FIFTH EDITION

PHYSICAL GEOGRAPHY

TO THE STUDENT:

A Study Guide for the textbook is available through your college bookstore under the title Study Guide and Lab Manual to accompany Physical Geography by Michael P. McIntyre, H. Peter Eilers, and John W. Mairs. The Study Guide can help you with course material by acting as a tutorial, review, and study aid. If the Study Guide is not in stock, ask the bookstore manager to order a copy for you.

FIFTH EDITION

PHYSICAL GEOGRAPHY

MICHAEL P. McINTYRE

Professor of Geography, Emeritus
San José State University
San José, California

H. PETER EILERS

Professor of Geography and Environmental Studies
Willamette University
Salem, Oregon

JOHN W. MAIRS

Professor of Geography
Southern Oregon State College
Ashland, Oregon

JOHN WILEY & SONS, INC.

New York Chichester Brisbane Toronto Singapore

ACQUISITIONS EDITOR	Barry Harmon
DEVELOPMENTAL EDITOR	Barbara Heaney
MANAGING EDITOR	Joan Kalkut
DESIGNER	Maddy Lesure
COVER PHOTOGRAPHS	Reginald Wickham
COPYEDITING SUPERVISOR	Gilda Stahl
PRODUCTION SUPERVISOR	Savoula Amanatidis
PHOTO RESEARCHER	Barbara Salz
PHOTO RESEARCH MANAGER	Stella Kupferberg
PHOTO RESEARCH ASSISTANT	Hilary Newman
ILLUSTRATION	Edward Starr
MANUFACTURING MANAGER	Lorraine Fumoso

Library of Congress Cataloging-in-Publication Data:

McIntyre, Michael P.
Physical geography / Michael P. McIntyre, John W. Mairs, H. Peter Eilers. — 5th ed.
p. cm.
Includes bibliographical references and index.
ISBN 0-471-62017-3
1. Physical geography. I. Mairs, John W. II. Eilers, H. Peter. III. Title.
GB55.M455 1991
910′.02—dc20 90-49420
CIP

Printed in the United States of America

10 9 8 7 6 5 4 3 2 1

Dedicated to
Kay, Brandt, and Anna
Nancy, Alicia, and Jeremy

PREFACE

For many Americans, geography brings to mind the study of place location, the memorization of names of state or national capitals, and the location of the longest rivers and highest mountains. This notion of geography as a compendium of places and physical or cultural features to be memorized and located on maps can be compared to memorizing the words in a dictionary. The correct recall of spelling or definition of words, while fundamentally necessary, does not constitute a command of language—and just knowing the location of a city, a mountain, or a nation on a map does not constitute the essence of geography. Knowledge of language is based on how words are integrated to develop and express thoughts, ideas, concepts, and ultimately understanding. So it is with geography. The discipline has always gone beyond simply investigating "what is where" and brought together for study the variety of physical and cultural phenomena that produces the earth's landscapes and allows understanding of the nature of place and regions, cultural diversity, and human interaction with the physical environment. Geography takes an integrated approach to understand "*why* it is there." Specifically, *physical* geography examines the interacting processes of the atmosphere, hydrosphere, lithosphere, and biosphere in order to understand the geographic aspects of the physical environment in which we live, as well as the role of humans in affecting that environment. This text was written for the purpose of promoting this understanding.

AUDIENCE

This fifth edition of *Physical Geography,* as with previous editions, is aimed at students with little or no academic background in physical geography. It is suited to lower division introductory college courses in which students do not necessarily intend to become geography majors but are seeking geographic knowledge as part of their overall education. At the same time, the text can help lay a foundation in physical geography for those who do proceed to a college geography degree program.

ROLE OF GEOGRAPHY

As colleges and universities throughout the United States revise their general education requirements and identify core curriculums, the importance of geography as an integrating discipline among the social and natural sciences is not lost on educators. Geography enjoys recognition as a subject that exposes the student to our world—its physical characteristics and processes as well as its human diversity and spatial interaction. In an increasingly complex, shrinking, and competitive world we are compelled to become more familiar with the connection between human action and the quality and quantity of environmental resources on this planet. Geography, when taught well, can provide the knowledge base and enthusiasm among students to generate interest in further detailed study of the physical, natural, and social sciences. This text was compiled with these thoughts in mind.

TEXT REORGANIZATION AND REVISION

The fifth edition, while retaining the philosophy and much of the style of the previous editions, has been significantly revised and updated. The collaboration of two additional authors has resulted in the reorganization of parts and chapters around a theme that allows students to progress, in logical succession, through the earth's atmosphere, hydrosphere, lithosphere, and biosphere. In this edition, Part One, "The Planet Earth," begins by introducing the earth in terms of its dimensions and motions, geographic coordinates, time zones, earth-sun relationships, and the fundamental conditions producing seasons. Material on maps and mapping has been moved to an appendix where the instructor and students can make supplemental use of the information. Part Two, "Earth's Atmosphere," examines the earth's atmospheric component and the interaction of incident solar energy with the atmosphere and earth's surface, presents the elements and controls of weather and climate, and discusses climate classification and climate regions. Part Three, "Earth's Water," brings together information on the hydrosphere: distribution, natural cycling, ocean currents, floods, groundwater, and the importance of fresh water and oceanic resources. The three chapters in this part consolidate, with some added material, the last four chapters of the fourth edition dealing with water and oceanic resources. Part Four, "Earth's Lithosphere and Landforms," includes an expanded section on plate tectonics and the geomorphic results of plate tectonic activity, introduces several new terms and concepts, and consolidates and adds material to chapters on tectonic and gradational processes and major landform types. Part Five, "Soils and Vegetation," presents soil and vegetation types as integral parts of the biosphere, at the interface with human occupation of the earth's surface. This final part introduces many important biogeographical terms and concepts and builds on the topics in preceding chapters to provide a synthesis of global physical geography. Soil characteristics, soil classification (U.S. Comprehensive Soil Classification System), global soil regions vis-à-vis climate regime, vegetation/environment dynamics, and global vegetation as geographic entities are the major topics in this section. In the final chapter, the global pattern of integrated physical phenomena, with consideration of human impact, is summarized with an emphasis on vegetation as the geographically unifying factor.

Topics which have been added or material which has been significantly revised in this edition include plate tectonics; freshwater and oceanic resources; "focus boxes" on hurricane Hugo, the southern oscillation, urban climate, Monterey Canyon, Mendocino triple junction, irrigation, and timberlines; U.S. Comprehensive Soil Classification System; and vegetation classification. The authors believe that the chapters on global soil classification and natural vegetation patterns and the concluding material integrating geographic components of the earth's physical environment are significant additions to the usefulness of the text.

PEDAGOGY

As part of the revision of the text, a short list identifying major objectives appears at the beginning of each chapter. Key terms are boldfaced in the text as before, but they are now listed at the end of each chapter. Although the technique was used in the fourth edition, new key conceptual sentences have been judiciously pulled out and highlighted within the text, emphasizing the importance of major points. Also, at the end of chapters are review questions and application examples regarding the chapter's subject matter. We believe these questions and applications can be useful to the student and to the instructor for classroom discussion of environmental issues or for expansion of lecture topics.

References appear at the back of the book under a general heading and by chapter headings. New titles have been added and some old ones dropped. These entries are suggestions for further reading and support material presented in the text. An attempt was made to include not only some introductory treatments of pertinent topics but to provide a few titles that are more advanced as well. This should allow for additional reading and study at more than one level.

TEXT DESIGN AND ILLUSTRATIONS

The text has been redesigned with a new typeface, highlighting color, and additional levels of subheadings. Numerous photographs, maps, and diagrams have been either replaced, updated, or revised and many new ones added. For example, explanatory diagrams have been added to the chapters on earth-sun relationships, climate, plate tectonics, vegetation, and soils. Although many of the fourth edition photographs still appear, in some cases the captions have been rewritten for clarity. New photographs, particularly of vegetation types and soil profiles, enhance the illustrative character of the text. Many topical photographs used as chapter and part openers have been added or changed. A new map of global soils has been added to the color plates. Overall, we believe the design work by the John Wiley & Sons staff has greatly improved the appearance and the function of the text.

ACKNOWLEDGMENTS

As with all textbooks, the final product is the result of the work of many: authors, editors, reviewers, researchers, designers, secretaries, students, family and colleagues. With all textbooks there is the potential for errors or mistakes which, despite the best efforts of all involved, somehow manage to appear. The authors accept full responsibility for any such errors and ask that they be brought to our attention. Within the scope of this work, the authors welcome comments and suggestions for improvement.

The staff at John Wiley & Sons has been extremely helpful in producing this revision of *Physical Geography*. Their patience and professionalism kept the project on track and running smoothly. Stephanie Happer, our editor, did much of the initial work to get the fifth edition off the ground; we also benefited from the guidance of her successor, Barry Harmon. Barbara Heaney, developmental editor, provided editing expertise and encouragement to the authors that resulted in much of the important organizational changes and pedagogical additions. The thorough and knowledgeable editing skill of Gail Hapke is sincerely appreciated. We thank Gilda Stahl, Stella Kupferberg, John Balbalis, and Maddy Lesure who worked on the copyediting, photo research, illustration, and design of the fifth edition. A special thanks to Savoula Amanatidis for coordinating the final production stages.

Many reviewers provided their suggestions for changes and improvement of the manuscript. We wish to thank

Professor Murray Abend
Jersey City State College

Professor Dan Gruber
West Valley College

Professor Ken Hinkel
University of Cincinnati

Professor Karl K. Leiker
Westfield State College

Professor Ralph Lewis
Eastern Oregon State College

Professor John Lyman
Bakersfield College

Professor Harold Meeks
University of Vermont

Professor Robert Petersen
San Bernardino Valley College

Professor James Powers
Pasadena City College

Professor Ted Schmudde
University of Tennessee

Professor Thomas W. Small
Frostburg State University

Professor Stephen J. Stadler
Oklahoma State University

Professor John Watkins
University of Kentucky

Professor Raymond Waxmonsky
SUNY–Buffalo

We also acknowledge the patience and contributions of our colleagues Claude Curran, Susan Reynolds, Ruth Monical, and Jim Rible. Special thanks to Brent Vanderpol, Scott Loban, Mark Leedom, Kirt Foster, and Megan Smith for their contributions to research and manuscript preparation. Any failings that remain are our own.

Michael P. McIntyre
H. Peter Eilers
John W. Mairs

BRIEF CONTENTS

CONTENTS

PART THREE

EARTH'S WATER

PART FOUR

EARTH'S LITHOSPHERE AND LANDFORMS

WORLD MAPS

APPENDIXES

PART ONE

THE PLANET EARTH

Grönland
(DENMARK)
Thule
Egedesminde
Godthab
Davis Strait
Denmark Strait
Greenland Sea
Reykjavik
ICELAND
SVALBARD
Hudson Bay
Hopedale
Battle Harbour
Gander
St. John's
NEW FOUNDLAND
Gulf Stream
NEW FOUNDLAND BASIN
PRINCE PATRICK

The globe. *Study of earth's physical geography begins with its size, spherical shape, and an understanding of global systems of time and location utilized by its inhabitants.*

CHAPTER 1

EARTH AS A SPHERE

OBJECTIVES

After studying this chapter, you will understand

1. The variations in the spherical shape of the earth.
2. The difference between earth rotation and revolution and how each affects our concept of time.
3. The definition of latitude and longitude and how each is determined.
4. What time zones are and why we have them.
5. Why daylight saving time is a form of self-deception.
6. What happens when a traveler crosses the International Date Line.

INTRODUCTION

For the most part, we perceive of the earth as spherical in shape. Of course, it is not really a perfect sphere—a geometrically perfect globe in which all radii from the center to the outer surface are of equal length. For one thing, there are significant irregularities all over its surface, and some of them seem very large. Mount Everest in the Himalayas is 29,029 feet (8848 m) above sea level, and the Mariana Trench in the western Pacific 35,810 feet (10,915 m) below. This is a vertical difference of about 12 miles (19 km). If we compare this 12 miles (19 km) with the earth's radius of approximately 4000 miles (6437 km), it becomes minuscule. For example, on a large classroom globe, where the earth's nearly 8000-mile (12,874 km) diameter is commonly represented by 16 inches (41 cm), the total maximum distance between the highest and lowest point would be little more than the thickness of a dime.

On a classroom globe, the vertical difference between the earth's highest and lowest point would be about the thickness of a dime.

In addition to the differences in relief of the earth's surface, there are other apparent deviations from sphericity. We know that the rapidly rotating earth has responded to centrifugal force and that a bulge occurs at the equator with a corresponding flattening at the poles. However, this amounts to a maximum difference between polar and equatorial diameters of only about 13 or 14 miles (21 to 23 km), again a tiny fraction of 4000 miles (6437 km).

The study and measurement of the earth's shape and size is called geodesy, and the word ***geoid*** refers to earth shape. It is a tribute to the complexity of earth dimensions that geodesists continue to acquire knowledge in their science. Additional knowledge helps establish an updated, accurate global reference base for new maps. Nobody yet knows the exact shape of the earth. Certain images from satellites suggest a pear shape; other studies based on gravity and magnetic measurements indicate indentations around the midlatitudes of both hemispheres. The people concerned with measuring the earth's surface—the geodesist, surveyor, and navigator—are very much aware of these variations. The key point, of course, is magnitude of variation. In the end, no matter what spherical deviations we consider, none is large enough to disturb our spherical perception.

The concept of the earth as a sphere goes far back into early Greek times when the classic logic of the Greek mind appears to have been tempered by a bit of inspired imagination. It was reasoned that the globe was the perfect shape, its surface equidistant from a central point and without beginning or end. Thus, since the earth was the handiwork of the gods, its shape must be spherical. Once the Greeks developed this theory they could easily prove it by the progressive "sinking" of a ship sailing over the horizon, the distinctive shadow on the moon during successive eclipses, the predictable change of the altitude of the stars and the sun with change of latitude, and so on. Columbus knew the earth was round even if his simple seamen did not. His error was assuming that it was smaller than it is. For our purposes we will assume the earth to be a perfect sphere as the Greeks did. This kind of generalization makes life much pleasanter for the student at problem-solving time by doing away with minor corrections. To avoid Columbus's error of too small an earth, we will adopt 25,000 miles (40,234 km) as a workable round number for the earth's circumference.

EARTH MOTIONS

The earth is constantly moving in two separate circular patterns at the same time. We are not immediately aware of this movement because we are part of the earth as is the atmosphere, both held securely to the planet by gravity. We can readily observe a number of phenomena, however, that are the direct result of these motions. Alternating day and night, the differing length of day during the year, and the progression of the seasons are some of these phenomena. The rising and setting of the sun and the shifting positions of the heavenly bodies are apparent motions rather than actual; these objects appear to migrate only because we observe them from a moving planet.

The earth is constantly moving in two separate circular patterns at the same time.

In everyday language the words "rotation" and "revolution" can mean the same thing. But they have separate meanings when used to describe the earth's motions. ***Rotation*** is the earth turning on its axis; ***revolution*** is its movement about the sun.

Rotation

Day and night are the result of rotation—every location on earth first turns toward the sun and then away as the earth spins on its axis. This is a constant and steady movement of about a thousand miles (1609 km) per hour at the equator.

CALENDARS

Pope Gregory XIII.

The calendar is a timekeeping device as surely as is the clock; both use regularly recurring phenomena in nature to regulate and organize our lives. Day and night (rotation) is the basis for our 24-hour clock, a relatively simple and straightforward arrangement. But the calendar attempts to combine rotation with the seasons/year (revolution) and the week/month (phases of the moon), none of which are exactly comparable. One year, or one revolution of the earth about the sun, is measured in fractional days, 365.25. One lunar month, or one revolution of the moon about the earth, averages 29.5 days, which when multiplied by 12 equals 354 days, a far cry from one full year. But how are we to make all these elements work together? This is the question that has vexed generations of would-be calendar makers. There is no easy or obvious answer.

The first modern calendar to achieve any extensive success was the Julian calendar, created by order of Julius Caesar in 46 B.C. It dealt with the extra quarter day each year by introducing the leap year, and it solved the short lunar year problem by arbitrarily lengthening each month except February. Our familiar months, their names, and varying lengths, have come down to us from the Julian calendar.

This remarkable calendar served very well for over 1500 years, even though, as it turned out, the Julian year was 11 minutes longer than the astronomical year. By A.D. 1582 this small problem had magnified into a 10-day error, so that on October 5 of that year Pope Gregory XIII felt it necessary to eliminate 10 days by declaring that the date should be October 15 instantaneously. An additional bit of minor tinkering with the formula took care of the difficulty. The resulting calendar became known as the Gregorian calendar and was adopted by Great Britain and the English colonies in America in 1752; it is the one in almost universal use today.

It isn't pretty but it works. We still have to go through the "30 days hath September, April, June, and November" routine; and it is still difficult to forecast what day Christmas will fall on several years hence—day and date are never the same from year to year. Calendar reform is always in the air; there are whole societies dedicated to it. Let's not hold our breaths, however. As long as the Gregorian calendar functions without error, newer, improved, slightly polished models are not likely to be adopted. Progress is not rapid in calendar circles. It took the British 170 years to make the move from Julian to Gregorian.

The movement is so regular that our system of time is geared to it and calculated on a 24-hour basis (one rotation) (Fig. 1-1). The sun, appearing to wheel around the earth once every 24 hours, rising in the east and setting in the west, is actually not moving around the earth at all. It only looks that way from a moving earth. The earth must therefore rotate in the *opposite* direction from the sun's apparent motion, or from *west* to *east*.

Day and night are the result of rotation—every location on earth first turns toward the sun and then away as the earth spins on its axis.

Revolution

Revolution, the movement of the earth in its orbit around the sun, is a different matter—variations are the rule. The earth moves in an ellipse about the sun, not a perfect circle. Thus the earth is constantly at different distances from the sun, ranging from 92 to 95 million miles (147 to 152 million km). In this less-than-circular orbit the earth is nearest the sun, or at ***perihelion*** ("near sun"), in January and farthest away, or at ***aphelion*** ("away from sun"), in July. It would seem that a couple of million miles more or less in distance would have no significant effect on the sun's ability to heat the earth and influence climate. However, many astrono-

Fig. 1.1. ***Rotation.*** *The sundial does not keep the same time as our familiar mechanical or electronic clocks and watches, which endlessly repeat a 24-hour interval. They would be identical if only rotation were involved, but the sundial also reflects the varying speeds and elliptical orbit of revolution. It therefore records all sorts of minor variations.*

mers and climatologists believe that shifts in the shape of earth's orbit may be a factor in periodic climate change. The speed at which the earth follows its orbit also varies. The most rapid advance is in January, and the slowest in July.

The earth is constantly at different distances from the sun.

Despite these variations, the earth manages to go around the sun in about the same length of time for each complete circuit, but this period does not come out even with our 24-hour days as determined by rotation. Our revolution measures about 365.25 days. That is why we have to add up the accumulated quarter days and splice them onto the tag end of February every four years.

LATITUDE AND LONGITUDE

Once the ancient Greeks had proven the earth to be spherical, they had to find a way to establish artificial coordinates on its spherical surface. Without such coordinates, it would be impossible to map or measure effectively. The idea of a coordinate measuring system is, of course, very simple on a flat piece of graph paper where all the X lines are parallel, all the Y lines are parallel, and X and Y lines cross at right angles. But how can we do this on a sphere? The Greeks worked out an ingenious system based on two important references: (1) the rotational axis of the earth with two fixed surface points at the poles and (2) the circumferential line dividing the earth in half, located exactly halfway between the two poles. This line is the ***equator,*** equating the earth into a Northern and a Southern Hemisphere. From this reference line and the two poles the geographic coordinate system of ***latitude*** and ***longitude*** surrounds the entire globe.

Latitude

The equator is the reference line for latitude; it is designated zero degrees (0°) latitude. It is also a ***great circle.*** A great circle is the circumferential line that is traced at the surface if an imaginary cutting plane were passed through the earth's center dividing it in half (Fig. 1-2). The imaginary plane, of course, could be passed through the earth's center in an infinite number of ways to create an infinite number of great circles (Fig. 1-2C). However, the equator is a unique great circle in that it is the trace of a cut that divides the earth in half equidistant from the poles, with the cutting plane at right angles to the polar axis (Fig. 1-2A). The equator serves as a baseline for a sequence of parallel lines that encircle the earth both north and south of the equator all the way to the poles. These lines are called lines of latitude or ***parallels.*** They are circles with their planes at right angles to the polar axis, but they cannot be great circles because only the equator divides the earth in half (Fig. 1-3).

Imaginary parallel lines encircle the earth both north and south of the equator all the way to the poles.

A line of latitude passes through every place on the earth. Therefore, when we speak of latitude we mean the angular distance north and south from the equator to any given location. Latitude is measured in degrees along the surface of the earth's sphere, with the equator at 0° and the poles at 90°. Because we number from 0° to 90° in *both* directions from the equator, each number except 0° appears twice, once in each hemisphere. We must add the suffix N (north) or S (south) to designate the hemisphere (Fig. 1-4).

MEASURING IN DEGREES Why are we measuring in degrees instead of miles or kilometers? This is because the earth is round and the circumference of a circle is 360°. We are

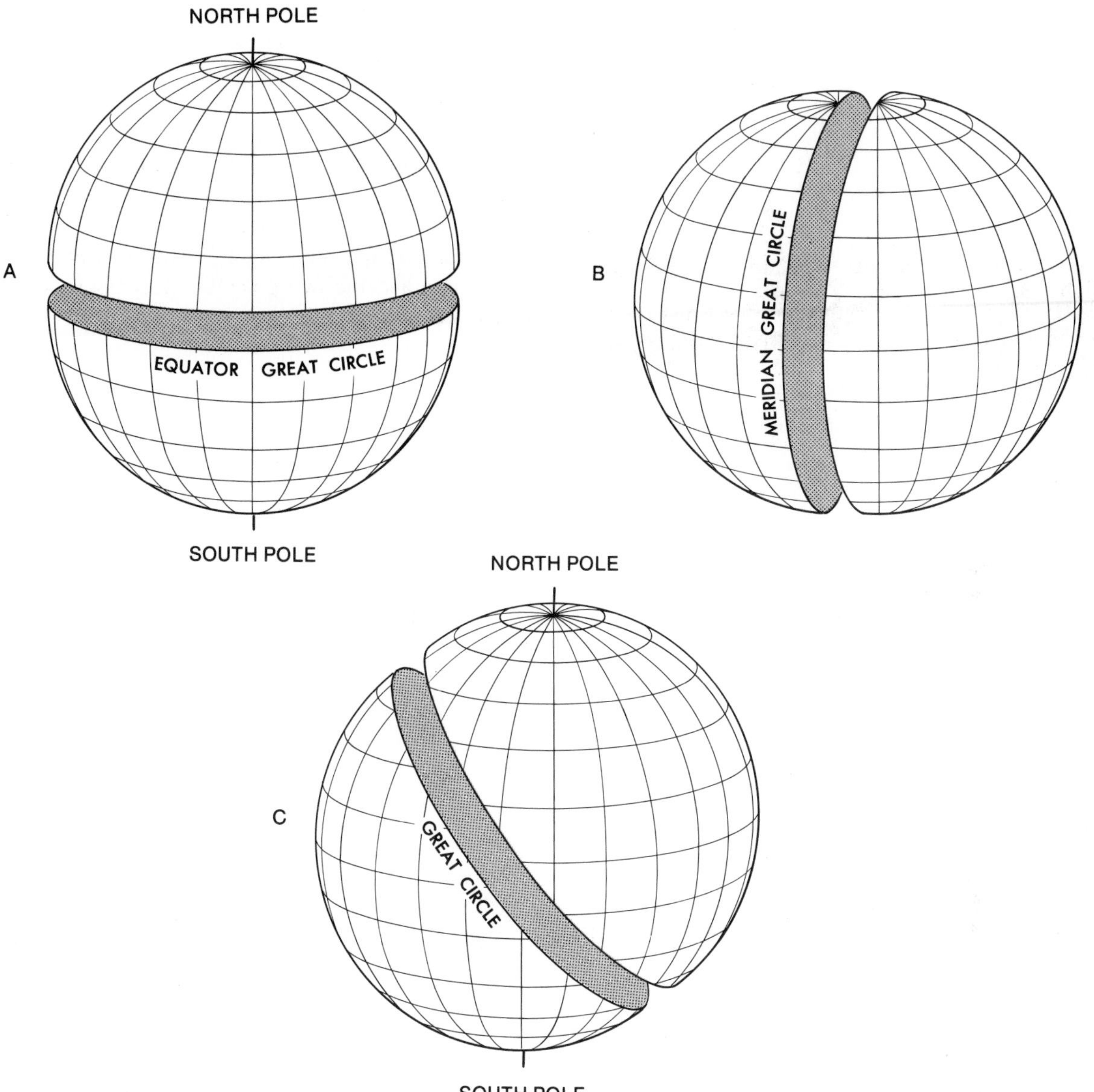

Fig. 1.2. ***A great circle.*** *No matter how it is sliced, any incision halving the earth describes a great circle on its surface.*

simply working through one-fourth of this circumference (0° to 90°) as we number the curved surface of the earth from equator to pole, both north and south. If necessary, latitude can be translated into miles or kilometers: 1° of arc on a great circle whose circumference is 25,000 miles (40,134 km) comes very close to 69 miles (111 km). A location at 10° N is somewhere on a circle 690 miles (1110 km) north of the equator. In addition, a degree of latitude breaks down into 60 minutes (1.15 miles or 1.7 km). A minute, in turn, breaks down into 60 seconds (0.019 miles or 0.03 km).

DETERMINING LATITUDE How do navigators figure out precise distance from the equator when this is an unknown? Nobody has drawn those critical circles right out there on the ground or over the water. Somehow they must determine the location of the equator as a point of departure for

Fig. 1.3. ***Parallels.*** *Although they appear to be straight lines on a map, lines of latitude (parallels) are actually circles, their planes parallel to that of the equator.*

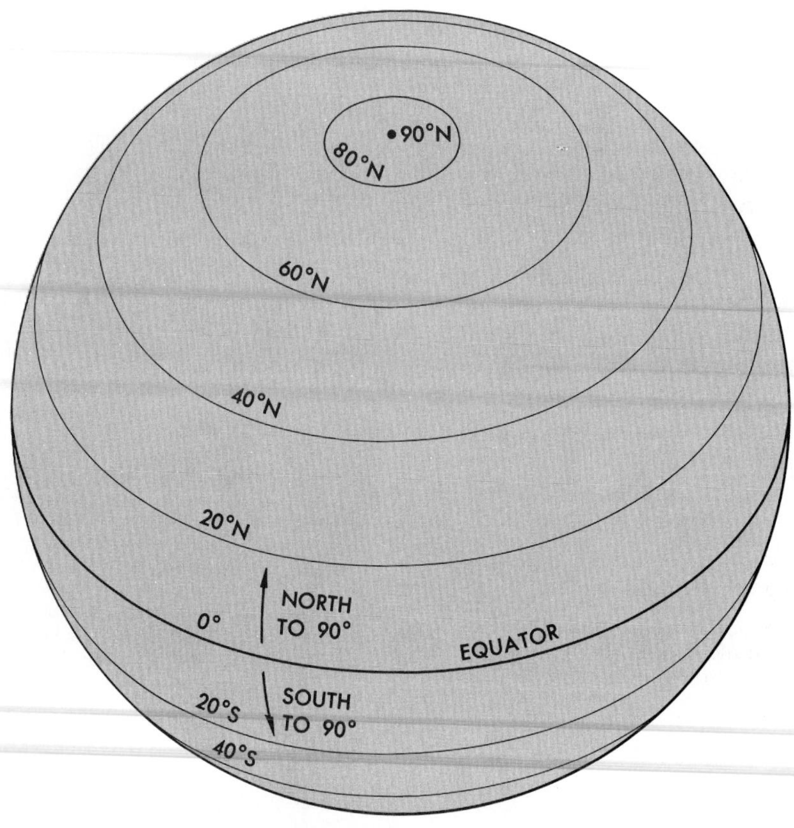

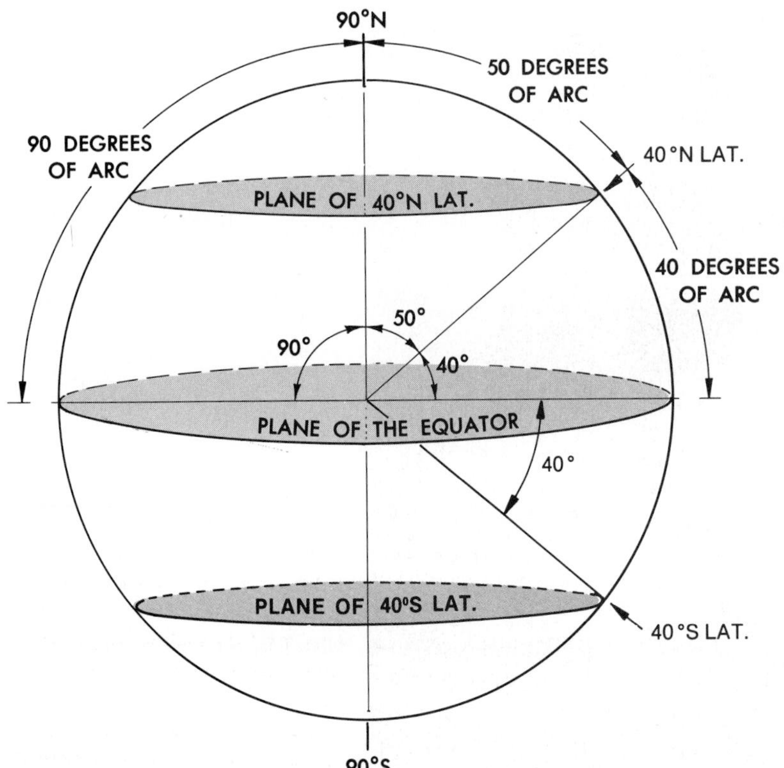

Fig. 1.4. ***Measuring latitude.*** *A north latitude of 40° represents 40° of arc measured northward from the equator along the curving surface of the earth, and also a 40° angle at the earth's center.*

reckoning. Strangely, or so it might seem, navigators look into the sky for help and observe the known behavior of the heavenly bodies. As an example, let us select the simplest of these heavenly bodies, one that cannot be mistaken for any of the others and one whose apparent motion follows a familiar pattern—the sun.

All navigators start with one important fact. On or about March 21 and on or about September 21, the sun at noon is overhead at the equator. The earth's axis is at right angles to the direct rays of the sun at these times; at other times of the year the noon sun's elevation in the sky varies in a predictable pattern. (Extensive discussion of earth–sun relationships follows in the next chapter.) Because we are trying to find a latitude of a given position, this knowledge helps us establish a reference for determining that latitude.

To proceed from this point, we simply use a sextant to measure the altitude of the sun above the horizon when it is highest in the sky, that is, the number of degrees in the angle between the noon sun and the horizontal (Fig. 1-5). Assume for the moment that we are standing on the equator at noon on either March 21 or September 21. The sun would be directly overhead, and its altitude would be 90° above the horizon. But 90° is obviously not the correct latitude. We know that latitude is measured as 0° at the equator and 90° at the poles. So we need to take one more step after measuring the sun's altitude above the horizon. We need to subtract the angle of the sun from 90°. Thus 90° minus 90° equals zero, which is the latitude of the equator.

If we move away from the equator, either north or south, the sun is no longer overhead and its noon altitude will be something less than 90°. The farther we go toward the poles, the lower the sun will appear at noon, until finally at the poles, the sun will appear to be resting on the horizon with no altitude at all. Therefore it is not difficult to see how the altitude of the sun at noon for at least two dates (around March 21 and September 21) lets us determine distance from the equator, or latitude. Once again, the altitude must be subtracted from 90°, or, stated differently, the latitude of any given location and the altitude of the sun at noon on March 21 or September 21 are complementary (when added they equal 90°). For example, if the sun's noon altitude measures 65°, latitude equals 90° minus 65°, or 25° north or south depending on whether the observer is looking south or north.

What about other days during the year? Since the apparent movement of the sun and other heavenly bodies follows

Fig. 1.5. ***Recording the sun's elevation.*** *The navigator lines up the horizon in the sextant eyepiece. Then by turning the knob with his left hand, he brings the mirrored image of the noonday sun down to the horizon. This maneuver will record the sun's elevation on a scale at the base of the instrument.*

regular patterns, we can reckon their positions relative to the equator on any date and at any time and apply corrections to compensate for variation. We know not only the specific dates when the noon sun is directly overhead at the equator, but also the dates when it is directly overhead at other latitudes. The latitude where the sun's rays are perpendicular to the earth's surface (or the sun appears overhead) at noon is known as the ***subsolar point*** (SSP). Navigators have long known the subsolar point for any day of the year, and from this information the latitude of a position can be found by adding or subtracting the latitude of the subsolar point.

We can reckon the position of the sun relative to the equator on any date.

For example, if we were in the Northern Hemisphere and the subsolar point for the day was, say, 20° N and the sun's noon altitude at our position measured 68°, then 90° minus 68° is 22° *plus* the 20° to compensate for the noon sun being directly overhead at 20° north of the equator. Thus the latitude is 42° N (Fig. 1-6). If the sun were directly overhead at noon at 20° S rather than 20° N, then the noon sun would be much lower in the sky from our observation point at 42° N. Its altitude would be 28° above the horizon. In this situation, 90° minus 28° is 62° *minus* 20° to compensate for the overhead noon sun being in the opposite hemisphere. The calculated latitude of the point of observation remains 42° N. Thus a simple formula can be stated for determining latitude when the subsolar point is known and the sun's noon altitude can be measured: subtract the observed altitude of the noon sun from 90° and then add or subtract, depending on hemisphere, the latitude of the subsolar point. This principle is relatively simple and has been understood and applied for centuries. Old-time mariners could always be sure of reasonably accurate latitudinal measurement even though their longitude was sometimes very questionable.

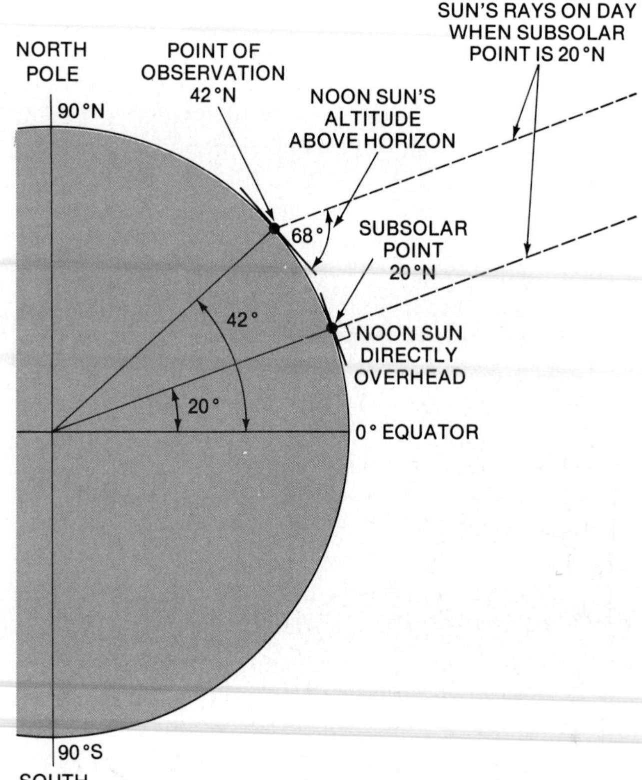

Fig. 1.6. ***Measuring the latitude of the observation point.*** *In this simple diagram the subsolar point, which varies in latitude throughout the year, is at 20° N latitude. Here the noon sun appears directly overhead or at 90° above the horizon, but the altitude of the noon sun at the same time at 42° N is 68° In this case, by subtracting the observed altitude of the noon sun (68°) from 90°, the difference (22°) when added to the latitude of the subsolar point gives the latitude of the point of observation.*

The measured altitude of Polaris requires no processing but is the correct latitude.

There is one other easy way to determine latitude, at least in the Northern Hemisphere, and that is by using ***Polaris,*** or the North Star. Polaris is so close to being permanently over the North Pole that the slight deviation can be generally ignored. Here the latitudinal numbering works in our favor, for the measured altitude of Polaris requires no processing but is the correct latitude. If we stood at the pole, Polaris would be directly overhead all night long, and its latitude would measure 90°, that is, latitude 90° N. As we move away from the pole, latitude decreases and the altitude of Polaris decreases at the same rate. There is no need to wait for a particular time or date, and we never need a correction factor. As long as we can see the North Star (Northern Hemisphere only), latitude determination is no problem. But isolating Polaris from its several million companions can be a difficult trick on a partially cloudy night (Fig. 1-7).

Longitude

To establish the other element in the coordinate grid—a series of lines crossing the parallels at right angles—we go

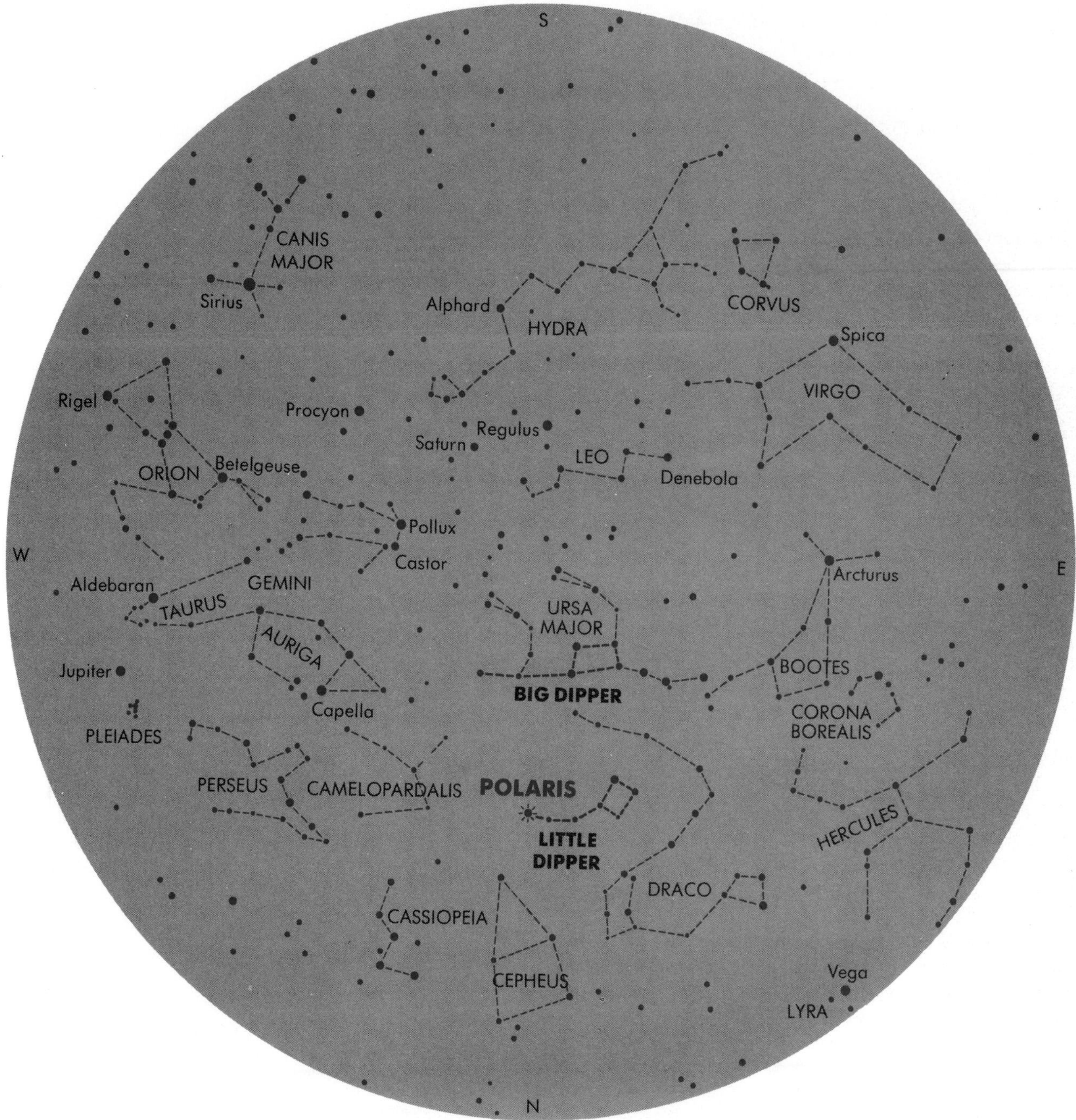

Fig. 1.7. ***Polaris.*** *A series of due north monthly observations will reveal the stars wheeling in arcs about a point well above the horizon. That fixed point is very close to the North Star. Around it the Little Dipper moves in a tight circle and the Big Dipper in a slightly larger one. The heavenly pattern in this illustration is as it appears in March.*

back to the great circle, this time to an infinity of great circles passing through both poles with their planes in the same plane as the axis. Thus, if the plane of the equatorial great circle was at right angles to the axis and the planes of all the parallels were parallel to it, then all the longitudinal great circles must cross all parallels at right angles (Fig. 1-8).

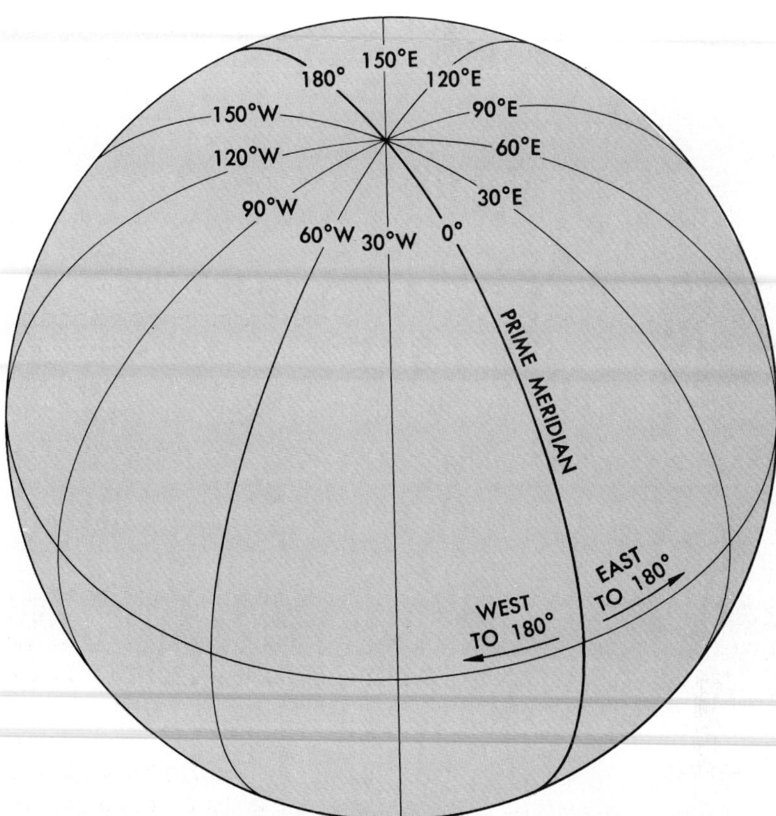

Fig. 1.8. ***Longitude.*** *All lines of longitude (meridians) are parts of great circles extending from pole to pole, their planes parallel to the axis. All meridians come together or merge at the poles.*

We have, after a fashion, duplicated graph paper. Of course, any two meridians come closer and closer together, or converge, as they reach the poles, introducing problems not present on the simple graph; this is the best we can do on a globe.

If the plane of the equator is at right angles to the axis and the planes of all the parallels are parallel to it, then all the longitudinal great circles must cross all parallels at right angles.

UNDERSTANDING MERIDIANS A line representing longitude is called a ***meridian.*** It is properly used to designate the total longitudinal great circle, but it more commonly refers to half a great circle, extending from pole to pole. This is because the numbering system designates the two halves of a total great circle by two different numbers 180° apart. Thus the half numbered 0° is really the half of a full circle containing the 180° meridian also—on the opposite side of the globe.

These lines of longitude, then, measure distance east and west of a baseline. However, arriving at a baseline from which to begin the numbering is not as simple as it is with latitude. There is no natural line corresponding to the equator, and many years ago every country thought that the meridian of its capital should merit the honor. After many years of international bickering, it became apparent that one meridian would have to receive worldwide approval, and the meridian of the Royal Observatory at Greenwich, England, was officially adopted in the late nineteenth century. Since then the Royal Observatory has been removed to the quiet countryside as suburban London engulfed Greenwich, but the prime or 0° meridian is still there in the form of a bronze marker (Fig. 1-9).

This line, then, from pole to pole, became the ***prime meridian*** or 0° longitude; all longitude is reckoned either east or west of it. The other half of this great circle running through Greenwich is athwart the central Pacific and, being halfway around the world from the prime meridian, is numbered 180° . Only these two meridians, 0° and 180° , require no directional suffix; all others must show a W or an E to indicate which side of Greenwich they are on. Thus, unlike

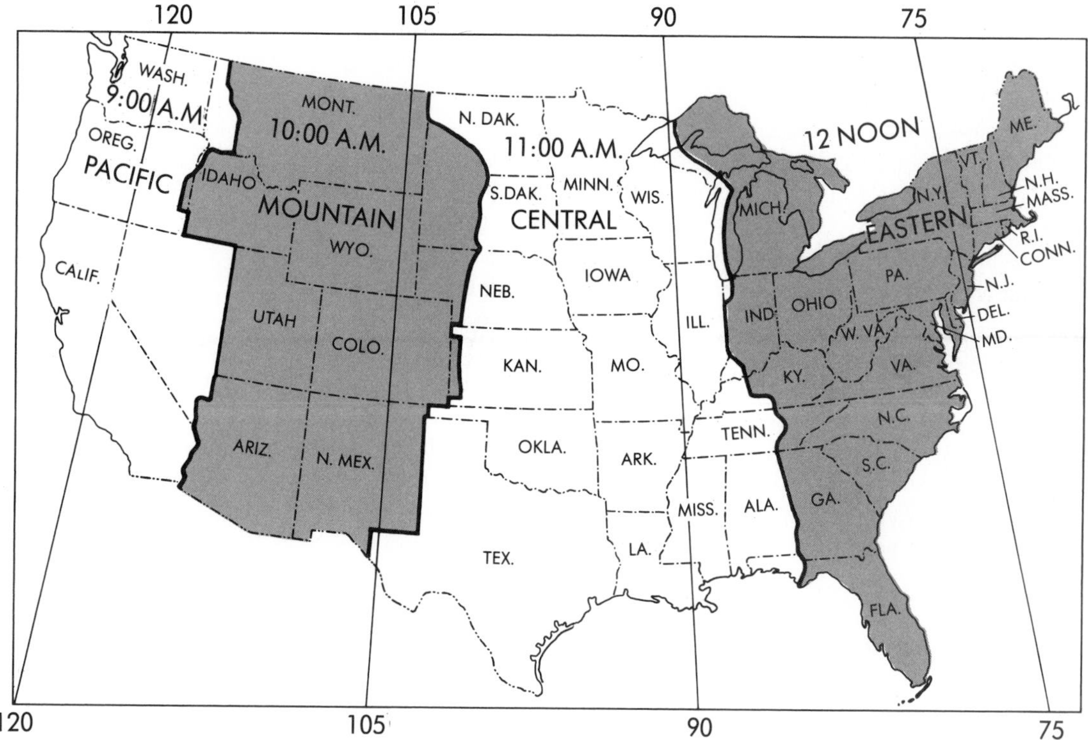

Fig. 1.11. ***Standard time zones in the United States.***

the day. Instead, it will indicate local ***zone time,*** which for most of us is not the correct sun time at all.

This has not always been the case. Not too long ago, people tried to keep local time according to the sun—but without a great deal of success. For an individual who spent a lifetime on one meridian, this was possible, but those who walked east or west, constantly changing meridians, would continually have to adjust their watches. To most, this situation was inconvenient but scarcely serious. Then railroads came along, and the problem became very serious indeed. Frequently, several trains, each attempting to keep its own local time, arrived at the same intersection simultaneously! So with the railroads leading the way, zone time was established, and everyone throughout a given area agreed to keep the *wrong* time but the *same* time for convenience and safety.

The entire world is divided into 24 times zones, each 15° wide.

The entire world is now divided into 24 time zones, each 15° wide; the time is based on the correct time of a meridian in the center of the zone. In actual practice the zones, except at sea and in unpopulated areas, deviate slightly at the margins to conform to state and national boundaries and, as in the United States, are frequently given descriptive names. In this country we have four such zones as shown in Figure 1-11:

Eastern Standard Time Zone, which keeps the time of the 75° W meridian.

Central Standard Time Zone, based on the 90° W meridian.

Mountain Standard Time Zone, centering on the 105° W meridian.

Pacific Standard Time Zone, which keeps the time of 120° W longitude.

So all clocks in the Eastern Time Zone read noon when the sun crosses 75° W, even though those near the eastern and western boundaries are almost a half-hour off, according to the sun.

This arrangement, of course, helps all business transacted within one time zone but causes the problem of sudden hour changes in time as a person leaves one zone and enters another. Suppose we were driving from Georgia to Alabama, the border of which is the change-over point from eastern to central standard time, and arrived at the Georgia side of the border at noon EST. This means that the sun is crossing the 75° W meridian on which EST is based. We have already spent the hour 11 to 12 on the road and progressed 50 miles (81 km), but as we cross the line into Alabama and CST, the time suddenly becomes 11 A.M. Central standard time is based on the 90° W meridian, and it will be another hour before the sun gets there. So we have the hour 11 to 12 all over again and can progress another 50 miles (81 km). We have gained an hour and do not have to give it back unless we return and the whole process is reversed. Those who travel all the way to the West Coast pick up a bonus of three hours. The general rule, then, anywhere on the earth (with the exception of the International Date Line, which will be discussed later), is that *as you go west you gain an hour every 15°, and as you go east you lose one.*

As you go west anywhere on the earth you gain an hour every 15°, and as you go east you lose one.

Daylight Saving Time

Daylight saving time has become a widely accepted practice that allows us an extra hour of daylight in the evening during the long summer days. Actually, it is a form of self-deception because all we do is get up an hour earlier than usual and go to bed an hour earlier. In other words, we exchange one of the early morning hours of light—which we normally waste in bed trying to sleep despite the peeping of the birds—for one of the early evening hours of dark when we are still awake but outdoor activity is restricted. Of course, we could easily do this without changing the clock, but therein lies the deception. We like to think that we are getting up at the usual time and still getting the bonus of an extra hour of daylight. So we adjust our clocks to read 7 A.M. when it is really 6 A.M. Now whose true local time are we keeping? Normally, we do not keep our own local time anyway but that of the central meridian of our time zone, so if we have changed the clock by an even hour, we must now be reading the time of a meridian 15° away from that central meridian. If we live in the Pacific Time Zone, our usual time is that of the 120° W meridian, and if that is 6 o'clock and we want the clock to read 7 o'clock, we must turn it one hour ahead; then we will be keeping the time of the 105° W meridian, or Mountain Time. Always, in daylight saving time, *the timepieces of any given zone are keeping the time of the next zone to the east.*

The Date Line

As we travel from zone to zone, people traveling west and gaining one hour every 15° of longitude could get carried away and pick up more free time than they should. A certain amount is acceptable, such as the three free hours gained by going from the East Coast to the West Coast of the United States and staying there. Even 23 hours is considered within bounds. A complete circuit of the earth, however, would give the traveler a full day's bonus, and here is where the line is drawn—literally—the ***International Date Line*** (IDL). When we cross this line in a westerly direction, we lose 24 hours.

There has not always been a Date Line because people didn't need one until they began to go around the world. The first circumnavigation of the globe by Magellan's party demonstrated the need for such a line, for despite keeping careful track of the date, the survivors of that voyage found that on their arrival in Spain, they were one full day in advance of the calendar. They should have had a day taken away to offset the one hour gained for each 15° of longitude in their westward journey. If they had gone around the world to the east, the accumulated one hour losses would have been rebated at a Date Line.

The International Date Line is essentially the 180° meridian running from pole to pole down the middle of the Pacific Ocean (Fig. 1-12). It is exactly halfway around the world from the prime meridian and together with the prime meridian it describes a great circle. In actuality, the Date Line deviates a bit from the 180° meridian so that it can run up the Bering Strait to separate Asia from North America, and it detours around islands and island groups to avoid the complication of differing dates on opposite sides of a single political unit. But, for our purposes here, we can consider the Date Line to be the same as the 180° meridian.

In this day of rapid communication we are constantly being reminded that there are two dates on the earth at any given time. News reports from Asia are frequently datelined a day ahead of the time of delivery. There is, however, one instant in every 24 hours when there is only one date on the earth—noon at Greenwich (midnight at the IDL). At any other time there are always two dates.

If we can imagine a midnight line sweeping around the earth, as it crosses each location the date changes.

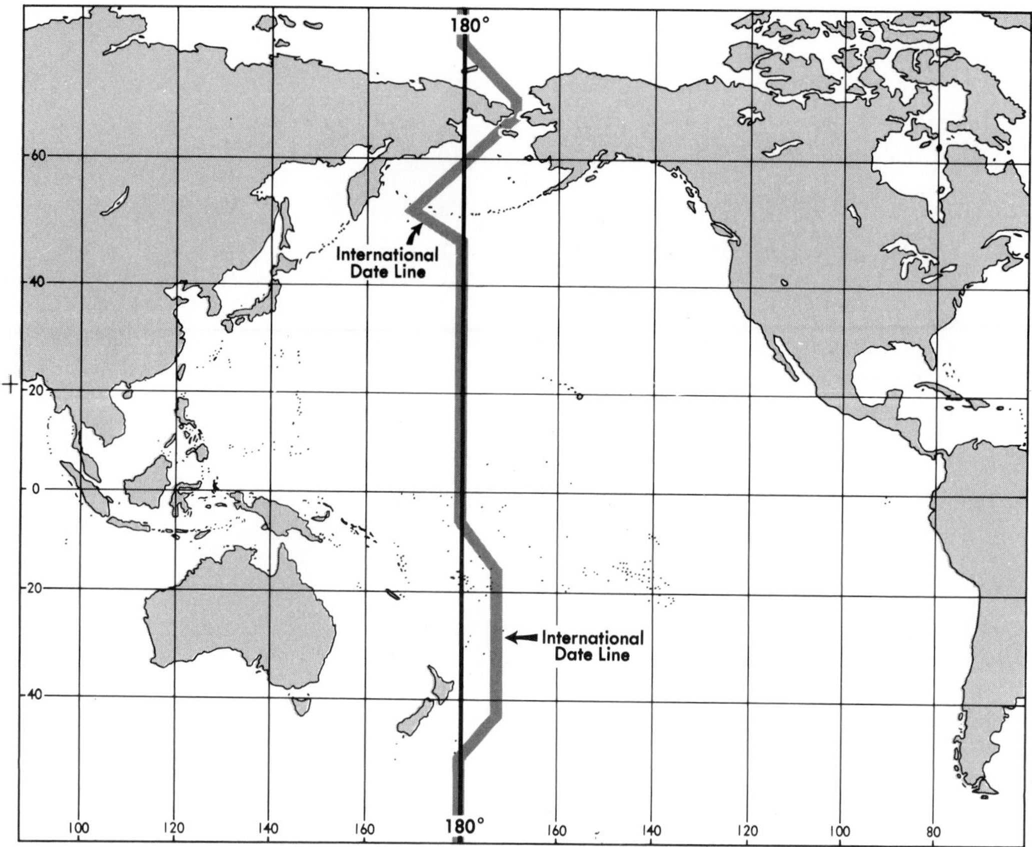

Fig. 1.12. ***The International Date Line.***

Midnight is the time when the date is advanced. If we can imagine a midnight line sweeping around the earth, as it crosses each location the date changes. This midnight line would be 180° (12 hours) away from the sun, so that when the sun is crossing the prime meridian, it would be midnight halfway around the world at the 180° meridian (Fig. 1-13). This is the instant when the entire earth has only one date. One hour later, when the sun has moved to 15° W, the midnight line has swept west 15°, from 180° to 165° E. Each spot within this area has had its date change, so if the date everywhere one hour ago had been the first of the month, one little 15° zone out in the Pacific would be the second. As the sun continued to move west, the wedge of new date would become larger and larger, until 12 hours later (noon at 180°, midnight at the prime meridian), one-half of the earth would be the second and one-half the first. Finally, when the sun had made almost a complete circuit of the earth and was again approaching Greenwich, all but a small part of the world would have experienced midnight and the date change, and only this narrow zone short of the Date Line would still be the first.

When we go west across the International Date Line we lose a day, and when we go east we gain a day.

Now if we are crossing the Pacific headed west toward Asia, crossing the Date Line will move us ahead one day (on the calendar). We lose a full day, and the only way to recover all or most of it is to go back across the Date Late or continue west all the way around the world, picking up 23 hours one at a time for each time zone and then stopping before crossing the Date Line again. Remember, the Date Line operates in opposition to the hour lines bounding the time zones: *by going west across the Date Line we lose a day, and by going east we gain a day.*

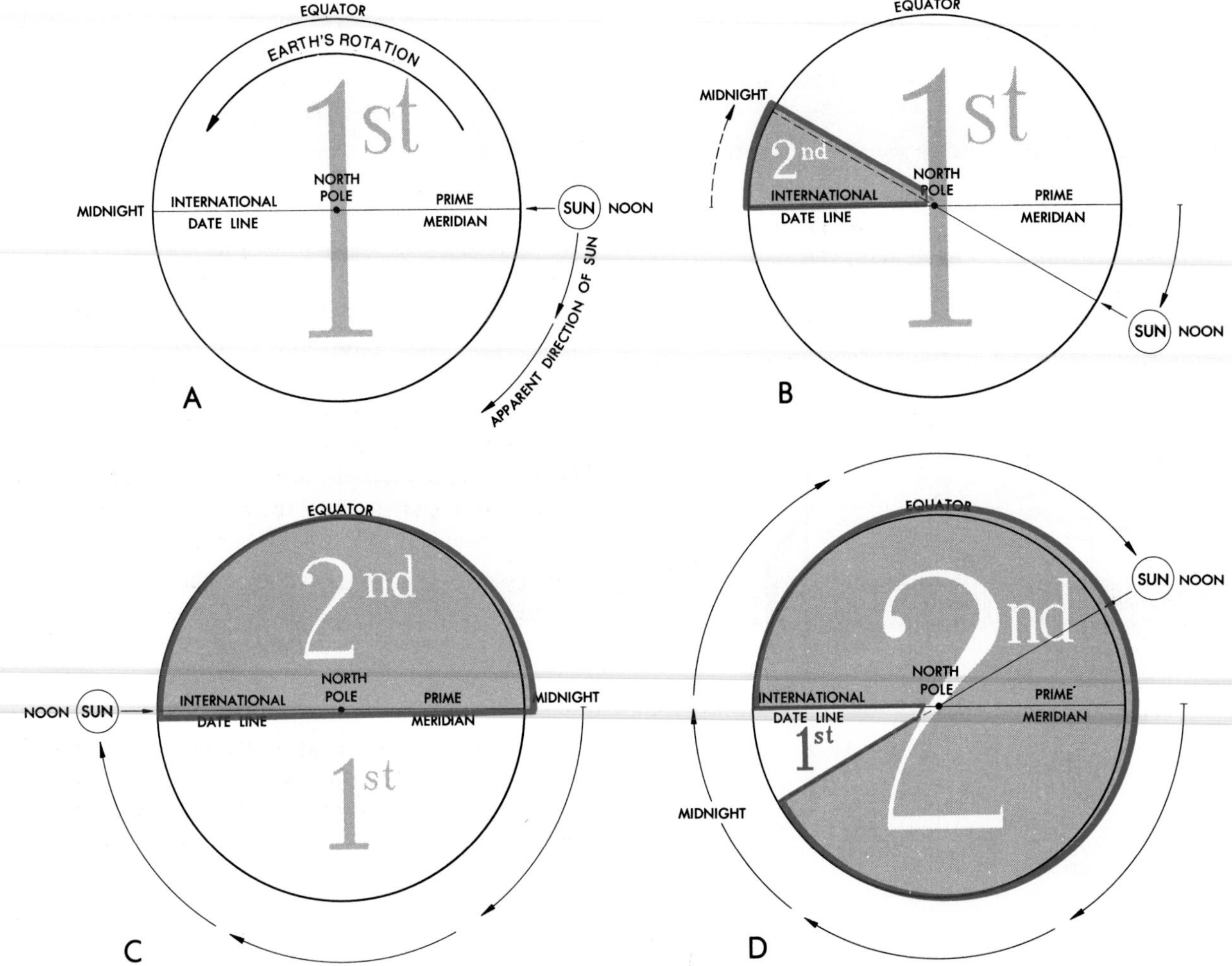

Fig. 1.13. ***Understanding the Date Line.*** *The apparent movement of the sun around the earth from A to D illustrates the function of the Date Line. The midnight line, sweeping around the earth opposite the noon sun, initiates a new day as it passes.*

CONCLUSION

The development of time zones and geographic coordinates, examples of systems we humans use to deal with the geographic peculiarities of our planet, can be traced directly or indirectly to the spherical nature of the earth and its motions in space. Our review of earth's shape, size, and motions in this chapter leads us to consider more specific interrelations between the earth and sun, the primary source of energy for the planet. We examine these relationships and their consequences in the next chapter.

KEY TERMS

geoid
rotation
revolution
perihelion
aphelion
equator
latitude
longitude
great circle
parallels
subsolar point
Polaris
meridian
prime meridian
Greenwich mean time
post meridiem (P.M.)
ante meridiem (A.M.)
zone time
International Date Line

REVIEW QUESTIONS

1. Why does the sun appear to rise in the east and set in the west every day?

2. If the equator is a great circle, why aren't all other lines of latitude great circles as well?

3. Why do we measure distance on the globe in degrees, minutes, and seconds instead of miles, yards, and feet (or their metric equivalents)?

4. If you are trying to find your latitudinal position on earth on a given day, why is it important to know the subsolar point for that day?

5. How is Polaris, the North Star, used as a navigational reference in the Northern Hemisphere?

6. What is the "zero" (0°) reference line for longitude? Why is it located where it is?

7. Why can't you use a pendulum clock to find longitude at sea? Under what circumstances could you use an ordinary wrist watch?

8. If zone time is not accurate, why do we use it?

9. What happens when a traveler crosses the International Date Line going west? Going east?

APPLICATION QUESTIONS

1. If you were an adviser to Julius Caesar circa 47 B.C., how would you explain the major defect in his proposed *Julian* calendar?

2. Take one orange and cut it into six evenly spaced parts along its lines of "latitude." Take another orange and cut it into six evenly spaced parts along its lines of "longitude." Study the sections. If you were presenting a lecture on the earth as a sphere, how could you use these orange sections to demonstrate the "great circle" concept?

3. If 1° of latitude on the surface of the earth is approximately 69 miles (111 km), at what latitude would you be after traveling 2967 miles (4773 km) directly north from 5° S latitude? If, on the day you arrived at your destination, the subsolar point was at 23° S latitude, what would be the altitude of the sun at noon at your new latitude?

4. Find out how many hours difference there are between your zone time and Greenwich mean time by using a world map showing lines of longitude.

5. Plan an itinerary from New York City to Sydney, Australia, including five stops along the way. Choose your own times of departure and either estimate your total traveling time or check with an airline guide. Describe the time changes you will encounter at each stop during your journey. You may have to check an atlas for time zones over the Pacific Ocean.

CHAPTER 2

EARTH AND SUN RELATIONSHIPS

OBJECTIVES

After studying this chapter, you will understand

1. How the tilt in the earth's axis of rotation causes changes in seasons.
2. The meaning of the terms ***solstice*** and ***equinox***.
3. The geographic significance of the Tropics of Cancer and Capricorn and the Arctic and Antarctic Circles.
4. How the sun's angle of incidence affects seasonal heating and cooling.
5. Why the lengths of day and night vary in the middle and high latitudes.

As photographed from the moon, the earth is illuminated by the sun as it rotates and travels around the sun.

INTRODUCTION

Our planet travels in an ellipse about the sun. As a result, its distance from the sun varies over the year. In a complete revolution the variation is little more than 3 million miles (4.8 million km), a small fraction of the average total distance of 93 million miles (149.6 million km) that separates the earth and the sun.

Some scientists believe that the shape of the earth's orbit changes significantly over periods of hundreds of thousands or even millions of years and that this factor may help to explain long-term climatic change. However, the present variation from a strictly circular orbit has little influence on the regular, annual heating and cooling we experience. In other words, how near or how far the earth is from the sun during its elliptical journey does not significantly affect the annual sequence of spring, summer, fall, and winter seasons.

What then are the major factors that account for the seasons? What earth–sun relationships explain why it is winter in the United States and the Soviet Union during January and February while it is summer in Argentina and South Africa? Why do nights become longer during the winter season and days longer during the summer? Why don't the seasons change near the equator? In this chapter we will see that the earth's aspect relative to the sun in its annual circum-solar journey has an effect on latitudinal heating and cooling, giving us seasonal change and climatic variation over much of the planet.

AXIS INCLINATION

The rotation of the earth causes such ***diurnal,*** or daily, occurrences as the progression from day to night. We are attuned to the cycle of light and dark and subsequent warmth and coolness which is attributed to rotation. But other longer term cycles are related to the axis of rotation and the way in which it is inclined, or tilted, relative to the

The inclination of the earth's rotational axis as it orbits around the sun causes the annual change in seasons at middle and high latitudes.

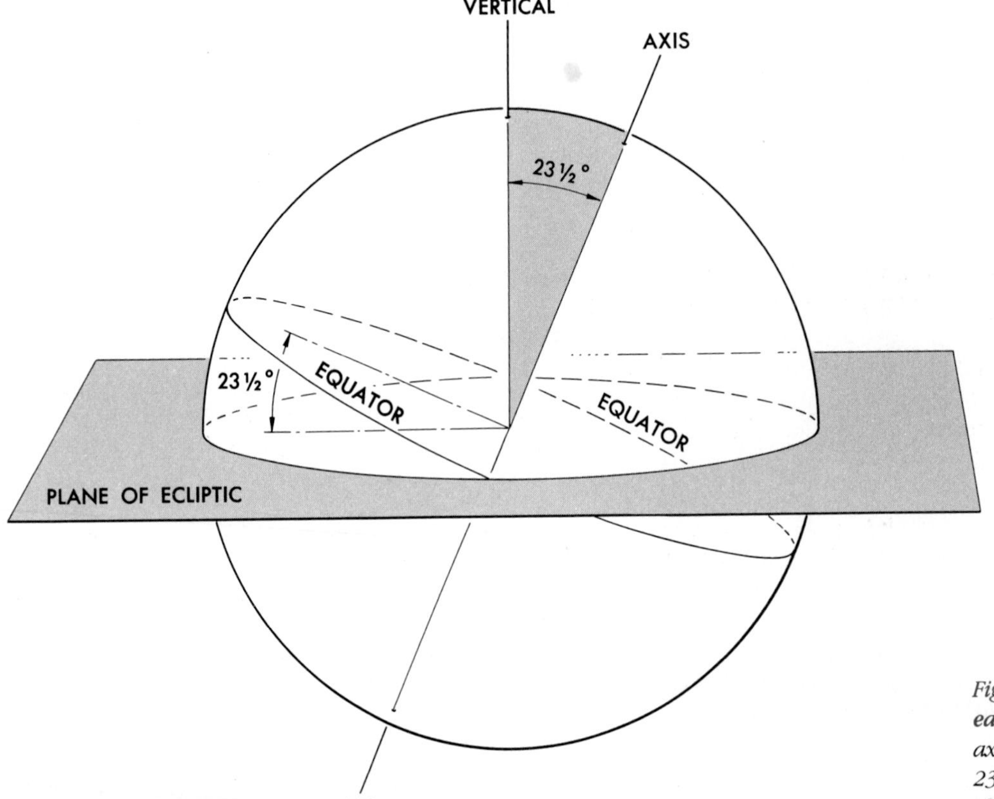

Fig. 2.1. ***Inclination of the earth's axis of rotation.*** *The axis maintains a constant tilt of 23.5° from a line vertical to the plane of the ecliptic.*

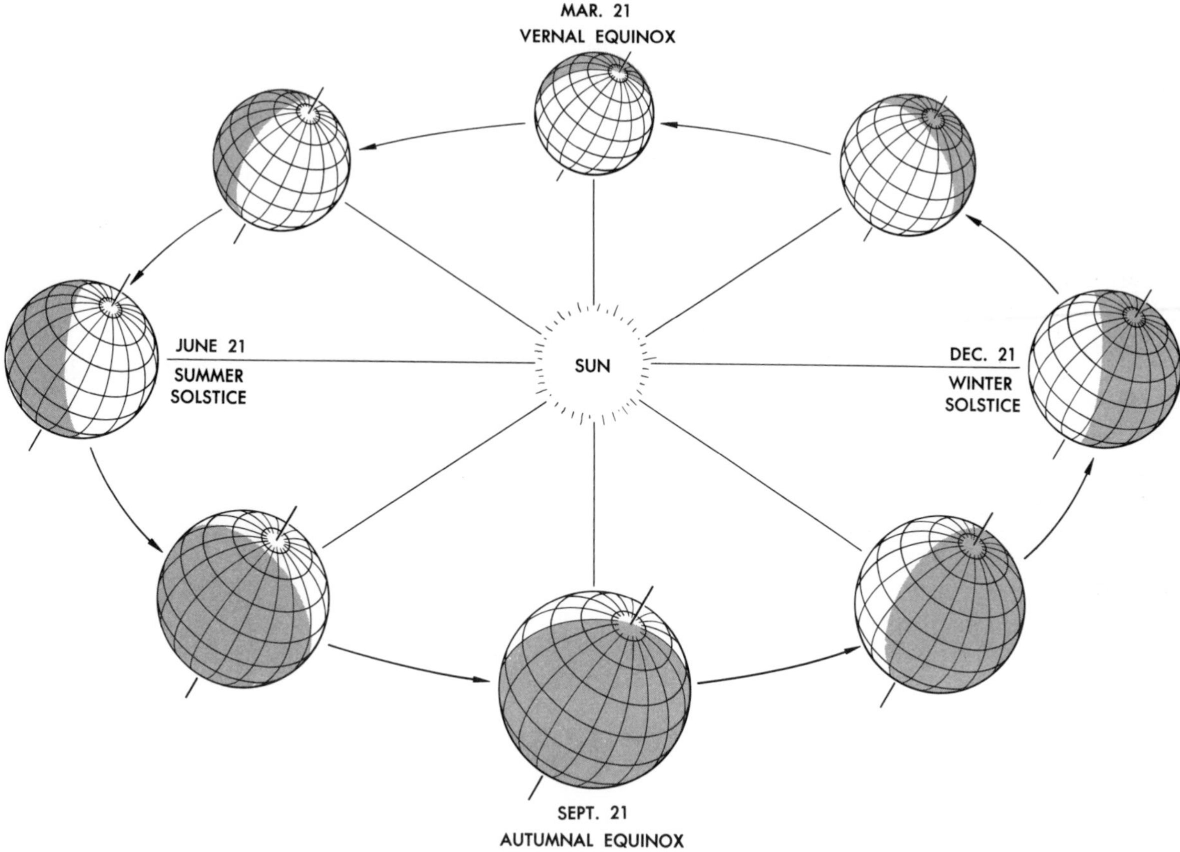

Fig 2.2 ***Revolution and axial inclination.*** *Both are required to complete this familiar annual sequence. The axis of rotation in any position is parallel to the axis in any other position (parallelism).*

sun. The inclination of earth's rotational axis as it orbits around the sun causes the annual change in seasons at middle and high latitudes.

The axis of rotation is tilted about 23.5° from the vertical, but everything is either vertical or horizontal relative to something else. How can we tell whether or not a sphere turning in space is vertical? Obviously, we need a horizontal plane from which to determine the vertical. The plane we use is called the ***plane of the ecliptic.*** It is the plane of the earth's orbit around the sun (Fig. 2-1). A line that intersects that plane at right angles can be considered to be vertical.

The axis, then, is inclined 23.5° from a line vertical to the plane of the ecliptic. It maintains this inclination, always pointing in the same direction in space, throughout every annual revolution. This is why the North Pole (an imaginary extension of the axis of rotation) continues to point to distant Polaris, the North Star, regardless of earth's position in its orbit. The night sky appears to rotate around this star. In other words, the axis of rotation is always parallel to itself throughout the earth's orbit around the sun. This is called ***parallelism*** (Fig. 2-2).

The axis of rotation is always parallel to itself throughout the earth's orbit around the sun.

Every year on about June 21 the northern part of the earth's axis tilts toward the sun, while the southern part tilts away (Fig. 2-2). As a result, the sun's most direct rays (high angle to the surface) strike a point well north of the equator, bringing seasonal heat to the Northern Hemisphere. At the same time, much less direct (low-angle) rays strike the surface in the Southern Hemisphere and solar heating is much less intense. The angle at which the sun's rays strike the earth's surface is significant. At a high angle relative to the surface, solar energy is more concentrated and the heating

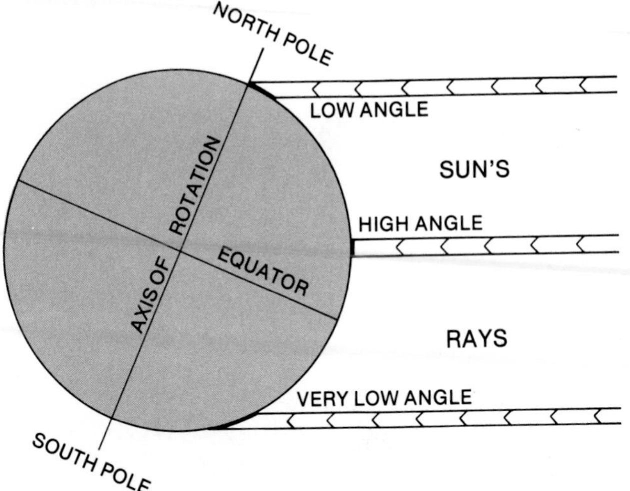

Fig. 2.3. ***Angle at which the sun's rays strike the earth's surface.*** *The three "beams" of sun's rays striking the earth's curved surface at different latitudes carry the same amount of energy. The angle at which each strikes the surface varies. The most intense heating is where the angle is highest, relative to the surface, and the energy most concentrated. Where the sun's rays are at a low angle the energy is diffused and heating less, since the energy is spread over a larger area. The case shown is that of the noon sun on or about June 21.*

greater; at a low angle, the energy is diffused and heating less (Fig. 2-3).

In the Northern Hemisphere June 21 is the longest day of the year and is called the ***summer solstice.*** Solstice literally means "the sun stands still." That is, on a solstice, the noon sun "pauses" before changing directions with respect to the horizon on subsequent days. For a midlatitude location in the Northern Hemisphere, the noon sun reaches its highest point above the horizon on June 21; on December 21 the noon sun reaches its lowest point above the horizon and begins its ascent in the sky, day by day, until the following June 21.[1] December 21 is called the ***winter solstice.*** The angles of the sun's rays at noon on the solstices is shown in Figure 2-4A and B.

In between the summer solstice and the winter solstice, on or about September 21, revolution has carried the earth through its orbit to a position where, although the axis remains inclined by the same amount relative to the plane of the ecliptic, the inclination is now at right angles to the sun's rays (Figs. 2-2 and 2-4C). In such a position the most direct rays strike the equator, concentrating the greatest heating midway between the poles. The September 21 date is considered the first day of fall for the Northern Hemisphere and is called the ***autumnal equinox;*** March 21 is the first day of spring and is called the ***vernal equinox.*** Equinox ("equal night") means that on September 21 and again on March 21 days and nights are of equal length at every latitude. These are the only two days of the year when this occurs.

If the axis of rotation were inclined to a greater or lesser degree, or not at all, relative to the plane of the ecliptic, we would experience a much different seasonal pattern.

If the axis of rotation were inclined to a greater or lesser degree, or not at all, relative to the plane of the ecliptic, we would experience a much different seasonal pattern in the middle and high latitudes. For example, if the tilt were 40° rather than 23.5° from the vertical, there would be greater contrast between the seasons. The winters would be colder and the summers much warmer, since the sun's angle in winter would be lower in the sky and in the summer much higher (see section on Sun Angle, p. 27). If the axis had no inclination at all, energy received daily would remain the same at any given latitude and there would be no seasonal variation.

Thus axis inclination is behind the pattern and intensity of the seasons. It also defines for us specific lines of latitude, which we now examine: the Tropics of Cancer and Capricorn and the Arctic and Antarctic Circles.

THE TROPICS—CANCER AND CAPRICORN

Most world maps and globes include a dashed line in each hemisphere, parallel to the equator and not far from it. Nearly everyone has carefully memorized at one time or another that the northern line is called the Tropic of Cancer and the southern one the Tropic of Capricorn. It is good to learn the names, but it is more important to know why these lines exist in their particular positions. You can find a clue in their latitude of 23.5° N and 23.5° S—the same as the angle of tilt of the earth's rotational axis! Clearly then, there is a direct relationship between the Tropics and the axial inclination, and in turn the seasons.

The ***Tropic of Cancer*** is defined as the northernmost point on the earth that ever experiences the sun directly overhead at noon, that is, at an altitude of 90° above the horizon. This only occurs on one day each year—June 21. The ***Tropic of Capricorn*** is the southernmost point on the

[1] These exact dates, the 21st of June, September, December, and March, may vary from the 20th to the 23rd. For our purposes here we will consistently use the 21st.

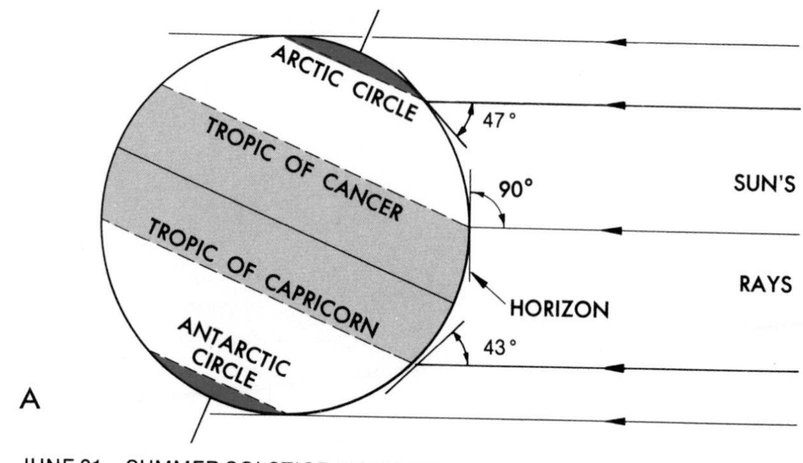

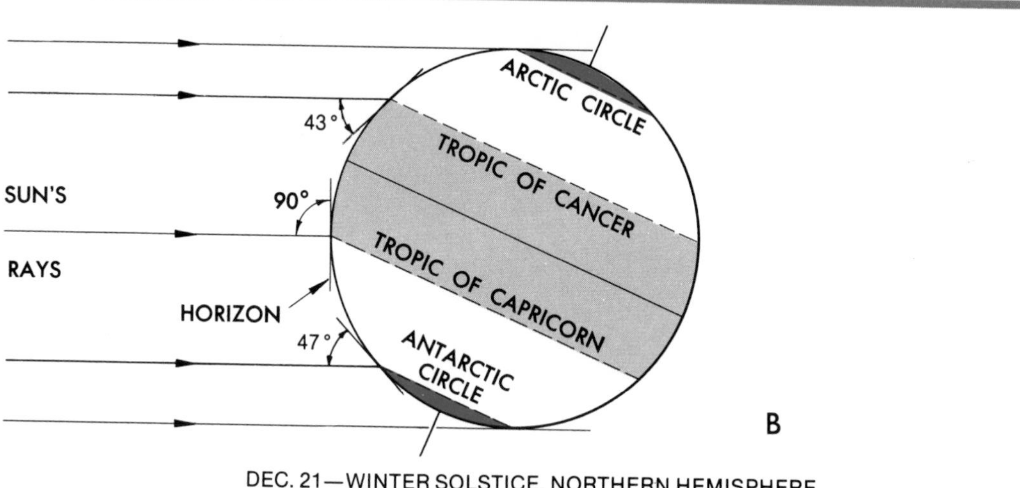

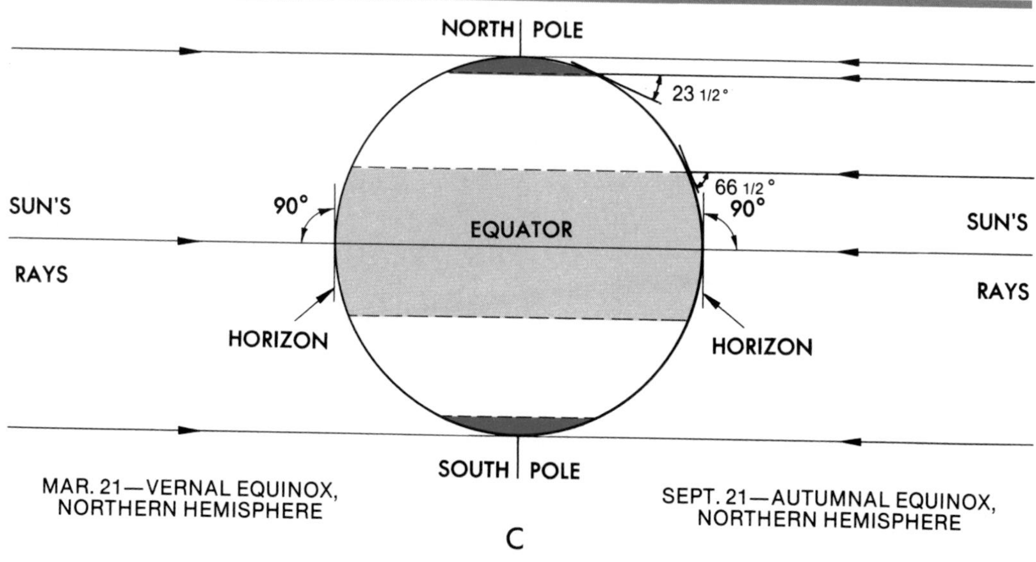

Fig. 2.4. ***Earth and sun at the solstices and equinoxes.*** *Note the changes in the altitude (angle above the horizon) of the noon sun at different latitudes on these dates.*

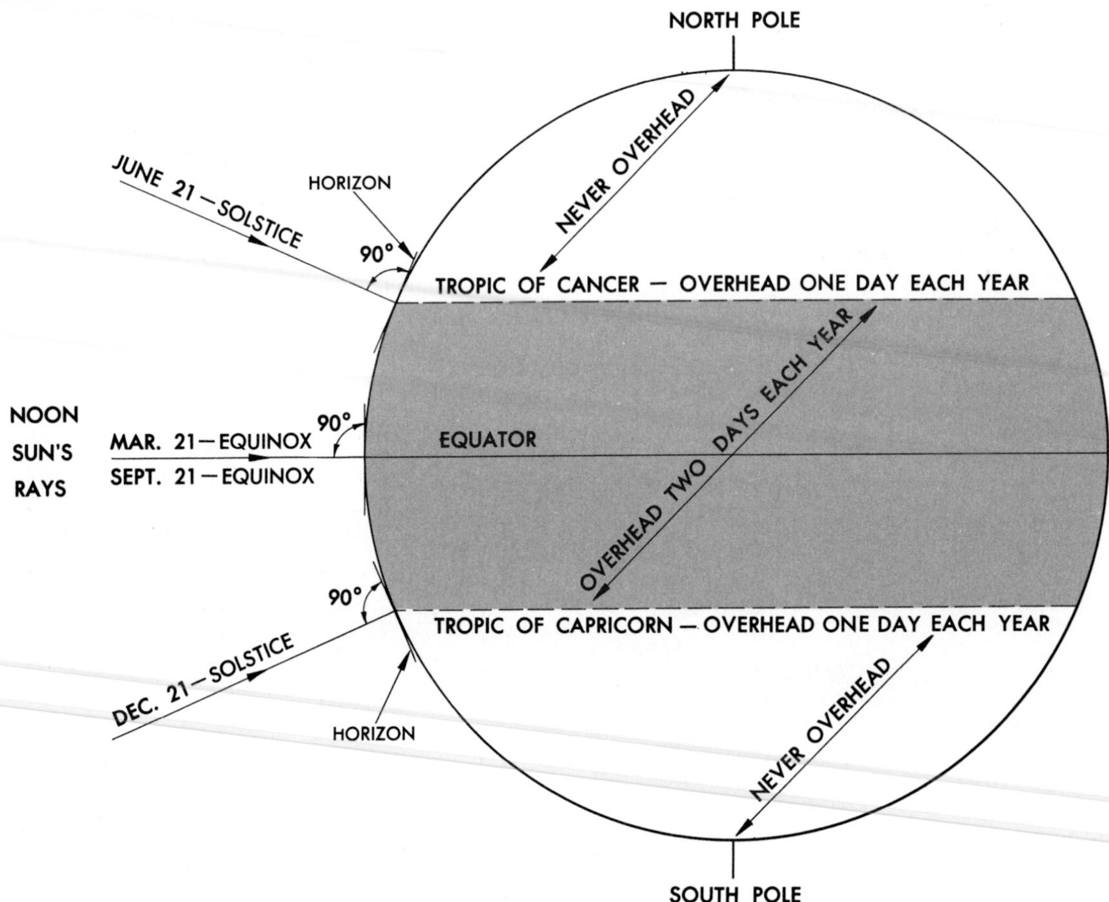

Fig. 2.5. ***The apparent annual migration of the noon sun.*** *Between the Tropic of Cancer and the Tropic of Capricorn the noon sun is overhead, or at 90° to the surface, twice during the year. This is because the earth is tilted on its rotational axis as it revolves around the sun.*

earth that ever experiences the sun directly overhead at noon—on December 21. These Tropics mark the outer limits of an important geographic region. Although the Tropic lines themselves experience the noon sun overhead only once a year, in between these parallels the sun is overhead at noon on *two* days out of the year (Fig. 2-5). This region is the only one on earth where this phenomenon occurs.

In between the Tropics of Cancer and Capricorn the sun is overhead at noon on two days out of the year.

At the equator, halfway between Cancer and Capricorn, these two dates are equally spaced between June 21 and December 21—on the equinoxes, September 21 and March 21. But the dates for the sun to be directly overhead for all other locations between the Tropic lines are unevenly spaced, depending on how close they are to the equator. The overhead sun appears to migrate from Cancer to Capricorn and back in one year, thus passing over each spot in between twice—once as it moves south and again as it moves north. The migrating high-angle sun delivers intense solar radiation to this region, which we often refer to as "the tropics."

THE ARCTIC AND ANTARCTIC CIRCLES

World maps and globes also include dashed lines in each hemisphere a short distance from the poles. These are the Arctic and Antarctic Circles. Like the Tropics, they mark the outer limits of a significant geographic area. Again, we find a relationship with axial inclination when we discover that the

Arctic and Antarctic Circles are 23.5° from their respective poles, or 66.5° from the equator (66.5° north or south latitude). The ***Arctic Circle*** is the farthest point from the North Pole that experiences at least one day when the sun never sets and one day when it never rises. The ***Antarctic Circle*** is the farthest point from the South Pole that experiences these phenomena. In other words, these lines are the equatorward limit where a person could experience at least one day (24-hour period) of darkness and one day of continual daylight. Let's examine the reasons for this situation in more detail.

At the Arctic Circle, the day when the sun fails to set is June 21. This day marks the summer solstice when the noon sun is overhead at the Tropic of Cancer. The one day when the sun does not appear above the horizon is December 21. This day marks the winter solstice when the noon sun is directly overhead at the Tropic of Capricorn. The situation is reversed at the Antarctic Circle. The polar regions enclosed by the Arctic and Antarctic Circles experience more than one 24-hour period of light and dark during each year. The number of these all-dark and all-light days increases the closer you get to the poles. The poles themselves have six months of light and six months of dark.

The polar regions experience more than one 24-hour period of light and dark during each year.

These peculiarities occur because each hemisphere tilts first toward the sun and then away as the earth revolves around the sun. During the Northern Hemisphere summer, when this hemisphere is tilted toward the sun, the sun's rays strike beyond the North Pole and, at the same time, fall short of the South Pole. Rotation, which causes day and night over the earth, is ineffective in the polar areas during this period because the north polar regions remain exposed to continuous sunlight, whereas those areas near the South Pole receive none at all. The distance that the sun's rays reach beyond the North Pole and fall short of the South Pole is at its maximum on June 21 and measures 23.5°. On December 21 the sun's rays fall short of the North Pole and strike beyond the South Pole by this same amount (Fig. 2-4A and B).

THE MIDDLE LATITUDES

The middle latitudes—between the Arctic Circle and the Tropic of Cancer in the Northern Hemisphere and the Antarctic Circle and the Tropic of Capricorn in the Southern Hemisphere—experience neither the sun directly overhead at noon nor a 24-hour period of continual darkness or of continual daylight during the course of a year. Those who live near the Arctic or Antarctic Circle will be familiar with extremely long summer days and winter nights. In each 24-hour period of the year, however, the sun will set and rise if only for a short time. Those who live on the poleward margins of the Tropics will never see the sun directly overhead, although the sun will appear very high in the sky at noon in the summer.

The Sun's Angle

The angle at which the sun's rays strike the earth's surface is called the ***angle of incidence.*** In our discussion so far, we have seen that this angle varies with latitude and time of year. If you lived at low latitudes—for example, in the Amazon River basin of South America or in Indonesia—the midday sun would be nearly overhead, at a high angle of incidence, all year. You would experience a very warm, unvarying climate. The earth in these locations continually receives direct solar radiation; that is, each day the noon sun's rays strike at or near a 90° angle to the surface. As a result, the heating throughout the year is intense and constant (Fig. 2-6).

At higher latitudes—in regions such as northern Canada or Siberia—the angle of incidence even during the period of highest sun (summer) is comparatively low. There are days or even months during the winter (depending on latitude) when the sun never appears above the horizon (Fig. 2-6B). This low angle of incidence is one reason for the lack of heating at higher latitudes; the other is day length.

Day Length

In the tropics, day and night are nearly equal in length throughout the year—about 12 hours of each (Fig. 2-6). Because the angle of incidence changes little in these regions and the amount of daily solar heating is nearly constant, the tropics are essentially seasonless.

At latitudes outside the tropics, the length of the day and night within a 24-hour rotation varies considerably between the summer and winter seasons.

At latitudes outside the tropics, the length of day and night within a 24-hour rotation varies considerably between the seasons of summer and winter. This condition is a direct result of the inclination of the earth's axis and is the reason

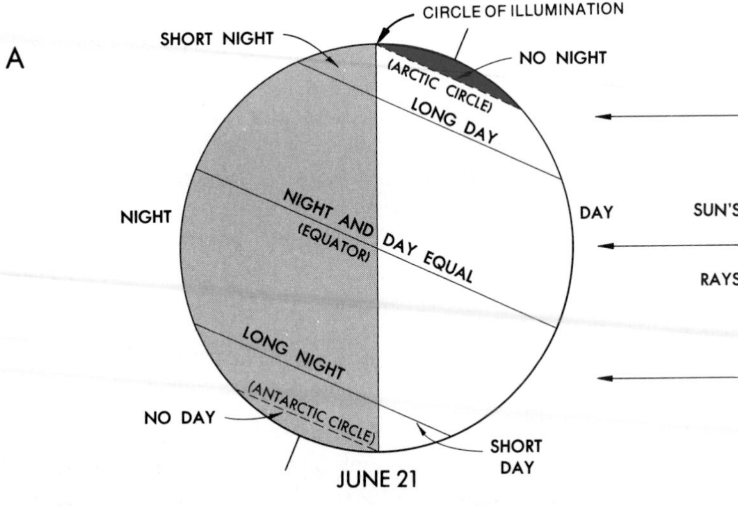

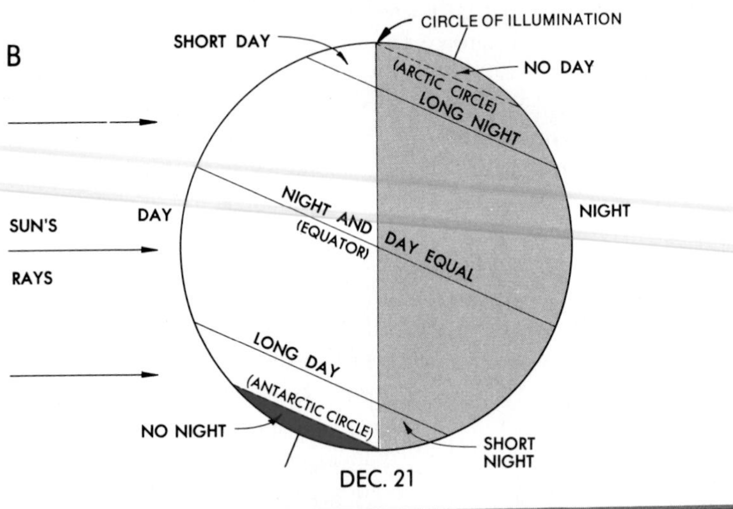

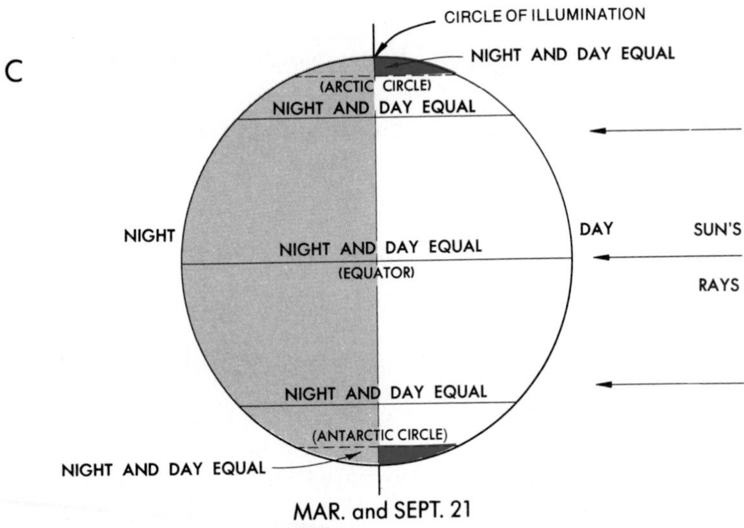

Fig. 2.6. ***The sun's angle.*** *The sun's rays strike any given location on the surface at different angles depending on the time of year. The circle of illumination, while always dividing the earth in half, shifts as the earth changes its position relative to the sun during the year. This results in seasonal changes in day/night lengths at middle and high latitudes.*

for the long summer days and the long winter nights of the middle and high latitudes (Fig. 2-6A and B). The length of day and night at a given latitude depends on the time of year. For example, around June 21 a midlatitude location in the Northern Hemisphere would have a relatively long day (sunlit period), but an equivalent latitude in the Southern Hemisphere would have a relatively short day (Fig. 2-6A). During a hemisphere's winter there is much less incoming solar energy at middle and high latitudes because of shorter days and lower angle sun. Most heat generated at the surface and in the atmosphere is readily lost to space during the long nights.

It may help to remember, that regardless of what day it is, one-half of the earth is always illuminated by the sun while the other half is in darkness. However, because of the axis tilt, the great circle marking the boundary between these two halves, the ***circle of illumination***, does not always pass through the poles as it does on the equinoxes (Fig. 2-6A, B, C). For this reason there are day/night length variations throughout the year everywhere except on the equator.

Summer days and winter nights are much longer in polar regions than in the middle latitudes.

As latitude increases, the length of days in summer and of nights in winter increases as well, so that summer days and winter nights are much longer in polar regions than in the middle latitudes. At or near the poles in summer, the lower angle of the sun's rays and the subsequent diffusion of energy across the surface does not allow heating on the scale of lower latitudes, even though day length is longer. In winter, there are days when the sun never comes above the horizon; there is no heating, and the earth's surface becomes colder and colder. Accordingly, very high latitudes are areas of cool summers and severe winters.

CONCLUSION

We have seen that the seasons are caused by the changing relationship of the earth and sun during the year. The consistent inclination of earth's rotational axis during its annual revolution results in changes in incident sun angle and changes in day length. These two variables, sun angle and day length, determine the intensity and duration of solar heating at a given latitude. Because sun angle and day length vary over the year in the middle and high latitudes, there are seasonal variations in heating and cooling of the surface and the lower atmosphere. If there were no tilt to the earth's axis of rotation, there would be no variation in incoming solar radiation and no seasons. In the next few chapters we will examine how the energy from the sun interacts with the earth's surface and atmosphere and how these interactions are related to global patterns of weather and climate.

KEY TERMS

diurnal	*autumnal equinox*	*Arctic Circle*
plane of the ecliptic	*vernal equinox*	*Antarctic Circle*
parallelism	*Tropic of Cancer*	*angle of incidence*
summer solstice	*Tropic of Capricorn*	*circle of illumination*
winter solstice		

REVIEW QUESTIONS

1. How important is the earth's elliptical orbit to seasonal change?

2. Earth's angle of rotation is 23.5° from the vertical. But how do we determine what is vertical?

3. Why does the North Pole always point toward Polaris regardless of earth's position in orbit?

4. What is the difference between a solstice and an equinox?

5. How are the Tropics of Cancer and Capricorn similar? How are they different? Why?

6. How are the Arctic and Antarctic Circles defined? How are they different from the Tropics of Cancer and Capricorn?

7. Why is it dark for six months and then light for six months at the poles?

8. Why is the noon sun never directly overhead in the midlatitudes?

9. Why are summer days in Norway very long but not very warm?

APPLICATION QUESTIONS

1. Suppose you are a planetary engineer in the distant future. It is your job to prepare planet X for colonization. Planet X is 90 million miles (144 million km) from its sun and has a rotational axis at exact right angles to its plane of the ecliptic and a perfectly circular orbit. Your company wants to build both summer and winter resorts to attract intergalactic tourists. Your budget will allow you either to tilt the planet or to change the shape of its orbit, but not both, to give the planet seasons. Which would you do? Why?

2. Does the sun really "stand still" during a solstice? What is a night equal to during an equinox? Why do you think solstices and equinoxes were so important to prehistoric people that they built monuments like Stonehenge to keep track of them?

3. Suppose you win a multimillion dollar lottery and are able to build two homes anywhere in the world. You enjoy summer activities but don't like to get overheated. Where would you build your two homes so that you could enjoy yourself year round? Explain your choice of locations.

4. You are planning a trip by dogsled to the South Pole. Your base camp on the Ross Sea is about 800 miles (1280 km) from your destination. You plan to travel six hours a day at an average speed of 3 miles (4.8 km) per hour. When is the *latest* you could set out in order to reach the pole *and return* to your base camp before the pole itself began its six-month long period of darkness?

PART TWO

EARTH'S ATMOSPHERE

CHAPTER 3

THE ATMOSPHERE'S COMPOSITION AND STRUCTURE

OBJECTIVES

After studying this chapter, you will understand

1. What materials make up the atmosphere.
2. How combustion and other human activities result in air pollution.
3. How the atmospheric layers differ.
4. What is meant by elements and controls.

The atmosphere is a mixture of many gases, and it is never clean. But humans have made it dirtier in so many different ways that it will probably never be normal again. Copper smelting doesn't do much for the countryside, even with high stacks to disperse the noxious emissions. Much of the smoke we see is particulate matter (solids and dust) cast aloft, yet invisible toxic gases are being introduced into the atmosphere as well. Neither is especially beneficial to plants or the human respiratory system.

INTRODUCTION

As the earth wheels through space, it is enveloped by an atmosphere held closely to it through gravity. A large majority of the planets in our solar system appear to have developed some sort of atmosphere in the early stages of planet formation. But because of the size of our planet, its distance from the sun, and its evolutionary history, our atmosphere appears to differ from that of any other planet in detail of composition.

Earth's atmosphere is a mixture of many gases (and some nongases) that we have always regarded as permanent. However, natural change of composition may be taking place very slowly through such processes as volcanism and radioactivity at the earth's surface, through seepage of certain atmospheric elements into space, and through the dynamic interaction of plants, animals, and atmospheric gases. In addition, it is becoming increasingly obvious that human activity is seriously altering the character of the atmosphere quite rapidly. This chapter will focus on the characteristics of the atmosphere—composition and structure—and will introduce the concept of elements and controls.

COMPOSITION OF THE ATMOSPHERE

Permanent Gases

A rather large variety of gases make up the total amount of air that extends outward from the earth about 6000 miles (9656 km). These gases are curious in both their uneven proportions and their tendency to stratify, or form into layers. In volume the atmosphere is dominated by two gases: nitrogen (78.1 percent) and oxygen (20.9 percent). Argon comes in third at 0.9 percent, and all the many others together make up less than 1 percent (Fig. 3-1).

Percentage of Gases by Volume		*Variable Components*
Nitrogen	78.1	Natural
Oxygen	20.9	
Argon	0.9	Water (vapor-ice-liquid)
		Dust (salt, spores, etc.)
		Carbon dioxide
Neon	Less than	Generally contributed
Helium	0.1%	by humans
Krypton	(trace	
Xenon	gases)	Dust (soot, etc.)
Hydrogen		Carbon dioxide
Ozone		Nitrogen dioxide
Methane		Sulfur dioxide
etc.		etc.

Fig. 3-1. ***Composition of the atmosphere.***

A large variety of gases make up the total amount of air that extends outward from the earth about 6000 miles (9656 km).

Variable Gases

Locally, these totals change slightly from time to time because of the introduction of variable gases. Two come from the earth and appear in the lower atmosphere:

1. ***Water vapor,*** the colorless, odorless, gaseous form of water.
2. ***Carbon dioxide,*** a gas derived from combustion (burning), organic decomposition, and respiration.

WATER VAPOR Water can exist as a liquid, solid (ice), or gas (water vapor) in air. As a gas, it can account for as much as 4 percent of the atmosphere mix over tropical oceans or as little as 0 percent over Antarctica in winter. In the long run a balance exists between the production and use of water vapor. Moisture evaporated from the sea as a gas is eventually delivered back as water or ice via streams, rain, snow, icebergs, and the like.

CARBON DIOXIDE A similar balance exists between the production and use of carbon dioxide. Processes that use up molecular oxygen and create carbon dioxide are matched by equivalent processes—plant photosynthesis and oceanic absorption—that give off oxygen and use carbon dioxide.

A temporary imbalance is normal; that is why carbon dioxide is classified as a variable. However, human action has upset the natural tendency toward carbon dioxide balance, as you will learn in the next section.

Gaseous Contamination and Smog

Humans have uncaringly or unknowingly become obsessed with combustion. Of course, combustion is normal; forest fires and volcanic activity were here long before people. We have, in fact, put the long-burning prairie and forest fire of the prehistoric past under very efficient control with the use of modern firefighting techniques and equipment. Smokey the Bear is not loafing on the job, nor are cigarette throwers and children with matches the main culprits in upsetting the carbon dioxide balance.

THE VAN ALLEN RADIATION BELTS

The powerful magnetic field of the earth exerts an influence well beyond the outer margin of the atmosphere. This zone, termed the ***magnetosphere***, may extend rather unevenly for more than 50,000 miles (80,500 km) from the earth. Satellites monitoring radioactivity in space as early as 1958 relayed the unexpected information that radiation of uncommon intensity was concentrated in two belts in the magnetosphere: one at about 10,000 miles (16,000 km) above the earth and a second lesser belt at 23,000 miles (37,000 km). They have been designated the ***Van Allen radiation belts*** after the physicist James Alfred van Allen (1914–) who first described them.

These radiation belts appear to be concentrations of highly charged particles from the sun, trapped by the earth's magnetic field. Thus here is yet another line of defense in protecting the earth and its inhabitants from dangerous radiation.

When we observe periodic solar flares and solar storms along the sun's periphery, magnetic disturbances in the magnetosphere follow faithfully as evidenced by short-term difficulties in radio transmission. During each solar storm a larger than normal amount of the sun's highly charged residue may find its way to earth.

Dermatologists have been attempting to correlate increased incidence of skin cancer with observed major solar flares, but so far they have had only varied results. Perhaps repeated exposure is more critical than exposure to single events. What if the earth's magnetic field should weaken, as it may very well have done each time it reversed itself in the geological past? Could the human race disappear abruptly because of bombardment by deadly radiation from outer space?

---

2. When they eventually find their way to the upper stratosphere and are exposed to ultraviolet energy, they decompose, freeing chlorine which in turn destroys the ozone.

By absorbing significant quantities of ultraviolet waves, the ozone layer serves as an important atmospheric barrier to potentially dangerous solar radiation.

The danger is that increasing ultraviolet radiation escaping the ozone trap may damage vegetation, step up the incidence of skin cancer, and even affect the long-run climate. Why take a chance? Chemical companies are now working on a replacement for CFCs, and the United States and the Soviet Union have agreed to drastically reduce their use.

Mesosphere and Thermosphere

Beyond the stratosphere is the ***mesosphere*** where temperatures decrease to their lowest reading in the entire atmosphere at the ***mesopause*** 60 miles (97 km) high. The mesosphere is dominated by oxygen in the form of single atoms resulting from separation of oxygen molecules by shortwave solar energy. Still farther out, to approximately 500 to 600 miles (805 to 960 km), the temperature once again increases, chiefly as a result of ionization,[2] giving rise to the term ***thermosphere***. It is important to understand, however, that the air is so thin at this level and the nitrogen and oxygen atoms so remote from one another that there is no real heat as we know it on the surface. At these heights the gaseous medium becomes so thin that satellites experience essentially no frictional drag, nor do they gather increased surface temperature from the surrounding gases despite the "heat" of the thermosphere. Surfaces exposed to direct sunlight, however, are warmed, and solar panels affixed to satellites can be used effectively to generate electric power. We must make a distinction between *sensible temperature,* which is tangible and personal, and the physicist's definition of *heat equals energy*. We would not feel warm in the thermosphere, despite the fact that the gases there have absorbed much solar energy and are moving rapidly.

We would not feel warm in the thermosphere, despite the fact that the gases there have absorbed much solar energy and are moving rapidly.

Those who select 600 miles (966 km) as the outer limit of the atmosphere must admit that there are gases beyond that point, chiefly helium. To call such thin air "atmosphere" is debatable, however. Above about 1500 miles (2414 km), helium gives way to hydrogen, the lightest of the gases. Perhaps the limit of trace hydrogen is about 6000 miles (9656 km), but the "top" of the atmosphere should not be thought of as a distinct boundary. Some people even dislike the term *interplanetary space* for the region beyond this, citing tenuous evidence that this greatest natural void of all is not absolutely without gaseous matter.

[2] At these high altitudes, ultraviolet waves and high-energy particles (cosmic waves) interact with oxygen and nitrogen atoms to expel free negative electrons. This process is called *ionization;* an alternative designation for most or all of the thermosphere is *ionosphere.*

ELEMENTS AND CONTROLS

The composition and behavior of the atmosphere, especially the troposphere, determine weather and climate. There are four basics or ***elements of weather and climate:***

1. Temperature
2. Pressure
3. Wind
4. Moisture (in all its forms—gas, liquid, and solid)

These four elements work together in endless variations, but at this stage, in order to understand them as precisely as possible, we will try to isolate and examine each element. In reality they cannot be totally isolated, as we will continuously discover, but it is worthwhile making the effort.

As we move along in our discussion, we make the transition from ***weather*** (the momentary state of the elements) to ***climate*** (a long-term average of the daily weather). With this change of emphasis will come an increasing concentration on a second set of factors called ***controls of weather and climate.*** How much rainfall does a given location receive on the average and why? What is the reason for extreme temperature variations within a short distance?

For example, latitude is probably the single most important control of temperature; it gets colder as you move away from the equator. But it also gets colder as you climb to the top of a snowcapped mountain or, often, as you approach the sea. Oceans and mountains are therefore temperature controls, too. There are moisture controls, and wind controls, and pressure controls—a virtually endless list. As we proceed with our discussion of elements, we will identify the important controls that help us to understand global weather patterns. Then, with the focus on individual climates, we will see how a particular set of controls acts to shape regional climatic conditions.

CONCLUSION

This chapter is an introduction to the earth's atmosphere—its composition and structure—and is preparatory to the discussion of the elements of weather and climate to be considered in the next three chapters. We can now conclude that most of the atmosphere is nitrogen and oxygen, but that it also contains other gases and even liquids and solids. Especially important are carbon dioxide, water vapor, and dust, for they are directly linked to heating and cooling. Although human activities are largely confined to the earth's surface, they influence atmospheric composition and air quality by additions of carbon dioxide, dust, CFCs, and other pollutants. We know not only that stratification of the atmosphere is mainly the result of temperature changes upward from the earth, but also that stratification is related to gaseous composition. Of the several atmospheric layers, the troposphere is most important to us for it is the air we breathe and the site of earthly weather. As we consider the elements of weather and climate, keep in mind that we are focusing mainly on the troposphere and that our major aim is to understand the fundamentals of weather that add up to the great diversity of climates over the earth's surface.

KEY TERMS

water vapor
carbon dioxide
dust
stratification of the atmosphere
tropopause
troposphere
stratopause
stratosphere
mesopause
mesosphere
thermosphere
magnetosphere
Van Allen radiation belts
elements of weather and climate
weather
climate
controls of weather and climate

REVIEW QUESTIONS

1. Which gases have a greater effect on our weather—permanent or variable gases? Why?

2. How has human action upset the carbon dioxide balance? Can natural occurrences upset this balance as well? What is the difference?

3. Why do we need dust in our atmosphere?

4. Why is our atmosphere stratified? How—and why—do the strata differ?

5. What is the difference between an element and a control?

APPLICATION QUESTIONS

1. We have said that our use of fossil fuels over the last century and a half has contributed to the pollution of our atmosphere. Go to the library and found out what the air was like in London, England, in the middle of the nineteenth century. Compare it with the air over Los Angeles today. Are we better off or worse off? In what ways? Are the results mixed?

2. Say an above-ground nuclear blast and a volcanic explosion both sent the same amount of dust into the air. Which "dust cloud" would be likely to remain aloft the longer? Why? Which would you rather be under when it descended? Why?

3. You already know that our increased wealth over the past century or so has increased pollution by throwing automobile exhaust, CFCs, and various other harmful substances into the atmosphere. But did you know that our increased consumption of hamburgers, steaks, milk, and cheese might be damaging our atmosphere as well? See if you can find out what all these things have in common and how they could possibly be pollutants. What are scientists trying to do about it?

4. As you know, there are both elements and controls of weather and climate. Tune in to a comprehensive weather report on your television or radio, and listen carefully for the discussion of each element. How does the treatment of elements contribute to the total weather picture? Are controls presented? If so, in what way?

C H A P T E R 4

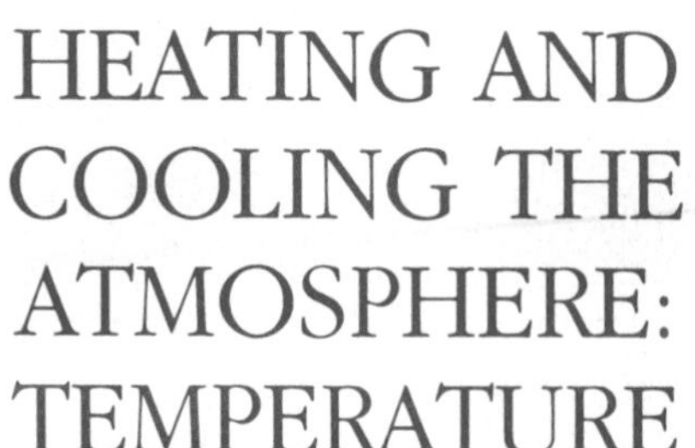

HEATING AND COOLING THE ATMOSPHERE: TEMPERATURE

O B J E C T I V E S

After studying this chapter, you will understand

1. How solar energy travels through the atmosphere to heat the earth.
2. How the atmosphere is heated.
3. How the global energy budget stays in balance.
4. How solar power can be used as a source of energy.
5. How and why temperature changes with altitude.
5. How isotherms can be used to show temperature patterns on a map.

The snow-laden trees proclaim that the season is winter, yet under the glass cover is a luxuriant profusion of plant growth—the greenhouse provides a demonstration laboratory of solar radiation behavior. Its function is as a transformer of wavelength from short incoming waves to longer heat waves. The earth and its atmosphere operate in similar fashion.

INTRODUCTION

Have you ever wondered why valleys are so much warmer during the summer than adjacent mountain tops? Why abnormally cold winter nights occur when the sky is clear? Why coastal areas have mild temperatures while inland areas at the same latitude suffer temperature extremes? Why the oceans don't freeze solid or boil over? In this chapter we begin our discussion of the elements of weather and climate by focusing on atmospheric temperature. Our first concern is with the means by which the atmosphere is heated. Then we consider heat transfer from one location to another. Our chief aim in this chapter, however, is to understand the global energy budget—how the temperatures that support life are maintained. In the process, we provide the background to answer questions about temperature like those above and pave the way for discussion of the other three weather elements: pressure, wind, and moisture.

TEMPERATURE

When we speak of temperature, we are concerned with relative hotness or coldness, that is, with the relative amount of molecular movement and how often molecules collide with one another—collisions produce sensible heat. Before we consider the mechanisms involved in heating the earth's surface and atmosphere, we must ask: "Where does the energy for heating come from?" The answer is, of course, from the sun. But the sun is merely one star among the heavenly legions and not a very remarkable one at that, in either brilliance or size. Moreover, the earth, being over 100 times smaller and 93 million miles (149,669,000 km) distant, intercepts an incredibly tiny amount of the sun's total radiation—actually less than one-half of one billionth. Yet the sun is our source of energy; the solar furnace, directly or indirectly, drives the earthly engine.

The sun is our source of energy; the solar furnace, directly on indirectly, drives the earthly engine.

What other sources could there be? Other stars, some of which are very large and hot, clearly emit enormous quantities of energy, but they are too far away to affect the earth. Internal heat released by vulcanism or radioactive decay can in no way be considered significant except perhaps in an extremely local situation. We must therefore depend on the sun. How much energy actually arrives at the earth's surface and how it is put to use are critical factors in determining the temperature of our planet.

THE SOLAR CONSTANT VERSUS INSOLATION

Even though we receive only a small fraction of the sun's energy, it is the only energy we have. By our earthly measurement standards, it is a very large amount. To measure this amount, we must visualize a plane at the far periphery of the atmosphere (wherever that might be) at right angles to the stream of solar radiation. The amount of energy received at the plane is called the ***solar constant.*** Theoretically, it amounts to about 2 calories[1] per square centimeter per minute. As the name implies, this figure should be constant. Actually, it tends to fluctuate 2 to 5 percent as a result of varying solar energy output caused by sunspots or solar storms, and the changing distance from sun to earth owing to our elliptical orbit.

Even though we receive only a small fraction of the sun's energy, it is the only energy we have.

The solar constant represents the potential amount of energy available to the earth, but the amount of energy that actually arrives at the ground is very different. This amount is called ***insolation.*** Don't call it sunshine. There is a tremendous difference between the total energy reaching the earth from the sun and that which is visible. Sunshine is merely the visible portion.

Now visualize an earthly plane parallel to the one at the outer margin of the atmosphere. The difference between the solar constant (energy received at the outer plane) and insolation (energy received at the inner plane) is on the order of 50 percent. In other words, there is a 50 percent casualty rate through the atmosphere (Fig. 4-1).

The difference between the solar constant and insolation is on the order of 50 percent.

We must remember, however, that at different seasons and at different latitudes the amount of atmosphere the solar energy must cross will differ, and the curved surface of the

[1]A calorie is a heat measure equal to that required to raise the temperature of a gram of water 1° C.

MEASURING TEMPERATURE

Grand Duke Ferdinand II, ruler of the Italian state of Tuscany from 1620 to 1670, was a significant figure in the advancement of scientific knowledge. He founded the Accademia del Cimento (Academy of Experiments), dedicated to the practice of experimental laboratory science and the testing of Aristotelian metaphysics. His resident genius was Galileo, long renowned as a scientist, if not generally respected in ecclesiastic circles.

Into this rich compost of intellectual ferment stepped the young mathematician, Evangelista Torricelli (1608–1647), first as Galileo's lieutenant and later as his successor. Within three years of his arrival at the academy, he was to preside over the experiment that resulted in the invention of a workable barometer, demonstrating that the atmospheric weight (pressure) could be measured and that the weight was not static but varied from time to time. Curiously, this critical knowledge opened the gate for the development of an accurate thermometer.

Some 40 years earlier Galileo had shown that air expanded when heated and that, in expanding, would draw liquid up a glass column. But this system was an open one, and the yet unknown effect of air pressure defeated early attempts at accurate calibration. The Grand Duke himself is given credit for realizing, in the wake of Torricelli's revelation, that a hermetically sealed tube could isolate the influence of temperature from that of pressure. As any good Tuscan would, he used wine as his liquid medium.

In the early eighteenth century, Gabriel Fahrenheit (1686–1736), a German physicist/chemist working in Holland, attempted to improve on the thermometer by (1) creating compactness utilizing heavier mercury as the liquid and (2) introducing a rational standard scale. He selected the freezing and boiling points of water as his bases but soon discovered that salty water did not freeze at the same temperature as fresh. After much experimenting, he concluded that 0° on his calibration should represent the lowest temperature at which he could supercool water. Selecting a purely arbitrary length for a degree, he found that 32° equaled the freezing point of pure water and 212° its boiling point at sea level.

The Swedish astronomer, Anders Celsius (1701–1744), followed up a few years later with what he thought was a somewhat simpler scale and called it *centigrade* (the prefix *cent* meaning 100). This system assigns Fahrenheit's 32° ice point as 0° and his 212° steam point as 100°, one degree thereby almost doubling the size of the F°.

Yet another scale used widely in the scientific world is the *Kelvin scale,* after Lord Kelvin (1824–1907) of Britain. It uses the concept of *absolute zero* defined as *that temperature at which molecules cease to move.* The technical definition of heat is related to the speed at which molecules vibrate: the more rapid the movement, the greater the heat. So Lord Kelvin, by simply using Celsius's degree spacing and ice/steam reference points, determined that absolute zero would be the equivalent of −273° C. Kelvin scale: absolute zero = 0° K, ice point = 273° K, steam point = 373° K.

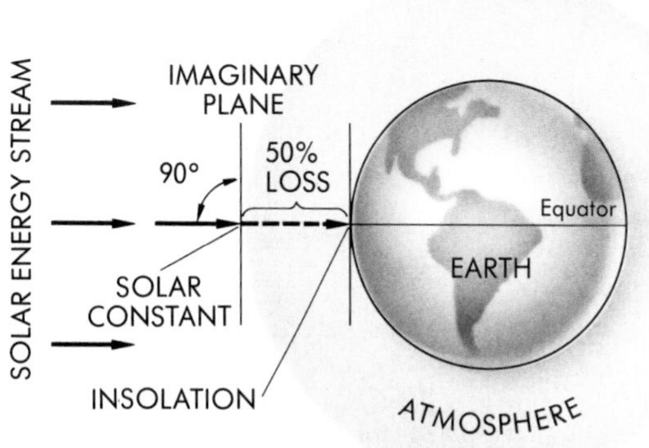

Fig. 4-1. ***The solar constant versus insolation.*** *There is an approximate 50 percent loss as the solar energy passes through the atmosphere.*

earth will accept the solar energy at something less than a right angle. So the 50 percent difference between insolation and solar constant is simply a working approximation.

ELECTROMAGNETIC RADIATION

Before we can consider the fate of solar energy as it passes through the atmosphere, we must be familiar with the ***spectrum of electromagnetic radiation*** (Fig. 4-2). Radiant energy travels through space in a series of waves. There is almost infinite variety in wavelength, but high-energy waves are very short—on the order of a billionth of a meter or less—and low-energy waves are very long—meters or even kilometers. The spectrum of electromagnetic radiation shows the variety of possible wavelengths and divisions or

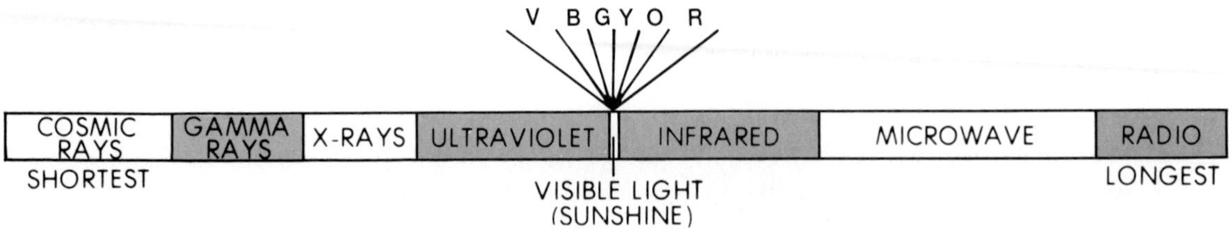

V B G Y O R

COSMIC RAYS	GAMMA RAYS	X-RAYS	ULTRAVIOLET	INFRARED	MICROWAVE	RADIO

SHORTEST — VISIBLE LIGHT (SUNSHINE) — LONGEST

Fig. 4-2. The solar spectrum of electromagnetic radiation. Visible light includes violet, blue, green, yellow, orange, and red wavelengths.

"bands." Thus the shortest wavelengths are referred to as gamma rays, those a bit longer but still very short are x-rays, and so on. Visible light, that is, the electromagnetic energy we can sense with our eyes, occupies a portion of the spectrum between ultraviolet and infrared wavelengths, and it may also be subdivided into a series of smaller bands from the shorter wavelengths of violet, to blue, green, yellow, orange, and finally to the longer wavelengths of red—the colors of the rainbow. By convention, we identify the part of the spectrum with wavelengths shorter than infrared as *short* wave and those of infrared and longer as *long* wave.

SOLAR ENERGY IN THE ATMOSPHERE

The sun is a very hot body. According to Stefan-Boltzmann's Law and Wien's Law, the *hotter* a body, the more energy it will radiate and the *shorter* will be the wavelength radiated. Thus, although the electromagnetic energy radiated by the sun includes a variety of wavelengths, nearly all are short. As solar energy passes through the atmosphere, air molecules, dust, and water droplets react selectively to the short wavelengths, as we will see in this section.

Solar Energy in the Upper Atmosphere

When the sun's radiation encounters oxygen and nitrogen atoms high in the ionosphere, it begins the process of heating the atmosphere. Heating cannot occur without ***absorption***. That is, the molecules of oxygen and nitrogen absorb some of the radiant energy and heating results. Some shortwave radiation is therefore used up and will not reach the earth. In the case of the high-energy waves intercepted in the ionosphere—x-rays and some ultraviolet rays—this is a good thing for they are deadly. Thus the ionosphere acts as a screen to shield us from harm.

Heating cannot occur without absorption.

At a lower level, in the stratosphere, ozone efficiently absorbs most of the remaining ultraviolet waves, some heating results, and again there is an important screen that protects life at the earth's surface. But, although ozone is receptive to ultraviolet, it is transparent as far as other wavelengths are concerned. When radiation passes through atmospheric gases without being absorbed, the process is called ***transmission***.

Some radiant energy waves, however, encounter infinitesimal gas molecules, and perhaps meteorite dust, but are not absorbed or transmitted. They are merely dispersed. This change of direction of energy waves is known as ***scattering***. The energy is not used or lost but is simply redirected. At these higher levels, scattering tends to be selective, and only the shorter visible light waves are affected. This gives the sky its blue color. Larger particles in the lower atmosphere also selectively scatter the longer red waves. So, when the air is polluted with dust and smoke, we are compensated with magnificent red sunsets.

Although ozone is receptive to ultraviolet, it is transparent as far as other wavelengths are concerned.

Solar Energy in the Lower Atmosphere

As the now reduced solar energy enters the lower and denser atmosphere (troposphere), dust particles and water droplets which cause scattering in the upper atmosphere react in other ways as well. Together with carbon dioxide and water vapor, they are excellent absorbers of certain wavelengths. However, these include only the very longest marginal wavelengths of the sun's emissions—red and some infrared. Therefore the amount of absorption and heating is limited.

Together with carbon dioxide and water vapor, dust particles and water droplets are excellent absorbers of certain wavelengths.

Fig. 4-3. ***Reflection.*** *A portion of the sun's short-wave energy stream is reflected off these cloud tops, redirected into space, and lost forever. A small part is absorbed by the tiny water droplets and transformed into heat. But a very large segment simply flows to the earth to be absorbed there.*

In the troposphere a more important factor in determining the amount of insolation the earth receives is the process of ***reflection***—the mirrorlike reversal of the energy stream back toward outer space where it is forever lost to the earth. Cloud tops are the most efficient reflectors, but dust and some gas molecules can also reflect. When flying above a cloud cover, we are immediately aware that reflection is occurring. Sunglasses are the order of the day to combat the painful brilliance of the white cloud tops (Fig. 4-3). Yet it is only the "sunlight" we observe; a certain amount of invisible short-wave energy is being reflected at the same time.

Clouds absorb a minimal amount of energy. Mainly, however, they reflect, and reflection, like scattering, merely redirects. The difference between the two is largely a matter of direction. Nonetheless, reflection from clouds, haze, or dust is never 100 percent efficient. Some energy is selectively transmitted. Beneath a dense cloud, it is not totally dark. A portion of light comes through, some being a product of scattering called ***diffuse daylight***, as well as much of the invisible short-wave spectrum.

Solar Energy on Earth

So now the earth has received its insolation—about 50 percent less than was present at the outer edge of the atmosphere:

1. Some energy was lost through absorption, which added heat to the atmosphere.
2. Some energy was lost through reflection and scattering, which sent it back into outer space.

But our topic is temperature, specifically the temperature of the troposphere—the part of the atmosphere in which we live. The question to ask is, *how much heat does the lower atmosphere gain as solar energy passes through it?*

The answer is, not much. A great deal of energy is turned back via reflection, and the atmosphere is largely transparent to the short sun waves that traverse it all the way to the earth. As we already mentioned, the ionosphere and the ozone layer in the stratosphere absorb and transform energy to heat, but these are in the upper atmosphere and are not involved in our direct personal experience of weather and climate.

Water vapor, water droplets, ice, dust, and carbon dioxide, because of their selective absorbency, combine to contribute *not more than 10 to 15 percent of the total heat gained in the troposphere or lower atmosphere.* Where does the rest come from?

TERRESTRIAL ABSORPTION AND RADIATION

If the sun is our sole source of energy and thus of heat, how does the troposphere acquire 85 to 90 percent of its heat if not by absorbing solar energy as it passes through? Actually, *the sun heats the earth and the earth heats the atmosphere.* The earth functions like a transformer in an electrical circuit. It receives energy in one state—mostly light—and transmits it in another state—mostly heat. An excellent absorber, the surface of the earth is highly receptive to the short-wave energy of sunlight. Molecular motion increases with inci-

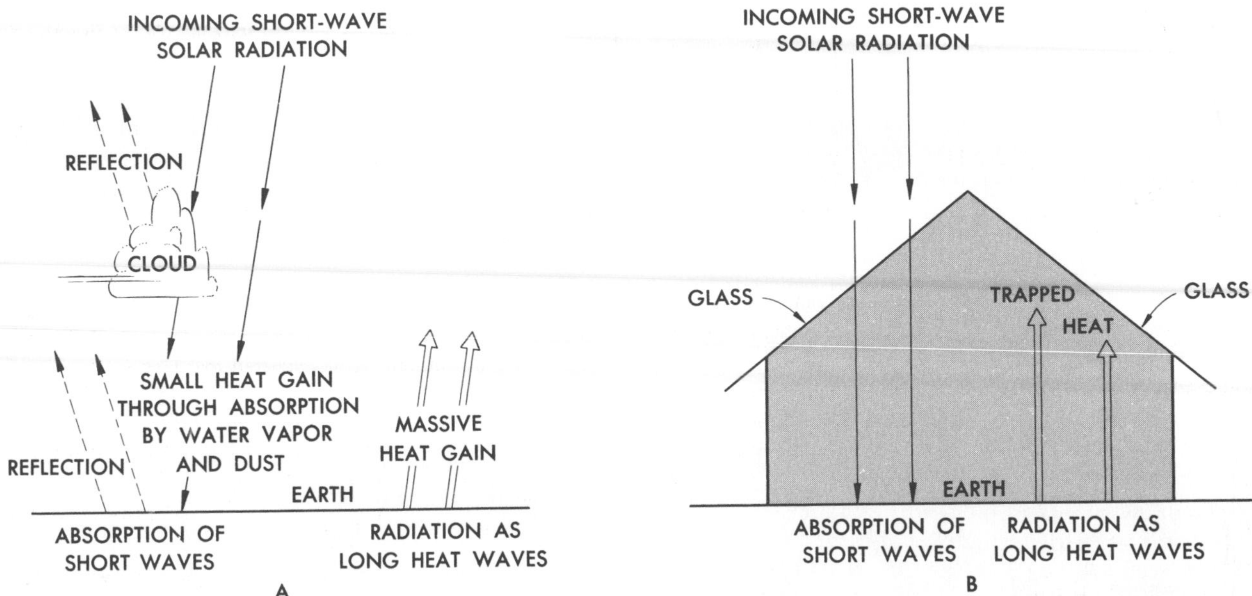

Fig. 4-4. ***Indirect heating of the atmosphere.*** *(A) Heating the lower atmosphere. (B) The greenhouse effect. Like the glass in the greenhouse, the atmosphere allows short-wave energy to reach the earth but traps the long heat waves as they are reradiated.*

dent sunlight at the earth's surface, and temperature increases. As we have seen, the wavelength of energy emitted is directly proportional to temperature (Wien's Law). Therefore the earth, at a much lower temperature than the sun, emits mostly in the infrared or *heat band* of the electromagnetic spectrum. Remember: the reason the troposphere did not absorb much of the sun's energy originally was that its gas molecules and particles tend to transmit rather than absorb short waves. The long-wave radiation that enters the atmosphere from the bottom after leaving the earth's surface is in the general range that water vapor, water droplets, dust, and carbon dioxide concentrated in the lower atmosphere can efficiently absorb. Heating is therefore both rapid and massive.

The sun heats the earth and the earth heats the atmosphere.

The Greenhouse Effect

The indirect heating of the atmosphere via the energy absorbed by the earth's surface is called the ***greenhouse effect.*** Gardeners can grow tomatoes in February by taking advantage of the fact that the glass roof in a greenhouse is transparent to light waves. The atmosphere works just like the glass greenhouse roof. Most light waves pass through the atmosphere, strike the earth, and are absorbed. The earth then reradiates them as long heat waves that find the "glass" a solid barrier, a trap. Create an instant greenhouse: park your car in the sun for a time with the windows closed and observe the change in interior temperature. You will be experiencing a demonstration of wavelength differences and selective absorbance (Fig. 4-4). It must be emphasized that, although the greenhouse effect is a natural phenomenon critical to the maintenance of temperatures suitable to life on earth, we are now so artificially enhancing the effect that global warming may result.

The atmosphere works just like the glass greenhouse roof.

Land Versus Water

The earth's surface is composed of a variety of materials, each with a different rate of warming and cooling. New snow, for example, will warm very slowly because, like the white cloud top, it reflects rather than absorbs. So will light-colored sand or soil and (to some extent) water, which makes up well over two-thirds of the surface of the earth.

The degree of reflection or ***albedo*** varies (Fig. 4-5). Snow is the best reflector, other light-colored materials are less so,

THE GREENHOUSE EFFECT AND CLIMATE WARMING

If the greenhouse effect is such a common experience as to be felt in a parked automobile, why are we so concerned about it? Without "greenhouse gases"—water vapor, carbon dioxide, methane, and others that absorb long-wave energy—all earthly radiation would go directly to space without heating the atmosphere. Fortunately, a mix of these gases holds the heat for a time before losing it to space. But one point must be kept in mind—the higher the concentration of greenhouse gases, the greater the absorption of long-wave energy and the ***higher the atmospheric temperature***. Throughout most of earth's history, the quantity of greenhouse gases has been relatively stable. There is even some evidence suggesting that the overall trend may be one of slight decrease in carbon dioxide through geologic time. Given the fact that the sun is gradually warming as it matures, we can see that the long-term stability of atmospheric temperature—so important to the evolution and maintenance of life on earth—is the result of a precarious balance between the quantity of incoming solar energy and the quantity of greenhouse gases to retain it. Over the eons, the warming sun has been balanced by the long-term decrease in greenhouse gases.

Today, however, we have become alarmed by evidence that our profligate use of fossil fuels and forest clearing are increasing atmospheric carbon dioxide and methane at a rapid rate. If the present trend continues, the concentration of carbon dioxide may soon double. With the balance tipped toward greater absorption, air temperatures will increase by an estimated 4° to 9° F (2° to 5° C). Although a major scientific controversy rages as to the full implication of climate warming, it is expected that higher temperatures will

1. Melt glacial ice, causing sea level to rise and drown some densely populated coastal areas.
2. Change world agricultural patterns, causing drier conditions in U.S. wheat regions, for example.
3. Affect river flows, with more flooding in some and less flow in others.

Has the expected warming begun? There is no clear indication yet, but most scientists say that it is just a matter of time.

and water is effective only when the sun is low. This means that during the middle of a summer day, or in the tropics where the sun is high in the sky all year, solar energy will penetrate water rather than reflect. Therefore we should probably regard water as a relatively poor reflector since the low-angle sun's rays that it does reflect are a good deal less effective heaters than direct rays. They not only lose energy by passing through a greater depth of atmosphere, but also disperse energy over a much wider area (Fig. 4-6). Dark plowed fields, evergreen forests, and asphalt parking lots (which increasingly cover our world) absorb much more solar energy than they reflect. You have probably noticed the heat-absorbent properties of blacktop if you've ever tried to walk barefoot across a parking lot in August.

	(Approximate Percentages)
New snow	90%
White sand	40
Light soil	30
Concrete	30
Forest or crops	15
Dark soil	15
Macadam	10
Water	5–90, depending on angle of receipt

Fig. 4-5. Albedo of selected earth surfaces.

With these endless diversities in mind, we can generalize to the extent of saying that only two basic materials make up the surface of the earth: land (29 percent) and water (71 percent). These two differ fundamentally in their ability to absorb and reradiate solar energy.

Land, on the whole, is a good absorber and in the summer heats up rapidly, whereas water, also a good absorber, increases in temperature only slightly over a three-month hot season. What causes this great difference to heat between land and water? The reasons are several.

1. Land is opaque. The energy arriving on one square foot is strongly concentrated at the surface. A similar square foot (0.09 m) of transparent water allows energy to penetrate to a considerable depth, thus dispersing the heating. The surface heats with only a fraction of the efficiency of a comparable area of land.
2. Water is moving, both vertically and horizontally, so that no given area of surface remains exposed to the

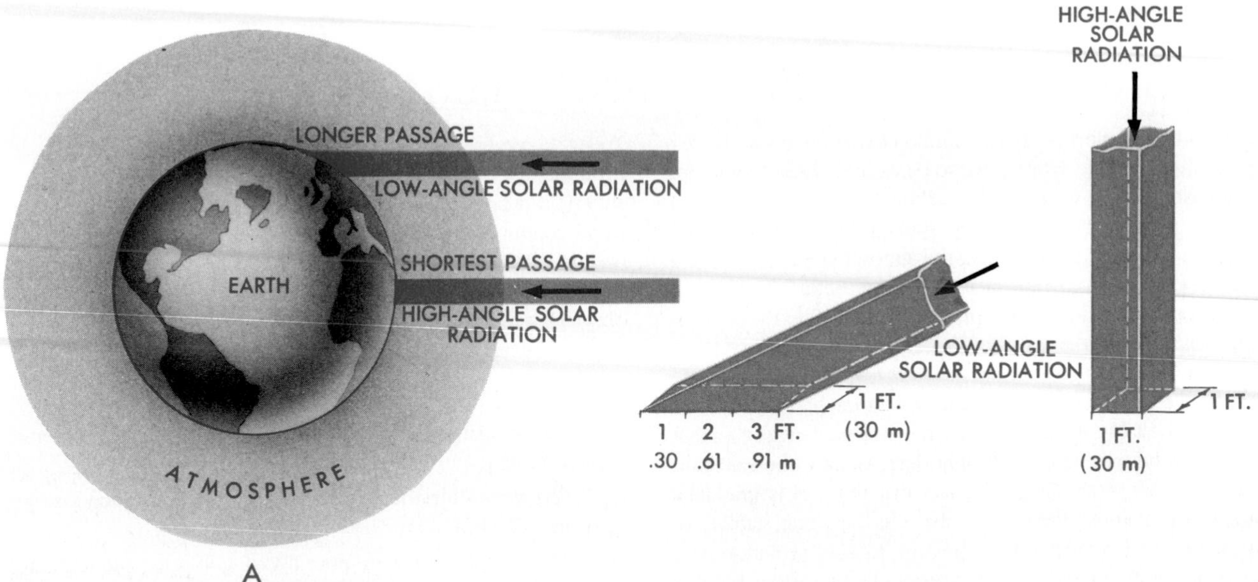

Fig. 4-6. ***Heating effectiveness of low-angle versus high-angle radiation.*** *Low-angle radiation loses energy as it passes through an exaggerated depth of atmosphere (A) and then disperses that lesser energy widely (B).*

same amount of energy as does a comparable land surface.

3. The cooling effect of evaporation over the water is more continuous and effective than over land.
4. Water has a higher ***specific heat*** than land. It is simply a physical law that, other things being equal, it requires almost five times as much energy to raise a gram of water one degree as it does a gram of dry earth.

Obviously, then, we can expect a continent to heat much more intensely during the summer than do the oceans.

Land is a good absorber and in the summer heats up rapidly, whereas water, also a good absorber, gets only slightly warmer during the hot season.

There is yet another dimension to specific heat: the rate of cooling. During the winter and at night, when the sun's energy is reduced or lacking, the earth will rapidly lose its heat, but water (with higher specific heat than land) will tend to maintain the heat it has, releasing it very slowly. Thus the temperature of the earth's land surface in the middle latitudes will vary from summer days to winter nights as much as 150° F (66° C), whereas that of the oceans may exhibit a change of only 3° to 5° F (2° to 3° C). Because the lower atmosphere takes its temperature characteristics from the earth's surface, we can easily see how oceanic air masses may differ radically from those that are normally found over land areas. The oceans and their air masses tend to maintain much the same temperature the year round, whereas the land and the air above it often show great seasonal extremes.

The oceans and their air masses tend to maintain much the same temperature the year round, whereas the land and the air above it often show great seasonal extremes.

Heat Transfer

There are several mechanisms by which the heat from the earth is transferred to the atmosphere and the heat in the atmosphere is transferred to various locations on the earth. We can divide these mechanisms into four pairs:

1. Conduction and radiation
2. Compression and expansion
3. Condensation and evaporation
4. Advection and convection

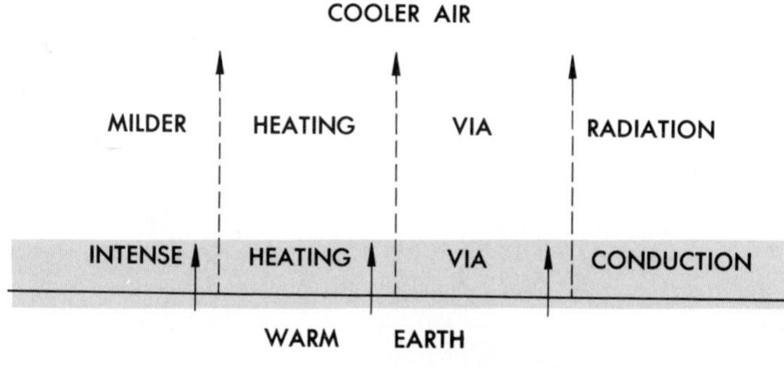

Fig. 4-7. ***Heat transfer.*** *Because conduction is much more efficient than radiation, the most intense heating occurs immediately above the surface.*

CONDUCTION AND RADIATION Two related mechanisms transfer heat from the earth to the air above it—conduction and radiation. *Conduction* is the movement of heat from a warm body to a cold one in *contact* with it. Heat will flow from the warmer to the colder until they are of equal temperature. As a simple illustration, place a sterling silver spoon in a cup of hot coffee and note how the handle heats to the temperature of the coffee. Silver is an excellent conductor and will rapidly assume the temperature of the liquid.

The warm earth will heat the adjacent cooler air through conduction.

The same principle applies to the earth and the atmosphere at their point of contact. The warm earth will heat the adjacent cooler air through conduction. At first this heated layer will be shallow, only a few feet deep, but as the day wears on it will become both warmer and deeper.

Radiation, on the other hand, is the transfer of heat through space. It is much like conduction except that contact is not necessary; therefore it can heat a wide zone of the lower atmosphere. Conduction is the more efficient of the two, however, and the air in contact with the earth will heat much more intensely. Putting your hand on the stove will result in blisters—intense heat transfer via conduction. In contrast, placing your hand several inches above the burner will make it pleasantly warm—demonstrating modest heat transfer via radiation (Fig. 4-7).

These processes differ because of differences in the air's ability to absorb heat. Water vapor (or humidity) is the most critical element, especially in maintaining the heat in the atmosphere. It may easily become excessively hot during the early afternoon in a desert climate simply from intense surface conduction and radiation; there are no clouds, and the earth receives its full complement of solar radiation. As soon as the sun goes down, however, that same clear air and lack of humidity allow the earth's accumulated heat to escape at a rapid rate. If, however, water vapor is introduced into the lower atmosphere, the long-wave terrestrial radiation is absorbed by the gas molecules, which in turn reradiate some of their heat back toward the earth. Water vapor not only closes the escape hatch for the earth's amassed heat of the day, but also holds it near the surface allowing only very gradual heat loss over a long night.

Conduction and radiation can also cool the atmosphere. We have found that the warm earth transfers heat to cold air through conduction. But on a long winter night with the ground snow-covered, the air still, and the day's heat dissipated rapidly via a clear atmosphere, the earth may be very much colder than the air above it. When this condition occurs, heat moves via conduction in the opposite direction and the lowest stratum of the air is cooled. The air at a higher level can also lose heat through radiation to the colder earth as well as out into space. Conduction and radiation may therefore function at any given time either to warm or to cool the lower atmosphere.

COMPRESSION AND EXPANSION Another common means by which the atmosphere may be heated or cooled is by volumetric change, that is, by compression and expansion. *Compression* occurs when an air mass loses altitude. As it descends, the greater pressure of air from above causes the same number of air molecules to compress into a smaller volume that results in heating. On the other hand, *expansion* of an air mass occurs when it gains altitude and air molecules occupy a greater volume. In this case the result is cooling. Such temperature changes are termed *adiabatic,* for they are accomplished by changes in volume only—without any outside heating or cooling applied. Adiabatic warming and cooling are common occurrences because sizable masses of air are constantly in motion in the troposphere.

Compression ***occurs when an air mass loses altitude;*** **expansion** ***of an air mass occurs when it gains altitude.***

CONDENSATION AND EVAPORATION Yet another means by which the atmosphere is warmed and cooled is by changes in the state of water. *Condensation,* the change of water vapor into a liquid, releases heat into the atmosphere. Where did the heat come from? It has been there all the time. In the process of becoming a gas (*evaporation*), the vapor absorbed heat. Let us suppose that a person is perspiring. As that perspiration evaporates from the skin, she feels cooler—heat has been subtracted. Where did it go? It is in the water vapor waiting to be released when the process is reversed and the gas becomes a liquid. It may not seem like a significant amount of heat to be considered, but when literally trillions of raindrops are formed instantaneously in a storm, a great deal of heat is released in a hurry.

Condensation, ***the change of water vapor into a liquid, releases heat into the atmosphere.***

We can carry this scenario a step further. Heat is required to change ice into water, so each drop of water has an increment of latent energy. Then heat is used in changing water to gas, and water vapor now has two increments of energy. If water vapor is suddenly changed directly into ice, as in the formation of snow (*sublimation*), two increments of energy are released and the air is heated rapidly (Fig. 4-8). These are much more than interesting phenomena; *latent heat of condensation* is an important element in the formation and perpetuation of most of our storms (Fig. 4-9).

Latent heat of condensation ***is an important element in the formation and perpetuation of most storms.***

We can now list four basic means of heating the atmosphere: conduction, radiation, compression, and condensation. The accompanying list shows each with its corresponding cooling mechanism:

Heating	*Cooling*
Conduction	Conduction
Radiation	Radiation
Compression	Expansion
Condensation	Evaporation

Fig. 4-8. ***Sublimation.*** *These delicate snow crystals, no two alike, result from the sudden change of gas (water vapor) to ice as the temperature goes below the freezing point. They have not gone through the water stage or they would appear as frozen raindrops.*

ADVECTION AND CONVECTION The air that is initially heated or cooled by the mechanisms just mentioned is seldom still. Once it has achieved its temperature in one place, it is likely to be transported elsewhere. Horizontal air movement of this type (*advection*) resulting from simple winds is normal enough, but vertical mixing or turbulence is also frequent, often as a result of *convection.*

To understand and visualize convection, we may set up a laboratory demonstration, using a tank of water with a Bunsen burner heating a point at the bottom. The heated water will rise above this spot, and the colder water from the top and sides will move in to take its place. This rising of the central column and sinking of the side columns is convection, triggered by heating at the bottom. The principle works equally well in gas as in liquid, so if the narrow band of atmosphere directly in contact with the warm earth is heated by conduction, it will rise and colder air from above will sink to replace it (Fig. 4-10).

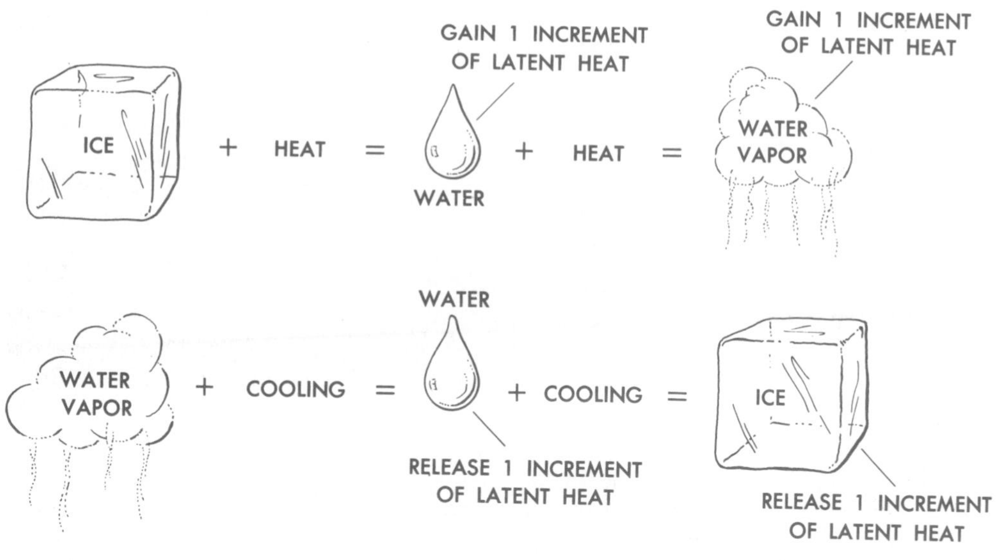

Fig. 4-9. Latent heat.

If the narrow band of atmosphere directly in contact with the warm earth is heated by conduction, it will rise and colder air from above will sink to replace it.

Convection is often explosive, as the trapped warm air suddenly breaks through the colder air above it during the hottest time of the day. This mechanism transports intensely heated air to great heights and exposes cool air aloft to the heating effects of the earth.

Both compression and expansion of sizable masses of air are involved here too, so the simple movement of warm and cold air vertically cannot be totally divorced from other processes. Generally, however, we can say that neither horizon-

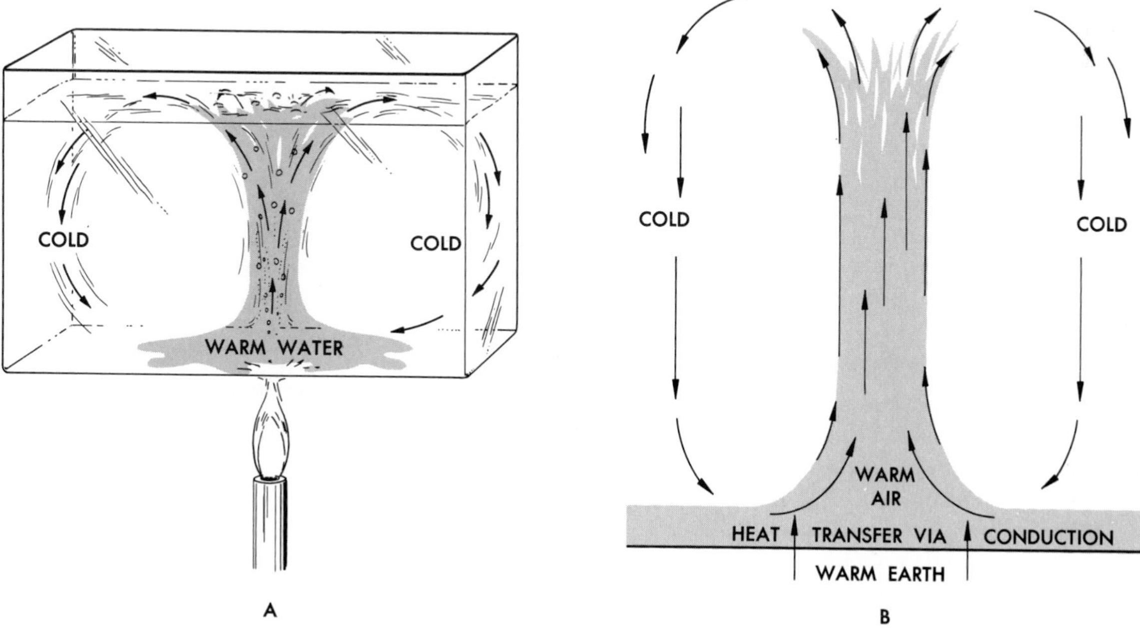

Fig. 4-10. Convection. Heat applied at the bottom of either a liquid (A) or a gas (B) causes a rising central column and sinking side columns.

tal winds nor vertical transfers of air in convection add or subtract any heat to or from the atmosphere that was not there already. They help distribute heat throughout the lower atmosphere and cannot be properly called heating or cooling mechanisms by themselves.

THE GLOBAL ENERGY BUDGET

Solar energy continues to arrive at the earth to be absorbed and changed into heat day after day, year after year, eon after eon. A heat buildup must take place unless an equal amount of cooling occurs. All our longtime climatic records reveal a strong continuity of temperature; a balance of heating and cooling must therefore exist. We use the term *global energy budget* because it implies balance. To avoid problems, the income column must equal the expenditure column in the household budget or the energy budget.

To avoid problems the income column must equal the expenditure column in the household budget or the global energy budget.

Vertical Exchange

We need not worry about the solar energy that is reflected from the atmosphere and earth surface since it is returned unused to space. The great amount of solar radiation that is absorbed by the earth is another matter, however. The condition of the atmosphere is important. On clear days when the humidity is low, the atmosphere intercepts only a small amount of the reradiated long-wave energy from the earth. Much of the energy that could have been absorbed by water vapor and clouds escapes to space. Conversely, on cloudy days with high humidity, more is absorbed and less escapes directly to space. Add to this the fact that water vapor, carbon dioxide, and other greenhouse gases are selective absorbers of long-wave energy. Thus certain wavelengths of the energy reradiating from the earth's surface go directly to space whether or not these gases and clouds are present. Such wavelengths are said to escape through "atmospheric windows"—parts of the spectrum not absorbed by air molecules. Even though these losses occur, much radiation from the earth is readily absorbed by the lower atmosphere where it is reradiated, some back toward the earth and some into outer space. In the course of this to-and-fro energy exchange from the earth to the atmosphere and back, a continual seepage of heat into space takes place, and it must be enough to balance our budget.

In the course of the energy exchange from the earth to atmosphere and back, enough heat must seep into space to balance our budget.

Energy transfer is not always instantaneous, and heat does build up over the short run. The afternoon is hot because energy receipt has exceeded loss for several hours, and the maximum temperature reading usually displays a lag of two to three hours after that of the highest sun. We usually experience the minimum temperature about sunup, at the end of a night of steady radiation loss with no compensating receipt. In most of the world the lengths of day and night vary with the seasons so that energy gain versus loss shows a seasonal as well as a daily cycle. The same kind of maximum/minimum temperature lag that shows up during a day also occurs on an annual basis. In the Northern Hemisphere July or August is usually the warmest month, not June at the solstice; and January or February is the coldest, not December.

Even this annual lag of a month or two before a balance is restored is not a long one, for solar energy is called on to accomplish many chores before it can complete its cycle. Solar energy evaporates countless gallons of water each day, melts great masses of ice and snow, sustains the complex food chain, and indirectly drives the great wind and ocean current systems. In addition, coal and oil are stored energy from the past waiting to be liberated when they are burned.

Horizontal Exchange

Heat moves horizontally as well as vertically, however. The tropics receive a surplus of solar radiation from a year-round high sun and lack of seasonality. Temperatures are continuously hot—but they are not getting hotter. Conversely, the high latitudes are never warm, reflecting the combined influences of low-angle summer sun and long winter nights—but they are not getting colder because of a long-term energy deficit. There must be transfer mechanisms (Fig. 4-11).

The ocean currents are one mechanism and the planetary wind system is another. These mechanisms can get a little complicated, as we will see later, but essentially they do the job. Imagine a mass of warm air from a tropical ocean being wafted poleward and carrying with it a cargo of water vapor from evaporation of the sea. The warm air takes its own heat

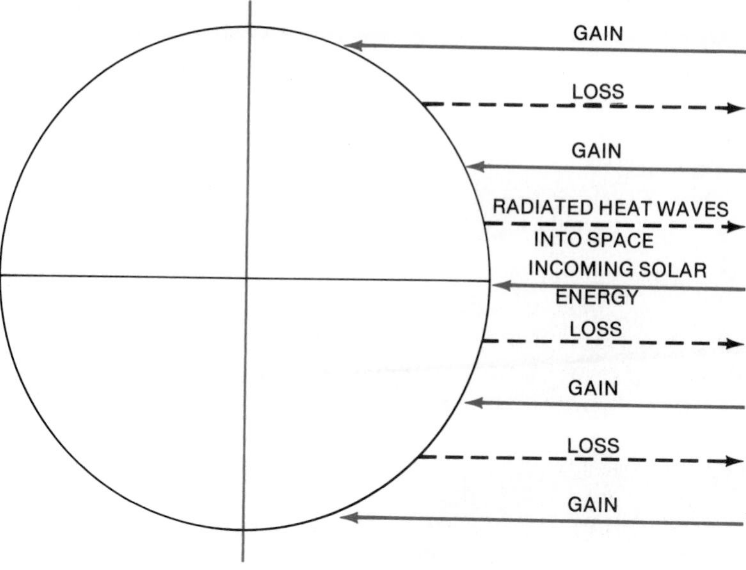

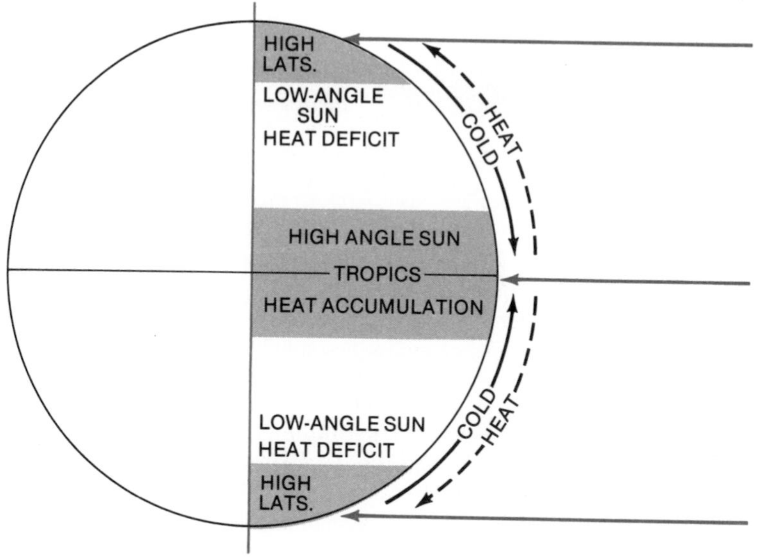

Fig. 4-11. ***The vertical and horizontal energy budget.*** *All the solar energy received and put to work at the earth's surface must be returned to space (A). Before it is returned, the surplus heat of the tropics must in some way balance out the lack of heat in the polar regions (B).*

poleward as it goes. Eventually, the water vapor condenses as it cools, releasing the latent heat far from its original tropical source. Such processes are occurring continually as cold polar air interchanges with warm tropical masses. It should not be surprising then that our middle latitudes, the point of contact and sharp contrast, should also be a region of storms and generally unstable weather.

The color map of oceans (Map IV) shows that all cold

FUSION

Tokamak Fusion Test Reactor, Princeton University.

There are two reasons why scientists are working hard to develop a practical way to use fusion energy: (1) the fuel is hydrogen, readily extracted from the ocean and hence virtually unlimited, and (2) radioactive waste materials are essentially nil. Following the most popular approach, all that has to be done is to heat ionized, electrically charged, hydrogen gases (deuterium and tritium) to temperatures rivaling those of the stars, in excess of 180,000,000° F (100,000,000° C) and then to compress and contain these gases so that the nuclei of the atoms fuse into new and heavier atoms. When this happens, energy is released to be captured and put to work for useful purposes.

We've already come a step along this route with the manufacture of the H bomb, but here the fusion was uncontrolled. The trick is to produce these immensely high temperatures while at the same time containing and using the energy that is released. *Plasma,* the ionized gases, must be confined for a matter of seconds to allow the atomic nuclei to fuse. Fusion then becomes self-sustaining, and no outside power is required beyond the original startup charge.

From the very beginning of fusion power research, it has been obvious that metal or plastic vessels were far too fragile to hold the superheated plasma. A significant breakthrough came with the concept of a controlled magnetic field to contain them. The Russians were the first to construct such a device, which they dubbed *Tokamak.* They are busily working on a much bigger one, as are the Japanese and the British. The U.S. version of this larger, new-generation, doughnut-shaped *Tokamak* is a $314 million model at Princeton University, financed by the Department of Energy. There in December 1982, American scientists achieved what has been billed as the first controlled fusion in history on their first try. The burst of plasma lasted just one-twentieth of a second and required far more energy input than it produced. Furthermore, the highest temperature applied to the gases was only in the 180,000° F (100,000° C) range, which is very cool by fusion standards. When 180,000,000° F (100,000,000° C) temperatures are produced, the energy burst will last several seconds, and the power production will equal the power expenditure. This equality is called the *breakeven point.* As of the beginning of the 1990s none of several test reactors had achieved it.

Recent research suggests that laser beams may be more effective than magnetism in containing the fusion reaction. There are even proponents of a totally different approach called "cold fusion" in which reactions may take place at room temperature.

Regardless of the approach taken, no one knows how long it will take to go from the breakeven point to the *burning point,* where the nuclear fusion reaction becomes self-sustaining and gives off more energy than is applied, and finally to large-scale commercial reactors. And who knows how much it will cost?

currents flow equatorward and all warm ones poleward. Interestingly enough, these flows of air and water are set in motion basically by the very temperature differences that require their existence. The surplus of tropical heat and the deficit of polar heat demand a budget mechanism, and the imbalance produces its own salvation.

The surplus of tropical heat and the deficit of polar heat demand a budget mechanism, and the imbalance produces its own salvation.

Climate Change

All this does not mean that the earth cannot experience long-term climate changes. There appears to be sufficient evidence to suggest that a number of these changes occurred in the past. A variety of possible factors might alter our world climates substantially.

We know little about sunspots. What if they multiplied sufficiently to cut down even a minor fraction of solar radiation? Or suppose the concentration of carbon dioxide and other manufactured atmospheric contaminants continues to increase as expected causing increased absorption, reflection, and scattering of selected energy wavelengths. Or, given the immense variation in the absorption and radiation ability of land and water, what would happen if their relative sizes change? Movements of the earth's crust could result in larger continents and smaller oceans; the Antarctic ice cap could melt and the sea flood all low-lying coasts. These are only some possibilities.

We work with the concept of an energy budget as a useful model in our attempts to understand the dynamics of atmospheric temperature. By thinking in terms of a complex system of energy input and loss, we may identify natural and human-generated changes in the budget in time to act accordingly.

SOLAR POWER

We have said that coal and oil are simply solar energy stored from the past. It would therefore seem wholly legitimate to look at the daily energy we receive from the sun as a source of practical instant power. The potential is there and has long been recognized. We constantly run across such statements in the press as, "if only 1% of the solar energy arriving in the Sahara Desert each day were converted to electricity, it would supply all of the world's projected energy requirements for the year 2000." Nonetheless, we did very little about it until the worldwide oil crisis convinced people to consider alternative energy sources seriously.

Of these alternative sources, much fevered argument has centered on nuclear fission with its many dangers. If only we could master the technology of the safer nuclear fusion, we could forget about fission and perhaps achieve the ultimate source of power. But where did the inspiration for fusion come from in the first place? The sun, which is in fact a mammoth fusion reactor.

The sun delivers huge amounts of energy on a daily basis to any who should harness it. Solar energy is clean, safe, and soundless; it does not alter the earth's heat budget and above all, it is free and infinite. Because it arrives at the surface of the earth in a diffuse form, the problem is to concentrate and focus it. We have mechanisms today that can do this, although undoubtedly they could be improved. Mirrors and heat collectors can put the solar heat to work directly, whereas power cells can transform electromagnetic radiation into electrical current (Fig. 4-12).

The sun delivers huge amounts of energy on a daily basis to any who should harness it.

The only essential drawbacks are darkness and cost. Darkness involves not only night but also clouds, fog, smog, and winter. Each of these factors limits—often for long periods of time—a steady and predictable flow of energy. We need to be able to store up the excess energy we receive on bright days for use when our energy supply is interrupted. But batteries for storing electricity, or water, rocks, and chemical banks for heat, have all proven to be immensely inefficient. A pile of coal covered with snow during a long winter loses no energy over time, but solar energy is lost at a rapid rate and so must be used immediately.

The excess energy we receive on bright days should be stored for use when our energy supply is interrupted.

In 1982 the Department of Energy in collaboration with the National Aeronautics and Space Agency (NASA) published a report indicating that it should be possible to develop a practical solar power satellite (SPS). The notion is to place giant arrays of solar cells into orbit 22,000 miles (35,406 km) above the earth, their movement synchronized with that of the earth so that the cells would face the sun permanently without any night or clouds. On earth, specialized receiving antennae would intercept the SPS energy stream in the form of microwaves, transform it into electric-

Fig. 4-12. ***Mirrors as heat devices.*** *We are beginning to think seriously about using the sun's energy these days. After all, it's free! This parabolic mirror device at Odeillo in the French Pyrenees, is properly called a solar furnace and is the largest in the world. The reflective surface measures 180 × 130 feet (55 × 40 m) and is a mosaic of 9500 individual facets.*

ity, and funnel it directly to regional power grids. Expensive dreaming? No one knows at this early stage in the game. At the very least it demonstrates creative imagination. Today's costs for much less ambitious schemes remain much too high for practical use, except in unique situations. However, no one doubts that they can be lowered with volume production of hardware while at the same time the costs of competitive energy sources keep going up.

As we project our energy needs into the immediate future, then, it appears that solar energy does indeed have some real potential and will very likely become a significant energy source in certain favored high-sun locations. Like mineral resources, there will be "have" and "have not" nations. Wherever practical, however, homes will be heated, cooled, and lighted, and industry powered by the same solar energy that continues to work in the old familiar ways nourishing the world's food chain, driving the wind and ocean current complexes, and tanning tourists.

TEMPERATURE CHANGES WITH ALTITUDE

Now that we have discussed the sources of heating on our planet, let's consider temperature changes in the troposphere in greater detail.

Normal Lapse Rate

Under usual circumstances, a thermometer attached to a balloon will indicate a constantly lowering temperature as it rises in the troposphere. The tops of mountains will support snow whereas the lower levels will not. That is, the closer we get to the sun, the colder it gets. This seems paradoxical until we remember that the lower atmosphere gets its heat from the earth. The farther we get from the radiator or heat source, the greater the cold (Fig. 4-13).

Visualize a stratified troposphere with the warmest air at the immediate surface and progressively cooler layers atop it. The change of temperature with altitude is called the ***lapse rate.*** If we assume the air to be reasonably still and the observer or instrument to rise through it, the rate of temperature loss is about 3.6° F (2° C) per 1000 feet (305 m). This loss is fairly predictable and is called the ***normal lapse rate.***

Adiabatic Lapse Rate

Another type of lapse rate is the ***adiabatic lapse rate.*** Where the normal lapse rate assumes moderately still air, the adiabatic lapse rate involves the lifting of a sizable mass of air. As the air mass rises, it receives less and less heat from the earth. Since we are dealing with a large mass of air, however, the temperature loss from simply being farther away from the source of heat will be slow and scarcely noticeable for some time. There is, however, an instantaneous heat loss caused by expansion. The rate of decrease differs depending on whether or not there is condensation in the air mass:

*Fig. 4-13. **Normal lapse rate.** The seeming paradox that the closer one approaches the sun the colder it gets is shown in this view of 8260 foot (2518 m) Mount Egmont in New Zealand. From the subtropical tree fern in the foreground, up past the timberline to the perennial snow, the temperature decreases visibly.*

1. If there is *no condensation* in the air mass, the rate of decrease is 5.5° F (3° C) per 1000 feet (305 m). This is called the ***dry adiabatic lapse rate.***
2. If condensation occurs in any of its forms during the ascent of an air mass, then the ***wet adiabatic lapse rate*** applies. This lapse rate will vary with situations from 2° to 3° F (1° to 1.7° C) per 1000 feet (305 m) but is *always less than the dry adiabatic rate.*

The adiabatic lapse rate involves the lifting of a sizable mass of air.

Let us use a ***chinook*** (United States) or ***foehn*** (Europe) wind to show how wet and dry adiabatic lapse rates work. These winds are warm, dry air currents blowing down from the mountains. The chinook, experienced along the eastern slope of the Rockies and Cascades in winter and early spring, melts the snow, thus opening up the range for grazing. It also frequently causes serious floods. If we set up a simplified and somewhat exaggerated example, the mechanics will become apparent (Fig. 4-14). We will place a 10,000-foot (3050 m) mountain range in the path of prevailing winds, thereby forcing the air mass on the windward side up the slope. If the temperature of this air mass is 60° F (16° C) at the foot of the mountain, as it rises the air mass will lose 5.5° F (3° C) per 1000 feet (305 m) because of expansion.

At the 2000-foot (610 m) level, it will have lost 11° F (6° C), and the original 60° F (16° C) air mass will now have a temperature of 49° F (9° C). At some point, any further cooling will cause condensation. This is called the ***dew point,*** which we will place at 49° F (9° C) for this air mass. From the dew point to the tip of the mountain, the *wet* adiabatic lapse

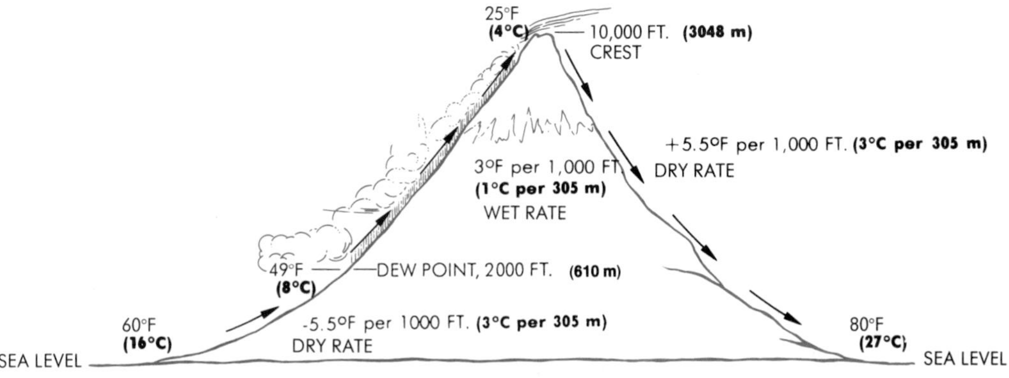

*Fig. 4-14. **Chinook.** The heat gained through condensation on the windward slope is reflected in the higher temperature at sea level in the lee (highly exaggerated).*

rate must apply. Since this rate is always less than the dry adiabatic rate, let us place it at 3° F (1.7° C) per 1000 feet (305 m). As the air mass continues to rise from 2000 feet (610 m) to the 10,000-foot (3050 m) crest, it loses 3° F (2° C) per 1000 feet (305 m) or 24° F (13° C) for the 8000-foot (2438 m) uplift. Thus the air mass arrives at the top of the mountain with a temperature of 25° F (−4° C).

Now if it descends the lee (opposite of windward) slope, the air mass will compress and warm as it goes. Condensation requires cooling, so condensation will cease at the crest. As the air mass comes down, it will warm at the dry rate of 5.5° F (3° C) per 1000 feet (305 m). The air mass descending 10,000 feet (3050 m) will warm 55° F (31° C), and this added to the 25° F (−4° C) at the crest gives it a temperature of 80° F (27° C) at the foot of the mountain.

By simply taking a 60° F (16° C) air mass over the top of a mountain and down the other side, we have increased its temperature by 20° F (11° C). Where did that heat come from? The only difference in ascent and descent is the zone of precipitation between the 2000 foot (610 m) and 10,000 foot (3050 m) levels on the windward side. The heat must therefore have been released by condensation. There was still a net drop in temperature as the air went up through this zone, but only at the rate of 3° F (1.6° C) per 1000 feet (305 m). In other words, the air continued to expand and lose heat at the usual rate on the way up, but condensation replaced part of that loss, and the 80° F (27° C) air mass at the foot of the lee slope is not only warmer but drier than when it started out.

The lapse rate, then, is simply the rate at which there is a decrease or increase of temperature with a loss or gain of altitude. The *normal lapse rate* assumes a still atmosphere with the observer moving through it; the *adiabatic lapse rate* applies when an air mass changes altitude and expands or compresses.

The* normal lapse rate *assumes a still atmosphere with the observer moving through it; the* adiabatic lapse rate *applies when an air mass changes altitude and expands or compresses.

Inversion

On occasion, the normal lapse rate may be reversed; that is, a gain in altitude will result in a gain in temperature. This situation is called a *temperature inversion,* or simply an inverted normal lapse rate. Inversions usually last a short time, but they are quite prevalent nonetheless. A long winter night with a clear sky and still air is an ideal situation for

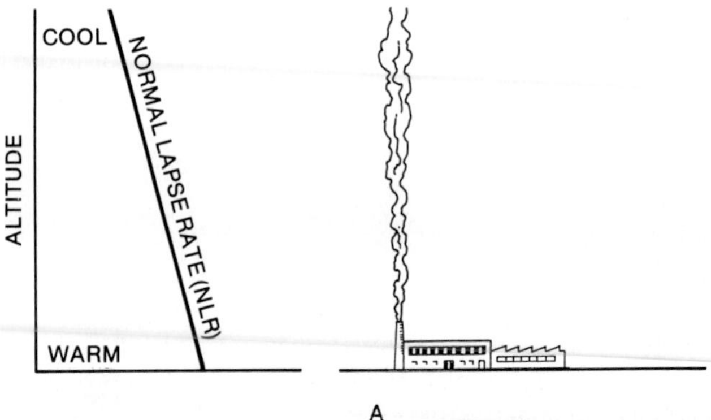

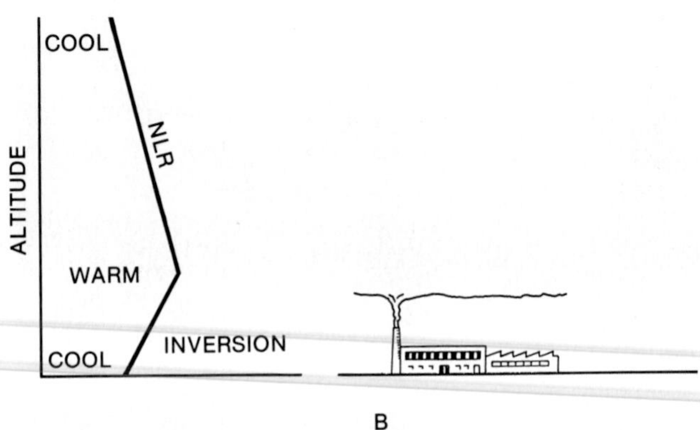

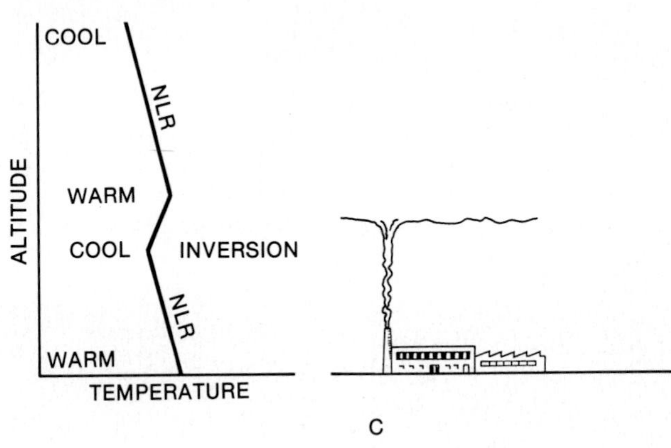

Fig. 4-15. ***Temperature inversions.*** *(A) No inversion. Temperature decreases with elevation. Pollutants from industries and vehicles disperse at high levels. (B) Surface inversion. Night-time cooling chills the air near the earth's surface while air remains warmer aloft. Pollutants may be trapped near the ground for short periods. (C) Upper air or subsidence inversion. Air descending from aloft is warmed by compression but does not penetrate stagnant and cool surface layer. Inversion between the two layers traps pollutants which may interact photochemically with sunlight to form smog. Under this inversion type, pollutants often persist in high concentration for long periods.*

the most common type of inversion, a ***surface inversion,*** so-called because the inverted lapse rate extends through a shallow layer upward from the earth surface. The heat of day radiates off the earth during the night, and by the early morning hours the surface of the earth is cooler than the air above it. Heat moves via conduction from the lower atmosphere to the earth, and a cool layer forms beneath warmer air above. This means that a thermometer taken aloft would indicate warming as it passed upward and out of this cool zone—just the reverse of the normal situation (Fig. 4-15).

Smoke rising from a chimney on a cool still morning is likely to continue upward for a short distance while it is still warmer than the air surrounding it. As the smoke rises, it cools and at the same time runs afoul of the warmer air strata above which it cannot penetrate. It flattens out abruptly here, making visible the sharp dividing line between warm and cold air. This inversion ceiling, which is usually very shallow but may extend to heights of 200 feet (61 m) or more, forms an effective barrier to the normal exhaust of combustion byproducts into the upper troposphere. Such a situation could be serious if it were long-lived, but most often this type of inversion lasts only a few hours until the sun comes up and begins to warm the earth. Conduction shortly restores the normal lapse rate, and all is well.

Some places such as the Los Angeles basin occupy a geographic position in which air often sinks and compresses for lengthy periods of time. With a stagnant layer of cool air near the ground, this sinking air will not continue all the way to the surface. The result is a ***subsidence inversion,*** also called an ***upper air inversion*** because the inverted lapse rate does not extend to the surface but remains at about 1000 to 1500 feet (305 to 458 m) above the ground (Fig. 4-15). Compression warms the descending air to a temperature that exceeds the temperature of the air in the cool surface layer. This means, of course, that no effluent gases or other combustion byproducts can escape from beneath the ceiling; because the sky is clear with a subsidence inversion, photochemical smog may develop. In places like Los Angeles where the inversion lid is slapped on top of a topographic basin, great concentrations of smog can develop (Fig. 4-16).

Fig. 4-16. ***Temperature inversion and smog over the Los Angeles Basin.*** *Warm air aloft confines pollutants from vehicles and industry to the cool surface layer. Because the inversion ceiling is below the crest of the basin perimeter, it forms a secure seal. This view is from the San Gabriel Mountains with the eastern edge of the basin on the right center.*

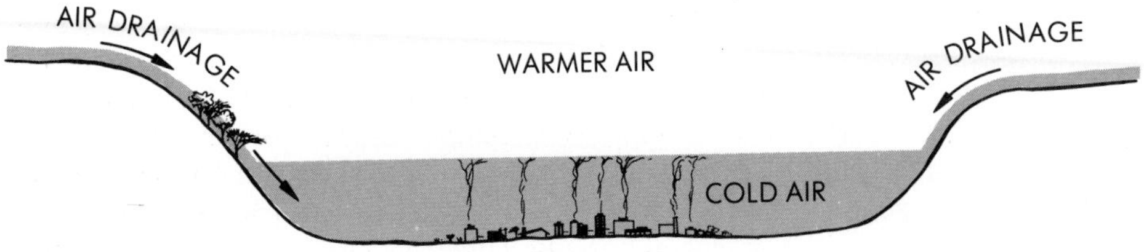

Fig. 4-17. ***Air drainage.*** *Cold air flows downslope like water, filling up terrain depressions to considerable depth.*

The worst feature of the subsidence inversion is its persistence. High winds or heavy rain can break it up, but usually the unhappy populace must simply wait for the high pressure (descending air) to move on. Without pollution there would be no problem, but since the pressure system cannot be changed, we must either accept worse and worse suffering or make considerable improvement in our current pathetic efforts to limit smog production.

The worst feature of the subsidence inversion is its persistence.

Air Drainage

Cold air at the surface, as in the common surface inversion, flows under the influence of gravity and the local topography. Being heavy and dense, cold air acts almost like water and moves down the slope to pile up deeply in pockets and valley bottoms. This is called ***air drainage*** (Fig. 4-17). If any significant altitude is lost, adiabatic warming occurs owing to compression, but on the relatively small-scale local scene there may be a good many degrees of difference in temperature between valley floors and sloping foothills. Air drainage intensifies the surface inversion in valleys, and during periods of stable weather, especially in the winter, the cold surface layer may persist for days or even weeks. Again pollutants from vehicles, factories, and wood-burning stoves may become concentrated and the air unhealthy to breathe. The practice is to plant frost-touchy citrus or peach orchards on slopes so that, as cold air develops at the surface, it will flow away. The moving cold air remains shallow on the slopes, affecting directly only the trunks of the trees, but

Fig. 4-18. ***Fighting off frost damage in the orchard.*** *Citrus is highly susceptible to just a few degrees below freezing, especially if the cold lasts up to six hours. The propeller is designed to produce ground-level turbulence and to discourage development of a fatal inversion. Old-fashioned orchard heaters and smudge pots have largely been replaced with the more efficient and cleaner propellers.*

deep accumulations of cold air in valleys often result in frost damage to delicate fruit and foliage (Fig. 4-18).

Cold air acts almost like water and moves down the slope to pile up deeply in pockets and valley bottoms.

ISOTHERMS

We often need to plot accumulated temperature data in map form to show the distribution of particular temperatures at a given time. One method would be simply to write the temperature at the locations of the various reporting stations. Imagine, however, how many figures would appear on a map of the world or the United States if all reporting stations were represented. It would require long and careful study for any sort of meaningful pattern to emerge from this method. The reading of such a map can be greatly simplified by the use of ***isotherms***, lines that connect all points of equal temperature.

Isotherms can be drawn on a map on which monthly temperature averages, daily means, yearly maxima, or any other temperature data set have been plotted. In connecting certain selected temperatures, the isotherms show a pattern that is not easy to see from the written figures alone, and they often reveal the marked influence of temperature controls.

Isotherms show temperature patterns not easily apparent from written figures.

For example, in Figures 4-19 and 4-20 (global distribution of isotherms for July and January), we see immediately that isotherms exhibit a general east–west trend. This expresses the influence of latitude as the major control of world temperature. Over the middle latitude continents, however, they tend to deviate sharply from this trend, especially along western coasts. For instance, the January 32° F (0° C) isotherm in the North Pacific is a fairly straight east–west line through the Aleutians, striking the west coast of North America at about Juneau, Alaska. Obviously, it cannot continue directly eastward into the continent, for we know that the winter temperature averages in northern Canada are well below 32° F (0° C). In order to find as mild a January in the interior of the continent as that of Juneau, we may have to drop as far south as central Kansas. The isotherm that connects these similar temperatures will therefore show a sharp break at the west coast and trend almost north–south.

In the North Atlantic and Western Europe this deviation in direction is even more apparent, reflecting both the larger Eurasian continent and the northward-probing warm Gulf Stream. By drawing a single isotherm, we see immediately that in this particular part of the world latitude as a temperature control is of considerably less importance than coastal versus inland location. Now check the 32° F (0° C) isotherm in the Southern Hemisphere for July (winter). It is simply a straight line because there are no great continents to introduce the factor of massive variation in seasonal heating between land and water. Isotherms are simply manufactured value symbols to help us interpret at a glance a whole panoply of temperature data plotted on a map. (For a discussion of maps and map symbols, Refer to Appendix B).

CONCLUSION

Solar energy is at the heart of the earth system. But there is a paradox here: the sun does not, for the most part, heat the atmosphere. We know now that to understand atmospheric heating is to understand electromagnetic radiation, that the short-wave energy emitted by the sun heats the earth and that the long-wave energy emitted by the earth heats the atmosphere. The greenhouse gases play a critical role in atmospheric heating by absorbing long-wave radiation. Our concern about climate warming is based on our understanding of this differential absorbency: higher concentrations of these gases, many of which are byproducts of human activity, mean greater atmospheric heating.

One point that is critical to the discussion of temperature is that the sun's rays do not heat the earth's surface evenly. We know that low latitudes absorb more solar radiation than high latitudes and that water, with its high thermal capacity, heats and cools much more slowly than dry land. Thus we are now aware of two interlocking aspects of the temperature picture that must be kept in balance if the earth is to remain livable. (1) The amount of energy entering the system must equal the amount returned to space, and (2) excess heat on the earth's surface must escape from warm areas to cold areas. With this knowledge we have the basis for understanding world weather and climate patterns. The need to equalize temperature is behind global pressure and wind systems, ocean currents, and storms. The heat of the low latitudes is largely transported to the high latitudes where it ultimately is lost to space by atmospheric radiation, and the global energy balance is thereby maintained. In the next chapter we look closely at the effects of unequal heating as we discuss pressure and winds.

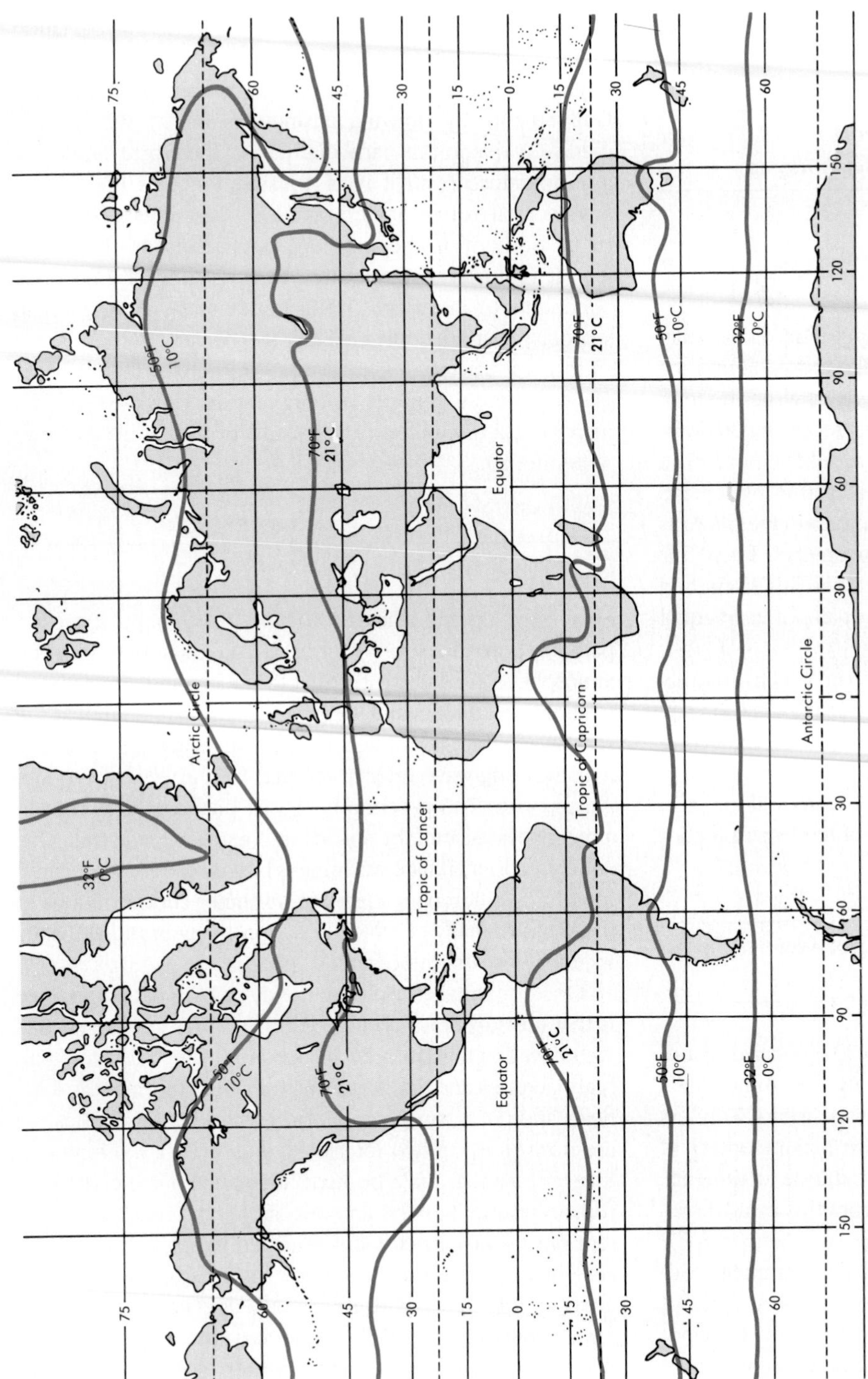

Fig. 4-19. ***World mean July isotherms (reduced to sea level).***

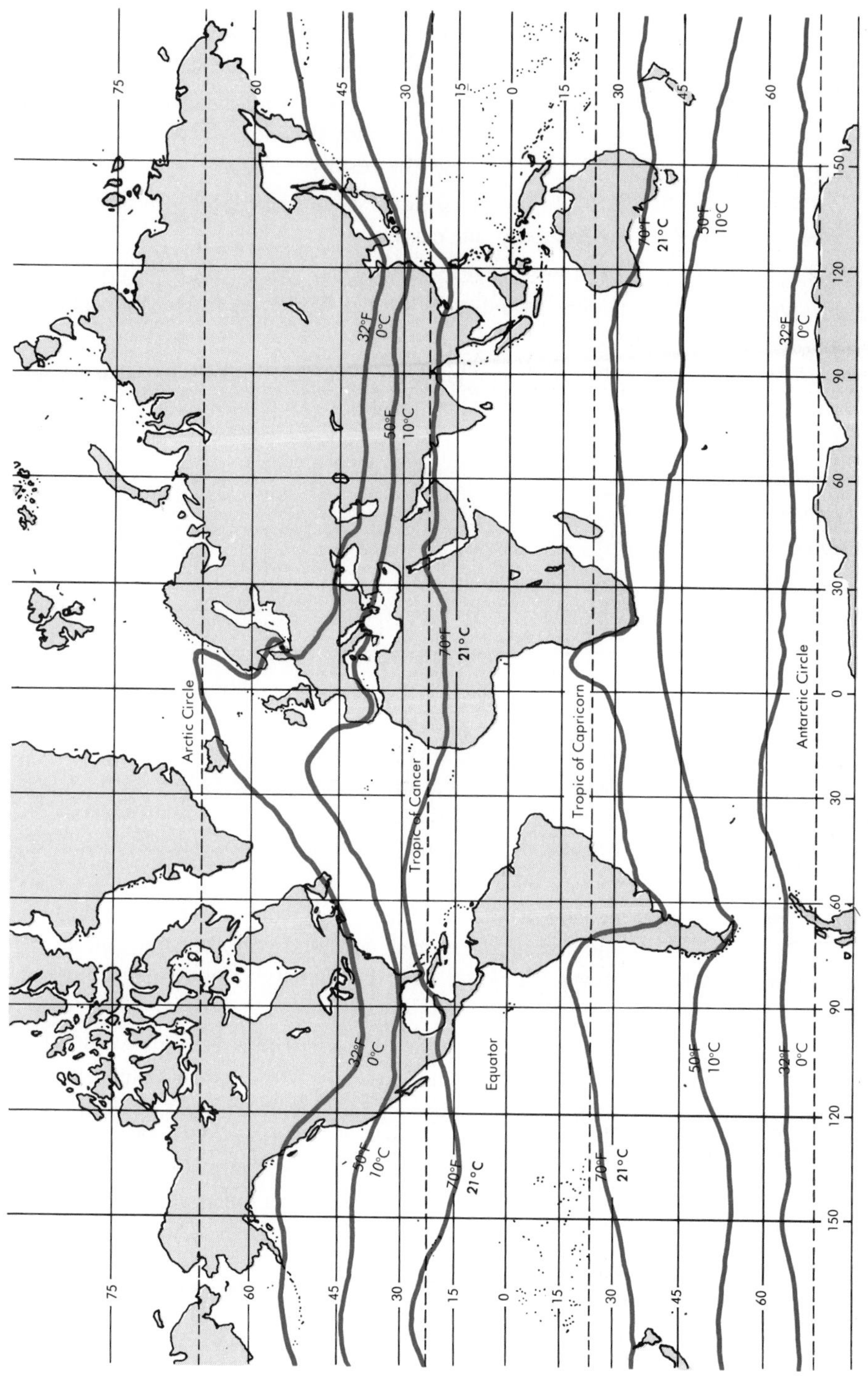

Fig. 4-20. **World mean January isotherms (reduced to sea level).**

KEY TERMS

solar constant
insolation
spectrum of electromagnetic radiation
absorption
transmission
scattering
reflection
diffuse daylight
greenhouse effect
albedo
specific heat
conduction
radiation
compression
expansion
adiabatic
condensation
evaporation
sublimation
latent heat of condensation
advection
convection
global energy budget
lapse rate
normal lapse rate
adiabatic lapse rate
dry adiabatic lapse rate
wet adiabatic lapse rate
chinook
foehn
dew point
temperature inversion
surface inversion
subsidence inversion
upper air inversion
air drainage
isotherms

REVIEW QUESTIONS

1. What is wrong with calling insolation "sunshine"? Why is insolation always less than the solar constant? Where does the "lost" energy go?

2. Why is the sky blue? Why are sunsets red?

3. Why is ozone a good thing to have in the upper atmosphere and a less desirable thing to have in the lower atmosphere?

4. Why does solar energy have to strike the earth *first* in order to heat the atmosphere?

5. If both water and land are good absorbers, why does water stay warmer in the winter and cooler in the summer than land does?

6. Why do we say that conduction, radiation, compression, and condensation can add to the heat that is already in the atmosphere but advection and convection cannot?

7. Why doesn't the sun's accumulated energy constantly make the equatorial regions hotter and hotter every year?

8. Why could we get more solar power from an orbiting satellite than we could from a ground-based collector?

9. Why does a thermometer rising through a still atmosphere usually show a steady decline? Why would a moving atmosphere change the results?

10. Why does a rising "wet" air mass cool more slowly than a rising "dry" air mass?

11. Under what circumstances would you expect air to be cooler on the ground than higher up in the atmosphere? Why?

12. "Isotherm" means "same heat." Why was this term selected?

APPLICATION QUESTIONS

1. Imagine that your next-door neighbor is in the habit of leaving her dog in the car when she goes to the grocery store. You have tried to explain that this practice can be dangerous, but she insists that she never does it when it's *really* hot out. Write a letter to her explaining why she is wrong and make up an experiment she could do to demonstrate your point.

2. Design an experiment that will show how conduction and radiation can both heat and cool the atmosphere. Be creative, and add other heating and cooling mechanisms if you can.

3. Imagine that a dry air mass with a temperature of 50° F has just reached the base of a 7000-foot mountain. Its dry adiabatic lapse rate is −5.5° F per 1000 feet. Dew point is at 2000 feet. The wet adiabatic lapse rate is −3° F per 1000 feet. Draw a diagram showing how and when the air mass will change temperature going over the mountain. What will the temperature be at the base of the mountain on the leeward side?

4. Most newspapers include a list of high and low temperatures for the country's major cities and a map showing the distribution of isotherms. Using an outline map of the United States (you can trace one from an atlas), plot the high (or low) temperatures by placing the number of degrees at the location of each city for which a record is given. Using these recordings as a guide, draw in your own isotherms. How do they compare to the published map?

5. The severity of air pollution is related not only to the number of automobiles and types of industries, but also to atmospheric conditions and topography. With these considerations in mind, assess your local area's susceptibility to severe air pollution in summer and winter. Does time of year affect severity?

CHAPTER 5

PRESSURE AND WINDS

OBJECTIVES

After studying this chapter, you will understand

1. What air pressure is and how we measure it.
2. What wind is and how it is related to air pressure.
3. How the Coriolis force affects the motion of the winds.
4. How we can use wind as a source of power.
5. What pressure and wind patterns exist over the earth's surface and why they behave as they do.
6. How the continents affect wind patterns.

Air pressure at the earth's surface is not easy for the human body to sense, but wind—which is certainly obvious—is the resultant of pressure differences. Wind can be put to work with the proper equipment. Sails to catch air currents and propel ships are "old hat" until we observe plastic, mechanically trimmed sails combined with diesel power on the modern Japanese vessel Shin Aitoku Maru. *These are called Sail Equipped Motor Ships, and the aim is to minimize the use of expensive fuel oil whenever possible.*

INTRODUCTION

Why should we be concerned with air movement and the differences in pressure that set air in motion? There are many reasons. Wind fills the sails of our recreational sailboats, turns giant rotors to generate electricity, and drives the waves that add to our enjoyment of the coast. Ascending air currents bring rainfall that nourishes the earth. On the other hand, the high wind in tornadoes and hurricanes can threaten life and property, wind chill can so lower temperature that winter travel is dangerous, and descending air can lead to inversions that trap toxic pollutants.

On a broader scale, winds and other air movements, together with ocean currents set into motion by wind, are central to the global energy budget—the warmth of the equatorial regions is spread north and south by moving air and water. In this chapter we discuss first the nature of atmospheric pressure and then the planetary and local wind and pressure systems. In the process, we will seek to understand these important elements of weather and climate.

PRESSURE

Pressure is the total mass of the atmosphere pressing down on the surface of the earth. At sea level in the middle latitudes, a column of atmosphere exerts an average pressure of about 15 pounds (7 kg) per square inch (3 cm). This amount of pressure is called ***one atmosphere.*** If this weight increases in a particular location, that is, if the air pushes down a little harder, the pressure is increased at the surface. This becomes a high-pressure center (***anticyclone***). On the other hand, the column may lift and push down less causing a low-pressure center (***cyclone***).

Pressure ***is the total mass of the atmosphere pressing down on the surface of the earth.***

Temperature is a frequent cause of changes in pressure (although by no means the only one). Air that is cooled at the bottom will sink and increase pressure. Warming air will raise and lower the pressure. If we climb a mountain or go up in an airplane, we are no longer at the bottom of the atmosphere, and pressures decrease as we gain altitude. A barometer may even work as a crude altimeter.

Unlike the other weather elements, pressure is something of an intangible. We cannot hear it, smell it, or, with some exceptions, feel it; in other words, the human body is a terrible barometer. It is not the best thermometer in the world for that matter, but we can tell general temperature changes, even if humidity variations and the like prevent us from being very accurate. Only a few individuals, however, are able to sense changes in pressure as their rheumatism and old injuries ache or their noses bleed when the barometer drops or rises. However, most of us feel pressure differences if we change altitudes rapidly by driving over a mountain pass or climbing steeply in an unpressurized airplane. Our ears "pop" or, if our ears are congested because of a cold, we may feel the pressure change as intense pain.

The Barometer

Because most of us cannot sense the minor pressure changes that occur with day-to-day weather, we depend strongly on instruments to keep us informed. The critical instrument is the ***mercurial barometer,*** which is not very different from the simple original invented by Torricelli in 1643. This first barometer was merely a glass tube with a vacuum in it, closed at one end. The open end was placed in a dish of mercury. As normal air pressure was exerted on the mercury surface, a column was forced up the tube to a height of roughly 30 inches (76 cm) (sea level and middle latitude). Variations in pressure caused changes in this height.[1]

It seems a little strange at first to measure pressure, a force, in linear units, but the height of the mercury column measured in inches is a direct indication of the pressure of the air. Today's modern barometers have only a few modifications on this original. Inches are subdivided into tenths or less for added accuracy, and a minor correction is made for the effect of temperature on the mercury. Because of their awkward dimensions, such barometers are usually firmly fixed in a permanent wall mounting.

A more convenient though somewhat less accurate instrument is the ***aneroid barometer.*** It is a small metal diaphragm with a partial vacuum inside. As the outside air pressure decreases, the sides bulge outward. This movement is shown on a dial that indicates the pressure change (Fig. 5-1).

Coming into much more general use today, and officially adopted by the U.S. Weather Bureau in 1940, is another unit of air pressure measurement called the ***millibar.*** It is a more logical unit than the inch since a millibar is a direct measure of force. A millibar is equal to a force of 1000 dynes per square centimeter, and a dyne is the force that will accelerate one gram of mass one centimeter in one second. (One dyne is approximately the weight of one milligram.) Since both the inch and the millibar are widely used,

[1]Water might be used. It is cheaper than mercury, but it is also lighter, and normal air pressure would push it up a tube approximately 33 feet (11 m).

Fig. 5-1. ***Aneroid barometer.*** *Compact and portable, the aneroid (right) is widely used, although it is not quite as accurate as the mercurial barometer. The instrument at the left, its face calibrated in inches rather than in millibars, also works as an altimeter. The inset at the bottom records the regular pressure decrease with altitude gained.*

the important thing to remember is their relationship. An inch of mercury is equal to 34 mb; that is, *one standard atmosphere is normally said to equal 29.92 inches or 1013 mb.*

Isobars

You may remember from Chapter 4 that we use isotherms to connect points of equal temperature on a map. When we want to view the pattern of pressure on a map instead, we use lines connecting points of equal pressure. ***Isobars,*** drawn at 0.1 inch (4 mb) pressure variations, take the form of roughly concentric circles indicating centers of high or low pressure. Remember, however, that pressure is relative—no specific number of inches is always low or high. If, for instance, we have a barometric reading of 30 inches (1016 mb) in one spot and the pressure rises in all directions from that spot, then it is a low-pressure center. This same 30 inches (1016 mb) could very well be a high-pressure center if the pressures on all sides of it were lower.

WIND

Nature seems to hate inequalities. It therefore tries to compensate for pressure differences by transferring air from one place to another. High pressure represents a surplus of air and low pressure a deficit. So nature sends air in the form of wind, from an area of high pressure to an area of low pressure. ***Wind*** is the result of pressure differences. Both its direction and velocity are determined by the relative location and intensity of highs and lows.

Wind blows from an area of high pressure to an area of low pressure.

Picture a low center as a lifting of air above a given place. This forms a "vacuum" of sorts at the surface that pulls air to it from all directions. The greater the lifting, the more effective the vacuum and the pulling effect of the low. A high, on the other hand, is at the bottom of a column of air being forced down, and the air at the surface is being pushed away from the high (Fig. 5-2). In general, wind is the movement of

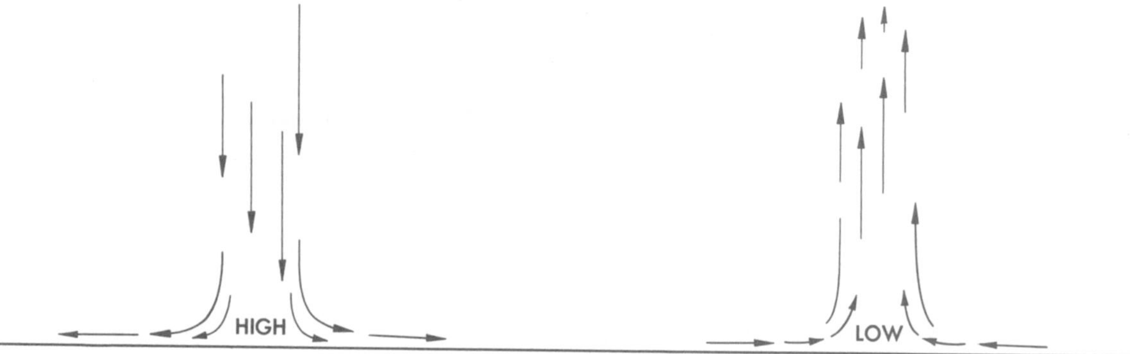

Fig. 5-2. ***At the surface, air flows into lows and out of highs.***

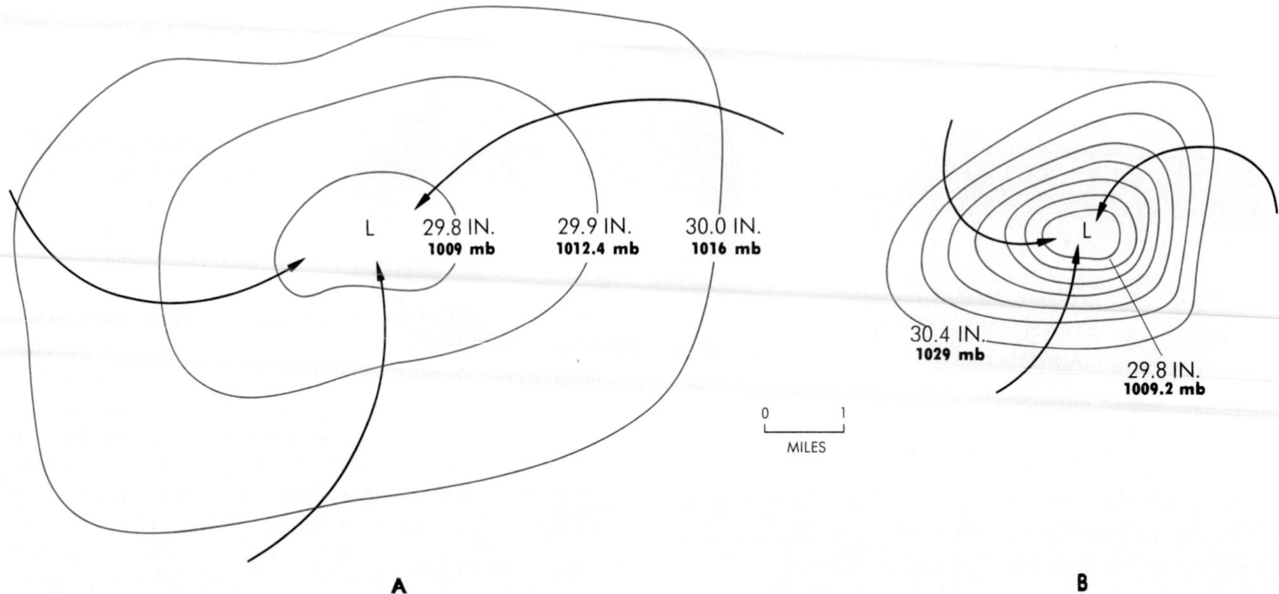

Fig. 5-3. ***Wind velocity.*** *The spacing of isobars shows the steepness of the pressure gradient. Wind velocities in (A) are much lower than those in (B).*

air parallel to the earth's surface. We describe the vertical movement of air as simply "ascending" or "descending."

Velocity

We see the intensity of high and low, and thus the velocity of winds, on the map by the spacing of the isobars (Fig. 5-3). If each isobar represents 0.1 inch (4 mb) pressure difference and the isobars are close together, there is a rapid change of pressure within a short distance, and wind will move across the isobars at a high speed.

This change of pressure represented by the spacing of isobars is called the ***pressure gradient.*** Wind will always flow across the pressure gradient *from high to low.* We measure wind velocity with an ***anemometer*** (Fig. 5-4).

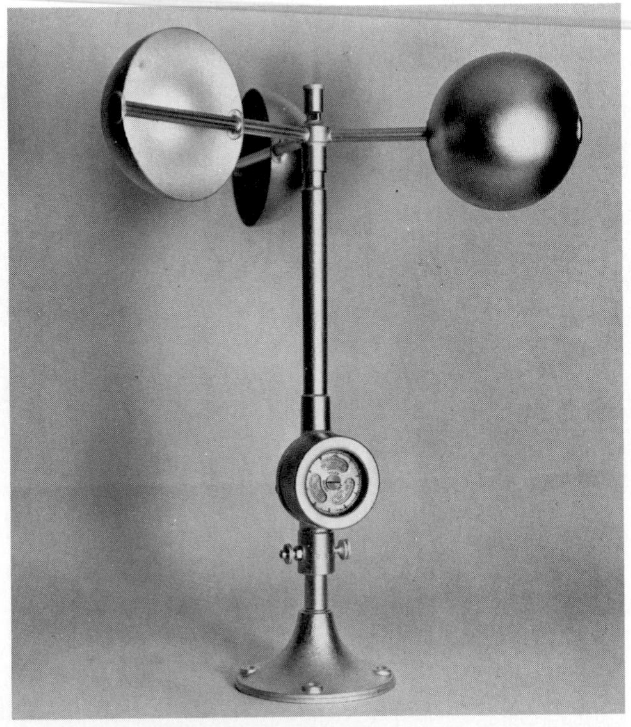

Fig. 5-4. ***The anemometer.*** *The anemometer measures wind velocity. Air currents catching the cups spin them at a rate that will be recorded in miles per hour on the dial at its base.*

Coriolis Force

Once the wind is set in motion by a difference in pressure, it does not flow directly from high to low as might be expected, but follows a somewhat devious course in arriving at its destination. This comes about through the effects of ***Coriolis force.*** Ferrel's law states the effect of this force: *Any horizontally moving object in the Northern Hemisphere will exhibit an apparent right-hand deflection and in the Southern Hemisphere an apparent left-hand deflection.* At the

GASPARD GUSTAVE DE CORIOLIS

The Frenchman G. Coriolis (1792–1843), as he signed his name, is often described as a mathematician or physicist. Perhaps he was a little of both, but his lifelong field of research was in applied mechanics. As a child of the early Industrial Revolution, he was interested primarily in how machines worked and concerned with improving their performance. Educated as a civil servant, he shortly turned to science. By 1829 he was teaching and publishing in the field of mechanics and had achieved membership in the Academy of Science.

In an 1831 address to that body, Coriolis commented on some new and provocative theories relative to inertial forces and accelerations in composite motions as they applied to hydraulic systems in machines. In his last book he explained these basic natural laws as "the momentum of relative velocity and rotation of the frame of reference." Although useful and well regarded, these concepts did not take the engineering world by storm. But the delayed and unintended application of the Coriolis force to elements in motion on the spinning globe is now regarded as his greatest and most significant contribution.

The earth rotates with greatest velocity at its largest circumference, the equator. Hence any horizontally moving projectile (air current, water current) that responds to the pull of gravity and runaway centrifugal force will, as it changes latitude, advance over a more slowly moving earth. When a moving object is viewed from the rotating earth, all these factors at work result in its apparent deflection from its original straight-line course.

Fluid mass application, borrowed directly from the Coriolis hydraulic systems theory, has aided our understanding of ocean current behavior. Belated recognition to the man who gave us a rational explanation of observable natural phenomena that had never been wholly understood was bestowed in 1963 when an official French oceanographic research vessel was named after him. It is perhaps ironic that Coriolis the scientist, not the engineer, was thereby honored.

---

equator no such force exists. Its effect increases with latitude and reaches a maximum at the poles. This applies to ocean currents, rivers, bullets, baseballs, and, of course, air currents.

Any horizontally moving object in the Northern Hemisphere will exhibit an apparent right-hand deflection and in the Southern Hemisphere an apparent left-hand deflection.

The Coriolis force is an apparent rather than an actual force. It actually depends on the point of observation. Picture a carousel and a child throwing a baseball at a fixed target off the whirling platform. If we station ourselves with the child on the carousel, we will be in the same kind of position we are in when observing the path of a horizontally moving object from the earth. If the child throws the ball directly at the target, the ball will curve away and miss badly.

This is only an illusion, however. When we view the same action from above (or observe the earth from outer space), we find that the ball's trajectory is perfectly straight and that the child's aim is faulty. She has failed to compensate for the motion of the platform. The ball only appears to curve. If the child never leaves the carousel, this illusion is very real to her. Deflected air currents are just as real to us as long as we use the earth's surface as a reference.

Once the air is set in motion, three forces affect its flow: the pressure gradient, which causes it to move from high to low; the Coriolis force, which causes it to veer off course; and friction, which causes it to spiral out of the high and into the low. Let us consider a Northern Hemisphere high-

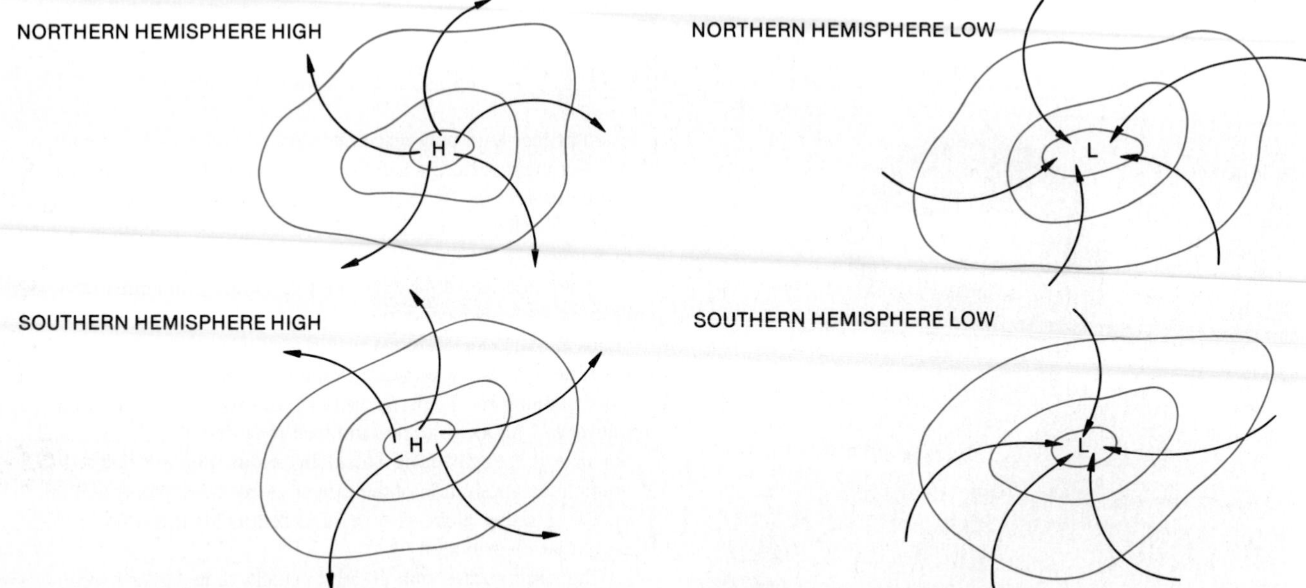

Fig. 5-5. ***Wind direction.*** *The attracting and repelling influences of lows and highs, respectively, set the air in motion but its ultimate course is determined by Coriolis force.*

pressure center with air moving away from it in all directions. *We must face with the wind* to determine which direction is right and then add a right-hand deflection to the general air flow. The result is that the air does not move directly out of the high but *spirals out clockwise.* On the other hand, the winds attempt to move at right angles across the isobars into the vacuum of the low-pressure center but are diverted to the right as they move. They eventually arrive at their destination but follow a *counterclockwise spiral.* In the Southern Hemisphere, these spirals are reversed (Fig. 5-5).

The degree to which air currents are forced to angle across the isobars or spiral is determined by latitude and friction—deflection increases with gain of latitude and decreases with increased friction. The rougher the terrain, the greater the frictional drag on air movement so that sea winds display greater deflection in response to Coriolis force than those blowing over land. Some upper air currents, which are essentially frictionless, are actually known to flow parallel to the isobars.

WIND POWER

Like solar energy, wind power is a free offering unique in this world of high and ever increasing cost. It is simply there for the taking. Wind is, of course, second-generation solar energy—a direct result of differential heating of the earth—and it displays most of the same advantages. Being reasonably soundless, pollution free, safe, and infinite, there is much to recommend wind as a large-scale power source.

Wind is, of course, second-generation solar energy, and it displays most of the same advantages.

People have been using wind, like the sun, since antiquity. Wind has been used to grind grain, pump water, and propel ships. Early settlers in the American West used windmills to pump groundwater for livestock and domestic use and, in the 1930s and 1940s before power lines were extended to farmlands, small wind turbines were used to generate household electricity. By the 1950s, however, with cheap power now available from fossil fuels and hydroelectric facilities, there was little interest in wind power.

As a result of the energy crisis of the 1970s, wind power was once again considered a viable alternative. Although we now have adequate supplies of oil for the immediate future, there is continued interest in wind power. Very rapid development of wind power took place between 1980 and 1986 under favorable state and federal tax credits. Nearly 13,000 wind machines with a combined generating capacity equal to an average nuclear or coal-fired power plant—enough for 200,000 homes—were placed in operation in the United States. Most of these machines were in California where

Fig. 5-6. ***Wind farm.*** *Scores of wind turbines, like these at San Gregorio, satisfy part of California's thirst for electric power.*

projections by the state's Energy Commission call for 8 percent of total electric energy needs in the year 2000 to be supplied by wind generation (Fig. 5-6). Since 1986, however, development of wind power in the United States has lagged somewhat owing to repeal of tax credits, cheap fossil fuel, and general lack of federal interest.

Suitable locations for "wind farms" can be found in many places, but the West Coast, Rocky Mountains, Great Plains, Hawaiian Islands, and Alaska are probably the most suitable locations. How about several thousand giant wind turbines installed in line across the central plains from Canada to the Gulf? Or these same turbines stationed on floating platforms off the coast?

Storage of energy becomes a problem on windless days. As yet we have no real solution, but ingenious methods have been suggested to supplement the inefficient methods already in use. Using flywheels, electrolyzing water to produce hydrogen, or pumping water into high reservoirs when the wind blows, and releasing it through power-generating turbines on calm days, may well have application to wind and other power sources where the present lack of storage inhibits development.

A SIMPLIFIED WORLD PRESSURE AND WIND PATTERN

The entire circulation system of the troposphere is driven by the solar heat engine with its striking energy contrasts between low and high latitudes. Without these contrasts there would be no need for horizontal heat transfer to satisfy the energy budget, and therefore there would be no complex of air and water currents. But this pressing need to get the surplus heat out of the tropics in exchange for cooler air is satisfied by a responsive circulation system.

In theory it should be simple. Cold air, heavy and dense, would flow along the ground toward the equator from both poles, while warm tropical air would, in turn, rise and move poleward at higher levels. In other words, there would be a giant convectional system. This is basically the way the system works, but nothing is ever quite as simple as it first seems. At least two major complications immediately arise: Coriolis force and differential heating of land and water.

To get a simplified overview, let's put the second of these aside for the moment and consider only the first. If we can visualize an all-water but rotating globe (to induce Coriolis force), we can see a predictable pattern of worldwide pressures and winds. Over the major oceans of the world is a reasonably predictable series of wind and pressure belts with names and terminology that are often nautical in flavor, reflecting the old-time mariner's reliance on a first-hand knowledge of the winds.

So let's imagine that we have temporarily erased the

If we can visualize an all-water but rotating globe (to induce Coriolis force), we can see a predictable pattern of worldwide pressures and winds.

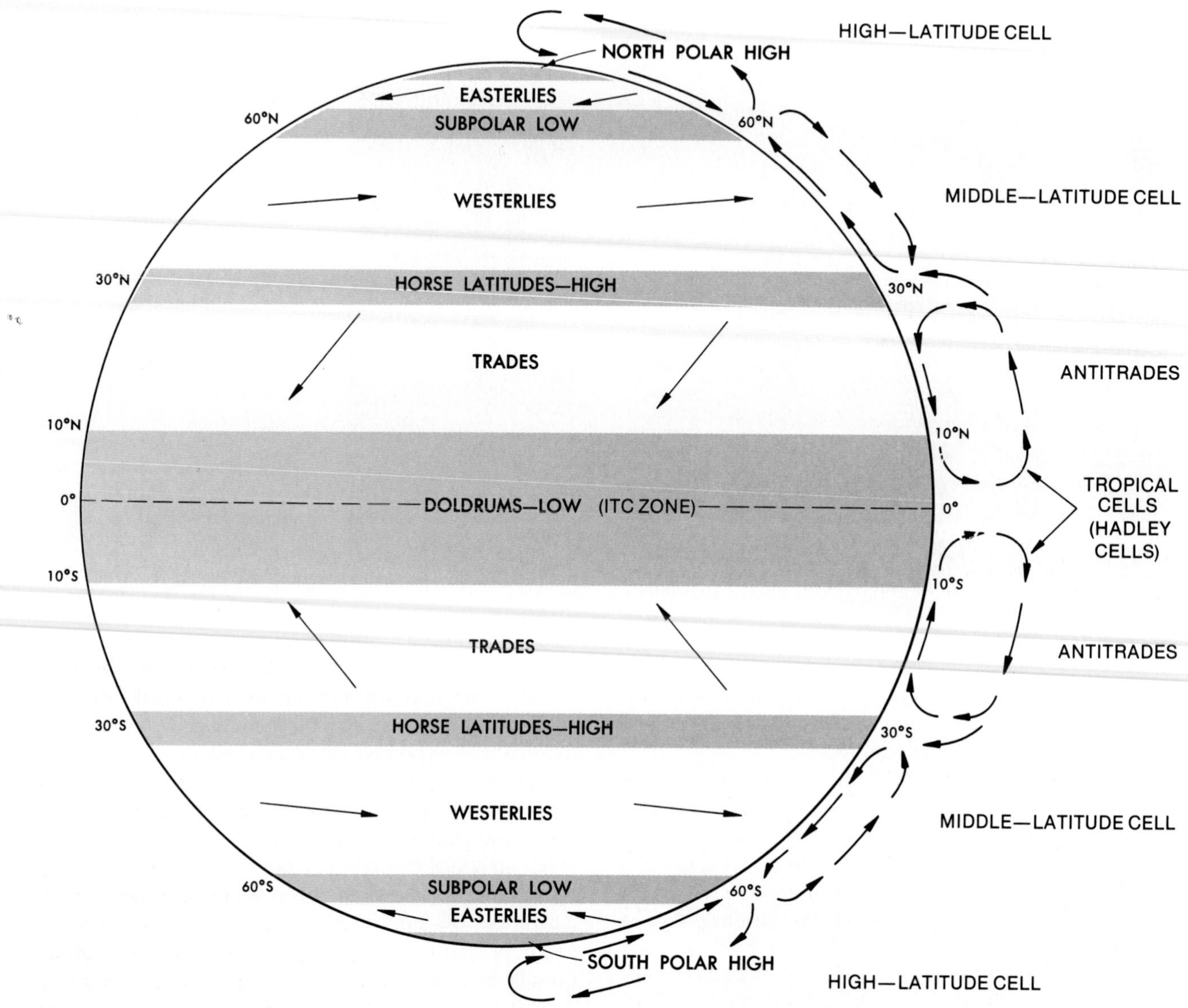

Fig. 5-7. ***Idealized world wind and pressure pattern.*** *Circulation cells are shown at right.*

continents with their seasonal temperature/pressure/wind changes and extended the elementary pattern of the oceans over the entire globe. There is a reason behind this procedure. Although the resulting system is both theoretical and simplified, the generalized outline can serve as a framework on which to hang some of the more complicated variations that we will develop later (Fig. 5-7).

Doldrums (Equatorial Low)

The belt immediately astride the equator and extending roughly 10° on either side is called the ***Doldrums*** or ***Equatorial Low.*** Its margins are not precisely at 10° N and 10° S because the entire zone, along with all the others on the earth shifts north when the overhead sun moves to the Tropic of Cancer and south when the sun moves to the Tropic of Capricorn. Nonetheless, the Doldrums generally coincides with the equatorial region and is the result of the constant high temperatures of this part of the world.

In the Doldrums the overall movement of the continuously heated air is *up;* thus the pressure is low. Since the air movement is mostly vertical, this zone is characterized by variable breezes and calms. The sailors who first recognized and named these belts dreaded the Doldrums, for they were often becalmed for long periods of time in equatorial latitudes. The expression "in the doldrums," meaning to feel depressed, has come down to us from those times, for there was nobody as unhappy as a sailor caught in the tropical calms.

Since the air movement is mostly vertical in the Doldrums, this zone is characterized by variable breezes and calms.

Horse Latitudes (Subtropical High)

A little distance out from the equator both north and south, at about 30°, is a second zone of calms. This is a narrow belt called the ***Horse Latitudes*** or ***Subtropical High.*** Again most air movement is vertical, but in this case the air sinks rather than rises. Temperature does not cause the air to descend. We are at the edge of the tropics and temperatures are high. However, the ***Antitrades,*** upper air winds flowing away from the equator, are increasingly deflected as they move (Coriolis force), and this deflection is not slowed by friction with the earth (Fig. 5-8). By the time the Antitrades have reached about 30° N and S latitude, they are flowing parallel to the equator. Air piles up and sinks of its own weight. The result is high pressure at the surface.

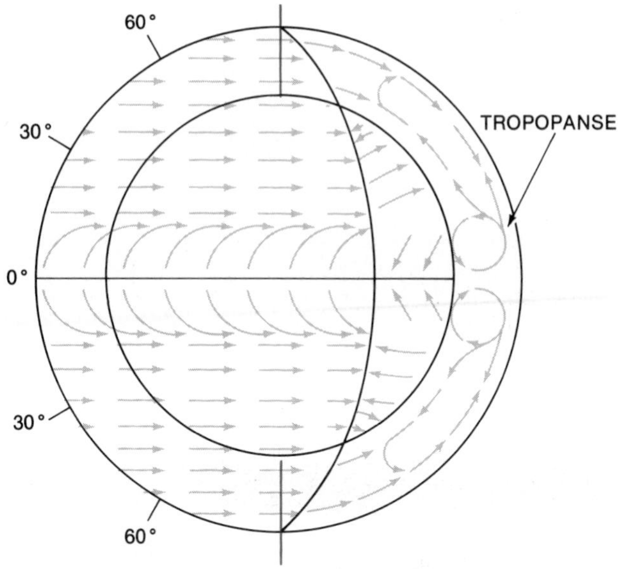

Fig. 5-8. ***Winds in the upper troposphere.*** *Warm air rises from the equator to the top of the troposphere, then flows north and south increasingly deflected by the Coriolis force but unaffected by surface friction. In the vicinity of 30° flow is westerly, parallel to the equator, and, with poleward flow no longer possible, air piles up and descends causing high pressure at the surface. Upper tropospheric flow at higher latitudes is also westerly.*

Trades and the ITC

Between the permanent high of the Horse Latitudes and the permanent low of the Doldrums is a broad zone in which air flows horizontally along the surface toward the equator (from high to low) (Fig. 5-7). In both hemispheres this belt is called the ***Trades.*** Because the Trades converge on the Doldrums, the Doldrums are said to occupy the ***Intertropical Convergence Zone (ITC).***

Before we continue, two points must be emphasized:

1. *Winds are named by the direction from which they come.* Thus a north wind, for example, blows *from* the north toward the south, and a west wind *from* the west toward the east. The terms *northerly, westerly, easterly,* and so on also imply the direction *from* which the wind comes.
2. Deflection—right-hand or left-hand—is *in the direction of air movement.* Thus with the wind at your back, right-hand deflection would be off to your right, and left-hand deflection off to your left.

Winds are named by the direction from which they come.

When we apply this principle to the Trades in the Northern Hemisphere, as the air moves from the Horse Latitudes toward the ITC, we see that it shifts to the right and takes a northeasterly course. Applying left-hand deflection to the Southern Hemisphere Trades, we find that they take a southeasterly course toward the equator. The trade winds, or more specifically the ***Northeast Trades*** (Northern Hemisphere) and ***Southeast Trades*** (Southern Hemisphere), characteristically flow consistently from the same direction at moderate velocities—perfect winds for sailing. Every westward-bound mariner took advantage of them if at all possible, and the term *trade* winds became truly descriptive (Fig. 5-9).

The trade winds flow consistently from the same direction at moderate velocities—perfect winds for sailing.

Tropical Circulation Cell

In the tropics we have already encountered a breakdown in the easy transfer of warm air toward the poles. Instead of direct flow in each hemisphere, there is a somewhat self-contained ***tropical circulation cell*** (also called a ***Hadley***

Fig. 5-9. ***The trade winds.*** *Fluttering and bowing before the brisk trades, coconut palms on Trinidad's north coast transpire huge quantities of moisture from their wind-blown crowns.*

cell after George Hadley who first postulated its existence): rising air at the equator, carried away at a higher level via the Antitrades, only to accumulate and subside at the Horse Latitudes and flow back at surface Trades toward the ITC (Fig. 5-7).

There is a somewhat self-contained* tropical circulation cell: *rising air at the equator, carried away at a higher level via the Antitrades, only to accumulate and subside at the Horse Latitudes and flow back at surface Trades toward the ITC.

The Westerlies

The tropical circulation system is not wholly closed, however. Some of the sinking air at the Horse Latitudes turns poleward along the surface in direct opposition to the fraction that sweeps toward the equator as Trades. The ever stronger Coriolis force causes this current to swing off toward the east so that finally the middle-latitude surface air flow (roughly, latitudes 35° to 60° north and south) comes from almost due west. This is where we get the term ***Westerlies.***

The Westerlies, unlike the Trades, are neither constant nor mild. Frequently gusty and boisterous, even violent on occasion, the winds are unpredictable and likely at any given moment to blow strongly from directions other than west. This behavior is highly typical of middle-latitude circulation in general, for here we have a broad zone of conflict between advancing warm tropical air masses going poleward and polar air currents attempting to push their way toward the equator.

We have a broad zone of conflict between advancing warm tropical air masses going poleward and polar air currents attempting to push their way toward the equator.

The Subpolar Low

An immediate result of this confrontation is the development of a narrow ***Subpolar Low*** (also called the ***Polar Front***) zone at the mean point of contact between tropical air masses and polar currents. The lighter warm air with its cargo of water vapor is forced to rise above the advancing cold along this line, and the released latent energy makes this a major region of storm generation.

Although a narrow belt at about 60° north and south is called the Subpolar Low and long observation has established it as a permanent zone of low pressure, the Polar Front itself will often move toward the equator. The character of warm/cold air conflict is one of erratic pulsation,

THE MANILA GALLEONS

Every year from 1565 to 1815, at least one Spanish galleon made the round trip from Manila, Philippine Islands, to Acapulco, Mexico. Their cargo was a rich mix of Far Eastern luxury goods destined for the grandees of Spain and the New World in return for Mexican silver and Spanish specialized manufactured products. Their route was unvarying—westward before the Trades, a smooth and pleasant journey; eastward via the Westerlies, the lumbering ungainly galleons pitching and yawing through some of the Pacific's nastiest weather.

Spain's colonial ventures into the Far East centered on the Philippines. At an early date, however, it became obvious that, although these islands supplied lost souls for the church to reclaim, they were not a productive source of gold for Spanish coffers as had been so many of the South and Central American colonies. It remained for China and to a lesser extent India and Southeast Asia to supply the riches in demand back home. A typical cargo outgoing from Manila might include cigars from the Philippines; silk, porcelain, lacquerware, and tea from China; spices from the Indies; gems from India; and ivory, camphor, and teak from Siam. Mexican silver, in both bullion and coins, paid for much of it. Until the mid-1930s the massive silver coin "one dollar Mex" still circulated and was the most valued coin in China.

These great ships, westward bound in the Gulf of Alaska, were amazingly seaworthy if thoroughly uncomfortable. Through all 250 years only a few were lost to the weather. Vaguely, through the drizzle and low clouds, the anxious lookout would make his first landfall. It could be Baranof Island in Alaska, Vancouver Island, or the dark and gloomy coast of northern California. Sailing before the squally Westerlies the captain had to stay well offshore and coast southward toward better weather and Acapulco. Here is where the English freebooters, Sir Francis Drake and Woodes Rodgers among them, sent their swift vessels to attack the battered giants. They didn't always win, but the enormous potential booty was reason enough to try.

Acapulco and Manila are 16.5° N and 14° N, respectively, both being well within the trade wind zone the year round. The southern tip of Hawaii is 19° N, which is also a trade wind latitude. Did Spanish ships sight the Hawaiian Islands? Almost a thousand Manila galleons plied this route over 250 years, and Pele, the Hawaiian goddess of fire, frequently signals from Mauna Loa. How could the Spaniards have missed it? Perhaps they didn't. Imperial Spain never made it a practice to publicize its secrets to the world. However, recent research in the archives at Madrid has turned up a map showing some sizable islands in the correct latitude, although the longitude was in considerable error. Because Spanish captains did not have practical chronometers until long after the English, accurate longitude measurements obviously suffered.

---

surges of cold air pushing the front temporarily far out of line, especially over the large Northern Hemisphere continents. On these occasions any orderly air flow in the Westerlies is disrupted, and the entire middle latitudes are affected by local squalls and storms coming from this ever restless subpolar arena.

Middle-Latitude Circulation Cell

By generalizing rather broadly, we can nonetheless visualize a moderately self-contained ***middle-latitude circulation cell*** in each hemisphere comparable to that of the tropics. A part of the sinking air at the Horse Latitudes spills poleward to form the Westerlies. Although in the course of their run across the middle latitudes the Westerlies are deflected sharply and interrupted by frequent storms, they eventually transport warm air into the Subpolar Low.

The Westerlies transport warm air into the Subpolar Low.

In this area the warm air encounters cold and is forced aloft. Part of this lifted air, at least theoretically, flows back at a higher level to help feed the downward movement of air in the Horse Latitudes. This upper air flow, above and counter to the Westerlies, appears to be as erratic and unstable as the Westerlies, but it is probably safe to say that there must be an overall drift of air toward the equator to complete the circulation cell.

Polar Highs and Easterlies

At the poles we encounter enduring and well-developed high-pressure zones. Like the Doldrums of the equatorial region where year-round high temperature results in low pressure, the cooling effects of Antarctica and the permanently frozen Arctic seas induce a thermal high. Air, chilled at its base, sinks and flows outward in all directions toward the equator, veering rapidly to the west (from the east) as it goes. These winds are called the ***Easterlies*** or ***Polar Easterlies.*** They bring cold air to the Subpolar Low to keep the warm/cold confrontation going. Again, theoretically, some of the warmer air from the Westerlies that lifts at this point flows back poleward to complete a ***high-latitude circulation cell.***

THE EFFECTS OF THE CONTINENTS

We have been discussing the patterns of winds and pressures that would occur if the world were without continents. It is reasonably accurate over the ocean basin, but once we include the land masses, variations immediately arise. However, these effects are not distributed evenly between the hemispheres. Not only does the Southern Hemisphere have a much higher proportion of ocean to land than does the Northern, but also its major land masses are in the tropics. The land masses in the Southern Hemisphere therefore lack strong seasonal temperature alteration. In contrast, Northern Hemisphere continents lie mostly in the middle latitudes. Therefore, they experience much more seasonal variation. The excessive cold of continental winters and the contrasting great heat of continental summers give rise to striking seasonal pressure differences. This condition significantly disrupts the simplified wind and pressure system (Figs. 5-10 and 5-11).

Once we include the land masses, variations immediately arise, but these effects are not distributed evenly between the hemispheres.

The Doldrums belt of our idealized all-water earth stays pretty much the same even over land. Temperatures remain high and pressure low. Note, however, that the ITC shifts north and south with the seasons. In the high-pressure belt of the Horse Latitudes, the heating of the land areas in summer causes continental lows to develop, thus breaking up the earth-encircling high belt into fragments. In the summer the Horse Latitudes become a series of isolated cells over the cooler seas. In the Northern Hemisphere there are two cells: one in the Pacific called the ***Hawaiian High*** and one in the Atlantic, the ***Azores High.***

There are three oceans in the Southern Hemisphere: the Pacific, the Atlantic, and the Indian, each with its high-pressure cell. During the winter, as continents cool and their lows weaken, these oceanic highs are connected across the land and the idealized belt reasserts itself. In the Northern Hemisphere where the continents increase their bulk in the higher latitudes, the winter cold is most severe in Canada and Siberia, causing intense high pressure. The high-pressure belt linking the Atlantic with the Pacific then becomes badly deformed with great northerly bulges. Winds that are, of course, the result of pressure differences exhibit seasonal variations in the vicinity of the continents.

The Subpolar Low Zone is a simple matter in the Southern Hemisphere. These latitudes completely lack sizable continents, and we have virtually an all-water earth again. In the Northern Hemisphere, however, the land masses are huge and the ocean basins severely restricted so that, once again, isolated cells at sea become the pattern. They are called the ***Icelandic Low*** and the ***Aleutian Low*** and are permanent in these areas. When the intense winter high asserts itself over Canada and Siberia, these lows, despite their small size, are very deep and well developed. In the summer when the continents are warmer (though not really hot at these latitudes), the lows are quite weak. The polar regions, which continue cold at all seasons, remain as permanent highs.

The Monsoon

Not surprisingly, except at sea, the orderly pattern established earlier can develop all sorts of complications. An excellent example of the kind of local winds that result from these complications are the relatively large-scale ***monsoon*** winds of Asia. Basically, their cause is differential heating of land versus water.

In summer the land mass of Asia heats up much more rapidly than the surrounding oceans, especially in its tropical portion. Local pressures exhibit a seasonal high over the cool sea, contrasting sharply with the heat-induced continental low that reaches its maximum development in the desert of northwest India/Pakistan. Reacting to this pressure

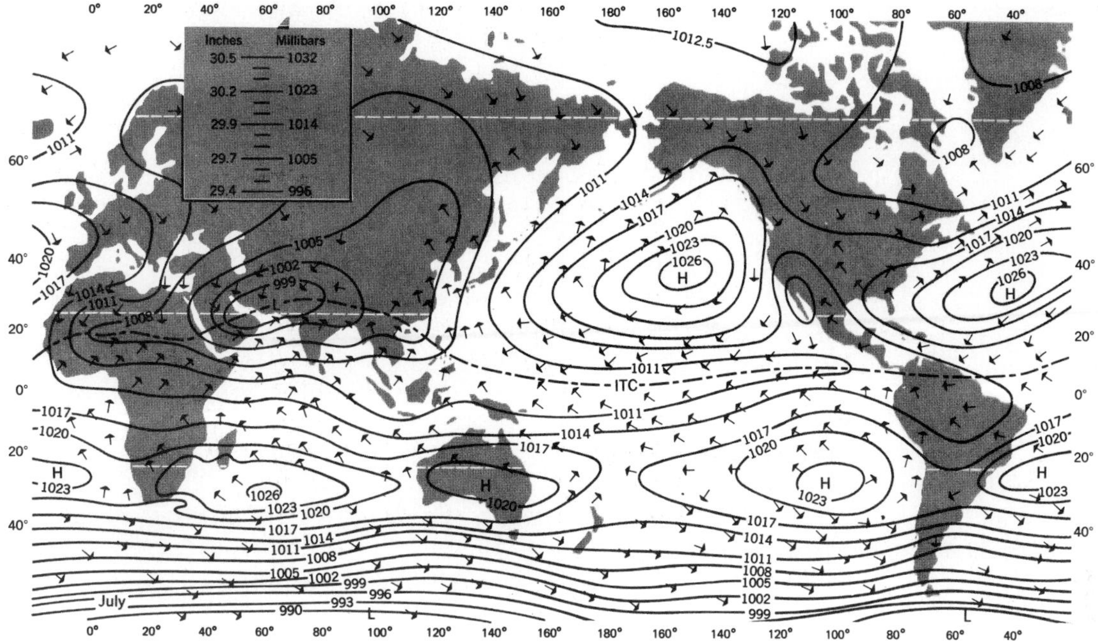

Fig. 5-10. World July isobars and wind patterns.

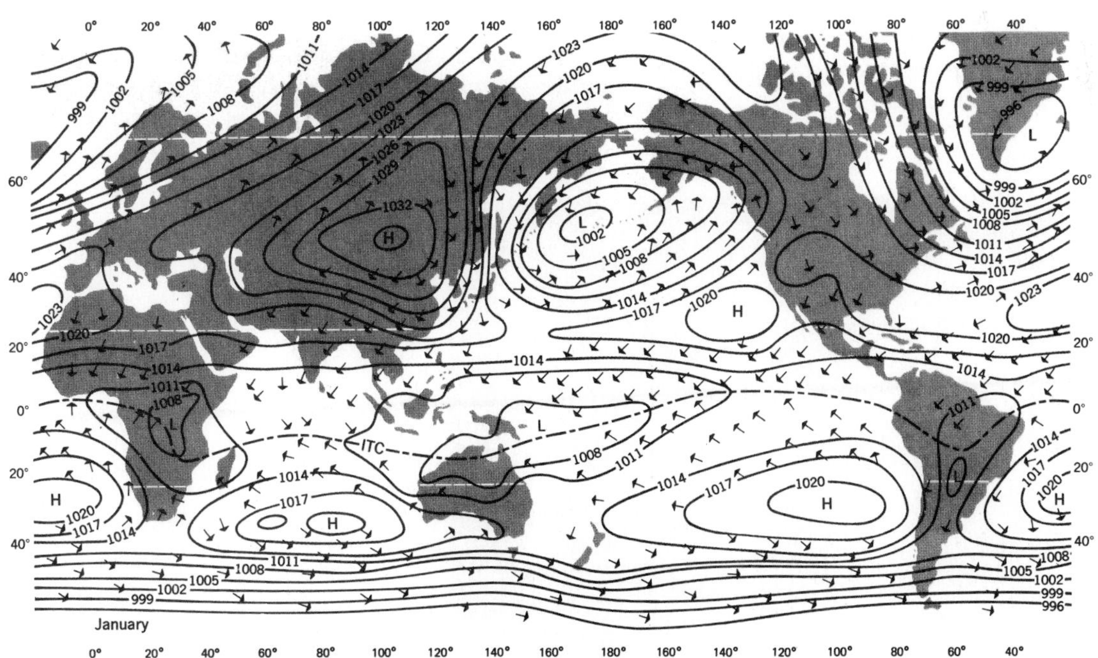

Fig. 5-11. World January isobars and wind patterns.

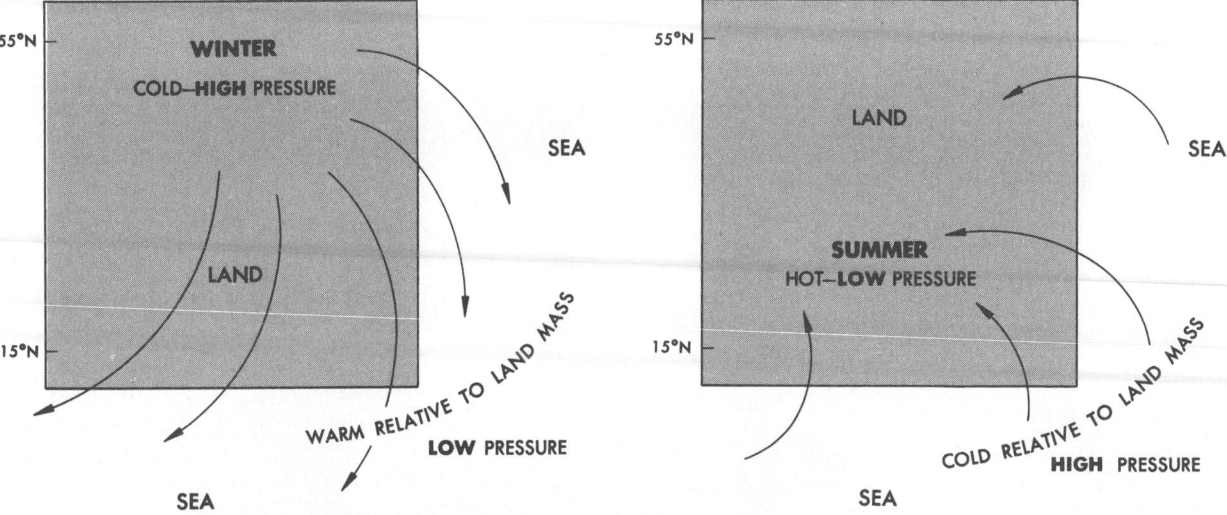

Fig. 5-12. ***Idealized Northern Hemisphere monsoonal circulation.***

difference, air flows onshore (from high to low), displaying as it moves a counterclockwise spiral. The entire procedure is reversed in winter. A cold continent develops high pressure, especially in northcentral Siberia, and the warmer sea, which has changed temperature very little from the summer, becomes a relative low. Now the air flows offshore, spiraling clockwise (Fig. 5-12).

In summer the Asian land mass heats up much more rapidly than the surrounding oceans, and air flows onshore, displaying a counterclockwise spiral.

This simple explanation of monsoon circulation, though correct as far as it goes, fails to take into account many of the intricacies of the local scene. For instance, the summer monsoon "breaks" with great suddenness in India; this situation would be unlikely if it were tied to the gradual progression of the seasons. But the great Himalayan barrier almost seals India off from the continent, thereby disrupting, channeling, or even blocking completely the easy flow of seasonal winds (Fig. 5-13).

Entering into this complex picture is a well-defined jet stream—an eastward-trending, high-velocity, upper troposphere air current that flows south of the Himalayas in sum-

Fig. 5-13. ***Monsoon.*** *Summer has officially arrived in Bombay with the "break" of the monsoon.*

mer and north in winter. (Later in this text more will be said about jet streams in general and their relationship to storms and precipitation.) Each year as our knowledge of their behavior increases, jet streams are taking on increasing importance in explaining monsoon phenomena.

There are also the generally southwest summer winds of tropical Asia, which have been explained as deflected Southern Hemisphere Trades rather than merely spiraling onshore breezes. The contention is that, as the heat equator (as opposed to the geographic equator) follows the summer overhead sun into the Northern Hemisphere, it pulls with it the southern southeast Trades. As they cross the equator, Coriolis force causes them to swerve to the right, thus becoming the southwest summer monsoon.

Without question, then, the monsoon is involved, and much is yet to be learned about its variations. Yet the underlying cause remains, simplified perhaps, but basic—the differential heating of land versus water. A huge land mass placed in the middle of the ocean naturally develops local winds. Out at sea, the Trades, Westerlies, and so forth, continue to blow in their predictable paths, but continental influences create the winds that are so characteristic of Asia.

THE NORTH AMERICAN "MONSOON" If the development of a monsoon requires a land mass of continental size and sufficient latitudinal spread to encompass both high latitudes and tropical to subtropical latitudes, then it appears possible that North America could become a site for monsoon activity. Interior North America can get almost as cold as Asia in the winter, and although the true tropics are almost lacking, the southern United States does become quite warm in summer. As a result, we could have a monsoon—not as well developed as that of Asia, which is larger, but a monsoon nonetheless.

Canadian cold air forces itself clear down to the Gulf on occasion, and sticky tropical air invades the Middle West. We do not have a name for this phenomenon as the Asians do, and sometimes its effect is overpowered by other more dominating weather factors, but the general tendency is there and we can see it operate close to home every year.

OTHER MONSOONS The rest of the world's continents are large enough to generate monsoons, but their mass is concentrated in the tropics. They get hot and pull air onshore in summer, but except for Antarctica no cold air masses are represented. Yet oddly, Australia, the smallest of the lot, displays a full-blown monsoonal circulation along its north coast.

This situation is caused by Australia's nearness to the huge Asian continent. During its winter there is no really cold air over the continent, and only a mild high develops—scarcely enough to cause significant offshore air currents. However, when it is winter in Australia, it is summer in Asia, and the attractive power of the Asian continental low pulls the air off the Australian north coast. In summer the desert heart of Australia develops the deep low required to complete the annual wind reversal.

If Australia were located far from Asia, it would have no monsoon. Similarly, the central east coast of Africa, from which the traditional Arab voyages to Asia and back took advantage of the seasonal wind reversal, is similar to northern Australia. If it were not near Asia, it would have no monsoon.

Sea and Land Breezes

On a much smaller scale, the same sort of air circulation that causes monsoons occurs along coastlines on a day-and-night rather than on a seasonal basis. The result is an alternating ***sea breeze*** and ***land breeze***. The circulation is characteristic of the tropics during the entire year and the summer season in many other parts of the world (Fig. 5-14). As the land heats up during the day, low pressure gradually develops in contrast to the high over the cooler sea. The mild pressure

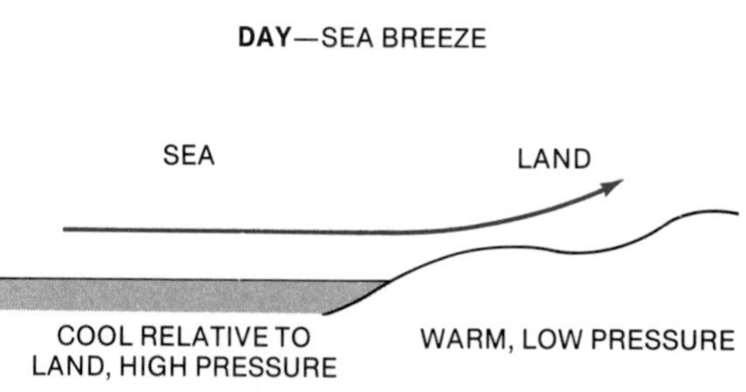

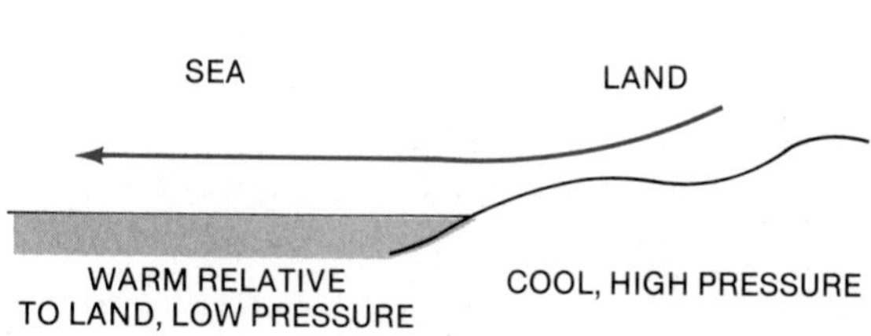

Fig. 5-14. ***Idealized land and sea breeze.***

difference causes gentle cooling breezes to move onshore (sea breeze). At night the reverse occurs, and the breezes blow in the opposite direction (land breeze). The entire phenomenon is extremely shallow and normally affects only a coastal strip a few miles wide. Frequently, however, it is the difference between reasonable comfort and oppressive heat.

As the land heats up during the day, the mild pressure difference that develops causes gentle cooling breezes to move onshore.

The monsoon and sea and land breezes are examples of local winds that defy the standard concept of worldwide pressure and winds. Occasionally, they are strong enough to completely erase the effects of these belts for a time; more often they merely modify them.

CONCLUSION

Overall, this chapter emphasizes the intimate relationship between atmospheric pressure, winds, and insolation. Here our knowledge of unequal heating and the need to equalize temperature differences is put to use. We can now see a sequence of events that begins with intense heating of the low latitudes and leads to the global circulation system. We know that the great contrast in the thermal capacity of land and ocean modifies the global pattern with monsoons and local sea and land breezes.

Understanding pressure and winds, as we will see, is central to the discussion of air masses, fronts, and storms in Chapter 7. First, however, in Chapter 6, we must turn to moisture, the remaining element of weather and climate.

KEY TERMS

pressure
one atmosphere
anticyclone
cyclone
mercurial barometer
aneroid barometer
millibar
isobar
wind
pressure gradient
anemometer
Coriolis force
Doldrums
Equatorial Low
Horse Latitudes
Subtropical High
Antitrades
Trades
Intertropical Convergence Zone (ITC)
Northeast Trades
Southeast Trades
tropical circulation cell (Hadley cell)
Westerlies
Subpolar Low
Polar Front
middle-latitude circulation cell
Polar Easterlies
high-latitude circulation cell
Hawaiian High
Azores High
Icelandic Low
Aleutian Low
monsoon
sea breeze
land breeze

REVIEW QUESTIONS

1. If a column of air is cooled at the bottom, what will it do? How will this affect atmospheric pressure? What if the column of air is heated?

2. How can a barometer measure pressure in inches when pressure is a force, not an object? How does a barometer that measures pressure in millibars differ?

3. Is a barometric reading of 30 inches high or low? How can the use of isobars help us find out?

4. What causes the wind? How can you tell in which direction the wind will move?

5. If the Coriolis force is an illusion, why do we need to know about it?

6. Why do pressure and wind patterns shift with the seasons?

7. What causes the main differences between the tropical circulation cell, the middle-latitude circulation cell, and the high-latitude circulation cell?

8. Why are there so many storms in the middle latitudes?

9. How—and why—does the true Asian monsoon differ from the North American "monsoon"?

10. What causes land and sea breezes?

APPLICATION QUESTIONS

1. Keep a daily log of the barometric pressure around 3 P.M. and 9 P.M. in your area for a week. Also note the weather conditions at those times. Can you see a pattern of relationships between pressure and weather conditions? How do your observations compare with your local TV meteorologist's observations?

2. If a barometer reads 29.5 inches, what would one read in millibars? How many dynes per square centimeter would this represent?

3. Prepare a set of directions, with illustrations, showing how to find out if the wind is moving in a clockwise or counterclockwise spiral. Imagine that your directions will be used by a pack of eight-year-old Cub Scouts on a camp-out. Be sure to make them *simple* and *clear*.

4. Photocopy a map of the world and plot a round-the-world sailing cruise beginning and ending in Boston Harbor. You must stop in at least five ports. How would you make sure your ship covered the least possible distance under the best possible weather conditions? Describe the winds you would be likely to encounter, and explain your reasons for choosing the latitudes in which you sail. Then compare your own imaginary trip with a real account of a recent circumnavigation by sail.

5. Imagine you are lost at sea in a small sailboat with nothing but a barometer to help you. How would you tell whether you were in the Doldrums or the Horse Latitudes? If you were in the Northern Hemisphere Horse Latitudes, in which direction would you paddle to pick up a trade wind? What could you expect to find if you went in the opposite direction?

6. Are there any wind generators in your area? What are they used for? Describe the site in terms of global and local wind patterns.

7. Consult the maps of pressure and wind patterns (Figs. 5-9 and 5-10) and identify places that have a potential smog problem owing to an upper air inversion.

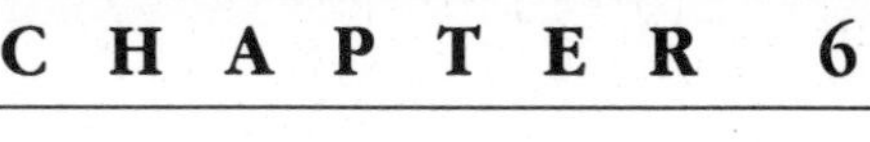

ATMOSPHERIC MOISTURE

OBJECTIVES

After studying this chapter, you will understand

1. How water becomes water vapor by evaporation and returns to its liquid state by condensation.
2. How dew, frost, and fog occur.
3. How clouds are classified.
4. How orographic, convective, and cyclonic precipitation occur.

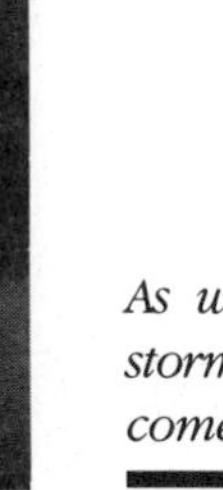

As waves crash on a coastal jetty, an isolated thunderstorm—a remnant of a fast-moving cold front—prepares to come onshore in western Oregon.

INTRODUCTION

Why does it rain? If you live in Oregon, why does it rain so much? If you are in Arizona, why does it rain so little? Why does dew form on some nights and not on others? Why is the air so wet and sticky in New Orleans during the summer, but neither wet nor sticky in Los Angeles during the same season? Before we can answer these and many other similar questions, we need to understand the behavior of moisture in the atmosphere—why condensation occurs and under what conditions precipitation takes place. By focusing on atmospheric moisture and precipitation, this chapter will continue our introduction to the four elements of weather and climate. When completed, we will be ready to consider how the elements work together, first in air masses and storms (Chapter 7) and then in world climatic regions (Chapters 8–11).

WATER VAPOR

Water vapor is odorless and invisible. The human body can sense it only in conjunction with air temperature—so-called ***sensible temperature***. When the air is dry, evaporation of perspiration from the skin cools us; when the humidity is high, evaporation is inhibited, perspiration is ineffective as a cooling mechanism, and we are not only hot but also uncomfortably sticky. (Don't point to a cloud and call it water vapor. It is not vapor at all but liquid droplets in suspension, just as is a plume of "steam" coming from the spout of a tea kettle. *Steam* is the term for gas that is superheated and therefore invisible.)

In terms of weather, water vapor is undoubtedly the most important atmospheric gas for three reasons:

1. It determines the precipitation and cloud potential of any given air mass.
2. It contains latent heat to be released on condensation.
3. It effectively absorbs radiated heat.

Water becomes water vapor by evaporation and returns to its liquid state by condensation. We will discuss both phenomena in this section.

Water becomes water vapor by evaporation and returns to its liquid state by condensation.

Evaporation

Water vapor finds its way into the atmosphere by evaporation, mainly from the oceans but also from plants (transpiration), soil, lakes, rivers, and ponds. The amount of water that changes form by evaporation varies widely from place to place. Because temperature is the major control, tropical seas and forests are by far the greatest contributors. The polar ice caps with their permanent low temperatures and the tropical deserts, despite their high temperatures, supply very little evaporated moisture. The middle latitudes fall somewhere between these extremes.

We have already seen how the planetary wind system transports moist air. Our particular problem here, then, is to identify and measure the water vapor taken into the air by evaporation and then to explore the ways and means of getting it back out again.

THE HYDROLOGIC CYCLE There is only a specific amount of water in this world, a closed system from which little is lost or gained. For the most part, water merely changes form regularly from ice to liquid to vapor. Of course, most of the earth's water stays in the oceans, but a small part is skimmed off the top by evaporation and is easily transported by winds to faraway places to form clouds and to fall as rain or snow. This water is returned to the sea via streams, springs, and icebergs as part of the ***hydrologic cycle*** (see Chapter 13), and the rest condenses and falls directly into the sea.

For the most part, water merely changes form regularly from ice to liquid to vapor.

RELATIVE HUMIDITY The water vapor that is in the air is known as ***humidity*** and we can express its measurement at a particular time and place in several ways. The most common type of humidity measurement is the percentage or ratio called ***relative humidity***. This is *the amount of water vapor in the air relative to the amount it is able to hold at a given temperature.*

"At a given temperature" is a critical part of the definition. The air's capacity to hold water vapor varies with temperature. Cold air can hold much less water vapor than warm. Thus air at 50 percent relative humidity is holding half the amount of water vapor that it is equipped to hold (Fig. 6-1).

Cold air can hold much less water vapor than warm.

If we had such a mass of air at sea level and caused it to ascend, the air would cool through expansion. Its volume would increase, but its moisture-holding capacity would decrease. Without any gain of moisture, the relative humidity would become greater than 50 percent. Essentially, we

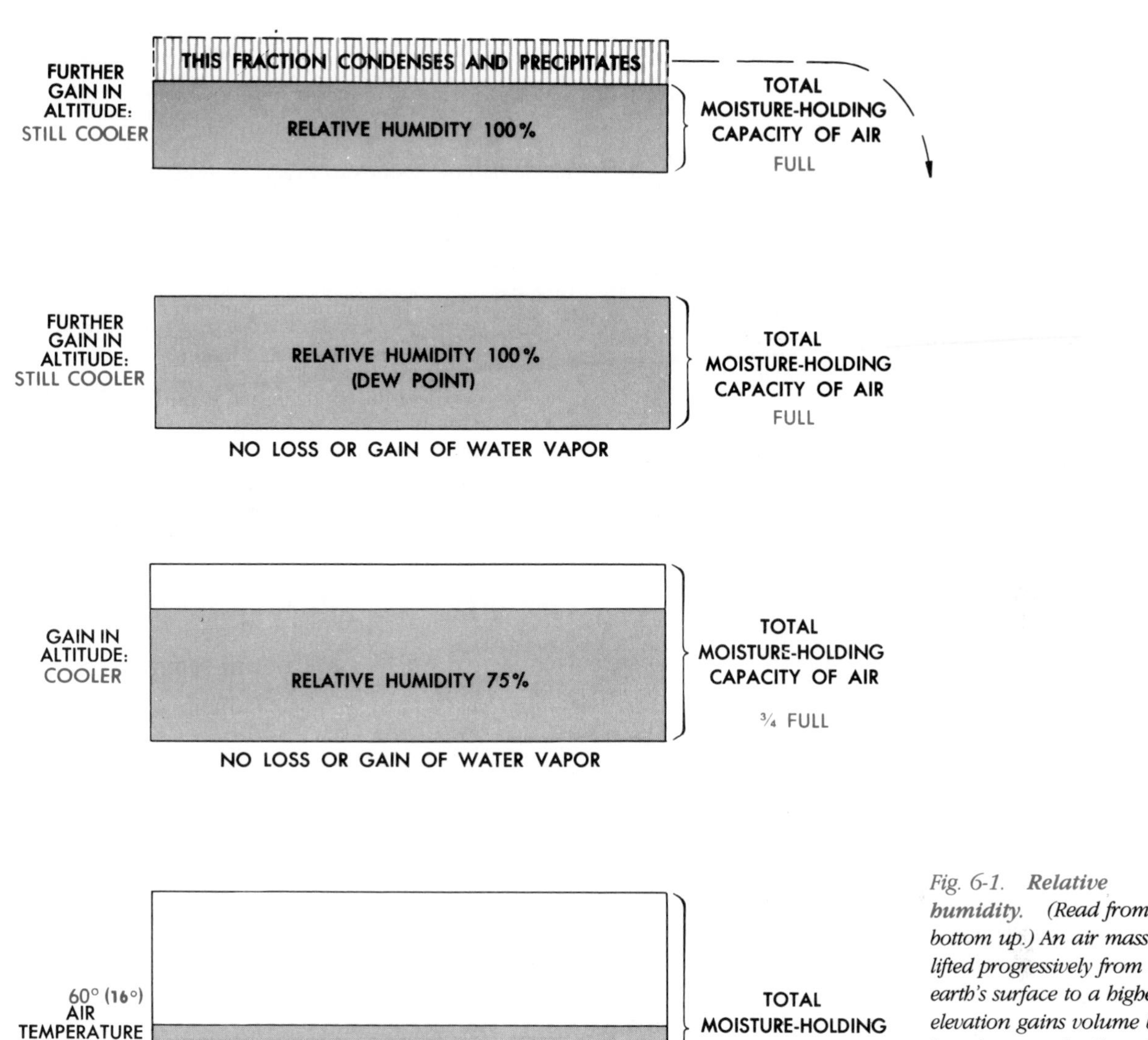

Fig. 6-1. ***Relative humidity.*** *(Read from the bottom up.) An air mass lifted progressively from the earth's surface to a higher elevation gains volume but loses heat, gradually reducing its ability to hold moisture. (The size of each quadrilateral represents moisture-holding capacity, not volume.)*

would be decreasing the capacity of the container. The same amount of water that half filled the large-capacity container now fills the small-capacity container almost to the top. When further cooling continues to decrease the moisture-holding ability of the air, the air mass will eventually become filled to capacity. This is ***saturation;*** the air is holding all the water vapor that it can hold, so the relative humidity is 100 percent. This is also the ***dew point temperature*** (or simply the *dew point*) because any further cooling will make the container too small to hold all the water and the surplus will spill out as condensation.

The dew point is where condensation begins as cooling continues. If this point is above freezing, condensation will be in the form of rain; if it is below freezing, snow will result. Continued cooling will produce continued condensation, and the relative humidity will be maintained endlessly at 100 percent. If we were to bring this same air mass back to its original temperature, the relative humidity would no longer be 50 percent; it would be less, for a portion of the original water vapor has been lost through condensation.

The dew point temperature is where condensation begins as cooling continues.

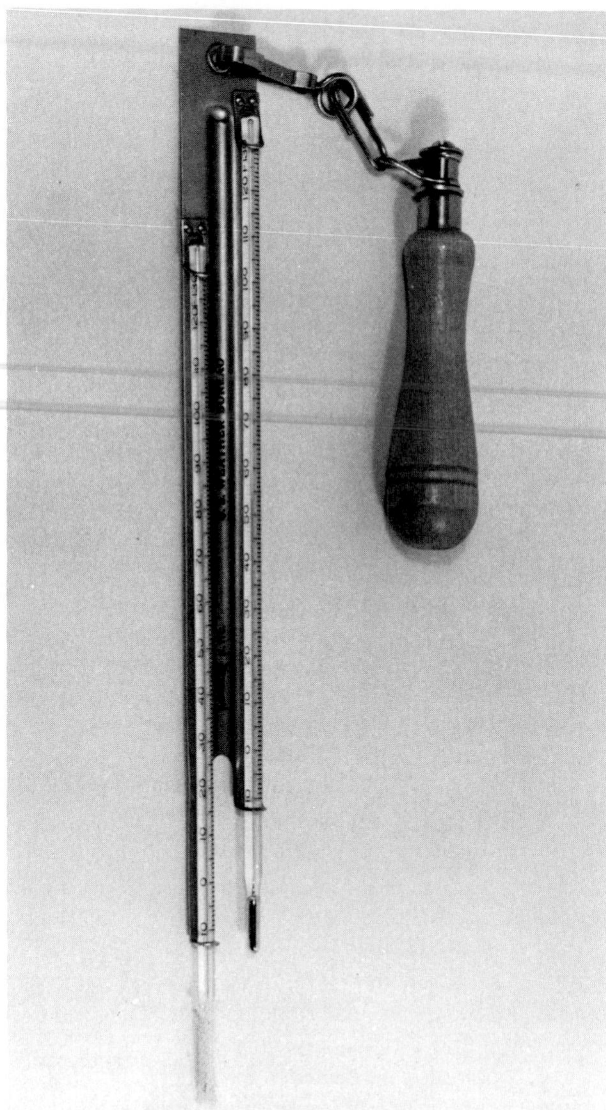

Fig. 6-2. ***Sling psychrometer.*** *Nothing more complicated than two mercurial thermometers, but note the cloth-covered wet bulb (left). The handle is for the operator to grasp and swing the entire mechanism overhead; this facilitates maximum evaporation and cooling.*

MEASURING HUMIDITY To determine the relative humidity, we use an instrument called a ***sling psychrometer*** (Fig. 6-2). This simple tool is made up of two standard thermometers mounted on a frame with a handle, one of which has a moistened piece of cloth wrapped about the bulb. We whirl the instrument around for several minutes and then read the difference of temperature between the two thermometers (if any). If the air is saturated, no evaporation of the moisture can take place in the cloth and so there will be no cooling. Under these circumstances, the two thermometers will register the same temperature. However, the drier the air becomes, the greater the evaporation and the greater the cooling of the wet bulb. A significant contrast in the temperature readings will be directly related to the amount of water vapor in the air. When applied to a table published by the Smithsonian Institution, it will give us an accurate dew point and relative humidity measure (Fig. 6-3).

The sling psychrometer is not practical for recording changes in humidity over time. Instead, we rely on the ***hygrograph*** which uses human hair to show changes in humidity. As we all know, our hair is sensitive to atmospheric moisture. It extends when moist and it contracts when dry. In the hygrograph, strands of hair are connected to a lever and a pen that traces humidity on a revolving drum for periods of a week or more.

Condensation

Condensation is the process by which water vapor returns to liquid water. There can be no condensation unless a surface is present on which liquid can condense. At ground level, trees, grass blades, parked automobiles, boats, and even animals provide suitable surfaces. We know that clouds form in the atmosphere, so there must be suitable surfaces there too, but what are they? From the discussion of the atmosphere's composition in Chapter 3, we know that the air contains numerous particles that we refer to generally as dust. Not all dust provides a suitable surface for condensation, but much of it does, and that part becomes the required ***condensation nuclei.*** Thus there are two requirements for condensation:

1. The air must be cooled to its dew point temperature.
2. Condensation nuclei or a suitable object at ground level must be available.

Condensation is central to the formation of dew, frost, fog, clouds, and precipitation.

Condensation cannot occur unless a surface is present on which liquid can condense.

TEMPERATURE OF DEW POINT[a]

Air Temp. Dry Bulb (°F)	*Depression of Wet Bulb (°F)*							
	1°	*2°*	*3°*	*4°*	*6°*	*8°*	*10°*	
20°	16	12	8	2	−21			
30°	27	25	21	18	8	−7		←Read
40°	38	35	33	30	25	18	7	Temperature
50°	48	46	44	42	37	32	26	of
60°	58	57	55	53	49	45	40	Dew
70°	69	67	65	64	61	57	53	Point
80°	79	77	76	74	72	68	65	←Here
90°	89	87	86	85	82	79	76	

PERCENTAGE OF RELATIVE HUMIDITY[a]

Air Temp. Dry Bulb (°F)	*Depression of Wet Bulb (°F)*							
	1°	*2°*	*3°*	*4°*	*6°*	*8°*	*10°*	
20°	85	70	55	40	12			
30°	89	78	67	56	36	16		←Read
40°	92	83	75	68	52	37	22	Percentage
50°	93	87	80	74	61	49	38	← of
60°	94	89	83	78	68	58	48	Relative
70°	95	90	86	81	72	64	55	Humidity
80°	96	91	87	83	75	68	61	←Here
90°	96	92	89	85	78	71	65	

[a]Excerpted from Smithsonian Meteorological Tables.

Fig. 6-3. ***Dew point and humidity meteorological tables.***

DEW AND FROST When moist air comes into contact with a cold surface, the water vapor will condense as it is suddenly brought to the dew point. This is ***dew.*** Any surface that is an especially good radiator will rapidly lose the accumulated heat of the day after the sun goes down and through conduction will cool the nearby air. If sufficient cooling takes place, dew forms as tiny water droplets.

The outside of a glass with a cold liquid in it will "sweat." The glass is not leaking but is removing moisture from the atmosphere by cooling. Similarly, the inside of windows in the kitchen and bathroom are likely to "fog" as the high humidity of these rooms contacts the cold panes. If the surface is below freezing, sublimation will occur and delicate ice crystals will form on that surface as ***frost.***

FOG On calm, clear nights when loss of the earth's heat through radiation is particularly efficient, dew will form rapidly. Then gradually as the night becomes longer, a deeper layer of air will be cooled below the dew point, water vapor will condense around any available condensation nuclei, and tiny droplets of moisture will appear throughout this zone. These droplets will be kept in suspension by normal minor air turbulence. This is ***fog.*** It may vary in thickness from a few inches to several hundred feet, depending on the degree of cooling and the moisture content of the air.

On calm, clear nights when loss of the earth's heat through radiation is particularly efficient, dew will form rapidly.

Most fog is formed in this manner and is properly called ***radiation fog*** because the lower atmosphere loses its heat by conduction and radiation to the cooler earth. However, if

CLOUD SEEDING: DOES IT WORK?

It all started in 1946 when Vincent Schaefer of the General Electric Research Laboratories serendipitously discovered that dry ice could create a snowstorm—in a home freezer. Schaefer was using the freezer as a cold chamber to test the ability of various substances to cause ice crystals in supercooled moist air. (Supercooled means that water droplets in the air are below freezing but still liquid.) None of the substances worked, but when he put in a bit of dry ice (frozen carbon dioxide) to lower the temperature, the fog of supercooled water droplets in the freezer crystallized into thousands of snowflakes. Later, in 1947, Bernard Vonnegut, one of Schaefer's coworkers and the brother of science fiction author Kurt Vonnegut, discovered that adding crystals of silver iodide (instead of dry ice) had a similar effect but worked more efficiently.

Because clouds often contain supercooled droplets, scientists reasoned that what occurred in the freezer could be duplicated in the atmosphere—rain and snow could be turned on at will.

In 1948 a group of government and private researchers tried the freezer experiment on a fully developed hurricane heading for Jacksonville, Florida. The storm abruptly changed course (perhaps on its own) and crashed into Savannah, Georgia. Despite such problems and with only limited knowledge of the exact processes involved, researchers continued to apply silver iodide to all types of clouds and storm systems.

By 1953 numerous private companies, operating mostly in the western states, offered cloud seeding services, and business with ranchers, orchardists, resort operators, towns, and public utilities was brisk. At middecade, nearly 10 percent of the United States was included in seeding programs with annual expenditures of more than $3 million. In 1959 cloud seeders claimed that they increased precipitation 10 to 15 percent annually in mountainous areas of the West.

The validity of this claim is clouded somewhat by the fact that the early 1950s was a period of drought, and the latter part of the decade was wetter everywhere in the West, with or without cloud seeding. Even various U.S. government-sponsored seeding programs in the 1950s and later, many of which were designed with strict statistical tests in mind, failed to clearly establish the efficacy of cloud seeding to increase precipitation.

The most famous of these programs is the Florida Area Cumulus Experiment (FACE) conducted by the National Oceanic and Atmospheric Administration (NOAA). After five years of massive seeding of isolated cumulus clouds with silver iodide (1970–1975), NOAA scientists declared that seeding can effectively increase rainfall by nearly 35 percent! The excitement quickly waned, however, when independent reviewers found serious flaws in the study. NOAA scientists responded with a new, tightly controlled followup experiment called FACE-2 (1978–1981), which was fashioned after clinical drug trials and designed to provide the definitive answer on the value of cloud seeding. Experimenters flying into the clouds didn't even know if they were applying silver iodide or just an inert sand "placebo." In the end, FACE-2 failed to support the positive results of FACE-1.

Atmospheric scientists now generally agree that cloud seeding has succeeded in only one set of experiments. The Israeli experiments (Israeli I, 1961–1967, and Israeli II, 1969–1975), which are considered statistically sound, indicate a consistent increase in precipitation of 15 to 30 percent when clouds over the drainage system supplying the Sea of Galilee (also called Lake Kinneret) are seeded.

Thus the jury is still out. It may be that cloud seeding has a positive effect but only in certain places and in certain cloud types. It is likely to require much more experimentation until we know for sure.

---

gentle breezes waft a normal radiation fog away from its place of origin to a region where conditions are not suitable for development of fog, it is called ***advection*** (or transported) ***fog***. The California coast often experiences this type of fog. A cold current offshore provides the cooling for saturated oceanic air masses that moves onshore with the daily sea breeze. These fog banks, formed over the cold current, are brought into the coast, where they remain until the nocturnal land breeze moves the moist air seaward (Fig. 6-4). Such fogs cool coastal forests (the coastal redwoods depend on them), but they often reduce visibility to the point that airports must be temporarily closed.

Radiation fog usually begins to evaporate as soon as the sun comes out. The top of the fog layer intercepts and absorbs the longer waves in the sun's spectrum while allowing the short waves to pass through and heat the earth. The fog is therefore "burned off" by being heated from both the top and the bottom, and it is usually gone by noon. The last remaining wisps of fog are not at the ground but at some intermediate level.

To most atmospheric scientists, fog is simply a low-lying cloud. However, the manner of its formation sets fog apart from other cloud types that result from an entirely different cooling process.

Fig. 6-4. ***Advective fog.*** *Looking north along the San Andreas Fault, San Andreas Lake in the middle foreground. Every summer afternoon a finger of sea fog creeps through the Golden Gate (from left to right at top) to envelop San Francisco and the Bay.*

CLOUDS Like fog, the other cloud types are composed of tiny drops of liquid water or ice held in suspension by air turbulence. Like fog, they are a result of moist air being cooled below the dew point. There is a major difference in the way they are formed. If the cooling and subsequent condensation result from moist air in *contact* with a cold surface—cooling through conduction and radiation—then it is called fog. If the cooling and subsequent condensation result from *adiabatic cooling,* the suspended water droplets make up a typical cloud.

Cloud Types. Clouds are a visual indication of the current weather conditions, and they have much to tell us about weather to come. We usually classify clouds by a combination of their appearance and the altitude at which they form. Although there is great variety in cloud shape and the clouds of one shape may change into another, we recognize three major forms:

1. ***Cumuliform*** clouds are relatively deep and fleecy, fluffy, or cottony. They indicate turbulent air movement but never cover the entire sky.
2. ***Stratiform*** clouds cover the entire sky, are usually gray (because they cut out a large part of the light), and show little vertical development. They indicate minimal turbulence.
3. ***Cirriform*** clouds are high-level clouds thought to be made of ice crystals, though some recent research suggests they may contain ***supercooled*** liquid water droplets with a temperature below freezing. They appear fragile and feathery. When covering the entire sky, they are so insubstantial that they are often described as a high-level haze.

Clouds are a visual indication of the current weather conditions and have much to tell us about weather to come.

In addition to the three basic cloud forms, four cloud families based on altitude can be recognized (Fig. 6-5):

1. ***Low clouds*** usually occur below about 6500 feet (2000 m). Here we find cloud types that cover the entire sky like ***Stratus*** or, if it is raining or snowing, ***Nimbostratus*** (the root *nimb* indicates precipitation). ***Stratocumulus*** are low, fluffy, grayish clouds with "sun breaks" between them.
2. ***Middle clouds*** are found between 6500 and 20,000 feet (2000 to 6000 m). One type is ***Altostratus,*** a smooth, grayish, blanketlike cloud that indicates the approach of a storm and through which the sun shows as a bright spot. The other type is ***Altocumulus,*** an indicator of fair weather with individual cumulus masses close together in a regular pattern.
3. ***High clouds*** are above 20,000 feet (6000 m) and in-

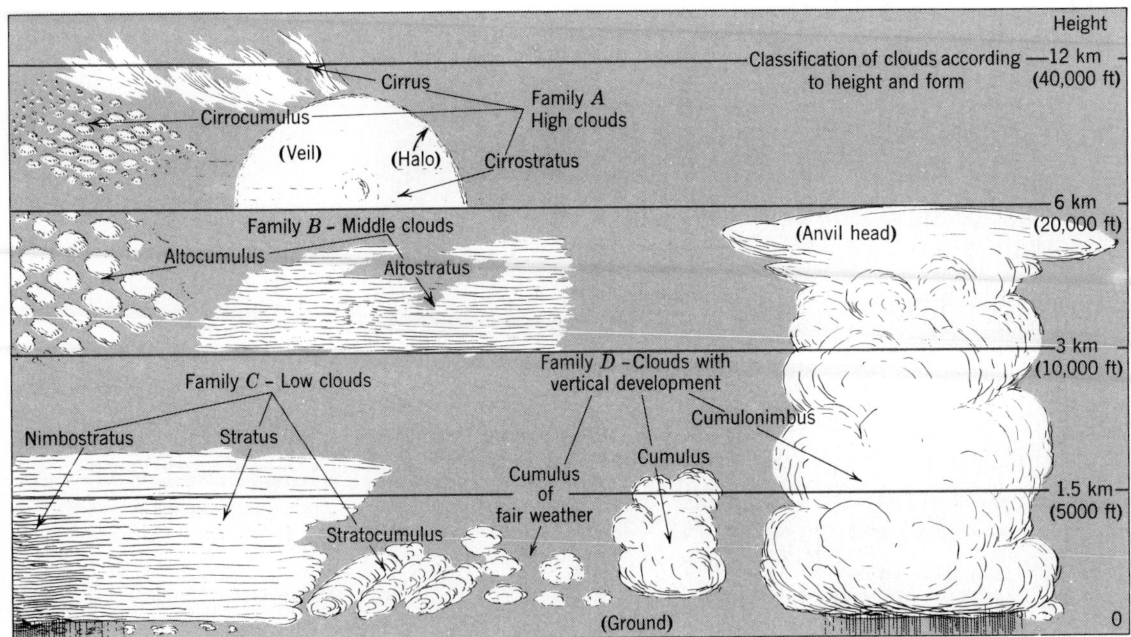

Fig. 6-5. ***Cloud families and types of clouds.***

clude ***Cirrus, Cirrostratus,*** and ***Cirrocumulus.*** Cirrus, the highest of the three and sometimes called "mares' tails," forms wispy trails that may show the position of the jet stream (Fig. 6-6). Cirrostratus is a light cloud that produces a halo around the sun or moon. Cirrocumulus, the "mackerel sky" that indicates the ap-

Fig. 6-6. ***High clouds.*** *The feathery wind-drifted character of ice crystals in suspension is evident in these cirrus clouds, called "mare's-tails."*

Fig. 6-7. ***Fluffy cumulus clouds near Page, Arizona.*** *These appear to be building up toward an afternoon shower or two.*

proach of a storm, is a layer of delicate cloud masses in a geometric pattern.

4. ***Clouds of vertical development*** make up the fourth group. These are the cumuliform clouds with little horizontal but much vertical spread. They include the small fair-weather ***Cumulus*** and the tall, bellowing ***Cumulonimbus*** that brings thundershowers and may extend to the top of the troposphere (Fig. 6-7).

PRECIPITATION Dew and fog provide very little usable moisture. Nowhere in the world does any sizable agricultural venture depend on dew or fog for its moisture. Somehow large masses of air must be forced to form clouds and to give up their moisture in sufficient quantities to cause the "real" rainfall (or snowfall). This we call ***precipitation.***

The principle remains the same whether we are dealing

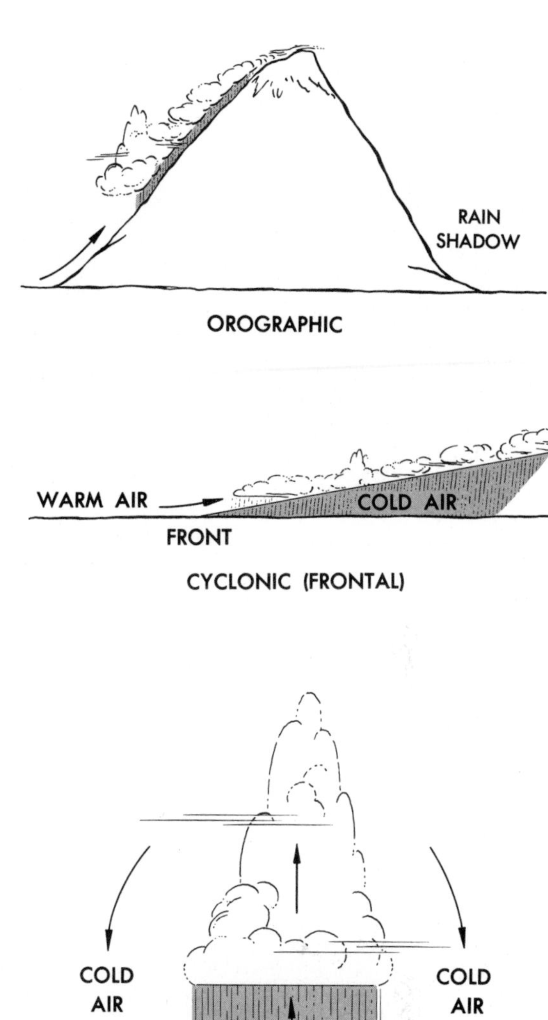

Fig. 6-8. ***Causes of precipitation.***

with dew, fog, cloud formation, or heavy rain. The air must be cooled below the dew point, but to cause rain, this must happen rapidly and on a large scale. The obvious way is to force a large mass of moist air to go aloft and to cool adiabatically (see Chapter 4).

All the world's significant precipitation occurs from one or a combination of three methods: orographic, convection, and cyclonic or frontal (Fig. 6-8).

THUNDERSTORMS

A cumulonimbus thunderhead, dark and brooding against the summer sky, is merely a malevolent variation of the common fleecy fair-weather cumulus cloud. Both are more often than not the result of hot-season surface heating and the development of a convective cell. The difference between the two is a simple matter of violence of uplift.

If a warm saturated air mass is abruptly lifted several thousand feet in a matter of minutes with great heat as an activator, huge quantities of energy are released through condensation and a predictable series of spectacular results will follow. Or much the same violent updraft can be initiated in an extended line along a cold front (frontal/cyclonic lifting) or a steep mountain wall (orographic lifting). In every case the top of the cloud reaches great height, usually well above the freezing point even in the tropics, and a full-fledged storm is under way.

The convective concept of a central rising column of air balanced by sinking side columns is correct as far as it goes, but the energy-rich thunderstorms normally contain several independently rising and sinking elements or cells. Each of these cells is an explosive entity in itself with savage wind-shear characteristics, so that thunderstorms are really complexes of internal churning. Caught up in all of this activity are supercooled water droplets carried alternately above the freezing line and back below to grow in size as each new layer of ice gathers. When they finally become too heavy to be kept aloft (or the force of uplift abates), down they come in a torrent of hailstones. "Hailstones the size of golf balls" may be a frequent exaggeration by the press, but in Coffeeville, Kansas, on September 3, 1970, a hailstone was measured at 6 inches (15 cm) in diameter and 1 ⅔ pounds (0.6 kg)—more like a softball than a golf ball.

The same internal turbulent fury that produces hail appears to trigger the onset of thunder and lightning. The rapidly rising air creates electrical charges that arrange themselves within a cloud, with the positive charges at the top and the negative at the bottom. On the earth immediately below, the usual ground negative becomes temporarily positive. This has been described as an "electrical shadow," faithfully following the cloud as it moves. Enormous electrical stresses between these opposing poles, as many as 100 million volts generated, eventually overcome the insulating effects of the surrounding air. Lightning flashes from the ground to cloud or within clouds, relieving nature's electrical imbalance. The instantaneous heat release by each bolt of lightning forces air to expand explosively. Sharp staccato reports, booming cannonades, and rolling, growling thunder off in the distance are all the indirect sounds of lightning's great heat.

Orographic. The word "orographic" means "relating to mountains." So ***orographic precipitation*** is produced when a topographic barrier such as a mountain range stands in the path of prevailing winds. This is a common means by which precipitation forms because such situations occur widely, as along the west coasts of North and South America.

Recall from our discussion of adiabatic lapse rates in Chapter 4 that moist air forced against a mountain front

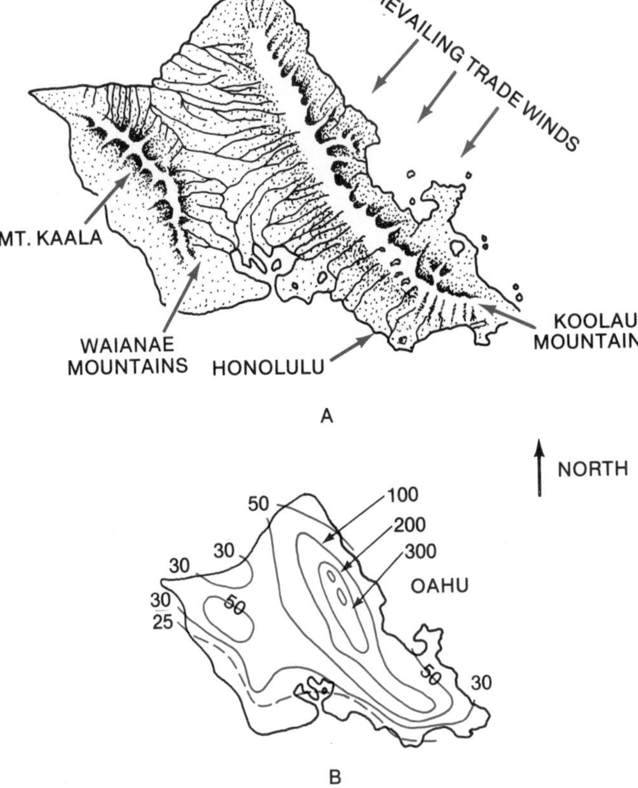

Fig. 6-9. ***Orographic rainfall.*** *Located at 21°N, the island of Oahu is in the zone of prevailing northeast trade winds. Moist air from the Pacific is lifted as it ascends the Koolau Mountains. The result is more than 200 inches (500 mm) of rain annually on the windward side. Honolulu, on the lee side of the range, is in a rain shadow with less than 25 inches (100 mm). Lines of equal rainfall, called* isohyets, *show the island's annual rainfall pattern (B).*

ascends. This results in expansion, cooling, and condensation. Clouds form as the temperature drops to the dew point, and, with further cooling, cloud droplets grow larger by joining together until they overcome the updraft and fall as precipitation. It takes roughly a million cloud droplets to make one rain drop. In this case the type and amount of precipitation are controlled by the slope and height of the mountain and the temperature and moisture content of the air mass.

Moist air forced against a mountain front ascends, resulting in expansion, cooling, and condensation.

In the lee of such a topographic feature, there is a relatively dry zone with descending and warming air called a ***rain shadow.*** Beyond this, the contrast between the windward and leeward sides of the mountain are often pronounced. The difference is very striking in many parts of the world. Oregon and Washington, divided north to south by the high Cascades, exhibit thoroughly split personalities: a cloudy and dripping west versus a sunny and dry east. Trade wind islands, be they West Indian, Hawaiian, or Madagascar, are similar in their sharp rainfall contrasts (Fig. 6-9). Orographic precipitation may occur in any season of the year or in any part of the world where the necessary conditions prevail. Wherever it occurs, the windward slope of the mountain will be wet and the leeward side dry.

Convection. *Convection* is not a new term. We dealt with it in Chapter 4 when we discussed the transfer of heat to high altitudes. It is also a major cause of precipitation. Remember that heat sets convective currents in motion—buoyant heated air at the surface trapped below cooler air aloft is a very unstable condition. As the hot earth on a summer day transfers its heat to the air above it by conduction, the increasingly warm air may at some point burst explosively through the cooler air above it. In this violent updraft the warm air cools rapidly through expansion as it rises several thousand feet in a matter of minutes, and condensation of a tremendous amount of moisture is virtually instantaneous. Tall cumulonimbus clouds form, and the resulting ***convectional precipitation*** is violent and usually short. In addition to the heat originally derived from the ground, the sudden and massive release of latent energy gives greater buoyancy to the air mass until its towering thunderhead may punch through the colder air above it to the level of the tropopause. Such rainfall is a tropical or summer phenomenon, and it is typical of many parts of the world.

As the hot earth on a summer day transfers its heat to the air above it by conduction, the increasingly warm air may burst explosively through the cooler air above it, almost instantly producing condensation of a tremendous amount of moisture.

Cyclonic. Cyclonic or *frontal precipitation* results when two differing air masses are side by side and one is forced against the other. (The details of air masses are given in Chapter 7.) The lighter air will ride over the heavier air and cool adiabatically, producing condensation and precipitation.

When one air mass is forced against the other, lighter air will ride over the heavier air and cool adiabatically, producing condensation and precipitation.

For instance, in the middle latitudes we frequently have warm and cold air in close conjunction. The warm air, which is lighter than dense cold air, will run over the top. This is like the warm air being forced up a mountain slope: it cannot invade the cold air, and it must move up over its sloping front.

Frequently, this slope is low angle, and thus the rate of expansion is slow, with gentle drizzle or snow flurries resulting. This type of precipitation occurs most often in the belt of the Westerlies where differing air masses meet (Fig. 6-10).

CONCLUSION

We now know that just because the air contains water vapor, it's not necessarily going to rain. First, there must be condensation, and this condition requires that the air be cooled below its dew point and that a suitable surface be available. Cooling of air at the earth's surface produces only condensation in the form of dew, frost, and fog. For precipitation to occur, moist air must be forced to rise and cool adiabatically. Thus, if conditions are right, ascending air means clouds and precipitation. We also know that descending air means clear skies and dryness. This is very important knowledge because, if we know where over the earth's surface moist air is likely to ascend and where air aloft is likely to descend, then we can understand world rainfall patterns. In this chapter we have begun to see how the elements of weather interact. In the next chapter we carry this theme further as we focus on air masses, fronts, and storms.

Fig. 6-10. ***Frontal precipitation.*** *Snow as well as rain can precipitate at the point of contact between warm and cold air. After the storm, travel on foot is often more practical than use of a vehicle.*

KEY TERMS

sensible temperature
hydrologic cycle
humidity
relative humidity
saturation
dew point temperature
sling psychrometer
hygrograph
condensation nuclei
dew
frost
fog
radiation fog
advection fog
cumuliform
stratiform
cirriform
supercooled
low clouds
Stratus
Nimbostratus
Stratocumulus
middle clouds
Altostratus
Altocumulus
high clouds
Cirrus
Cirrostratus
Cirrocumulus
clouds of vertical development
Cumulus
Cumulonimbus
precipitation
orographic precipitation
rain shadow
convectional precipitation
cyclonic (frontal) precipitation

REVIEW QUESTIONS

1. Which atmospheric gas is most important to our weather? Why?

2. What does "relative" mean in the term *relative humidity*? Why is it important to understand?

3. What forms of precipitation do *not* require condensation nuclei?

4. What is the difference between a cloud "form" and a

cloud "family"? What value is there in knowing the types of clouds?

5. Name the three causes of precipitation. How do they differ? What mechanism do they have in common?

APPLICATION QUESTIONS

1. What do you think would happen if the relative humidity in a completely dust-free room reached 100 percent?

2. Design an experiment to show changes in relative humidity using a strand of your own hair.

3. The next time it rains—or snows—see if you can determine the cause of the precipitation. Was it orographic, convective, cyclonic, or some combination of these? Find clues on the local TV weather reports.

4. Describe the kinds of clouds you see for the following week, at two or three times a day. Can you classify them as to form, family, and type? How do you think they relate to the weather you experienced during that time?

CHAPTER 7

AIR MASSES, FRONTS, AND STORMS

OBJECTIVES

After studying this chapter, you will understand

1. What an air mass is and how the different air masses function.
2. What a front is and how the different major fronts function.
3. How storms are classified, what causes the various types, and why they behave as they do.

Air masses in conflict breed storms and it isn't difficult to see on this satellite image where the action is. Dry, bitterly cold air from interior Canada is being driven down across the central plains, midwest and far into the south to encounter warm, moist, tropical air. The line of contact is the sharply defined cloud line called a Cold Front. One may be certain that atmospheric turbulence and severe precipitation of one kind or another is rampant along the front, from New England to the West Indies.

INTRODUCTION

Mark Twain observed, "If you don't like the weather, wait a minute." Of course, he lived as we do in the middle latitudes, and there is something about this area that does not lend itself to a monotonous chain of daily weather events. Intense cold and blizzard can be followed in short order by warmth and drizzle. Even within the tropics where the weather is noted for being the same day to day, major storms do break the pattern occasionally. This chapter, our last with a primary focus on weather elements, begins with a discussion of air masses and how they form. We then turn to the various interactions between air masses that result in storms. Storms, as we will see, vary in type and severity, but all are important to the hydrologic cycle and to the earth's energy balance.

AIR MASSES

By ***air mass*** we mean specifically ***a great body of air, continental in size, that is relatively homogeneous in both temperature and moisture.*** Since the earth itself transmits these characteristics to the lower atmosphere from its surface, we

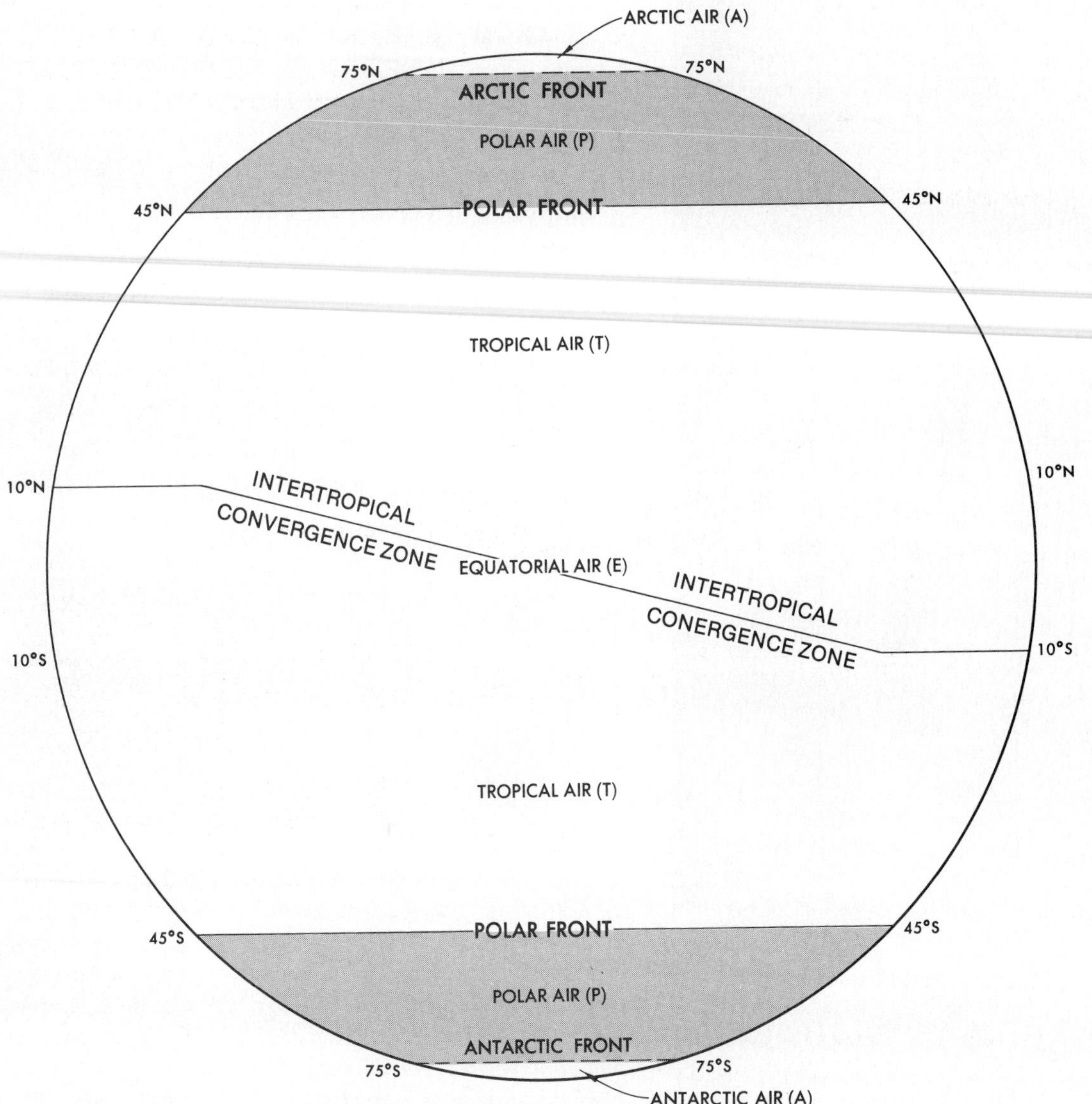

Fig. 7-1. World air masses, fronts, and the intertropical convergence zone (highly diagrammatic).

are, of course, also talking about a considerable earthly region or *source area* that shows more than a little homogeneity of its own. Another element to be considered is *time,* for a massive body of air cannot truly reflect the character of the surface it overlies without some reasonable time in residence. That is, the source area should be a quiet haven over which the air mass may linger.

Air masses do not exist just anywhere, and they are not just any random bit of atmosphere. Their basic requirements appear to be satisfied most effectively in the permanent high-pressure regions of the world. These areas are characterized by a tranquil, downward, vertical movement of air with dispersing surface flow in strong contrast to the turbulence of the lows. The *high latitudes* with their generally low temperatures and the permanent high-pressure cells in the *subtropics* would seem to be ideal, and it is here that we isolate the basic sets of air masses (Figs. 7-1 and 7-2).

The basic requirements of air masses appear to be satisfied most effectively in the permanent high-pressure regions of the world.

Because the source area is so important in identifying and giving character to the air mass, our classification scheme is as much one of source area as of air mass. The two sets of air masses—polar/arctic and tropical/equatorial—obviously exhibit a strong difference of temperature. Each set has three types for a total of six basic air masses. Figure 7-3 lists the air masses and their important characteristics. Let us see precisely how they differ.

Polar/Arctic Air

CONTINENTAL POLAR The classic source areas for the ***continental polar* (*cP*)** air masses are the broad high-latitude stretches of Siberia and Canada. Typically frozen and dusted with snow through a long winter season, any air mass overlying these regions must become thoroughly chilled. Even

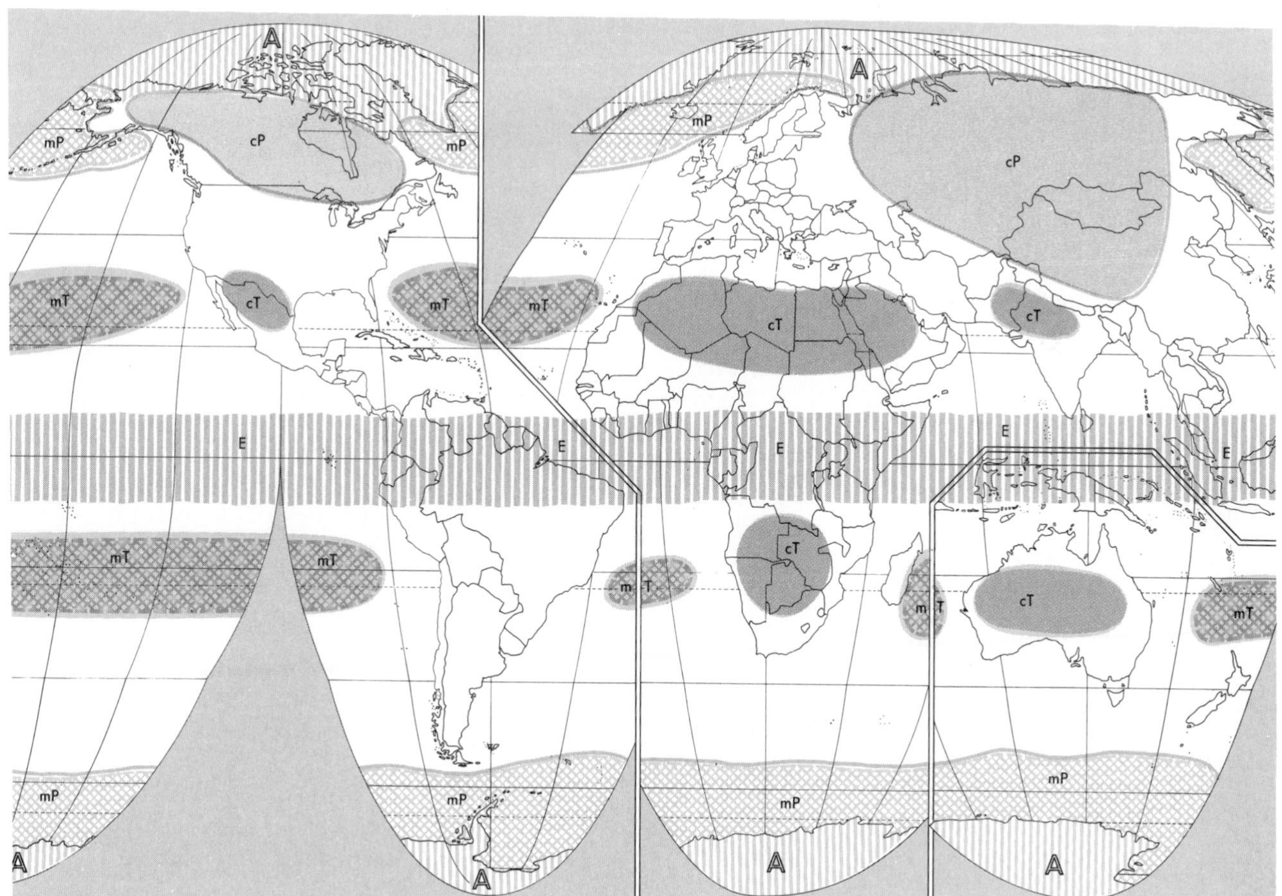

Fig. 7-2. ***World air masses and source regions.***

Major Division	*Air Mass*	*Symbol*	*Source Area*	*Characteristics*
Polar/Arctic	Continental polar	cP	High latitudes of Siberia and Canada	Cold and dry
	Maritime polar	mP	High-latitude oceans: North Pacific/ Bering Sea, North Atlantic/Norwegian Sea, oceans fringing Antarctica	Cool and moist
	Arctic	A	Frozen Arctic seas and Antarctica	Very cold and very dry
Tropical/Equatorial	Continental tropical	cT	Low-latitude deserts	Hot and dry
	Maritime tropical	mT	Oceanic high-pressure cells of the Horse Latitudes	Warm and moist
	Equatorial	E	Warm equatorial seas	Warm and very moist

Fig. 7-3. ***Summary of major air masses.***

the short summer is not actually hot. Thus one fundamental characteristic of cP air is cold—exceptionally so in winter, modified slightly in summer.

Because chilled air cannot hold moisture even if the frozen continent could offer any for assimilation, the cP air is dry as well as cold. Again summer is different. As the snow melts, river ice goes out and marshes and tundra bogs become common features. The somewhat warmer air gains a shallow layer of humidity through increased evaporation. For the most part, however, we can accurately describe cP air as *cold* and *dry* (Fig. 7-4).

MARITIME POLAR Now we come to ***maritime polar (mP)*** air masses that could not be called *polar* if they were not

Fig. 7-4. ***The cP air mass.*** *Winter totally dominates each year in the far Canadian interior. This ice fog is representative of the still and cold cP.*

cold. Their source areas are the warmer high-latitude oceans (not including those that are habitually frozen): the North Pacific/Bearing Sea, the North Atlantic/Norwegian Sea, and the entire sweep of Southern Hemisphere oceans fringing Antarctica. This air, overlying unfrozen seas with frequent probing warmer currents, simply cannot be as cold as the continental masses. As a result, it is a far more efficient evaporator of readily available moisture. Thus mP air is *less cold* and *more moist* than cP air.

Air overlying unfrozen seas with frequent probing warmer currents is a far more efficient evaporator of readily available moisture.

ARCTIC Many students of high-latitude meteorology identify an *arctic (A)* air mass. Its source areas are the permanently frozen Arctic seas and Antarctica. Note that this is *poleward* of polar air source areas. Such an air mass does, of course, have much in common with polar air masses, but there are recognizable differences between A air and the somewhat warmer and more humid cP summer air mass/mP air mass. Arctic air is described as *very cold* and *very dry,* with no seasonal change.

REGIONS AFFECTED In the United States we experience two of these air masses—cP and mP. Both appear only in the winter when there is a strong tendency for polar air to push out and invade neighboring territory. In the interior of the country and on into the northeast coastal region, cP air from the north Canadian plains is a regular winter visitor. It overlies the Dakotas and parts of Minnesota, Montana, Wisconsin, and Michigan for long periods of time. These states might even be regarded as marginal fragments of its source area. Elsewhere in the northern half of the country, cP air is experienced as cold spells when the air mass surges outward in waves or lobes. On occasion such surges will even reach the Texas Gulf Coast and northern Florida.

The West Coast is normally protected from invasion by the generally prevailing westerly circulation and by the Cascade mountain range. Cold air moving along the ground in a moderately shallow layer has difficulty overcoming high continuous chains of mountains—although it can flow around compact masses or through the gaps and passes of broken chains. However, most of North America, with its north/south trending mountain ranges; is open to invasion of cold air from the Canadian interior.

Continental polar air brings with it the cold dry character of its source area, producing sunny crisp days when one can dash from the house to the garage and back in shirtsleeves without great discomfort, only to discover frozen ears.

Continental polar air brings with it the cold dry character of its source area, producing sunny crisp days.

Maritime polar air sweeps in along both coasts of the northern United States. It is a great deal warmer than cP and is also usually close to the point of saturation. A blustery day in Boston or Ketchikan, with the temperatures in the 20s (F) and a cold sleet coming down, chills one to the marrow and for most people is distinctly more uncomfortable than frozen ears in eastern Montana.

On rare occasions, Arctic air escapes its source area, bringing with it a severe cold that is long remembered. The cold snap during the winter of 1988–1989 is one example. Temperatures of −40° F (−40° C) and below were recorded in northwestern states like Montana. In southern Alaska, where temperatures dipped below −50° F (−46° C), newspapers reported that boiling water thrown into the air would freeze before it hit the ground.

Europe is exposed to the same three air masses as North America but on a slightly different scale. Eastern Europe is closest to the Siberian cP source area and experiences the coldest temperatures. Despite the east/west trending character of European mountain ranges and thus a seeming invitation to cP and A invasion, this "grain" of the land instead allows easy access to strong Westerlies, supplying a continuous mass of mP air from the North Atlantic. Along with warming and humidifying influences of the Mediterranean and Baltic, mP air generally keeps the colder air at bay.

Nonetheless, Western Europe is open to occasionally devastating cold spells of some duration when this balancing system temporarily breaks down. Interior Asia and the Far East feel the brunt of the greatest of all cP air masses during the winter. Only Japan, Korea, and the immediate coast of eastern Asia experience the slight moderation which North Pacific mP air brings.

There is no cP air in the Southern Hemisphere, although a great A mass overlies Antarctica. The lightly populated lands of the southern ocean normally experience only mP. South America is closest to Antarctica and in addition thrusts its high Andes across the path of the Westerlies, so that mP air masses prevail much of the year. One has only to read Darwin or the accounts of many ships' logs as they describe the weather of the Strait of Magellan to appreciate the chill winds and high humidity that typify this part of the world. Even as far north as the Argentine Pampas occasional equatorward surges of mP air masses have been recorded.

Australia and South Africa are barely within range, but isolated New Zealand experiences mP air as a frequent winter visitor at its southern extremity.

Tropical/Equatorial Air

Tropical and equatorial air masses are warm, the precise opposite of the biting polar and arctic cold. As a result, they have the capacity (if not always the opportunity) to carry immense quantities of water vapor.

Tropical and equatorial air masses are warm, the precise opposite of the biting polar and arctic cold.

Tropical air has two source areas: the permanent but somewhat migratory oceanic high-pressure cells of the Horse Latitudes and the great low-latitude deserts. Both are warm, but there are some fairly important differences. ***Continental tropical* (*cT*)** is the product of the desert source area, and ***maritime tropical* (*mT*)** overlies the seas.

The ***equatorial* (*E*)** air mass is the product of warm seas along the equator.

CONTINENTAL TROPICAL The cT air mass is basically *hot* and *dry*. Heat may vary a bit with season, but the dryness remains the same. A typical example is the extensive Northern Hemisphere air mass that develops over the greater Sahara source area, which is the epitome of heat and aridity (Fig. 7-5). However, since worldwide tropical deserts are frequently limited to narrow coastal strips, only two other source areas can even approach the Sahara in size or climatic extremes: central Australia and southern Africa.

The cT air mass is a drying air mass that "burns up" the populace and vegetation when it ventures outside its source area. The dreaded ***Sirocco*** that periodically blights Mediterranean Europe is cT air drawn out of the Sahara by a passing northerly cyclone. The United States is seldom afflicted by the invasion of true cT air because of both the elevation and lack of bulk of the northern Mexico source area. Southern Californians nonetheless know the ***Santa Ana*** wind that occasionally brings small-scale quasi-cT air from the southwest

Fig. 7-5. ***The cT air mass.*** *By far the world's greatest cT air mass develops over the 3 million square mile (7,769,964 km^2) superheated Sahara.*

desert. This wind comes with mixed blessings. Its high velocity damages property, and its dryness and dust cause human discomfort. On the positive side, it sweeps the Los Angeles basin free of smog, making it possible to see the surrounding mountains and offshore islands.

The cT air mass is a drying air mass that "burns up" the populace and vegetation when it ventures outside its source area.

MARITIME TROPICAL Maritime tropical air is considerably more widespread than cT because of the greater size and prevalence of its source areas. The Pacific Ocean is by far the world's largest single physical feature. In the Southern Hemisphere, it is all encompassing at every latitude, and in the Northern Hemisphere, although it is limited in breadth at the far north, it reaches almost half the earth's circumference in the tropics. The North and South Atlantic as well as the Indian Ocean are also extensive source areas for mT air.

We would expect mT air masses, which overlie tropical seas, to be moist to the point of saturation, but this is not always the case. There is a difference between the eastern and western sections of the Horse Latitude high-pressure cell in each ocean basin. As world pressure maps (Figs. 5-10 and 5-11) show, the center of these high-pressure cells is offset to the eastern side of the basins (toward the west coasts of the continents). This suggests that downward-moving air will be most dominant there. As the air descends into the high, it is warmed adiabatically. As the warm air spirals outward from the high on these eastern sides, it is chilled at the base by cold ocean currents. The result is drier and more stable air.

At the other end of the same mT air mass, downward motion is less marked and a warm current there contributes to warmer and less stable air (Fig. 7-6). Despite these differences between east and west, we classify mT air as *warm* and *moist.* When elements of it push beyond the source area, they carry these characteristics with them.

***Despite these differences between east and west, we classify mT air as* warm *and* moist.**

One variant of mT, called locally ***Gulf air*** (gT), after the Gulf of Mexico, regularly flows each summer up the Mississippi valley to induce an unwelcome sample of the maritime tropics far into the continental interior. Shanghai, Sydney, and Buenos Aires, all located at the western extremities of ocean basins, also experience the hot stickiness of saturated mT air as oceanic highs shift poleward with the summer sun.

EQUATORIAL The equatorial air mass is a product of the Doldrums and is at all times *warm* and *very moist.* Obviously, such an air mass is difficult to separate from western mT, but it carries slightly more water vapor than eastern mT and twice as much as cT air.

Modified Air Masses

Because an air mass is so large, it takes a fair amount of time to achieve a temperature/moisture equilibrium with its source area. When some of it probes beyond the source area proper and invades surrounding territory, it carries with it the basic characteristics of its home region and tends to resist change. When we talk of air masses departing the

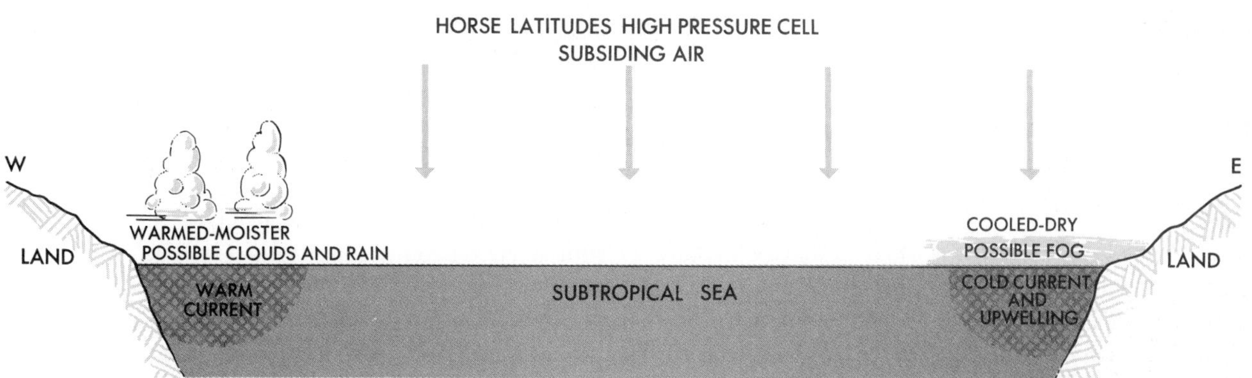

Fig. 7-6. ***Maritime tropical air.*** *The tropical oceanic cell displays a drier, more stable eastern end as opposed to the greater humidity and cloudiness of the west.*

source area, we never mean that the entire mass, or even a substantial part of it, flows off and completely abandons its home. Instead, we are dealing only with these pulsations on the edge of the mass, advancing and retreating on a relatively short-term basis. Although much of the theory of air mass interaction is based on the recognized ability of the mass to retain certain reasonably stable temperature/moisture characteristics as it travels abroad, change is possible. Infrequently, such extreme change occurs that the visiting air mass is completely altered.

The margin of an air mass is generally shallower than the main body, and if it surges too far and stays too long, it will eventually change character. Two of the most common types of air mass modification occur when

1. The invading air is *colder* than the surface over which it is flowing and the mass is being heated at the bottom. We add a "k" (from the German word for cold, *kalt*) to the air mass symbol to indicate this modification—as in mTk.
2. The invading air is *warmer* than the surface over which it is flowing. For this we add a "w"—as in mTw.

The margin of an air mass is generally shallower than the main body, and if it surges too far and stays too long, it will eventually change character.

In the first instance, moist maritime tropical air moves in from a cooler sea over a summer-heated continent. The air mass is warmed at its base, begins to exhibit instability, and thundershowers may well develop. A "k," then, means instability. When the "w" is added, light and buoyant maritime tropical air may encounter a cold surface with chilling at the bottom and the development of an inversion. This will be a stable air mass with the possibilities of fog and concentrated air pollution.

AIR MASS BOUNDARIES

Fronts

In the process of locating and labeling air masses, we have identified high-latitude cold air masses and low-latitude warm air masses, but no moderate, lukewarm, or tepid air masses. This omission is understandable because *there simply are no middle-latitude air masses.* In this critical region in both hemispheres we find continual conflict between the advancing and retreating margins of the masses on either side. This region is where intensely differing masses meet, a line of juncture called a ***front.***

One feature of any front is the "unwillingness" of different types of air masses to mix. Like oil and water, each maintains its individual identity over frontal zones of 50 to 100 miles (80 to 160 km). A front, then, is just the opposite of the usual experience in nature where gradual transition is the rule. Here we have essentially a very sharp line between strongly divergent elements. As such, it is a line of conflict and atmospheric dynamism.

One feature of any front is the "unwillingness" of different types of air masses to mix.

POLAR FRONT It is significant that the Norwegians originated the concept of air masses and fronts, for their weather, coming to them off the stormy Norwegian Sea, is usually bad and highly unpredictable. Their concern with comprehending the ingredients of weather and establishing a scientific means of forecasting is understandable. When they finally got their whole weather picture together and all their basic air masses established, it turned out that Norway (and the middle latitudes in general) was directly astride the Polar Front much of the year.

The Polar Front is the most decisive of all the world's fronts. It separates polar and tropical air, and anyone who establishes permanent residence within its zone of influence is automatically going to encounter "interesting" weather. Warm, moist tropical air is forced to skid up over cold, dry polar air, thereby releasing a great deal of latent energy through condensation. This is the beginning of a whole sequence of events leading to storm formation.

Along the Polar Front, moist tropical air is forced to skid up over cold, dry polar air, thereby releasing a great deal of latent energy through condensation.

ARCTIC–ANTARCTIC FRONTS There are other fronts too. Remember the Arctic/Antarctic air masses and their somewhat subtle differentiation from the various kinds of polar air. The so-called ***Arctic Front*** is the zone that separates cold or cool air and very cold air. It is most clearly defined around the periphery of Antarctica and at the permanent ice line of the restricted Northern Hemisphere ocean basins. In summer it shows up to some degree along the north coasts of Canada and Siberia as well.

Intertropical Convergence Zone

Air masses also meet within the tropics where the Trades from both hemispheres converge. Yet, because both air masses are very similar in temperature and moisture content (both are a combination of mT and E air), this is not a front in the true sense of the word and is better thought of as a zone roughly equivalent to the Doldrums. As you learned in Chapter 5, this is the Intertropical Convergence Zone or ITC. This zone coincides with a tropical trough (low pressure) marked by precipitation, cloudiness, and apparent storm genesis.

STORMS

Just about any low-pressure center with cloudiness and precipitation can be called a storm, if

1. It is localized on something less than a continental-size scale.
2. It establishes its own spiraling air circulation.
3. It can be measured on a barometer.

The most frequent and important storms are related to converging air masses along the world frontal zones and along the ITC.

It need not blow the roof off the sun porch before it becomes a legitimate storm; wind velocity is not the main criterion. ***Cyclone, low, depression,*** and ***storm*** are all good terms and mean pretty much the same thing. Within the category of storm, however, other names describe more specific types. A hurricane or typhoon is certainly capable of taking the roof off the sun porch, and a tornado will blow the house away and all the furniture, pets, and human occupants as well. A storm can also be much milder with modest breezes and a little drizzle as its most violent features.

Converging Air Masses

Not all storms are strictly the result of converging air masses. The ordinary thunderstorm, which is sometimes violent, is an example. Nevertheless, the most frequent and important of our storms are in some way related to converging air masses along the world frontal zones and along the ITC. These include:

Polar Front Zone (Extratropical Storms)

1. Middle-latitude cyclones
2. Tornadoes

Intertropical Convergence Zone (Intertropical Storms)

1. Easterly waves
2. Weak Equatorial Lows
3. Hurricanes/typhoons

On or near the Polar Front (and at times along the Arctic Front also) we find the ***middle-latitude cyclones.*** Occasionally, we use the term ***extratropical storms,*** simply meaning that they occur outside the tropics. Middle-latitude cyclones are large [up to 1000 miles (1600 km) in diameter], slow-moving, relatively mild (as far as wind velocities and general destruction are concerned), and quite common throughout the year.

Tornadoes are extratropical storms but are different from, though in some cases related to, the middle-latitude cyclones. The most violent and destructive storms known, they occur in limited regions at certain times of the year, are extremely small in size, and move very rapidly.

The three types of ***intertropical storms*** include ***easterly waves,*** which occur within the Trades; ***weak equatorial lows,*** centered on the ITC; and ***hurricanes*** or ***typhoons*** (the different names apply to the same type of storm in various parts of the world), which are small, seasonal, violent, and occur in certain limited locations.

Jet Streams and Upper Air Waves

Involved in the generation and movements of these various storms to some degree are the jet streams, which were first discovered during World War II by high-flying aircraft whose pilots were greatly chagrined to find that they were making almost no progress despite full power ahead. ***Jet streams*** are high-velocity upper air currents usually flowing from west to east at the troposphere/stratosphere boundary. It is wrong to speak of a single jet stream, for there are many, but only the ***Polar Front jet stream,*** which flows westerly above the Polar Front, has been studied in great detail. Nonetheless, enough reliable work has been done to support the idea of a considerable relationship between the various jet streams and the moisture and temperature characteristics of surface climates, particularly of storm genesis. They are being introduced here because of the almost certain influence of the Polar Front jet stream on the storms of the Polar Front and vicinity, and a very probable relationship

Jet streams ***are high-velocity upper air currents usually flowing from west to east at the troposphere/stratosphere boundary.***

between the lesser known ***tropical easterly jet stream*** and disturbances on or near the ITC.

The Northern Hemisphere Polar Front jet stream varies in both height and velocity with season. It is lowest [4 to 5 miles (6 to 8 km)] and strongest [200 to 400 mph (320 to 650 km/hr)] in winter and highest [7 miles (11 km)] and weakest [50 to 75 mph (80 to 120 km/hr)] in summer. It is a narrow twisting band of air that meanders wildly at times and often breaks down into several parallel elements rather than a single current. Its main course is much farther north in summer than in winter, suggesting a relationship with the similar advance and retreat of the Polar Front below it or the seasonal surface temperature variations that control that front.

During the wild winter of 1976–1977, which was excessively dry in the west and bitterly cold in the east, the jet stream was credited with directing Pacific storms on a far northerly course into western Canada. Once clear of the Rockies, however, these storms came swooping southeastward into the United States bringing some Pacific moisture with them and drawing more from the Gulf of Mexico (Gulf air). Part of the problem seemed to be a major winter-long high-pressure cell persisting off the West Coast, the apparent result of a huge oceanic pool of cold water. At the same time a comparable warm pool was identified farther out to sea. Precisely why these waters differed so greatly in temperature remains unexplained, but the result was an unusual jet stream pattern that directed a series of storms through the eastern United States that dumped snow 12 feet (3.7 m) deep in places like Buffalo, New York. The position of the jet stream and associated Polar Front and storm tracks was a reaction to unusual surface temperatures.

Another wide swing of the Polar Front jet stream was responsible for the scorching summer heat of 1988. In this case, the jet abandoned its normal summer position along the Canadian border and extended from California well north to Hudson Bay. This allowed a vast high-pressure cell to invade and dominate the entire Middle West, blocking the inflow of moist Gulf air that usually brings high humidity and summer rains. The drought was so severe that civic leaders, and some scientists, warned that the dreaded greenhouse warming of the planet had begun. The role of greenhouse warming is still being debated, but by fall the jet stream had returned to a more normal position.

A similar Polar Front jet stream exists in the Southern Hemisphere middle latitudes. It presumably acts much as its Northern Hemisphere counterpart, although the lack of large continents and resultant differing air masses undoubtedly causes variations. Intermittent jets are also known. Thirty miles (48 km) above the poles we find the ***polar night jet streams,*** which are active during six months of bitter winter cold when the sun remains below the horizon but disappear when it rises. Over the Indian Ocean is a reverse jet stream that changes its direction 180° as surely as do the surface monsoon air currents in response to seasonal heating and cooling of the Asiatic continent.

In addition to jets, other kinds of air motions seem in one way or another to affect storms at the earth's surface. The upper Westerlies, for instance, feature a series of recognizable long waves below the level of the jet. They move about in a quasipredictable pattern and appear to be an element in the basic atmospheric heat exchange between the polar and equatorial regions. Recent research points to a relationship among these long waves, the jet stream above them, and the formation and movement of low-level cyclones. A weaker but somewhat similar wavelike pattern has also been identified in the Trades, which may play some part in the development of hurricanes.

Polar Front Storms

MIDDLE-LATITUDE CYCLONES Of the five storms previously mentioned, the middle-latitude cyclone is by far the best understood. Because it has been the major weather maker in the United States and Western Europe where most of the world's scientists are concentrated and because of its frequent occurrence and mild, slow-moving character, it has been carefully studied and analyzed for many years.

According to the Norwegian frontal wave theory, middle-latitude cyclones may form along the Polar Front wherever a wave or protuberance of one air mass advances against the other. If cold air advances against warm air or vice versa, the warm air is forced aloft and a low-pressure center comes into being. Around it, the air begins to circulate in a typical Northern Hemisphere counterclockwise spiral and a storm is born.

Around the low-pressure center, the air begins to circulate in a typical Northern Hemisphere counterclockwise spiral, and a storm is born.

Certain locations along the front appear to be particularly active in cyclogenesis (the formation of cyclones): the Texas Panhandle, the Gulf of Lions (on the southern coast of France), and the upper Yangtze valley (northern China), to name but a few. Outstanding in this regard, producing more and bigger storms than all other Northern Hemisphere sources combined, are the two permanent Subpolar Low cells, the Aleutian and Icelandic centers. These spew out a

SOVIET AGRICULTURAL PLANNING

Wheat—USSR.

An attempt at practical application of wave theory has been going on in the USSR for some time now. During 1954–1960, 100 million acres (40,468,000 ha) of new wheat fields were plowed and planted east of the Volga River and across the steppes of Kazakhstan—Khrushchev's "Virgin Lands." Traditionally, the semiarid Ukraine, 1500 miles (2400 km) to the west, had been the Russian breadbasket. A combination of increasing population and diversification of Ukrainian farms to crops other than wheat forced the opening up of new farmland, and despite its marginal rainfall (comparable to the American "Dust Bowl"), the decision was made to go ahead. The planners knew very well that there would be failures, perhaps even 50 percent over the years, but they reasoned that the 1500 miles (2400 km) between the Ukraine and Kazakhstan was exactly one-half the width of an upper tropospheric wavelength. Therefore, in theory at least, both regions would not be subject to drought at the same time—and, of course, both would not achieve maximum rainfall at the same time either. As it has worked out over the last 30 years, the facts have at least roughly supported the theory. There has never been a national bumper crop, nor has there yet been an across-the-board failure.

steady stream of cyclones, especially in the winter. In the Southern Hemisphere where the Subpolar Low is a belt in the uninterrupted sea, storms typically form almost at random anywhere astride the Polar Front.

Once the cyclone is formed on the Polar Front, its generally slow movement is to the east as it is pushed by the Westerlies, its course determined by the front itself and the jet stream above it. As long as the center of the storm "rides" the front, it can maintain itself, for each cyclone depends on the interaction of both cold and warm air to sustain its vigor. If it were to "jump the track"—that is, move to either the warm or cold air side of the front—the interaction of warm and cold air would cease and the storm would die.

Imagine a cyclone astride the Polar Front in the Northern Hemisphere. The air would spiral counterclockwise about the low-pressure center. On the western side of the storm this would cause cold air to move against warm at the front, forcing the warm air aloft, and very likely precipitation would result. This segment of the Polar Front within the storm where cold air is advancing against warm is called the *cold front*.

On the eastern side of the storm, warm air would attack cold at the front. Again, the warm air would flow over the cold, expand, and cool as it rose, causing precipitation. This portion of the Polar Front where warm air is forced against cold is called the *warm front*. Thus the cold front, the warm front, and the center of the storm where air is rising vertically are the precipitation zones within a cyclone. However, the pattern and character of precipitation differ slightly between the two fronts (Fig. 7-7).

The cold front, and the warm front, and the center of the storm where air is rising vertically are the precipitation zones within a cyclone.

At the warm front the advancing light and buoyant warm air has difficulty pushing cold air aside and instead flows up and over it. Because the cold air is retreating at the same time—though less rapidly than the advancing warm air—its shape is drawn out by surface friction. Thus the frontal slope over which the warm air flows is very gentle and, consequently, it cools slowly. Stratiform clouds form along the line of contact, and a broad zone of drizzle or snow flurries develops on the cold air side of the front.

On the western side of the storm where cold air is advancing against warm at the cold front, a somewhat different situation prevails. Since friction now acts to slow that part of the advancing cold air in contact with the earth's surface, the depth of the cold air may become considerable, and its leading edge will be close to vertical as opposed to the thin wedge slope at the warm front. This means that, as the warm air is forced aloft, it follows up this steep leading edge,

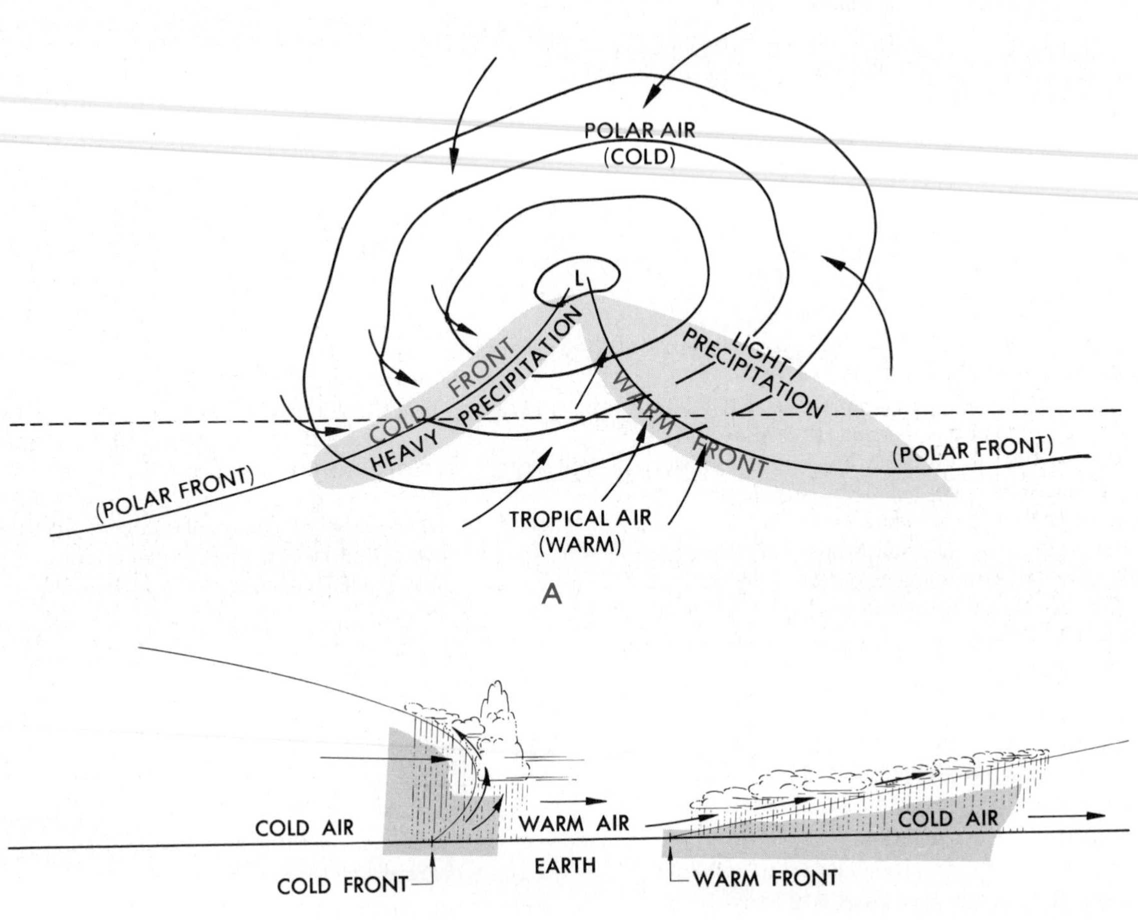

Fig. 7-7. Warm and cold fronts—middle-latitude cyclone. (A) The fronts as they appear in a typical Northern Hemisphere storm. (B) A cross-section of the same storm (along the dashed line).

cooling rapidly and producing high cumulonimbus clouds and heavy precipitation. Frequently, the cold air above the surface moves forward more rapidly than that which is in contact with the ground and slowed by friction. When the trapped warm air bursts aloft suddenly, thunder, lightning, and torrential precipitation result. Note that the precipitation zone is rather narrow and in advance of the cold front. There may also be precipitation of a much less vigorous character behind the front (or in the cold air sector) as the warm air swings up over the cold air.

The chief differences between warm and cold front precipitation are these:

1. Warm front precipitation is a light rain or drizzle, whereas cold front precipitation is likely to be much heavier.
2. Warm front clouds are usually stratiform, whereas cold front clouds are cumuliform.
3. The warm front precipitation zone is broad and wholly on the cold air side, whereas the cold front precipitation zone is quite narrow and is astride the front with the greater part of the precipitation occurring in the warm air sector.

When the television meteorologist shows us the daily satellite image of winter storms marching in from out of the North Pacific, the changing cloud pattern reveals two constants: (1) the long, trailing southwesterly cold front and (2) a counterclockwise cloud spiral about the low-pressure center of the storms (Fig. 7-8). Warm front clouds, a result of not very decisive lifting of warm moist air, may or may not be there on all occasions. When they are apparent, they look

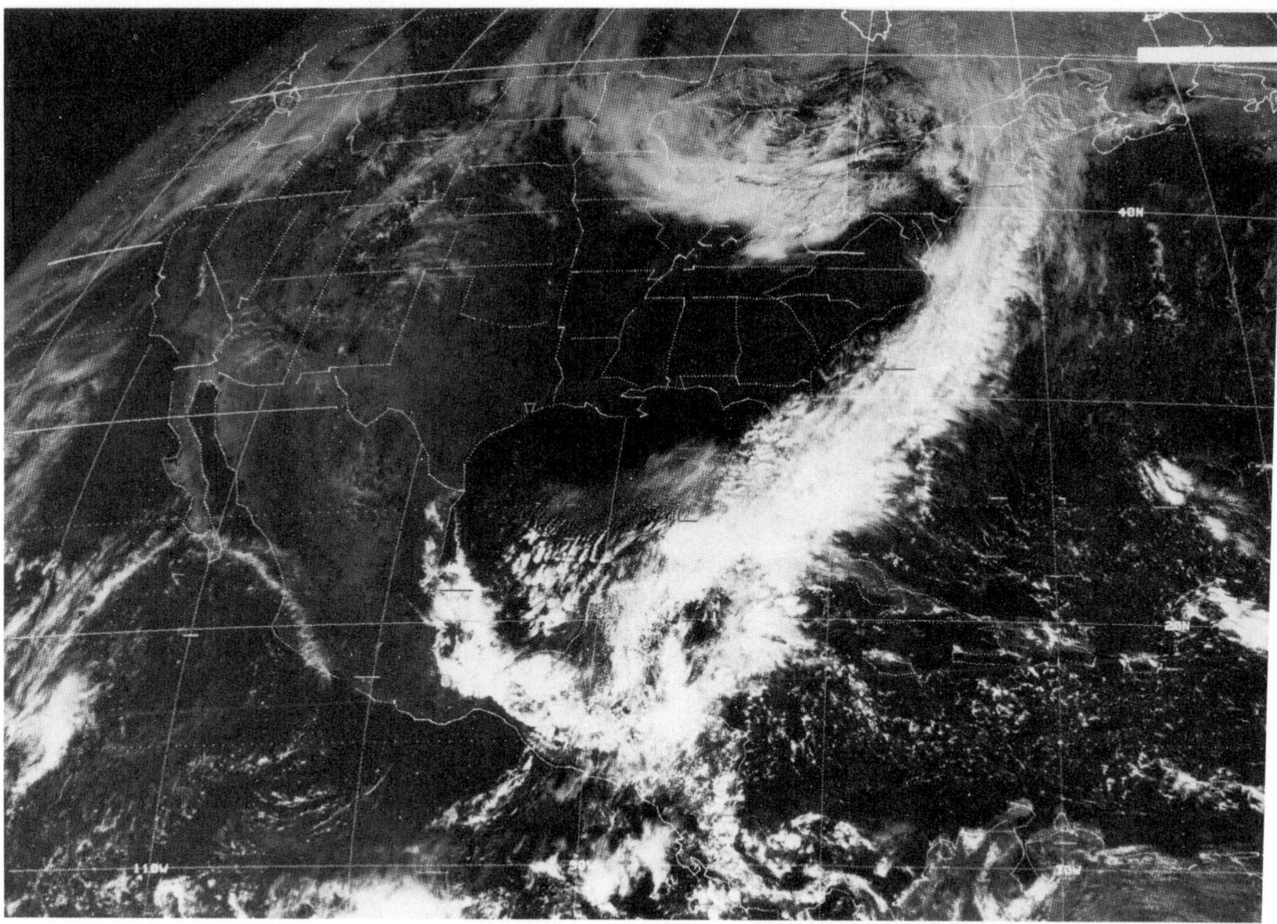

Fig. 7-8. ***Air masses in conflict.*** *These air masses breed storms, and it isn't difficult to see on this satellite image where the action is. Dry, bitterly cold air from interior Canada is being driven down across the central plains, the Midwest, and far into the South to encounter warm, moist tropical air. The line of contact is the sharply defined cloud line called a cold front. Atmospheric turbulence and severe prec'.ation of one kind or another are rampant along the front, from New England to the West Indies.*

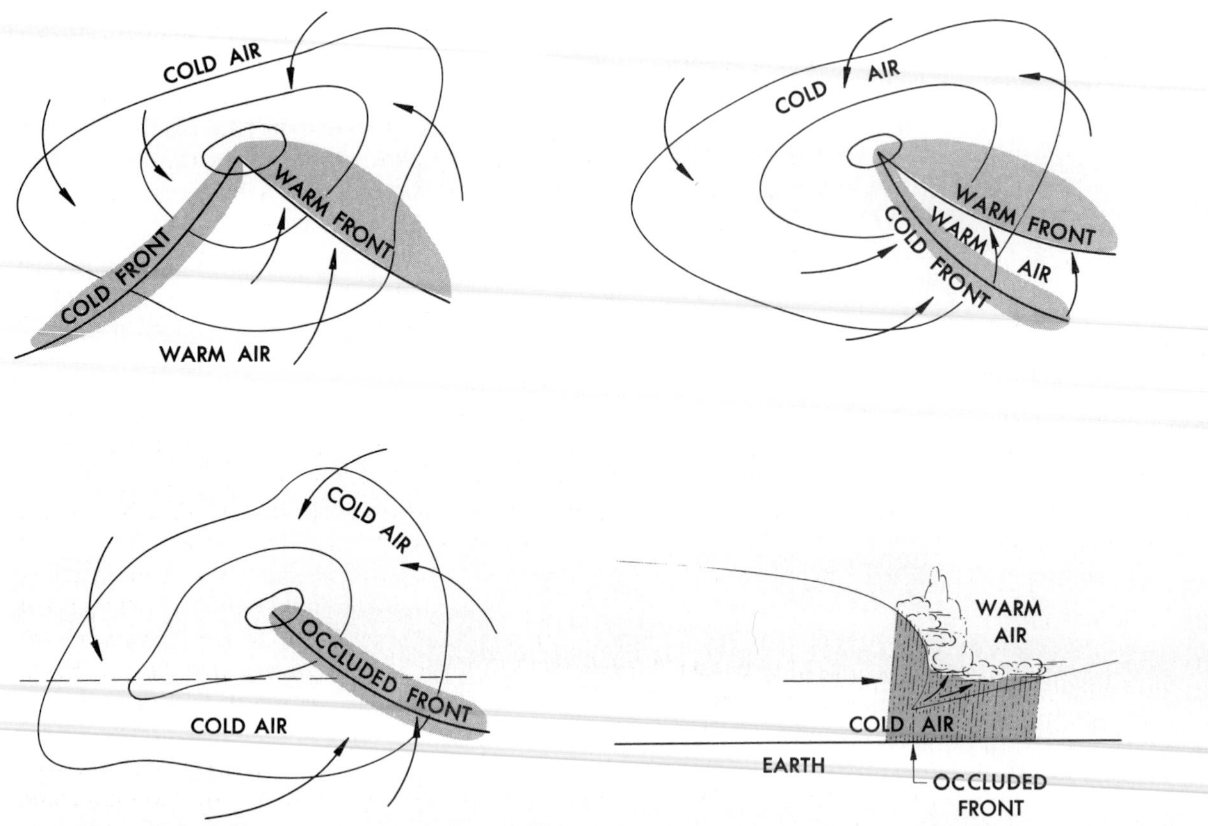

Fig. 7-9. ***Occluded front.*** *In (A) and (B) the cold front is advancing more rapidly than the warm front, culminating in an occluded front (C); (D) is a cross-section along the dashed line in (C).*

to be an undistinguished and disorganized blob at the cyclone's advancing margin. The cold front, on the other hand, is crisply defined, a narrow band of cumulus and cumulonimbus clouds reflecting a great deal of light for the satellite sensor to record. Surely there will be precipitation when that element of the storm passes over.

Because the cold front advances more rapidly than the warm front, it can catch up with the warm front after a storm has gone on for several days. When this occurs, it is called an ***occlusion*** (or ***occluded front***). As the cold front gradually moves in on the warm front, the warm air sector becomes progressively smaller until finally, when the fronts come together, the last of the warm air is forced aloft. At the surface only cold air remains. The rising warm air above the front gives off precipitation, and the vortex at the storm's center may maintain itself weakly. In essence, however, the storm has "jumped the track"—it is no longer astride the Polar Front (Fig. 7-9). The important interaction of warm and cold air has ceased, and the storm usually dies out gradually. Not every cyclone develops to the occluded stage, but the possibility is always present.

When the cold front catches up with the warm front after a storm has gone on for several days, an **occlusion** ***(or*** **occluded front*****) is said to occur.***

It is important to recognize that two basic air circulations are involved in these middle-latitude cyclones:

1. The primary Westerlies forcing the general movement of all storms in the middle latitudes from west to east.
2. The local spiral winds about each individual low center.

This steady parade of slow-moving storms migrating through the middle latitudes of both hemispheres gives the basic character to our day-to-day weather.

TORNADOES The second kind of storm associated with the Polar Front, the tornado, is known for its sudden appearance without much advance warning, its violent character, its small size, and its rapid movement. Nevertheless, it is related to the middle-latitude cyclone, for it tends to develop from strong thunderstorms. It does so only under particular conditions, however.

Tornadoes are largely all-American storms, occurring mainly in the central plains states from the Gulf to Canada. They also occur, though somewhat less frequently, east of the plains in the South, the Middle West, and even now and then along the eastern seaboard into New England. West of the Rockies, tornadoes are virtually unknown. The rest of the world is not immune; Australia reports the largest number. But the greatest tornado frequency by far is in the United States.

Most of these storms occur in late spring to early summer, although the entire warm half of the year is subject to at least occasional tornadoes. The greatest number appear in the South in the spring and gradually move north with the onset of summer. By late summer the number of storms has decreased, but some of them may be experienced as far north as the Dakotas or the Canadian prairie provinces.

Before the development of radar as a tracking device (radar does not usually pick up the tornado itself, just the rotary thunderstorm that indicates its presence), we were unaware of tornadoes at all unless they tore up the countryside in transit. Although they often presented an immediate danger, no warning was given. As a frustrated Iowa meteorologist once observed, "Many of our weather reporting stations are only 9:00 A.M. to 5:00 P.M. operations, but many of our tornadoes are on the night shift." So today when fewer storms are slipping by undetected, there is the illusion that more tornadoes are taking place than ever before. At least we hope it is an illusion. In any case the largest number continues to occur in May with June second and April third; the south central plains states from Texas to Kansas are center stage in this annual drama.

Like the common middle-latitude cyclones, tornadoes move in a general easterly direction, but they are a great deal smaller and advance with greater speed. A typical tornado may be only a few hundred yards across. Tornadoes have been known to wipe out all the houses on one side of the street completely, while scarcely affecting those on the other side. Furthermore, they move in a skipping pattern. Although the funnel remains active and visible and torrential rains accompany it, the bottom of the funnel lifts off the ground for a while before coming back down to carve out a path of destruction. Luckily, the storm is normally short, and although it may run a great distance, it generally disappears abruptly after only a brief period of activity.

Almost all tornadoes (90 percent of the average 400 per year in the United States) appear to originate somewhat in advance of the cold front of a middle-latitude cyclone. Spring is the turnaround point in the annual North American "monsoon" circulation. Cold dry continental polar masses are still far south of the Canadian border, but warm superhumid Gulf air is just beginning its summer surge northward. At the cold front their sharp diversities are magnified. Temperature contrasts on opposite sides of the front seem to be less important than the moisture differential. If the warm temperature is high enough to hold a large amount of water vapor and if its humidity is near saturation, then the advancing cool air, with its upper air moving more rapidly than that at the surface, will cause a violent updraft accompanied by extremely heavy and rapid condensation. Given these conditions in Kansas in May or June, a tornado or even whole clusters are likely. This is the best that the meteorologist can do—forecast tornado likelihood. Just how many will develop, if any, or exactly where is beyond the forecaster's knowledge. Once a tornado has shown itself, then we know that it will quickly outrun the front heading east–northeast, and developments from then on are in the lap of the gods.

Almost all tornadoes appear to originate somewhat in advance of the cold front of a middle-latitude cyclone.

Thunderstorms, too, are formed in this same zone ahead of the cold front, and some tornadoes appear to spring from them as they do occasionally from isolated thunderstorms elsewhere. Once again, we must credit the jet stream with an assist in storm formation, although general agreement has not been reached as to its exact role relative to tornadoes (Fig. 7-10).

The destructive power of a tornado is enormous, and it is without doubt the most violent storm in nature. Wind velocities may exceed 250 mph (400 km/hr), but the few direct measurements that have been made to date suggest that most tornadoes have wind speeds somewhat less than 200 mph (320 km/hr). However, the results of even these velocities can scarcely be exaggerated, and weird stories are legion—straws driven through trees, babies snatched from mothers' arms, and so on ad nauseam. In conjunction with the wind are other equally powerful forces. The lifting ca-

WEATHER FORECASTS—HOW GOOD ARE THEY?

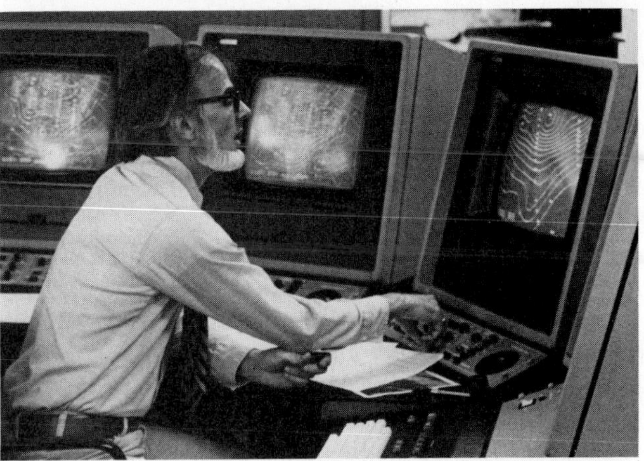

Weather Forecasts: National Weather Service computer readout room.

Weather forecasts provide a basis for many of our decisions—how to dress, whether to grab an umbrella, whether or not to cut hay, plant seeds, or wash the car. How accurate are these forecasts, especially in light of the millions of dollars spent in the last 30 years for satellites, high-resolution sensors, and sophisticated mathematical models of the atmosphere that run only on state-of-the-art high-speed computers? The answer is, it depends how far in the future the forecasts extend. In general, however, forecasts are not much better than climatology, that is, predicting weather simply on the basis of what has happened in the past.

Surprisingly enough, the National Weather Service keeps track of its own batting average. The agency reports that, since 1966, its precipitation predictions for up to 48 hours for all seasons across the country have increased in accuracy from 20 to 30 percent over what could have been done by the climatological record alone. Prediction of severe storms like tornadoes continues to be difficult. Alerts issued for the U.S. Middle West have about 40 percent accuracy in predicting a twister somewhere in the region, but determining where in the region funnels will occur remains an elusive goal.

As might be expected, general short-range weather forecasts of 12 hours or less are the most accurate. Accuracy falls off rapidly beyond 24 hours, and the usefulness of predictions from even the most advanced computer models takes in a period of only five or six days at best. Although a slow increase in accuracy is likely, weather forecasting continues to be one of the toughest and most challenging problems facing science and technology.

pacity of a tornado as the violent winds spiral upward can, for example, pick up houses off their foundations. The dark appearance of the storm funnel may be at least partially caused by great volumes of soil, fences, and assorted livestock sucked into the vortex. In addition, a tornado center is probably the deepest low-pressure center experienced anywhere on earth. As it suddenly envelops a building, the normal air pressure trapped inside is many times that outside, and the building explodes. Only a steel-reinforced construction with a high percentage of windows is relatively safe from this type of destruction as the glass blows out to relieve the pressure.

In the eyes of the general urban public, tornadoes often seem remote and slightly rural in character. Generally, dra-

Fig. 7-10. ***A clear-weather dust devil.*** *This cyclonic wind has the look of a tornado about it and is a particularly big one spiraling tightly about a compact low-pressure center. It is the product of surface heating rather than frontal lifting and is probably closely related to the short-lived but spectacular water spout at sea. Dust devils and water spouts are capable of local damage but are not in the same league with tornadoes.*

Fig. 7-11. ***Tornado aftermath at Inverness, Mississippi.*** *Something violent has just gone through here.*

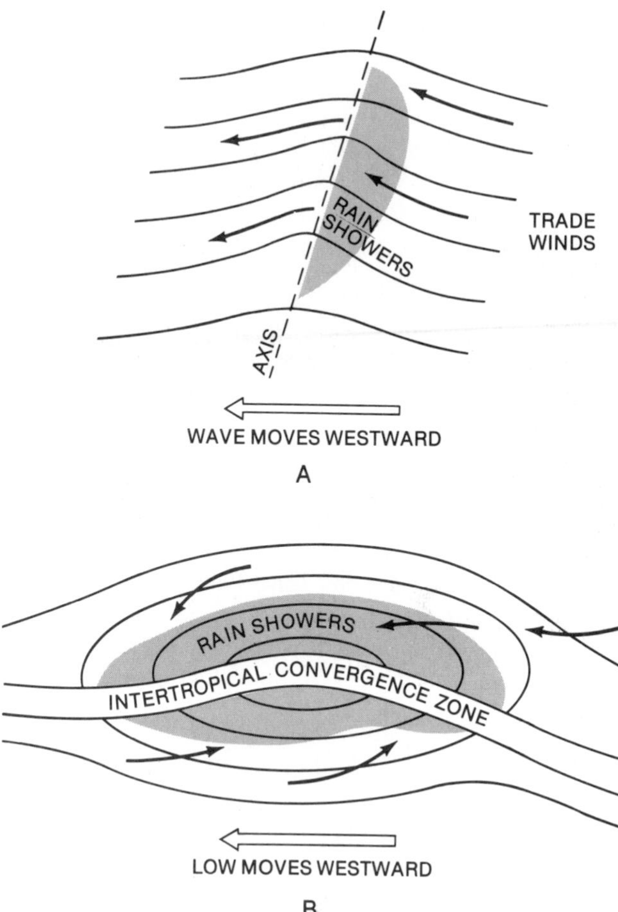

Fig. 7-12. ***Weak intertropical storms.*** *(A) Easterly waves are characterized by a band of convergence in the Trades leading to uplift and many convective showers. (B) Weak Equatorial Lows develop along the ITC. Both storms move from east to west.*

matic reports of tornadoes come in from the countryside and crossroad hamlets, but that is only because the United States, as most countries, has a lot more countryside with small towns than cities. However, there is nothing at all to keep a tornado out of any town or city. In recent decades several sizable cities have witnessed devastation as they found themselves directly in the path of one or more tornadoes: Flint, Michigan, 113 deaths; Waco, Texas, 114 deaths; Worcester, Massachusetts, 94 deaths; Wichita Falls, Texas, 54 deaths; Tulsa, Omaha, and Xenia, Ohio, all damaged severely (Fig. 7-11).

Statistically, the likelihood of a tornado occurring over any given square 50 miles (130 km) of area is about once in 1000 years. Yet Gainsville, Georgia, has been visited three times in less than 80 years: 1903, 28 deaths; 1936, 203 deaths; and 1944, 44 deaths. This is the nature of probabilities—many other towns in what we regard as prime tornado country have never been "honored."

Intertropical Storms

EASTERLY WAVES The dramatic hurricane has largely overshadowed any other type of tropical storm both in the popular mind and in organized research. But other storms are significant elements in day-to-day tropical weather. One—the easterly wave—is the seed, so to speak, from which hurricanes flower. Easterly waves occasionally appear within the streamlines of the Trades—north or south of the ITC between about 5° and 30° latitude (Fig. 7-12A). Warm, moist air converges in the eastern side of the wave resulting in lifting and precipitation. Air flowing out of the wave to the west diverges and descends giving exceptionally clear skies. The wave moves slowly westward bringing with it a rainy period that lasts one to two days. Little is known about the origins of easterly waves, although they appear to be related to temperature anomalies in the higher levels of the Trades. Fortunately, fewer than 10 percent of these storms become hurricanes.

The easterly wave is the seed, so to speak, from which hurricanes flower.

WEAK EQUATORIAL LOWS Convergence of the Trades along the ITC brings together two very similar air masses. But instead of a general lifting all along the convergence zone, as might be expected, lifting is focused on weak low-pressure centers that, once developed, tend to move westward (Fig. 7-12B). Warm, nearly saturated air flowing into the lows requires little lifting to produce rain and clouds, and rainfall tends to occur from numerous individual convective storms.

Lifting along the ITC is focused on weak low-pressure centers that, once developed, tend to move westward.

HURRICANES/TYPHOONS As unspectacular as the easterly waves usually appear to be, some of them, at certain places and at particular times of the year, develop for unknown reasons into spectacular and dangerous hurricanes or typhoons. These storms are limited to summer and fall and occur almost exclusively in the western part of major ocean basins in both hemispheres. These regions include:

- The southeast corner of the North Atlantic (Caribbean and southeast U.S. coast).
- The southwest corner of the North Pacific (Philippines to Japan and southeast Asiatic coast).
- The northwest corner of the South Pacific (northeast Australian coast and adjacent islands).
- The northwest corner of the South Indian Ocean (Madagascar and the east coast of South Africa).

The South Atlantic is an exception, probably because the ITC and associated trade wind system does not move south of the equator as it does in other oceans for reasons we do not know. In addition, less frequent storms may occur in the Bay of Bengal, the Arabian Sea, off the west coast of Baja California, and the northwest coast of Australia.

Despite the difference in names, hurricanes and typhoons are the same type of storm. Typhoon is simply the Far Eastern name, and hurricane is the Caribbean name. Other local names also refer to the identical phenomenon, such as cyclone (India), ***baquio*** (Philippines), and ***willy willie*** (Australia).

Whatever it is called, the storm seems to develop from an easterly wave that simply deepens. The low center becomes deeper and deeper, the winds spiral about it with higher and higher velocity, and the general dimension of the storm becomes more compact—100 to 200 miles (160 to 320 km) across. The National Oceanic and Atmospheric Administration's Hurricane Warning Center in Florida has decided, rather arbitrarily, that when winds reach 75 mph (121 km/hr), such a storm is to be classified as a hurricane, although as it continues to grow, winds commonly exceed 100 (160) or even 150 mph (240 km/hr). In spite of these high winds, the storm itself advances quite slowly, on the order of 5 to 10 mph (8 to 16 km/hr).

Like the tornado, the hurricane/typhoon is somewhat too robust to study easily, but hurricane–hunter planes have been flying into the centers of these storms for several years in an effort to unravel their mysteries. They have been studied by satellite since the early 1960s.

Initially, these storms appear 8° to 10° off the equator because they need enough Coriolis force to allow the violent spiraling of the winds about a low-pressure center. The equatorial regions are thus exempt. Heavy concentrations of hurricanes/typhoons on the western sides of oceans may be a result partly of the generally exaggerated poleward shift of the ITC here, and partly of somewhat higher air and water temperatures. The vigor and driving force undoubtedly come from the latent heat that is released as warm saturated oceanic air is drawn up the vortex to be condensed into torrential showers. Therefore, the warmer the air and the higher its moisture content, the greater the energy potential.

Initially, hurricanes/typhoons appear 8° to 10° off the equator because they need enough Coriolis force to allow the violent spiraling of the winds about a low-pressure center.

One curious feature of the hurricane/typhoon is the *eye*—a warm, calm, cloudless region at the very center that has no counterpart in any other cyclone. First-hand accounts describe the sudden cessation of the wild wind, warmer temperatures, and the clearing of the sky, with only a heaving sea to hint of storm conditions. This is the eye and its passage is of short duration, with the wind picking up abruptly and this time blowing 180° from its previous flow. Only a column of sinking air could account for all these characteristics of the eye. Yet from our perspective in the middle latitudes, it is unusual, to say the least, to encounter sinking air in the midst of a brisk vortex that is obviously accomplishing mass uplift of huge volumes of moist air (Fig. 7-13).

Once the storm is fully formed, it begins to follow a typical path—reasonably predictable in general outline but wholly unpredictable in detail. Pushed along by the easterly Trades, hurricanes/typhoons, regardless of hemisphere, begin a slow movement westward. At this stage they may very well gain in strength and intensity, for they are still over tropical seas and tap a constant supply of warm, moist air.

Fig. 7-13. ***The eye of the hurricane.*** *Filling the Gulf of Mexico from the Yucatan to the mouth of the Mississippi, a huge hurricane, with eye clearly visible, sidles slowly toward the northwest. Although its forward progress may be leisurely, it spirals about its eye at well over 100 miles per hour (161 km). This one finally came ashore near the U.S.-Mexican border, dripped mightily, carried some things off downwind, and finally blew itself out far inland.*

Sooner or later, however, a high percentage of these storms begin to curve away from the equator, generally following the western end of the permanent oceanic Subtropical High cell. No two are exactly alike, which makes forecasting difficult. On rare occasions, one may not curve at all; others will curve at varying rates; and some may even stop or form a loop in their course. The disastrous Cuban hurricane of 1963 crossed the southern end of that island and then turned and recrossed it before resuming its northward curve (Fig. 7-14).

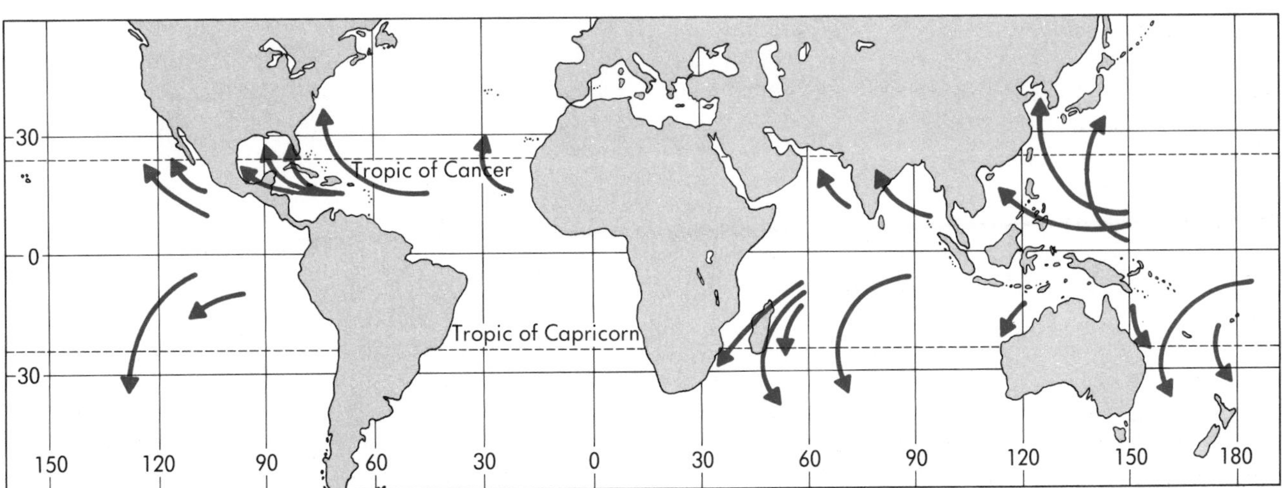

Fig. 7-14. ***Typical hurricane tracks.***

Pushed along by the easterly trades, hurricanes/typhoons, regardless of hemisphere, begin a slow movement westward.

There are two ways a hurricane can stop; both involve losing the moist air that is its fuel. If it invades a large land mass, the air drawn in at its base will rapidly decrease in moisture content and the storm will fade out, often very rapidly. If it survives at sea to complete its poleward curve, it will finally recurve at higher latitudes and move to the east. Here the ocean is colder and the air above it less capable of holding moisture. A tropical storm at these latitudes will gradually become a mere westerly squall.

A hurricane can stop in two ways, both of which involve losing the moist air that is its fuel.

The high winds accompanying violent tropical storms are the obvious destructive force that accounts for the awe in which they are held, especially in those regions of the world where such insubstantial construction materials as nipa and bamboo are the rule. The wind is not all that is to be feared. Heavy seas kicked up by wind and torrential rains are widespread for many miles, far from the center of the storm. The combination of overtaxed surface drainage channels attempting to carry off the rainfall and excessively high tides and storm waves (called the ***storm surge***) running up the river mouths from the sea can result in disastrous floods. One of the great natural calamities of all time occurred in this manner on the Ganges Delta in 1737 when over 300,000 persons were killed outright and uncounted others starved later because of flooded fields and subsequent crop loss (Fig. 7-15). In May 1965, again in November 1970, and still again in 1977, almost exact duplicates struck Bangladesh, killing many thousands. The newspapers were full of accounts of "tidal waves" engulfing the low delta islands, but the storm surge is what they actually meant. A tidal wave, or more precisely a *tsunami,* is something different (see Chapter 23).

Fig. 7-15. ***Typhoon aftermath.*** *Grain in the field, ready for harvest, is particularly susceptible to damage by both high winds and pounding rain. These Oriental rice farmers are attempting to salvage a bit of their crop, which represents eight months of labor and next year's food for the family.*

HURRICANE HUGO

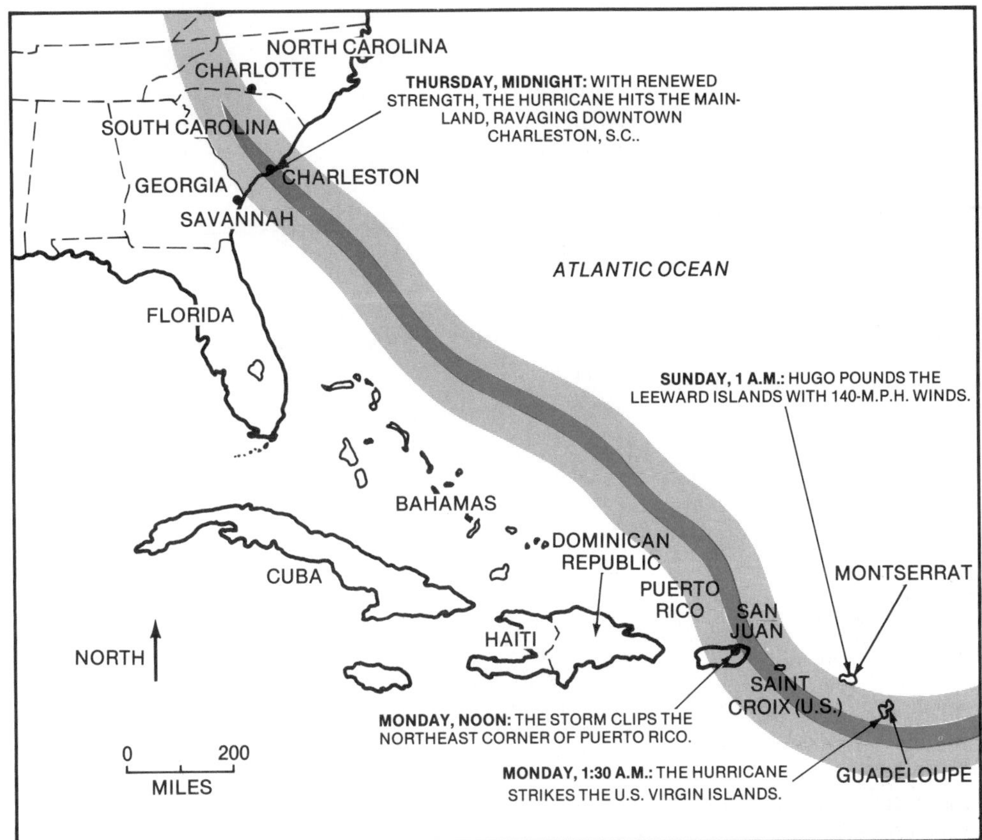

Late Thursday night, September 21, 1989, hurricane Hugo, packing winds of 135 mph (215 km/hr) and accompanied by a 17 foot (5 m) storm surge, came on shore near Charleston, South Carolina. Low-lying barrier islands and coastlands were severely battered. In the next few hours, Hugo deluged the city of Charlotte where flood-waters ran up to 5 feet deep, left 21 people dead, totally destroyed hundreds of homes, and otherwise left a path of destruction from the coast inland before spending its last energy over eastern Canada.

The powerful storm, one of the 10 worst to reach the U.S. mainland this century, began as an easterly wave over open water off the coast of Africa. As the storm moved westward with the Trades, wind velocity increased and a typical cyclonic flow developed. Recognized as a hurricane and named Hugo, the storm made its first landfall on the Leeward Islands of Guadeloupe and Montserrat with winds exceeding 140 mph (225 km/hr). Most of the islands' buildings were damaged, and many were destroyed. Traveling to the north–northwestward, the storm then hit the islands of Saint Croix and Puerto Rico leaving thousands of people homeless with little food and no fresh water.

At this point meteorologists, who had been watching Hugo's progress with the aid of satellites, were unsure where it would go next. A strong high pressure over the North Atlantic then sending air toward a low center over the Appalachians could steer the storm toward the southeast coast of the United States. If these pressure systems moved or the winds abated, Hugo might head for the more populous northern coastline. In addition, tropical storm Iris, paralleling Hugo some 700 miles (1100 km) to the east, might draw from Hugo's warm air source and weaken the storm.

Hugo weakened as it traveled over the Caribbean islands. Wind velocities dropped to about 100 mph (160 km/hr), and some forecasters predicted that it would not regain its former strength. This was not the case, however. The high- and low-pressure systems remained in place, Hugo overpowered Iris, and a revitalized Hugo set a course for the southeast coast of the United States.

When it was all over, Hugo's week-long binge caused an estimated $2 billion in property damage and at least 51 deaths.

Still another type of destruction was displayed when, on September 17, 1988, hurricane Gilbert, described as one of the strongest storms in history, plowed into the Gulf Coast near Brownsville, Texas. A dancing vanguard of nearly two dozen tornadoes preceded it, overturning cars and mobile homes, and tearing the roofs from buildings.

In the United States our hurricane warning system has become increasingly effective in allowing the populace adequate advance time for evacuation and "battening down." Luckily, hurricanes move slowly, and although they cannot be deterred, a careful daily to hourly check on their movements does make a warning system practical. Even in the United States, however, where hurricane warnings are best developed, there was heavy loss of life in Cameron, Louisiana (1957), and near Biloxi, Mississippi (1969), when hurricane veterans refused to believe that they were in danger and attempted to sit it out. This viewpoint persists; it was estimated that only 20 percent of Brownsville's residents had left town prior to Gilbert's landfall. Fortunately, although many were injured, there was no loss of life.

KEY TERMS

air mass
continental polar (cP)
maritime polar (mP)
arctic (A)
continental tropical (cT)
maritime tropical (mT)
equatorial (E)
Sirocco
Santa Ana
Gulf air (gT)
front
Arctic Front
cyclone
low
depression
storm
middle-latitude cyclone
extratropical storms
tornado
intertropical storms
easterly wave
weak Equatorial Low
hurricane
typhoon
jet stream
Polar Front jet stream
tropical easterly jet stream
polar night jet stream
cold front
warm front
occlusion
occluded front
baguio
willy willie
eye
storm surge

CONCLUSION

This chapter continues the previous discussion of atmospheric moisture with its consideration of storm precipitation. To the discerning eye, the topics covered—air masses, fronts, the ITC, and storms—bring together all that we have learned thus far in Part Two. Air masses develop in response to differential heating and cooling of the land and oceans, and they move in response to world pressure and wind patterns. Air masses carrying moisture and solar energy interact along fronts and the ITC to produce precipitation, which, besides being an important aspect of the hydrologic cycle, plays an important role in the earth's energy balance. To the geographer, however, knowledge of moisture and the other three elements and how they interact is not an end in itself. Instead, understanding weather elements is a prelude to studying the broader picture of global climate and the great variety of patterns within. Climate, depending as it does on long-term averages, is more important to the way we live and work than is day-to-day weather. In the rest of Part Two, we focus on major climates and what they mean to human use of the earth. Later, in discussions of hydrology, landforms, soils, and vegetation, we will draw on our knowledge of weather and climate to understand a variety of other global patterns.

REVIEW QUESTIONS

1. What makes a particular area of atmosphere an "air mass"? Why don't air masses form over the United States?

2. How are mP and mT air masses *similar?* How are cP and cT air masses *similar?* Why?

3. How do the eastern and western sections of an mT air mass differ? Why?

4. When a portion of an air mass moves out of its source area, why does it tend to retain its original characteristics? How, then, can it change?

5. Why are there no middle-latitude air masses? How does this help to explain the "stormy" character of the middle latitudes? Of Norway, in particular?

6. How could the behavior of a jet stream cause a bad winter in 1976–1977 and a bad summer in 1988?

7. If the cold front and the warm front are both zones of precipitation in a middle-latitude cyclone, how can you tell the difference between them? What happens if the cold front catches up to the warm front?

8. How do the conditions that create a tornado differ from those that create a hurricane/typhoon?

9. If hurricanes in the Northern Hemisphere move from

the east to the west and then curve northward, which way do hurricanes in the Southern Hemisphere move?

10. What are the two ways in which a hurricane can stop? How are they similar?

11. Why do many people believe that tornadoes never go through the middle of a city?

12. How does unequal heating of the earth's surface lead to middle-latitude storms? To tropical storms?

APPLICATION QUESTIONS

1. Find out what air masses and fronts most affect the weather in your area and use that information to explain the kind of weather you have been experiencing recently. What can you expect in the near future?

2. Imagine you live at 10° N latitude and are tired of battening down for hurricanes, but you want to move the least possible distance from your present home. Which way would you go to avoid these storms? Would it matter if your choice were a hotter or cooler place?

4. What hurricanes or tornadoes have been the most violent in the past 10 years? What caused them? What damage did they do?

5. Most daily newspapers in the United States include a current weather map showing the positions of high and low pressure and associated cyclonic storms (if they are present). Select a seven-day period beginning with the appearance of a new cyclonic storm in the west and follow the progress of the storm eastward. Plot the path of the storm through the period and estimate its average speed. Would it be possible to avoid the storm by driving away from it in your automobile (assuming that you do not exceed the legal speed limit)?

CHAPTER 8

CLIMATE CLASSIFICATION

OBJECTIVES

After studying this chapter, you will understand

1. Why weather records make it hard to describe a climate accurately.
2. Why the use of averages can be misleading.
3. Why the classification of climate must avoid both too little and too much generalization.
4. How the ancient Greeks classified climate.
5. How the Köppen (and modified Köppen) system of climate classification works.
6. How to read and prepare a climograph.

Reviewing daily weather maps at the National Weather Service.

INTRODUCTION

To this point in Part Two we have focused on the dynamic character of the earth's atmosphere. We have seen that incoming solar radiation results in surface and atmospheric heating. The degree of heating varies with latitude, altitude, continent, and ocean. With such temperature variation come the global patterns of barometric pressure, wind, moisture, and the conflict of air masses we see in storms.

The focus on weather, which involves the instantaneous condition of the atmosphere, tells us little about the long-term ranges and averages that characterize specific places and broad regions. For this information we turn to climate.

We have dealt with the word ***climate*** briefly in an earlier section, but it seems to be one of those everyday terms that become a little sticky when an exact definition is needed. Fundamentally, it is a mix or average of the ever changing, momentary weather at any given place. Theoretically, then, every region in the world has its own climatic personality made up of the interaction of the four weather elements. In this chapter, we will discuss how we can classify these climatic personalities, but first let us consider the nature of weather data and some problems in climatic definition.

WEATHER DATA

One variable that makes it difficult to define climate with great precision is our reliance on observed weather data and the lack of dependability of some of these data. Time, for instance, is a critical factor in weather observation. Generally speaking, the longer the record the more reliable our conception of a given climate. This is because a long-term average tends to iron out the freak year and short-run oscillation. Occasionally, however, we are forced to operate with a very brief record. When this becomes necessary, the location of the record station helps us assess the validity of the data.

The longer the record, the more reliable our conception of a given climate.

In most parts of the world only a five- to ten-year transcript of continuous weather observation could seriously distort our attempt to visualize climatic character. In places like New York and the Netherlands, where the weather changes constantly, a short-term record would never do. The hallmark of climate in the equatorial and polar regions, however, is deadly monotony, and thus the short-run record need not be fatal. For example, although the accumulated observations at the South Pole cover little more than a 30-year period, they show such a day-to-day and year-to-year similarity that long-range projections are being based on them with a reasonable expectation of only negligible error.

We should be aware, however, that even records amassed over many years are likely to include at least minor errors. Even in 50 to 75 years, official observations may have been transferred from somebody's backyard to the city hall to the airport tower and finally back downtown to the smogbound roof of the weather bureau. During this same time observers have used increasingly efficient instruments, producing a considerable range of recordings, while simultaneously changing definitions of "killing" frost, "trace" of rain, rainfall year, moisture equivalents of snow, and the like.

Even records amassed over many years are likely to include at least minor errors.

PROBLEMS IN CLIMATIC DEFINITION

Problems may also arise in simply averaging weather data to arrive at climate. A climate that regularly features great extremes cannot be described by an artificial mean "somewhere in the middle," no matter how long the record. Both Seattle and Chicago, for example, display an annual average temperature of 50° F (10° C), but to cite 50° F as typical of both cities would imply a totally false picture of the temperature realities of one of them. In Seattle, day after day from summer to winter, the temperature is never very far from 50° F. In Chicago, most days of each year are far above or below 50° F; the average simply does not make this fact clear.

A climate that regularly features great extremes cannot be described by an artificial mean "somewhere in the middle."

Other kinds of local peculiarities lend character to nearly every climate. When we compare Seattle to Chicago again, we see that both receive approximately 33 inches (84 cm) of precipitation annually. Seattle's, however, is delivered mostly in the form of a soft drizzle over 250 days of gloomy, gray cloud cover each year, whereas Chicago's arrives in part as explosive hour-long summer thundershowers or as wind-driven, deeply drifted winter snow. Another distinguishing factor could be fog. In most parts of the world this phenomenon is limited and harmless, yet on the occasional west coast it is a daily occurrence with a significant influence. Similarly, in some regions the frequency of killing frost, above and beyond the mere monthly temperature average, alone determines the kind of agriculture that can be practiced.

All these elements are part of the total climatic image, and all are the product of climate controls that in one way or other influence the fundamental elements. Seattle is affected by the control of a moderating sea whereas Chicago is continental; recurrent fog is frequently the product of a cold ocean current; and killing frosts, even far into the subtropics, may well result from the control of topography over air drainage from a massive, cold source region.

Thus, if we say that climate is essentially the long-term mean of the daily weather conditions, we are on the right track. The climatic controls are at work in our definition of the average condition, and the result is a *basic* climate picture. All climates, however, have special individual details that make them distinctive from all the others. The climatic definition, then, must point up and explain individuality. It is, therefore, a complex of many things, not the least of which is a measure of insightful interpretation and analysis.

CLASSIFICATION

No two places in this world have exactly the same climate. Witness the home garden featuring a wide variety of plants, each with its own specific climatic requirement. The azalea must have shade, the hibiscus full sun; the camellia requires constant moisture, the hollyhock virtually none; the oleander needs great heat, the rhododendron as little as possible; the rose is totally frost hardy, the geranium moderately so, and the poinsettia not at all. Yet every one of these plants may exist in the same garden, none more than 50 feet (15 m) from the other; change their places and each will expire. Admittedly, mechanical sprinklers and chemical fertilizers give the gardener an edge over nature, but unquestionably there are many microclimates here. Train a peach tree to grow against a south-facing white wall in Scandinavia and it will produce a bountiful crop, or plant an orchard no more than 3 miles (5 km) from the east shore of Lake Michigan for equally spectacular results (Fig. 8-1). Again, these are microclimates: the fruit trees are reacting to purely local situations.

No two places in the world have exactly the same climate.

Fig. 8-1. ***Cherry orchards near Traverse City, Michigan.*** *Each fall the lake acts as a heat reservoir to extend the growing season. The same lake, cool in April, discourages the early blossoming occasionally caused by an early warm week. This common "false spring" is inevitably followed by one last freeze of a dying winter. A few miles from the lake, fruit trees do not do nearly as well in these high latitudes.*

If the world's climatic pattern is merely an infinitude of slightly differing climates, how do we even begin to classify? The key is to *generalize.* If we cannot find any extensive regions with *identical* climates, we will generalize a bit and classify regions with *similar* climates. By setting some relatively broad limits beyond which a given climate cannot trespass, we can arrive at an acceptable if not perfect system of classification, the reasonable alternative to dealing with climatic chaos.

If we cannot find any extensive regions with identical climates, we will classify regions with similar climates.

Climatic Concepts of the Ancient Greeks

The ancient Greeks were the first to give serious attention to the concept of climatic similarities. They had long been aware of the character of the Sahara Desert, which travelers encountered (no matter what the route to the south) immediately next to the Mediterranean basin. Here was an uninhabitable world of shifting sands and wildly extreme temperatures. They called it the ***Torrid Zone*** and totally shunned it. Across the mountains to the north was another sharply alien environment of cold winds and snowy forests, the home of primitive tribes. This had to be the ***Frigid Zone*** and no self-respecting Greek felt any particular desire to invade it. As a matter of fact, the Greeks felt singularly blessed to live in the moderate Mediterranean world, which they called the ***Temperate Zone*** (Figs. 8-2 and 8-3). Given the shortage of hard data in those times, the Greeks can hardly be faulted for their climatic naïveté. They were at least recognizing the basic control of latitude on temperature. In a modern context, however, it would be somewhat extreme to generalize the climate of the whole world into three temperature zones. We therefore need to find other factors besides latitudinal temperature on which to base our classification.

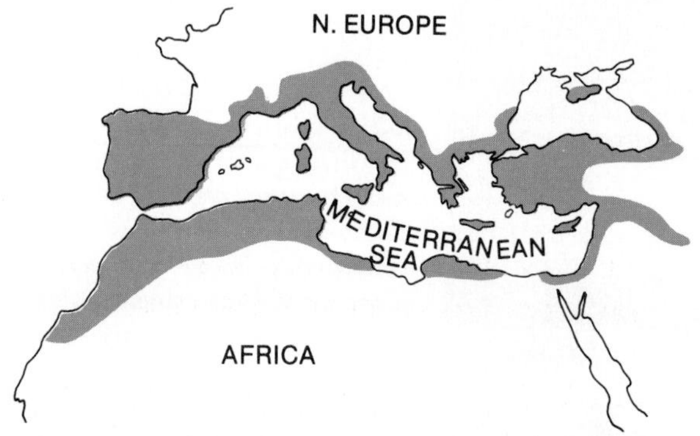

Fig. 8-2. The temperate world of the ancient Greeks.

The Greeks were guilty of climatic naïveté, but they at least recognized the basic control of latitude on temperature.

The Factors of Classification

Any responsible climatic classification must consider, among other things:

- I. **Moisture**
  - A. Precipitation
    1. Amount
    2. Distribution (seasonality)
    3. Type (rain or snow)
    4. Cause(s)
    5. Reliability
    6. Extremes
  - B. Humidity
  - C. Clouds
  - D. Fog
- II. **Temperature**
  - A. Average
  - B. Range
    1. Annual
    2. Diurnal (daily)
  - C. Extremes
  - D. Cause(s)
- III. Winds and Pressure
- IV. Storms

How do we pull all this together? Very early it dawned on somebody that the natural vegetation of any particular region was a visible indicator of climate. It tends to reflect strongly the complex interrelationship of all the factors on our list above and is a great deal easier to observe and classify than the limited data derived from unevenly distributed weather instruments. Thus, although vegetation regions and climatic regions do not coincide precisely, they are similar enough (as we point out in Chapter 27) that many of the standard classification systems are based heavily on their close relationship.

Natural vegetation tends to reflect strongly the complex interrelationship of climatic factors and is easier to classify than limited data from weather instruments.

The Köppen System

The best known and generally accepted standard of the twentieth century has been the system devised by Wladimir

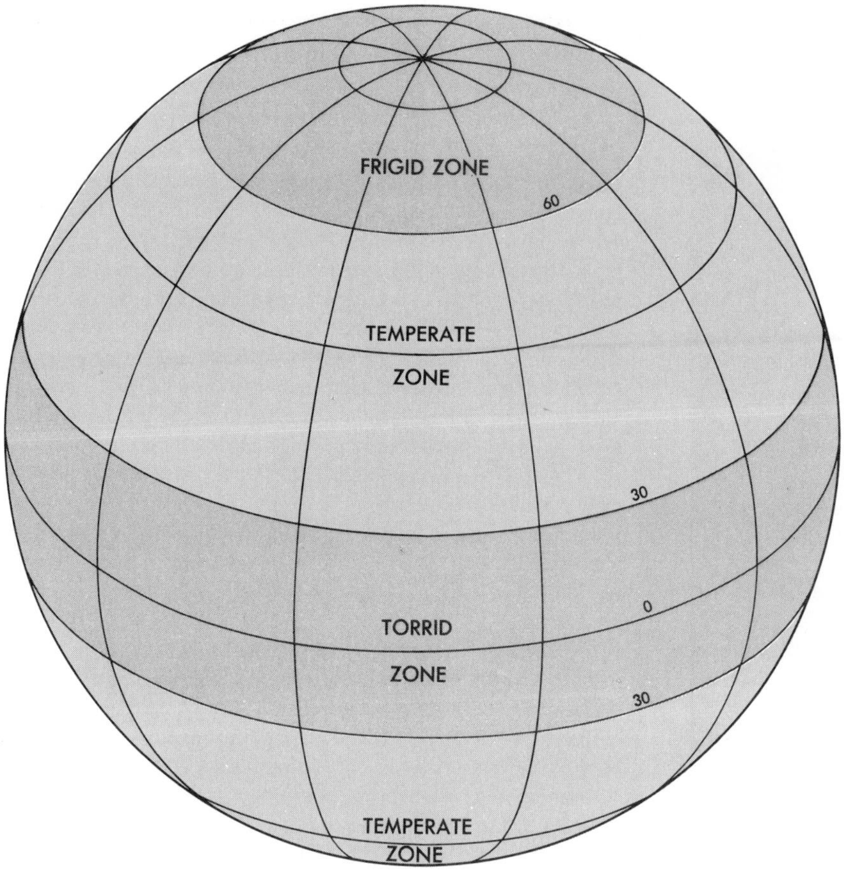

Fig. 8-3. ***The Greek classification.*** *This slightly modified version of the ancient system continues to appear in some elementary textbooks. It points up the strong influence of latitude on temperature but is badly flawed by failing to bring out other temperature controls and especially by completely ignoring moisture as a climatic element.*

Köppen in 1918 (Appendix C). He worked with natural vegetation as a guide and then, applying quantitative methods, arrived at a climatic classification. The ***Köppen system*** gives critical boundary lines specific values or arrives at them through the use of formulas. Boundaries are constantly revised as new data become available. Each region is assigned letter symbols.

The Köppen system gives critical boundary lines specific values or arrives at them through the use of formulas.

This approach is valuable for the advanced student, but at the introductory level pure unmodified Köppen can become a little turgid. When decoded, his symbols are based on German words, which must be translated, and for those clinging to English units, the Celsius degrees and metric scales decipher into awkward fractions. Nonetheless, most commonly used classifications bear a strong family resemblance for in some manner or other they are merely Köppen simplifications.

A Modified Köppen

In the classification about to be introduced

1. We substitute descriptive titles for Köppen symbols.
2. We generalize rather broadly where Köppen is specific.
3. We set limits of most climates somewhat wider than Köppen does, thereby combining a number of his divisions.

Wladimir Köppen

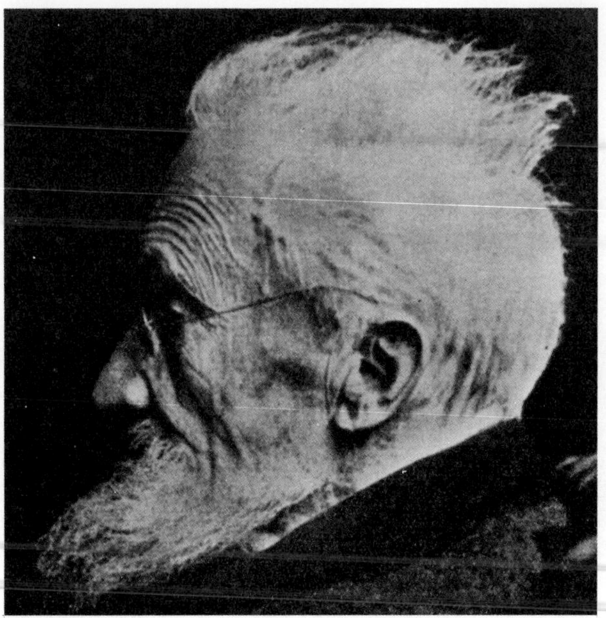

Wladimir Köppen

Born in Saint Petersburg (Leningrad), Wladimir Köppen (1846–1940) spent his youth on a Crimean estate which the Czar had presented to his father, a distinguished historian. The early influence of these years in a region of benign climate and subtropical vegetation was to mark his professional life. In moments of nostalgia Köppen pointed to the Crimean experience as the spark that ignited his curiosity about the relationship between climate and vegetation and led to his specialization in botany.

At Heidelburg and Leipzig in Germany Köppen studied all the sciences in addition to botany, and shortly thereafter he embarked on a career as a meteorologist at the German Government Naval Observatory in Hamburg. Forty-four years later he retired, but never during his service as a government technician did he become a bureaucratic drone. Throughout those years he remained an active scholar and publisher. Köppen is renowned for his pioneering contributions to the mapping of ocean winds, the use of devices for exploring the higher atmosphere, the melding of atmospheric physics and synoptic meteorology, and the brilliant anticipation of atmospheric fronts. At the age of 75, he even became involved with his son-in-law Alfred Wegener in advocating the theory of continental drift.

Geographers, however, remember Wladimir Köppen for his climatic classification based on the relationship of vegetation and climate. Inspired by August Grisebach's map of vegetation regions of the earth (1867), he worked off and on at expressing his theory in map for over 30 years, finally publishing a heavily botanical "Classification of Climate" map in 1900. It was moderately well received by his fellow scientists, but a substantial revision of 1918, in his own words "freer from botanical geography and more closely adjusted to pure climatology," was of major significance. The climax of his work was the joint editorship of the great five-volume *Handbuck der Klimatologie (Handbook of Climatology)*—at the age of 84.

The result is four general groups with a total of eleven separate climates, each with its own peculiar set of characteristics:

Tropical Climates

Tropical Wet

Tropical Dry

Tropical Wet and Dry

Subtropical Climates

Mediterranean

Humid Subtropic

Marine West Coast

Midlatitude Climates

Middle-Latitude Dry

Humid Continental

High-Latitude Climates

Taiga

Tundra

Polar Ice Cap

Map I displays these climatic regions. If this color map seems frighteningly complex, be assured that it possesses both logic and a large element of predictability. Note, for instance, that the color representing a particular climate is repeated over and over again but always in the same relative position on each continent. The reddish shade is strongly concentrated near the equator, and the bright green is always on middle-latitude west coasts.

This approach simply means that an identical set of controls is at work in the same place on each continent. Therefore, if we learn the personality of a climate in one location, we automatically understand that of every other spot where the same color appears on the map. If we were to establish a huge hypothetical land mass in the mid-Pacific, it would be relatively easy to predict its various climates (at least in rough fashion), for they would be mere repetitions of those found elsewhere.

An identical set of controls is at work in the same place on each continent.

We will make no attempt to classify the oceans and the mountains. The reason is not that they do not have climates but that longtime weather records are not available at sea as they are at land stations. In addition, mountains break down into what is essentially a mass of microclimates. Every 1000 feet (300 m) of elevation changes the temperature roughly 3° F (1° C), whereas shady slope versus sunny slope, windward slope versus lee slope, and so on, introduce so many variations that it becomes impossible to integrate the climate of each individual valley into the world pattern.

THE CLIMOGRAPH

We can visually display a number of basic characteristics of each of these climates by plotting the data of a representative station on a graph called a ***climograph.*** This is a condensed version of the main features of each climate region (Fig. 8-4).

A climograph is a condensed version of the main features of each climate region.

The climograph is made up of vertical columns corresponding to the 12 months of the year. They are always labeled starting at the left with January and following in sequence so that December is the right-hand column. In the Northern Hemisphere, the middle columns will be summer and the outside ones winter. In the Southern Hemisphere, the reverse will hold.

Along the right margin is a precipitation scale. We plot the normal monthly precipitation by blacking in each column from the bottom to the proper height as the scale indicates. When each month has been plotted, the total annual precipitation becomes visible at a glance as does the typical distribution of that precipitation throughout the year.

Along the left margin of the graph is a temperature scale. Using this scale, we plot the average monthly temperature

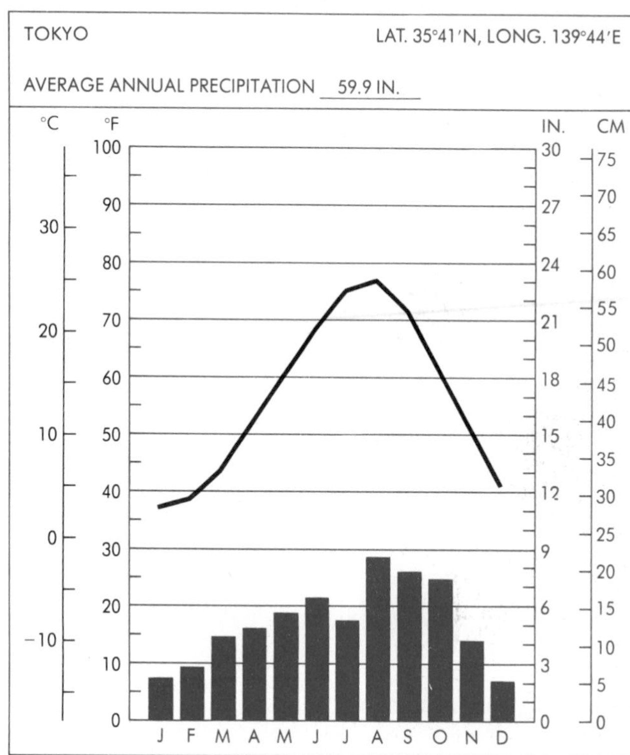

Fig. 8-4. ***The climograph.*** *In chart form, the longtime weather record of an actual station becomes readable. Monthly temperature averages are connected to form a curve, and monthly precipitation averages are represented by shaded bars.*

by placing a dot in the middle of each column. When we have plotted all months, we draw a line connecting the dots. This curve shows the average temperature for each month and the annual temperature range.

Because of their usefulness as visual representations of the more important features of each climate, we will include typical climographs of actual stations for each region discussed in the following chapters.

CONCLUSION

Each location on the earth's surface has a unique climate, that is, a unique combination of weather elements interacting over time. To view climate on a regional and global scale we must generalize, although overgeneralization, as characterized by the ancient Greek system, is of little value. The Köppen system of climate classification was initially based on world vegetation patterns and is now, in modified form, the system most used by geographers.

In this chapter we explored the rationale and problems related to climate classification and introduced the Köppen system. The scene is now set for a descriptive treatment of the earth's major climates. In the next chapter we focus on tropical climates. In subsequent chapters we cover subtropical, mid-latitude, and high-latitude climates.

KEY TERMS

climate	*Frigid Zone*	*Köppen system*
Torrid Zone	*Temperate Zone*	*climograph*

REVIEW QUESTIONS

1. Given the nature of weather records, why is it often difficult to present a truly accurate description of climate for a particular location?

2. Why might simple averages of temperature and precipitation give a false impression of an area's climate?

3. Why must we generalize when discussing the climate of a large region? Why are gross generalizations about climate such as those of the ancient Greeks of little value?

4. Why is vegetation a good indicator of climate?

5. Maps showing world climates include very little, if any, information about climatic patterns in mountainous regions and over oceans. Why?

6. What is the value of a climograph?

APPLICATION QUESTIONS

1. Contact your nearest National Weather Service office and request a list of daily maximum and minimum temperatures, and precipitation amounts for the past year for your town. Calculate the average annual temperature and the average daily precipitation. Are these averages useful, misleading, or both? Explain your reasoning. How would you prefer to characterize your climate? (Consider *seasonal* averages, for example. Do they give a more accurate picture?)

2. List the factors that you think should go into an accurate description of the climate in your area. How does the vegetation in your area serve as a climatic indicator for each of these factors?

3. Prepare a climograph for your area using weather data gathered from the National Weather Service or back issues of your local newspaper. Do the same for another area, perhaps one where you've always wanted to live or travel. Are you surprised by the results?

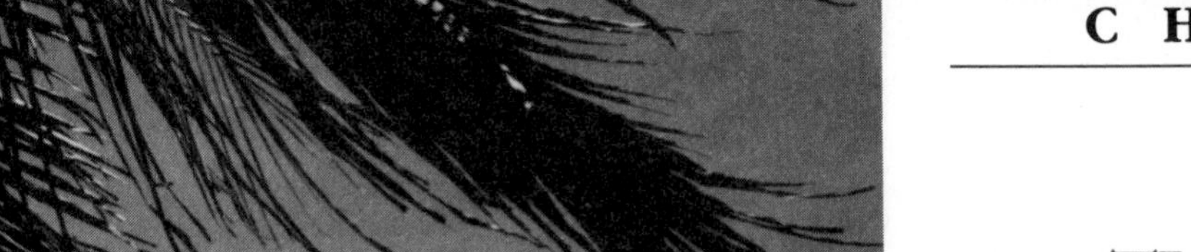

CHAPTER 9

THE TROPICAL CLIMATES

OBJECTIVES

After studying this chapter, you will understand

1. Where the Tropical Wet climate occurs, what its conditions are, and what causes these conditions.
2. Where the Tropical Dry climate occurs, what its conditions are, and what causes these conditions.
3. Where the Tropical Wet and Dry climate occurs, what its conditions are, and what causes these conditions.

The dense growth of palms on the west coast of Trinidad, an island in the Caribbean Sea, reflects the warm and wet climate under which they thrive. We often associate the tropics with warmth and moisture. However, while the tropics are warm throughout, there are great contrasts in moisture from perpetually wet and seasonally wet to the driest of deserts.

INTRODUCTION

Understanding the earth's major climates is crucial to understanding most other physical and cultural patterns. For example, climate strongly influences the development of mature soils and thus agriculture and, in part, the way people live. The rate of erosion, the presence or absence of forest, the size of rivers, and the abundance of fish all reflect the nature of climate. Armed with the modified Köppen system of climate classification presented in the last chapter, we now begin an exploration of world climate patterns beginning in the low latitudes with the tropical climates.

The tropical climates are located in a wide belt astride the equator—approximately the area that the Greeks called the Torrid Zone. As the Greeks looked south from the Mediterranean, the northern margin of the Sahara was obviously the outer edge of the excessively hot land and the beginning of a more moderate temperature region.

Others have suggested that the Tropic of Cancer and the Tropic of Capricorn are the limits of the true tropics, but actually the Greeks were nearer to the truth when they selected the Sahara margin some 30° to 35° from the equator. Köppen suggested that the tropics should be characterized by a cold month averaging 64° F (18° C or warmer), and the region the Greeks described comes close to this average. It is also the poleward limit of coconut palms.

Within this zone are three climates, all generally warm as the word "tropical" would indicate, but differing markedly in their precipitation characteristics. They are (1) *Tropical Wet,* (2) *Tropical Dry,* (3) *Tropical Wet and Dry* (Fig. 9-1).

TROPICAL WET

The major representatives of the ***Tropical Wet*** climate occur within 10° of the equator and therefore are located in roughly the same place as the Doldrums and at times the Intertropical Convergence Zone. The Amazon basin in

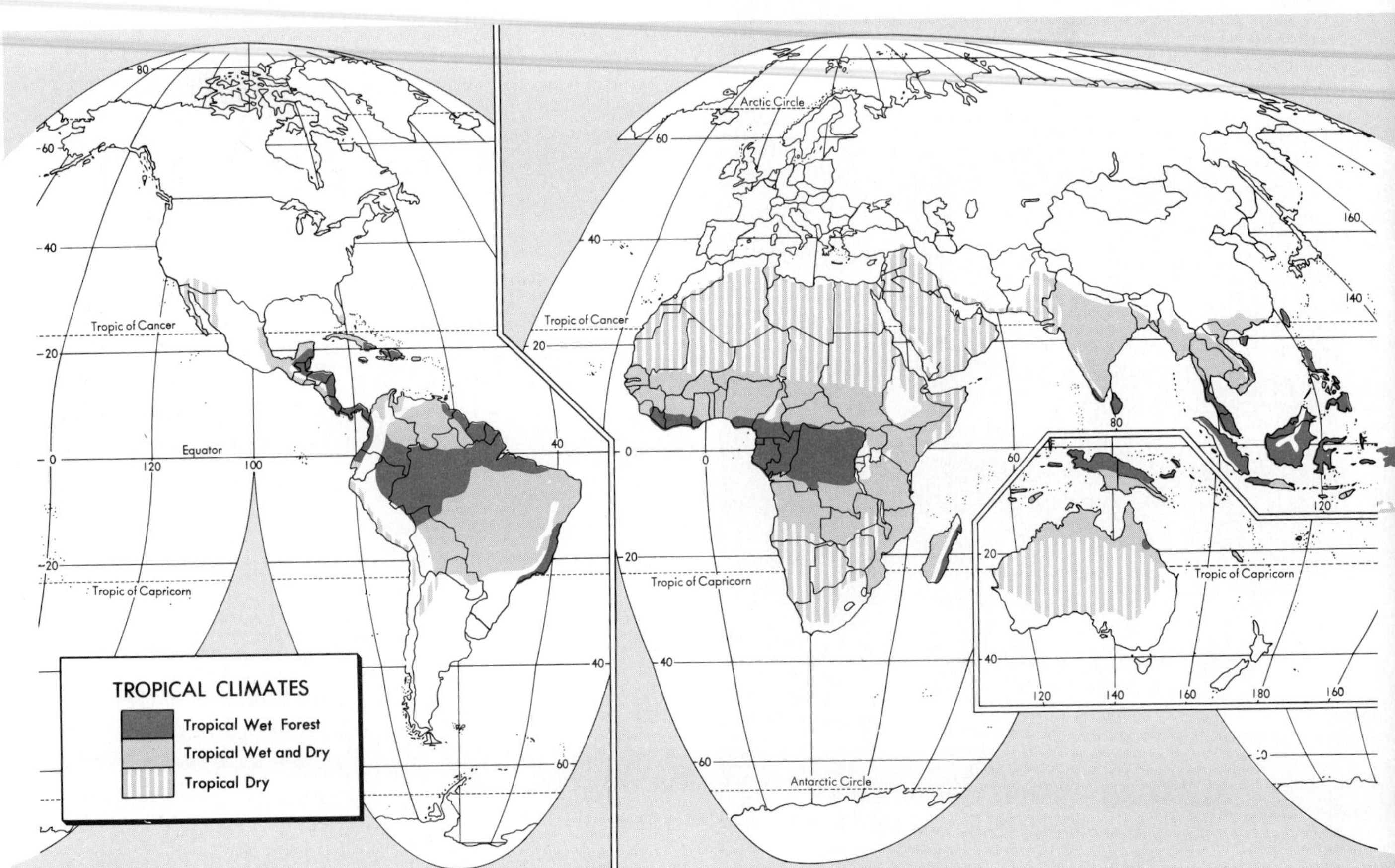

Fig. 9-1. Tropical climates.

South America, the Congo basin and closely adjacent area in Africa, and most of the East Indies are the chief land areas of the Doldrum belt, and each exhibits a Tropical Wet climate.

The major representations of the Tropical Wet climate are located in roughly the same place as the Doldrums.

Tropical Wet is synonymous with "rain forest," which is a key to the precipitation characteristics of this climate. Rain forest vegetation requires heavy and frequent rainfall. Probably 65 inches (165 cm) would be a minimum requirement, and 100 to 150 inches (250 to 380 cm) would not be uncommon. The distribution of this moisture throughout the year must be even, at least to the extent that no real dry season is experienced. Köppen suggested that, because the tropics have a high rate of evaporation, any month receiving less than 2.4 inches (6 cm) of rainfall must be regarded as dry. Therefore any tropical area with at least 65 inches (165 cm) of rain in a year and no less than 2.4 inches (6 cm) of rain in any month will support rain forest vegetation and must be classified as Tropical Wet.

The local populace may speak of the wet season and the dry season, and it is entirely possible that one season may experience double the precipitation of another. A true Tropical Wet, however, has no dry season. What the local people are referring to is a wet season and a *less* wet season (Fig. 9-2, Sandakan).

Most of this moisture comes from daily convectional showers. Saturated equatorial air heated by a sun that is nearly always overhead gives rise to this phenomenon that occurs in many places. The rain will come virtually "on schedule" every day during the period of maximum heating—2 to 3 o'clock in the afternoon. In general, mornings are bright and fresh, but as the day wears on, the lower atmosphere, heated by conduction and radiation, becomes increasingly warm and sticky until relief arrives in the afternoon in the form of clouds and torrential showers. The violently ascending convectional column is topped by towering thunderheads and lightning, and thunder frequently accompanies the short but heavy rainstorm. For a while at least, as the rain descends in sheets and the runoff floods the ground, there is a welcome cooling, but the storm is short and within an hour the sky clears, the vegetation steams under a high sun, and the heat returns, with even higher humidity than before the storm.

In tropical climates, mornings are bright and fresh, but as the day wears on, the lower atmosphere becomes warmer and stickier until torrential showers bring relief.

Rarely, a weak equatorial low will move through, giving rise to several days of cloud cover and rain. By far the most important moisture source is the convectional shower. Since the heating is constant the year round as a result of equatorial location, and the showers occur regularly, it follows that no dry season will exist and the landscape will be heavily forested.

Temperatures are, of course, generally high, but despite the proximity to the equator, Tropical Wet climates do not record world record highs. The daily clouds during the early afternoon cause the heat curve to level off so that 90° to 95° F (32° to 35° C) is the normal daytime maximum. This temperature is, of course, warm, especially when combined with high humidity, and afternoons are enervating and uncomfortable. Nonetheless, it is a far cry from the over 100° F (38° C) typical of many locations much farther away from the equator. The worst thing about these 90° F (32° C) temperatures from the standpoint of human comfort is that there are no seasons, so every day is the same. The only real relief comes at night when temperatures may drop to 65° to 70° F (18° to 21° C). These are the coolest temperatures ever experienced here. Lack of seasonal temperature change means that there is more than a kernel of truth in the old saying, "Night is the winter in the tropics."

There is more than a kernel of truth in the old saying "Night is the winter in the tropics."

As the climographs (Fig. 9-2) show, the average monthly temperature (both day and night) is never far from 80° F (27° C), and the temperature curve is virtually a straight line. The annual temperature range never varies more than a degree or two. The daily range, which does not show on the climograph, is considerably larger, averaging as much as 25° F (14° C).

Tropical Wet in the Trades

Although the bulk of this climate occurs in the Doldrum belt, there are several narrow coastal strips of Tropical Wet some distance from the equator (Fig. 9-1). These are all the same latitude (20° to 25° N and S) and all are on east coasts: the eastcentral coast of Brazil, the east coast of Madagascar, the northeast coast of Australia, the east coasts of Central America, Hawaii, most West Indian islands, and the east coasts of Vietnam and the Philippines. (This last region in Southeast Asia is confused by local monsoon circulation but generally fits the overall world pattern.)

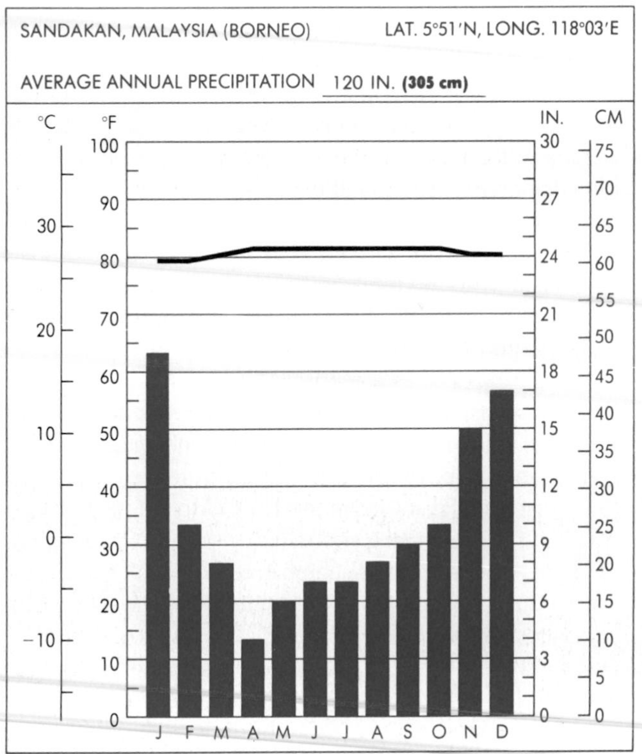

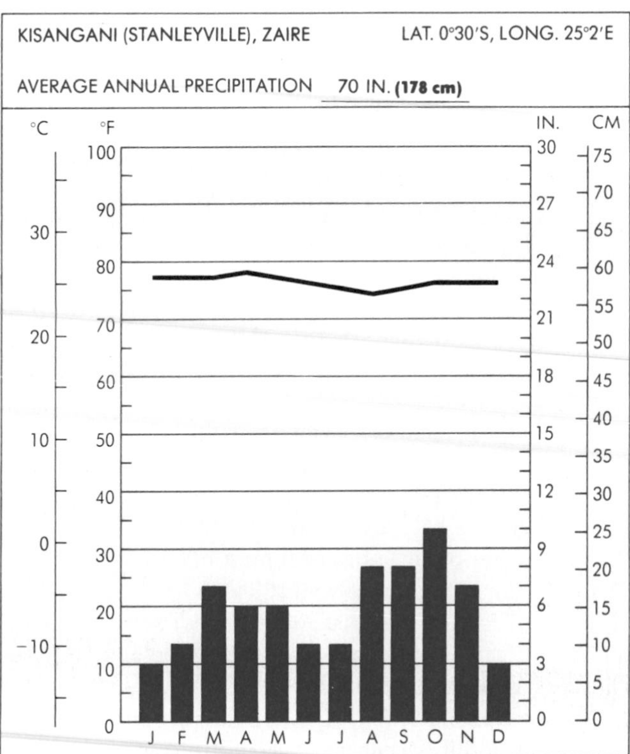

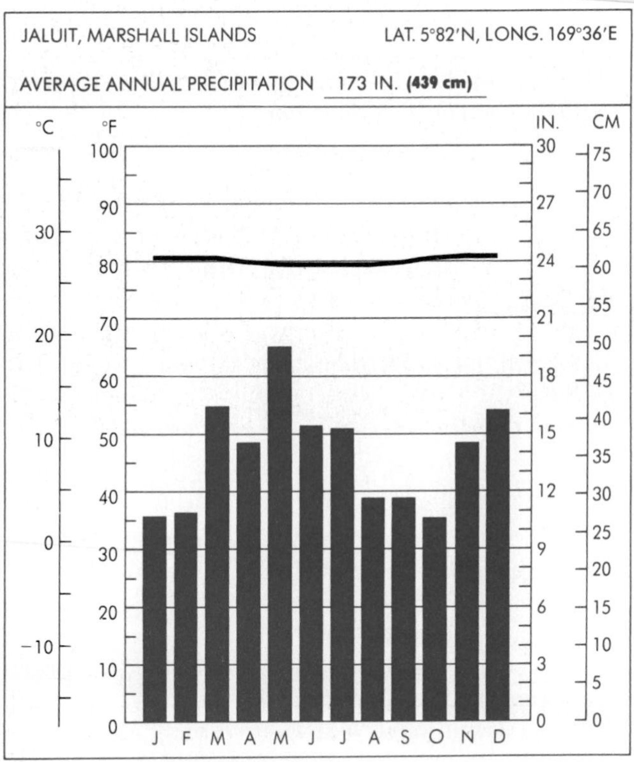

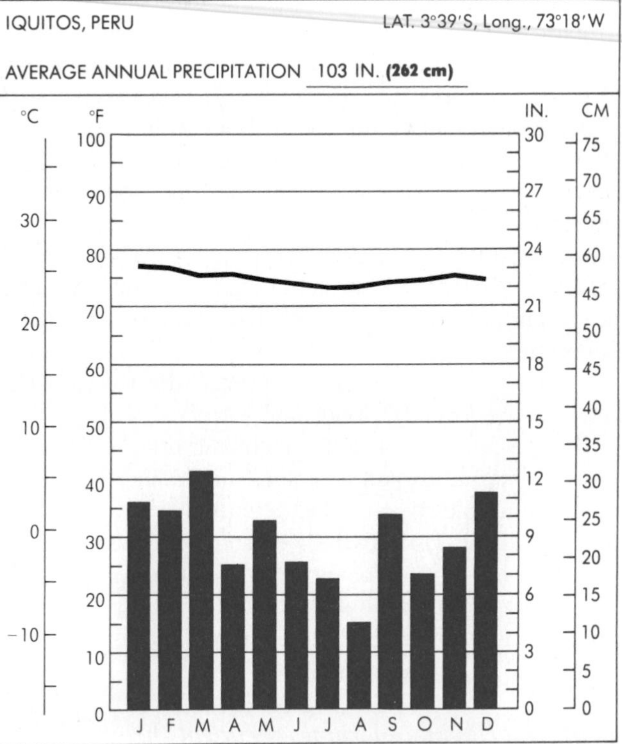

Fig. 9-2. Tropical Wet stations.

These areas have three climatic controls that are responsible for their similarities.

1. Despite their distance from the equator, they are still well within the tropics and thus experience continuously high temperatures.
2. They are all in the trade wind zone and therefore are constantly exposed to moisture-bearing winds from off warm seas.
3. Each of the regions features uplands rising abruptly in back of the coast. So as the oceanic air masses reach these highlands, they are forced aloft and orographic precipitation bathes the mountain slopes. On the coast proper the heated oceanic air most often gives up its moisture in convectional showers, but the seaward-facing slopes are perpetually cloud-shrouded and drippy.

As the oceanic air masses reach these highlands, they are forced aloft and orographic precipitation bathes the mountain slopes.

The end result is a climate characterized by continuously high temperatures, constant heavy rainfall the year round, and dense rain forest vegetation, all of which is very similar to that of the equatorial regions. We have no alternative, then, but to classify this climate as Tropical Wet even though these trade wind coasts are separate from the Doldrum regions. The causes are different but the result is the same, and in classifying climates we are concerned primarily with results.

Tropical Wet Populations

Human beings have not sought out Tropical Wet regions as a preferred habitat. Along with the deserts and polar regions, throughout history the tropical rain forests have shown up as great blanks on the world population map. This is a difficult region, as many an American or European who has been forced to live there for any length of time will attest. The moist heat day after day can be enervating, and diseases are difficult to control, especially those carried by hosts, such as yellow fever, malaria, filariasis, and liver fluke. Mildew and mold flourish, attacking and decomposing textiles and leather in a remarkably short time. Furniture secured with glue falls apart, and termites may honeycomb wooden houses and foundations almost overnight. Yet people do live and prosper in this climate, and even the reluctant immigrant adjusts at least physiologically, after a week or two.

Small numbers of wandering tribes occupy most of these Tropical Wet regions. Their economy is based on hunting, fishing, and gathering what the forest provides. Some clear small sections of the forest by fire and carry on a rudimentary agriculture among the stumps and snags. Soon, however, they are forced to move on and repeat the operation, for without the protection of the mature forest, soluble soil minerals are quickly washed away and weeds, beyond the control of primitive tools, soon crowd out the crop plants (Fig. 9-3).

Some limited locations within the Tropical Wet have become the home of large numbers of people and have achieved an important economic position because of certain

Fig. 9-3. ***Traditional life in the tropics.*** *The tropics have long supported human habitation, but in relatively recent times resource development has marked the land with urbanism, industry, and organized production ventures. "Small society" represented by the village and the aborigine may be on its way out, yet some vestige of traditional life persists in remote areas.*

The Hoti Indians of southern Venezuela subsist on primitive agriculture (main crops: plantain, bananas, maize, and sweet potatoes), hunting (peccary, tapir, monkeys, and birds), and collecting wild fruits (mostly from palm trees) and honey. They move frequently to new agricultural fields and more abundant game and fruit resources.

factors offsetting or complementing the basic difficulties of the climate. Some of these are:

- Superior accessibility
- Occurrence of valuable minerals
- Above-average soil fertility
- Application of capital and technical skills

The island of Java, during the Dutch colonial regime, is an example. Because Java is an archipelago (a group of islands), it is easily accessible from the sea. In addition, its location off the coast of Southeast Asia places it near important shipping routes. Its mountain backbone of active volcanoes have supplied fertile ash showers that have rejuvenated the soils more rapidly than they can be leached. Economic deposits of minerals, petroleum in particular, have been found. All these factors, combined with Dutch capital, direction, and development, has made Java a highly productive region capable of supporting a population of many millions. Today the nation of Indonesia has gained a measure of individual freedom at the expense of economic solvency, but the potential remains.

Sri Lanka with its tea and coconut plantations, various West Indian islands with their sugar, Central America with its bananas, Malaysia with its rubber and tin, Surinam with its bauxite, and Nigeria with its cacao plantations and petroleum are other examples of Tropical Wet areas that have progressed above the average because of one or more special factors.

On the whole, however, such development occurs at the expense of the earth's most naturally productive and diverse forest. Vast tracts of tropical rain forest are now being cleared each year as emerging nations in these latitudes make way for more modern agriculture and a strengthened export economy. Much former rain forest in Brazil, for example, is now planted in cattle pasture, and the meat produced is served outside the tropics in fast-food restaurants. Although the Tropical Wet climate is likely to remain a dominant feature of equatorial regions, the forest for which it is known may soon be a thing of the past. (See Chapter 27 for more discussion of rain forest vegetation.)

The forest for which the Tropical Wet climate is known may soon be a thing of the past.

TROPICAL DRY

All the world's ***Tropical Dry*** zones are found in a broad belt at about the same latitude—roughly astride the Tropics of Cancer and Capricorn. Thus they are in the region dominated by the trade winds and are often called trade wind deserts. In addition, they all have west coastal frontages. Some, notably in North and South America, are narrow coastal ribbons hemmed in by mountain ranges closely paralleling the coast. Others, as in Australia and Africa, lack this topographical control. These regions extend long distances inland (Fig. 9-1).

All the world's Tropical Dry zones are in the region dominated by the trade winds and are often called trade wind deserts.

The largest of the Tropical Dry regions by far is a continuous zone that extends from the Atlantic Coast across North Africa, Arabia, and into northwestern Pakistan and adjacent India. Traditionally, three different names have been applied to the various sections of this extensive tropical desert. In North Africa it is called the ***Sahara***; between the Red Sea and Persian Gulf it is the ***Arabian*** Desert; and in India/Pakistan it is the ***Thar***. In virtually identical latitudes, the west coast of North America is also Tropical Dry. Baja California and the coastal strip across the Gulf of California are included here as well as the Imperial valley and lower Colorado River section of California and Arizona. Frequently, the name ***Sonoran*** Desert is used to refer to the entire area of North American Tropical Dry.

The Southern Hemisphere also exhibits the Tropical Dry climate, with every continent represented. In the northern third of Chile and the entire coast of Peru is the ***Atacama*** limited to a strip between the Andes and the sea. Astride the Tropic of Capricorn and on the west coasts of Africa and Australia are sizable dry regions extending far into the heart of each continent. These are the ***Kalahari*** and ***Australian*** deserts, respectively.

Tropical Dry Characteristics

Tropical deserts are hot and dry; this is scarcely news. But how hot is hot and how dry is dry? More importantly, why does this situation exist in these particular locations?

Except for the very high latitudes where evaporation is limited, any place receiving less than 10 inches (25 cm) of precipitation per year is normally regarded as a true desert. Under these circumstances, there is little or no natural vege-

---

Any place that receives less than 10 inches of precipitation per year is normally regarded as a true desert.

EL NIÑO, LA NIÑA, AND THE SOUTHERN OSCILLATION

Rain can be a disaster along the desert coast of Peru.

Air and water conditions in the equatorial Pacific Ocean shift between two extremes—El Niño and La Ninã—which in turn are responsible for such weather anomalies as severe droughts, torrential rains, icy winters, and the breakdown of monsoons. To appreciate the extremes, remember that under normal conditions the Trades flow outwardly from the Subtropical Highs over the Pacific Ocean to converge near the equator at the Intertropical Convergence Zone.

Along the coast of Peru, the Trades blowing offshore cause upwelling of cold, nutrient-rich water that supports abundant marine life and thousands of fish-eating birds. Warm ocean currents propelled by the Trades flow toward the western Pacific where they continue poleward as the Japanese and East Australian Currents (Map IV). Cold water flowing equatorward in the California and Humboldt Currents completes the circulation. A strong Subtropical High pressure center is usually in place off the coast of South America, and a low-pressure center persists along the ITC off northern Australia.

Under the normal conditions just described, the western side of the Pacific basin tends to be moist and rainy, and the eastern side arid. However, with an average frequency of every four years and for a period lasting a year or so, ***El Niño*** and the ***Southern Oscillation*** (together labeled ***ENSO***) wreak havoc on these patterns. El Niño, meaning the Christ child—so named for its frequent appearance around Christmas—is a marked increase in ocean temperature (2° to 5° F or 1° to 3° C) in the eastern Pacific. It occurs with the Southern Oscillation, a sort of seesaw by which the normal low and high atmospheric pressure patterns switch places across the southeastern and western tropical Pacific. Now westerly winds replace the normal easterly flowing Trades. This wind reversal moves warm water eastward to the coast of South America. Warm water signals the return of El Niño—upwelling ceases, ocean production drops, and fish-eating birds starve. Particularly intense ENSO events, such as those of 1976–1977, 1982–1983, and 1986–1988, bring global weather anomalies such as heavy rain in deserts of coastal Peru, drought in Australia, absence of the wet monsoon in India, and drought in the U.S. Midwest.

With another reversal in the Southern Oscillation, the Trades once again resume easterly flow, upwelling returns, and cold characterizes the coastal waters off South America. However, in certain years (1975 and 1988 were the last two), the sea surface off South America becomes abnormally cold. Such cold extremes, for which the term *"La Niña"* (the daughter) has been proposed, have only recently been recognized. La Niña is thought to bring another set of global weather anomalies such as wetter-than-normal monsoons in Asia and colder-than-normal winters in Canada.

What causes the Southern Oscillation in the first place? Researchers are currently using an array of sensors including satellites to monitor ocean temperatures and global wind patterns. They have developed promising mathematical models. Yet the mystery remains. We are still a long way from understanding the cause, let alone being able to predict the occurrence and effects of El Niño and La Niña.

---

tation, only drought-resistant shrubs at best, and this cover is widely spaced among considerable areas of bedrock, sand, or gravel.

Actually, in the tropics where evaporation is exceptionally high, even 15 to 20 inches (38 to 50 cm) of rain each year is scarcely enough to support more than a skimpy bunch grass cover. Although this area may have some small value in good years for nomadic grazing of small flocks, it is so unreliable that we can legitimately classify it with the true desert as too dry for more than sporadic human habitation.

Unreliable rainfall is a characteristic of the world's dry regions. As a general rule, *the smaller the total precipitation, the greater the variability.* Lima, Peru, which averages a mere 1.8 inches (5 cm) of rain per year, recently went over 13 years without any measurable precipitation and then had several storms that brought the average back to normal. Sudden violent storms are typical in the dry regions, resulting in heavy surface runoff on the hard baked earth. Gullies (also called ***arroyos*** or ***wadis***) are cut by short, raging torrents, adobe buildings melt, and people drown.

Unreliable rainfall is a characteristic of the world's dry regions.

In some parts of the Tropical Dry, the rainfall is concentrated in the winter, as in the northern Sahara or the southwestern United States. In other places such as the southern Sahara and the Thar, rainfall comes in the summer. In all Tropical Dry regions, however, the total rainfall is low and usually unreliable.

The Role of West Coast Exposure

Why should west coastal exposures at these particular latitudes be so dry? Kalama in the northern Chilean nitrate fields has never recorded even a trace of rain since records were begun 100 years ago, and many other places are almost as dry (Arica, Chile, 0.02 in. [0.1 cm]; Iquique, Chile, 0.6 in. [1.5 cm]). One cause appears to be the trade winds. Notice the Tropical Wet trade wind coasts on maps (Fig. 9-1 and Map 1). Here the Trades are forced up over highlands, precipitating most of their moisture at the east coast. In almost every case the Tropical Dry regions are immediately opposite these trade wind coasts and thus are in the lee or rainshadow of the highlands and the continental masses. As we learned in Chapter 7, Tropical Dry regions are also strongly affected by the permanent Subtropical High cells offshore where the air is usually sinking and warming—a condition that discourages precipitation.

Nearly all the Tropical Dry regions are immediately opposite the trade wind coasts and thus are in the lee or rainshadow of the highlands and the continental masses.

It may seem strange that the world's driest regions should be along the ocean. To be sure, the subsiding air masses of the Horse Latitudes discourage precipitation, but certainly local sea breezes must introduce moist oceanic air onto the heated land from time to time. So they do, except that an invisible barrier exists to intercept this air and remove its moisture just short of the coast. This is a cold ocean current, and there is one off the west coast of every continent in these tropical latitudes. Furthermore, the offshore drift of the general trade wind pattern tends to pull surface waters out to sea, allowing colder water from the depths to rise along the coast—a process called *upwelling*. This process is discussed in more detail in Chapter 13.

Because all Tropical Dry regions front on west coasts, cold water is bound to exist offshore. A cold current can be almost as effective as a mountain range in intercepting moisture. Saturated air masses moving over a broad cold current will be cooled from the bottom, and condensation will occur in the form of dense fog. As this fog drifts in over the desert coast, it brings about the curious anomaly of the air being filled with moisture much of the year in a place that has never recorded more than a fraction of an inch of actual precipitation.

Fog that drifts in over the desert coast brings in moisture-filled air much of the year in a place that hardly records any actual precipitation.

Chilling of the air at the surface along with warming by subsidence at the eastern end of the Horse Latitude anticyclone produces an inversion that makes normal convection virtually impossible. In South America the Humboldt Current (the largest of the cold ocean currents), together with the Andes paralleling the coast, cause the Tropical Dry zone to extend to just a degree or two south of the equator.

Patterns of Tropical Dry Heat

Heat is the other typical feature of the Tropical Dry climate. Here is where the world's record high temperatures occur, the current record being a shade reading of 136° F (58° C) at El Aziziao, Libya, in the Sahara.[1]

Yet the Tropical Dry region is at the outer edge of the tropics, a considerable distance from the equator. In the Tropical Wet region at the equator the temperature rarely exceeds 100° F (38° C) even on the hottest day.

The reason, of course, is cloud cover and rain. The sun is directly overhead in both regions at some time or other each year, but in the Tropical Dry summer, the overhead sun and the lack of any clouds allow temperature to build up continuously well into the late afternoon. Only along a narrow band of fogbound coast is there any moderation. This fact has helped to make the west coasts the preferred locations for permanent habitation in the desert (see Fig. 9-4, Swakopmund).

In the Tropical Dry summer, the overhead sun and the lack of clouds allow continuous temperature rises in the afternoon.

[1] Whenever dealing with world records of temperature, precipitation, and so on, keep in mind that places with such extremes are very uncomfortable spots in which to live and therefore are lightly inhabited. The cited world records simply mean that for some reason an observer lived there and had the official instruments. It is entirely possible that there are hotter places than El Aziziao, but for that very reason they are good places *not* to be and their temperatures go unrecorded.

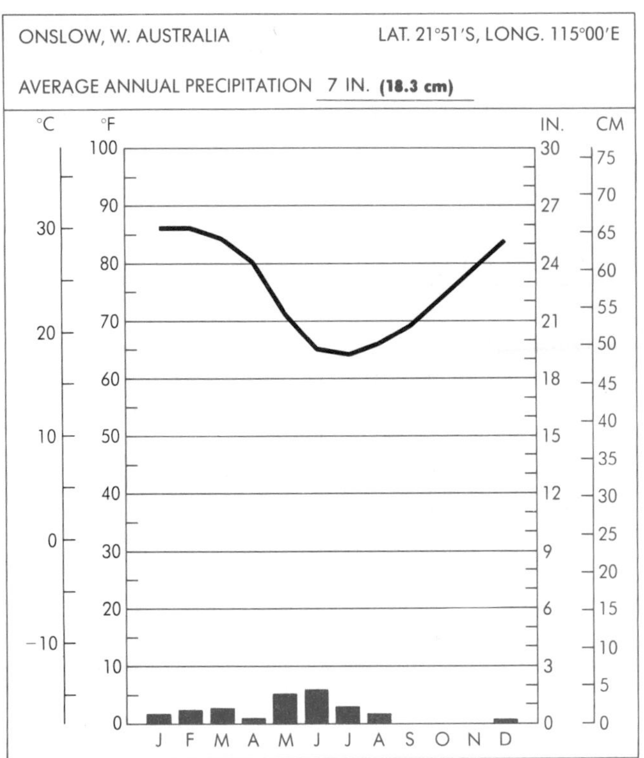

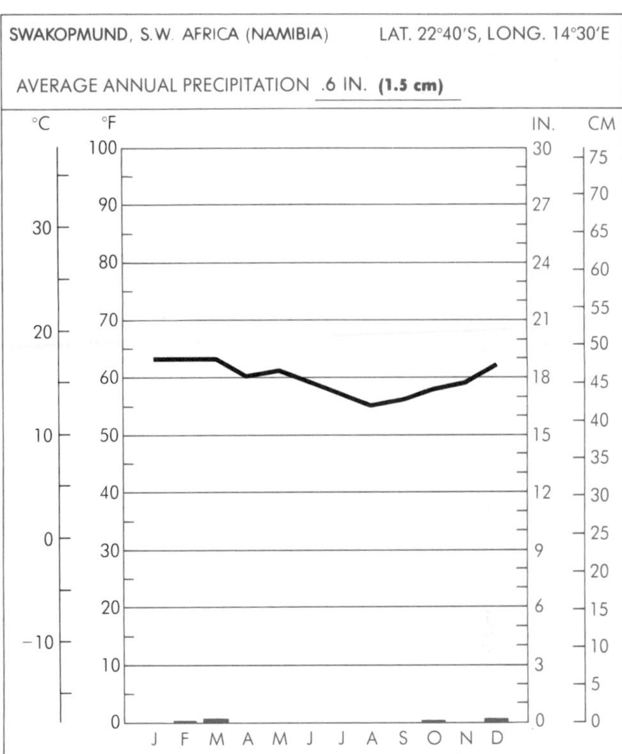

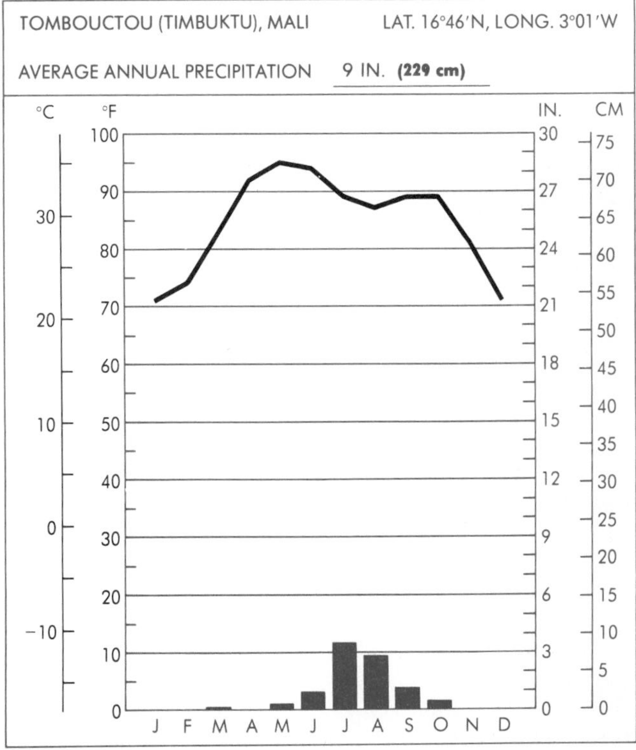

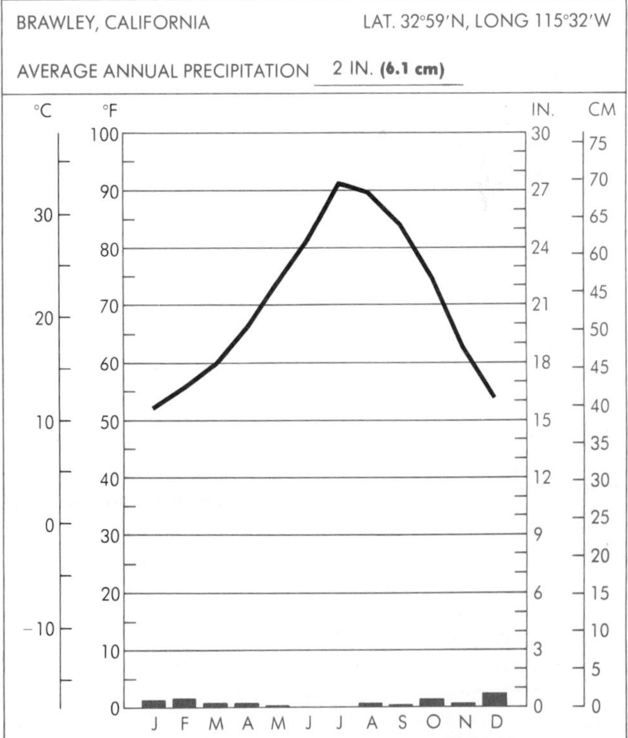

Fig. 9-4. Tropical Dry stations.

These heat extremes occur in the summer when the sun is high; daytime maximums are regularly over 100° F (38° C) for two to three months. In winter the overhead sun is 45° to 50° of latitude away, temperatures are mild, and clear sunny days and crisp nights prevail. Actually, where Tropical Dry locales are easily accessible to populated regions with more severe weather, they have become winter resorts—witness Palm Springs, Phoenix, and Tucson.

The annual range of temperature, then, is sizable, being the largest of any tropical climate. The daily range is even larger. Remember, the control of daily range is moisture in the air. If the atmosphere is heavily charged with water vapor, the radiation of heat from the surface of the earth will be intercepted and absorbed, with the moisture acting as a blanket to hold in the day's heat. Erase this blanket and the daytime accumulation of heat will be radiated off into space in only a few hours after sunset. The Tropical Dry regions with their predominantly dry air will therefore have a very large daily temperature range. Where summer daytime temperatures may go as high as 110° to 115° F (43° to 46° C), they will drop to 60° to 65° F (16° to 18° C) at night. In winter, something on the order of 80° F (27° C) (daytime) and 45° F (7° C) (at night) is typical.

As seen in the climographs (Fig. 9-4), the average for the hottest month is in the neighborhood of 90° F (32° C). This is the mean of both day and night for the entire month. Cool nights act to pull down the daytime temperatures of 100° F (38° C) or warmer to this lower average. Similarly, the coldest month temperature is an average of the more extreme day and night maximums.

Tropical Dry Populations

The Tropical Dry regions do not possess such climatic charms that people have flocked in to live there, and now, having checked some of the details, it is not difficult to see why. Water is the critical factor. Wherever water is available in surface streams, springs, or underground strata, the desert sustains life (Fig. 9-5). A year-round growing season, maximum sunshine, ***and water*** make the oasis agriculturist a small-scale but extremely productive operator.

Traditionally, the oasis has been one of two attractions for people in the desert; the other has been minerals. Given rich mineral deposits, a person will go anywhere and live under the most difficult conditions. On the whole, the deserts are no more productive of minerals than any other climatic region, climate having relatively little to do with the occurrence of ore-bearing formations. Such water-soluble minerals as nitrates and borax could be found at the surface only in dry regions and are usually the result of high evaporation, but the popular image of the desert prospector being more common than prospectors elsewhere is erroneous. Why do people assume great numbers of prospectors go to the desert? Simply because the desert prospector is the

Fig. 9-5. The ancient Saudi oasis of Oatif. Brilliant green date palms dot a dry Arabian landscape. Wherever there is water in the desert, as here, people will gather round it in tight little settlements.

only person out there and thus commands attention. No large numbers of others have willingly selected the Tropical Dry regions as their home. This may not always be the case, however. Air conditioning, new mineral strikes, or cheap ocean water desalinization could mean new opportunities in the desert.

TROPICAL WET AND DRY

Located roughly between the excessively moist Tropical Wet and the arid Tropical Dry are the regions of ***Tropical Wet and Dry*** climate. As the name implies, these regions have characteristics of each of their neighbors.

Located between the too moist Tropical Wet and the arid Tropical Dry are the regions of Tropical Wet and Dry climate.

The map (Fig. 9-1) shows this intermediate location. Examine Africa first, where the regions run quite straight, unimpeded by major mountain ranges or coastal indentations. Between the Congo basin and the Sahara is a narrow band of Tropical Wet and Dry climate broadly designated as the ***Sudan–Sahel***, whereas in the Southern Hemisphere a comparable zone (usually called the ***Veldt***) shows up between the Congo and the Kalahari. In South America, despite the variations brought about by the Andean range and the Caribbean, the same general pattern again occurs. North of the Amazon basin, interior Venezuela and Colombia (***Llanos***), and parts of Central America and the West Indies make up the Northern Hemisphere Tropical Wet and Dry region, whereas most of southern Brazil (***Campos***), Paraguay, and sections of Argentina, and Bolivia represent the Southern Hemisphere. Much of Southeast Asia from South China to and including India is Tropical Wet and Dry, and the entire north coast of Australia and adjacent Indonesian Islands possess this climate type.

Rainfall Seasons in Tropical Wet and Dry

The term *Wet and Dry* places the emphasis on exactly the right element—the annual rainfall seasons. In the middle latitudes, we tend to break the year up on a temperature basis—that is, warm season (summer) versus cold season (winter)—because this is the most prominent seasonal change. However, near the equator where temperature changes are slight, the dramatic difference in seasons is wet versus dry, and the local inhabitants invariably refer to them in this way (Fig. 9-6). How do we explain this phenomenon? The reasons are varied, but generally there are two basic causes: one accounts for the rainfall seasons in Southeast Asia and Australia, and the other for those in most of Africa and Latin America.

Near the equator the dramatic difference in seasons is wet versus dry.

SOUTHEAST ASIA AND AUSTRALIA Let us begin in Southeast Asia. Why does this area have alternating wet and dry seasons? This is the region of the monsoon, the seasonal winds whose characteristic circulation is chiefly responsible for the distinctive rainfall in this part of the world. Whether it is a result of jet streams, upper-air frontal activity, deflected Trades, differential heating of land versus water, or all of these factors operating together, the end-product is a summer rainy season and a winter dry season.

Saturated air drifting onshore from tropical seas each summer readily gives up its moisture on the mountain flanks by convection. The amount varies with distance inland, latitude, exposure, and the like, but in general oceanic air tends to invade the land during the warm months. Colder and drier continental air begins to move offshore with the onset of winter, and although the details vary from place to place, precipitation ceases in large part and the cold season becomes the dry season.

As colder and drier continental air moves onshore with winter, precipitation largely ceases and the cold season becomes the dry season.

In the Southern Hemisphere, but still within the greater Asiatic realm, namely, northern Australia and parts of the African east coast, monsoon circulation also operates as the dominating force. This monsoonal influence, then, produces rainy summers and dry winters throughout a sizable part of the world.

AFRICA AND LATIN AMERICA In the Tropical Wet and Dry regions of Africa and Latin America the essential cause of rainfall seasons is something else again. Recall that as the overhead sun makes it apparent annual migration from Cancer to Capricorn and back again, all the world's standard wind and pressure belts tend to follow. Their change in position lags somewhat behind the sun and the shift is small (on the order of 5° to 10° of latitude), but to regions along the margins of these belts, even such minor changes can

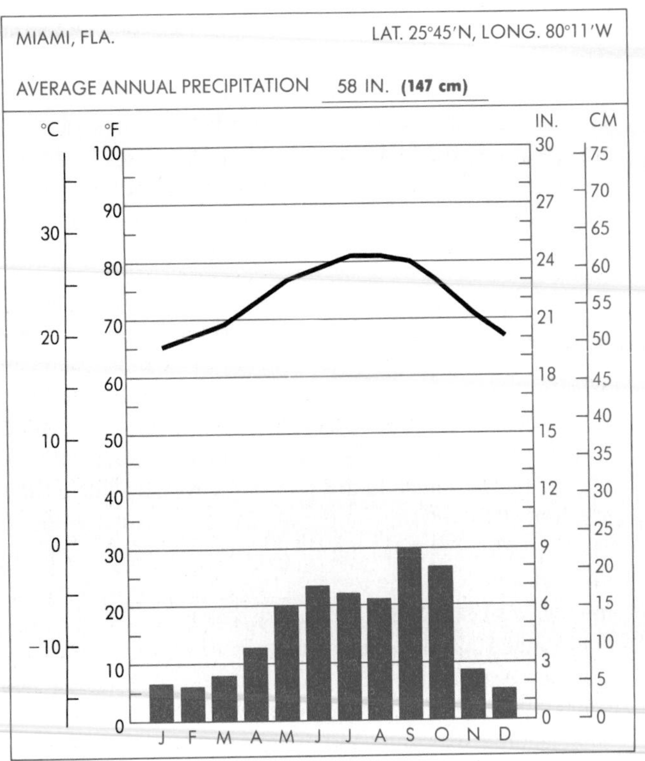

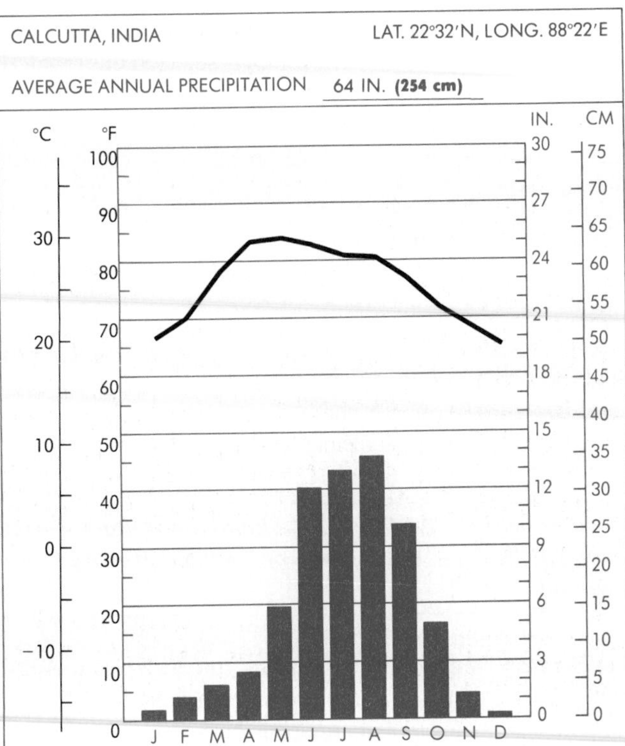

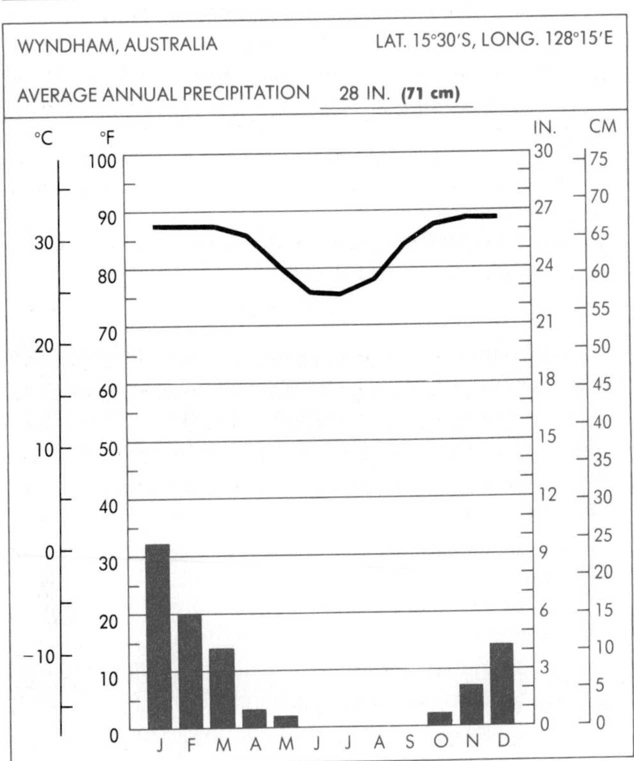

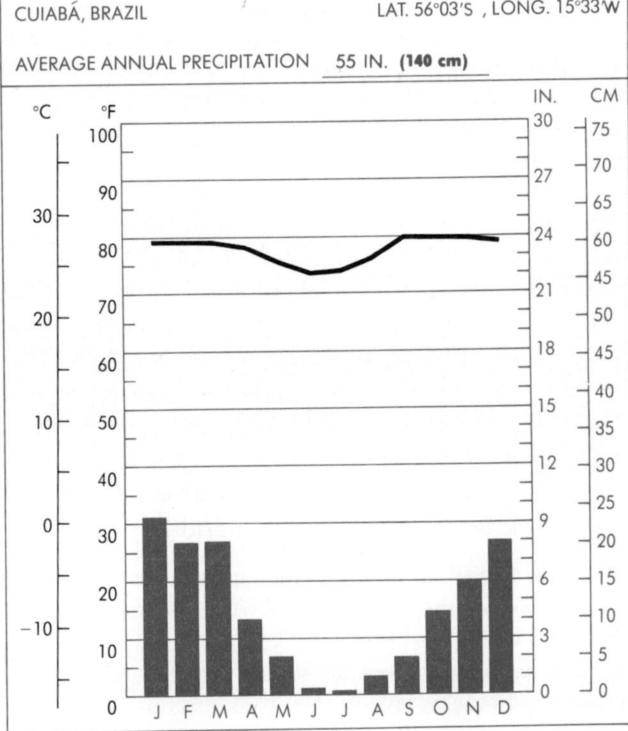

Fig. 9-6. ***Tropical Wet and Dry stations.***

RAINFALL RECORDS

Rainfall Records: Mount Waialeale from Hanalei Valley, Kauai.

The wettest place in the United States is also billed as the wettest in the entire world. Mount Waialeale's observatory on the Hawaiian island of Kauai has been recording rainfall data for 50 years, and during that time the average annual receipt has been 472 inches (1199 cm)—which is almost 40 feet (12 m)! We're dealing here with orographic precipitation in the trade winds. Kauai crests at about 5000 feet (1524 m), and the weather station is on the windward slope not far below. All trade wind islands and coasts display this same tendency, but they must be thrust up to some height. Trades are not likely to give up their moisture over an open sea or a small tropical atoll.

Almost as wet on the average and a good deal more spectacular in its extremes is the 4389 foot (1340 m) Indian hill station, Cherrapunji, where a continuous record has been kept for well over 100 years. The annual average is 451 inches (1145 cm), but in 1861 the total measured was 905 inches (2299 cm) and during the 12-month period from August 1860–July 1861, 1042 inches (2647 cm)—just short of 90 feet (37 m). But the annual average is not a sure thing; only 12 years later a mere 283 inches (719 cm) accumulated.

The remarkable part of the Cherrapunji records is that all the rain fell in the summer, with the four midwinter months bone dry. In this part of the world the monsoon is responsible. Again, the cause is orographic lifting, as the saturated summer monsoon sweeps in off the Bay of Bengal only to run afoul of a Himalayan outlier, the Khasi Hills.

If it were not for the cooling effect of the approximately 4000-foot (1220 m) elevation, these two locations would be classified, respectively, as Tropical Wet (trade wind coast example) and Tropical Wet and Dry (monsoon example). But the lowland below reflects these same rainfall patterns, if not amounts, and the constant heat is unmodified by elevation. Thus all of windward Hawaii is Tropical Wet and the Ganges delta typical Tropical Wet and Dry.

Cherrapunji is the site of a former British tea plantation, and the villagers who reside there do indeed endure these climatic peculiarities. In contrast, the Mount Waialeale station sits in a permanently cloudbound upland swamp, not an environment that would support any kind of permanent settlement. So it is essentially an automated climate observatory. This brings up a problem—should we regard its records as official? The accepted rule through the years has been that somebody has to live there continuously with the official instruments. Otherwise we could blanket the world with automated stations—the top of Mount Everest, central Greenland, or the middle of the Sahara.

We have to admit, however, that there is a lot of moisture at Mount Waialeale, "legal" or not.

---

determine climate. The Tropical Wet climate is almost wholly within the Doldrum zone, and the Tropical Dry is dominated by the Trades and Horse Latitudes, but the Tropical Wet and Dry has no such wind or pressure belt of its own. Instead, it is located astride the boundary separating the Doldrums from the Trades. Thus even slight latitudinal changes of the Trades or Doldrums will affect the Tropical Wet and Dry.

Even slight latitudinal changes of the Trades or Doldrums will affect the Tropical Wet and Dry.

Consider the Northern Hemisphere summer when the overhead sun is at the Tropic of Cancer. The Doldrums, in shifting slightly northward, has lapped over the Tropical Wet and Dry region and is now under the influence of all the factors that give rise to the Tropical Wet climate. At the same time in the Southern Hemisphere the Trades have moved slightly north and are dominating that hemisphere's Tropical Wet and Dry region, bringing Tropical Dry conditions. As the sun moves to the Tropic of Capricorn six months later, this movement is reversed.

In essence, then, the Tropical Wet and Dry zones have no real climate of their own but instead borrow that of their neighbors. The Sudan–Sahel of North Africa, for example, experiences an imported Congo climate in the summer with daily showers and high humidity, whereas in the winter arid Sahara conditions move in. The result is a summer rainy season and a winter dry season very much like that of the monsoon areas. Undoubtedly, this same set of circumstances should prevail in Southeast Asia and Australia except for the local monsoon currents. The monsoon currents are so assertive as to wipe out any traces of the standard wind and pressure system. Climatically, however, the results are the same, and we can logically class all these regions as Tropical Wet and Dry.

Precipitation, Vegetation, and Heat

In determining precisely which areas will be included as Tropical Wet and Dry, it is more important to consider the characteristic distribution of precipitation throughout the year than the total amount. However, if a region receives less than 20 inches (51 cm), it must be classified as Tropical Dry. Any place in the tropics receiving any amount less than 20 inches (51 cm) in one year and experiencing a winter dry season will be Tropical Wet and Dry; there is no limit as to total. Some locations exceed 150 inches (380 cm).

The natural vegetation will closely reflect the total rainfall and the length of wet and dry season. If we were to travel from the northern edge of the Congo to the southern margin of the Sahara, we would observe a regular progression of vegetation. Just north of the tropical rain forest the trees are more widely spaced, and occasional open glades result from the short dry season. Near the desert, where the dry season is dominant, short grass turf and stunted thorn bush grow. In between these two extremes is the savanna, extensive grasslands where the vegetation sometimes reaches a height of 8 feet (2.4 m). All these regions are classified as Tropical Wet and Dry because they have a winter dry season and a total precipitation of more than 20 inches (51 cm).

Temperatures, of course, are generally high in any tropical region, and as might be expected they are closely related to those of neighboring climates. The summer temperatures are similar to those of the Tropical Wet climate. Continually high humidity and afternoon cloudiness combined with 90° F (32° C) days and 70° F (21° C) nights are the average. Heat is greater near the desert margins, and the daily range increases proportionately. Winter, being the dry season, displays temperatures somewhat like those of the desert winters, roughly 75° to 80° F (24° to 27° C) days and 55° to 60° F (13° to 16° C) nights.

Tropical Wet and Dry Populations

The wide range of conditions within the Tropical Wet and Dry regions leads to variations in human occupancy and land use. In the Orient this is a highly productive agricultural region. Some places in southern China and India support as many as 3000 people per square mile; these people depend directly on the land for their livelihood. Paddy rice and the monsoon are highly compatible as long as the rainfall is over 40 inches (102 cm) per year, and the climatic rhythm becomes the pulse of life (Fig. 9-7). Sugar cane, as in the West Indies and India, also adapts to the alternation of wet and dry seasons and allows these areas to produce a cash crop and sustain sizable numbers of people. Sisal and pineapple can also withstand partial drought.

Regions where the natural grasslands have been put to less intensive use as pasture support much smaller populations. The savannas have not proved to be the ideal grazing land that some have supposed them to be. They are isolated, alternately parched and flooded, and plagued by disease. Only a special breed of livestock can survive here, and more often than not such a rugged animal does not produce particularly tender steak. Only in Australia has any sort of commercial grazing proved successful, and even there, despite persistent efforts, losses are high and profits small.

There are always economic possibilities, however. Mineral strikes such as copper in Zambia, iron in Venezuela, and phosphates, uranium, iron, and bauxite in Australia draw large numbers of people and pay for the construction of modern transport, which helps to defeat isolation. New crops and new agricultural techniques may also help. At Humpty Doo in northern Australia, a joint U.S.–Australian venture in large-scale mechanized rice culture has been attempted, although with less than complete success.

In these areas and elsewhere, peanuts, cotton, and other tropical crops have shown varied results. With world population continuing to rise at a rapid rate, the relatively empty portions of the Tropical Wet and Dry do have a potential as a productive home for humans.

CONCLUSION

The three climates of the low latitudes—Tropical Wet, Tropical Dry, and Tropical Wet and Dry—share the warmth of high-angle sunshine but differ greatly in amounts of precipitation. The consistently wet and warm equatorial regions of the Tropical Wet climate have sparse but increasing human populations. As the population grows, more rain forest will fall before spreading agricultural and urban development. The earth's driest and warmest lands are Tropical Dry. Here descending and diverging air keeps skies clear, and daytime temperatures may soar to 130° F (54° C) or more. Human occupation is limited to scattered oases. Tropical Wet and Dry occupies the transition between the extremes. Precipitation tends to come with the high summer sun in Africa and

A

B

Fig. 9-7. ***Paddy rice and the monsoon.*** *(A) The monsoon is the pulse of Asia. Its annual rhythm with assured summer rain signals the teeming millions to assemble at the paddy. Each stalk of rice is hand planted, and everyone participates because everyone eats. (B) Those who worry less about the serious relationship between hard physical labor and minimal food on the table enjoy the carefree lassitude of trade wind beach resorts.*

South America and with the warming continent and monsoon winds in southern Asia. Human use of Tropical Wet and Dry varies from marginally productive grazing of domestic animals to high-yielding rice paddy agriculture. In the following chapter we continue our consideration of world climates with a look at the subtropics.

KEY TERMS

Tropical Wet
Tropical Dry
Sahara
Arabian
Thar
Sonoran
Atacama
Kalahari
Australian
arroyos
wadis
El Niño
Southern Oscillation
ENSO
La Niña
Tropical Wet and Dry
Sudan–Sahel
Veldt
Llanos
Campos

REVIEW QUESTIONS

1. List the climatic characteristics of each of the three tropical climates. How are they similar and how are they different?

2. What is meant by the old saying, "Night is the winter of the tropics"?

3. Why do some Tropical Wet regions appear in the Doldrums and others in the Trades?

4. What is the main source of moisture in the Tropical Wet climate? If the Tropical Wet is so close to the equator, why doesn't it record the world's highest temperatures?

5. Why do many of the world's hottest regions occur on the west sides of the continents? Why are desert nights so cold?

6. How does the seasonal shift of the sun north and south influence rainfall in the tropical climates?

7. In what way is Tropical Wet and Dry a transitional climate?

8. How are the Tropical Wet and Dry climates of Southeast Asia and Australia different from those of Africa and Latin America? What would happen if Africa and Latin America experienced monsoons?

APPLICATION QUESTIONS

1. Photocopy a map of the world and draw in the boundaries of the three tropical climates. Shade each with a different colored pencil.

2. Choose *one* area in each tropical climate and explain why it experiences its particular climatic conditions. Is it typical of its climatic region? Why or why not?

3. Account for the heavy rain at Mount Waialeale and Cherrapunji. How are they different? If they are the two wettest places in the world, what are the two driest? Account for these in the same way.

4. Photocopy the climographs in Figures 9-2, 9-4, and 9-6. Remove place names and practice identifying the climatic region in which each of these areas occurs.

CHAPTER 10

THE SUBTROPICAL CLIMATES

OBJECTIVES

After studying this chapter, you will understand

1. Why the subtropical climates are considered transitional, and how they differ from each other.
2. Where the Mediterranean climate occurs, what its conditions are, and what causes those conditions.
3. Where the Humid Subtropics are, what their conditions are, and what causes these conditions.

Some subtropical climates are humid, some are distinctly lacking in moisture during the summer, but all have long, sunny frost-free periods. In southern Spain, the hardy, drought-resistant olive is planted in groves that cover the hills. Winter rains nourish the trees to bare the fruit that ripens in the hot summer sun.

INTRODUCTION

The discussion of world climate patterns now shifts from the low-latitude tropical climates to those immediately poleward—the ***subtropical climates.*** In a word the subtropical climates are transitional, reflecting many characteristics of both the tropics and the middle latitudes. We would expect them to have a little frost as a limit to agriculture and, at the opposite swing of the seasons, somewhat milder and more "livable" temperatures than are generally found in the tropics. Large numbers of people, happily occupying the subtropics, proclaim that their climate combines the best features of both the tropics and the middle latitudes. They may be right (Fig. 10-1). In the United States we have popularized the term *Sun Belt* with all its positive connotations, to refer to the subtropics and the desert. Our demographers and economists point to a steady migration from the more northerly *Snow Belt* climate.

The subtropical climates are transitional, reflecting many tropical and middle-latitude characteristics.

There are two subtropical climates: (1) the Mediterranean in regions facing out on west coasts and (2) the Humid Subtropic on the east coasts. This opposing coastal orientation introduces some sharp differences in climate despite generally identical latitude.

Fig. 10-1. ***The subtropics.*** *Not as hot as the tropics, not as cold as the middle latitudes, the subtropics are just right for many folks. Not everyone can afford to move there permanently, but many visit and vacation. The French Riviera is representative of world-renowned playgrounds that sell sunshine and "atmosphere."*

MEDITERRANEAN

The west coast subtropical climate gets its name from its typical location along the margins of the Mediterranean basin. Some people have criticized use of the term ***Mediterranean*** for this particular climate in that it is merely a geographical location and, unlike the other climatic names, does not refer to vegetation or some outstanding feature of the climate. Nonetheless, through long use it has become the traditional term, and it does aid the student in location.

Regions of Mediterranean Climate

Mediterranean climatic regions are always west coastal. Because they are subtropical and immediately next to the tropics, they must share a boundary with the Tropical Dry climates. The Tropical Dry are the most poleward of the tropical climates and always appear on west coasts too. In the Old World Mediterranean the west coast immediately north of the Sahara Desert is breached by the Mediterranean Sea and the coast is extended greatly. Thus almost the entire fringe of this inland sea represents the Mediterranean climate.

Mediterranean climatic regions are always west coastal and most share a boundary with the Tropical Dry climates.

Normally, if the continental west coast is straight, regions with a Mediterranean climate will be rather limited, for they must be near the sea and their latitudinal extent is not great. So if there were no Mediterranean Sea, only the coasts of Morocco and Portugal would be "Mediterranean." In fact, however, the Mediterranean Sea multiplies the length of the coastline many times. This region is therefore the largest area of Mediterranean climate anywhere in the world. Involved are the coasts of Spain, southern France, peninsular Italy, Yugoslavia, Albania, Greece, Turkey, the Middle East (Levant) states, Libya, Tunisia, and Algeria. the Sahara reaches the sea from southern Israel to Libya, but elsewhere the Mediterranean climate encircles the Mediterranean Sea and even pushes into the Black Sea coastal areas in Turkey and the Crimea (USSR).

On the west coast of North America, in the same latitudes as the Mediterranean Sea, is another Mediterranean climate region. It includes the California coast from the Mexican border where it meets the Sonoran Desert northward to just beyond San Francisco Bay, and also the greater part of the Central valley. In South America, the central one-third of Chile occupies a similar position, and in South Africa, the southern margin of the Kalahari Desert, in the vicinity of Cape Town, is also Mediterranean. Presumably if Africa were longer, the region would be greater, but South Africa barely reaches subtropical latitudes, and its Mediterranean representative is the world's smallest. An unusual situation exists in Australia. The Great Australian Bight (bay) indents the south coast, presenting two west coasts in these latitudes. Thus Australia has two Mediterranean climate areas, one in the Perth region of West Australia and one involving the Adelaide region in South Australia (Fig. 10-2).

Rainfall Pattern of the Mediterranean Climate

One outstanding feature of the Mediterranean climate is its rainfall pattern. Total annual rainfall is low, averaging about 15 to 20 inches (38 to 51 cm), but it is highly concentrated in the winter whereas the summer is almost absolutely dry. Frequently, the arid season is dominant, reflecting the nearness of the tropical deserts. As in the Tropical Wet and Dry, then, the year is broken up into two distinct seasons on the basis of rainfall, except that this time winter is rainy and summer is dry.

Total annual rainfall in the Mediterranean averages only about 15 to 20 inches, most of it concentrated in winter.

This distinctive rainfall pattern is the result of the wind and pressure systems of these particular latitudes, which shift slightly north/south with the seasons. Wind and pressure systems affecting the Mediterranean regions are

1. The Trades, which dominate the tropical deserts and exert a particularly drying influence on the west sides of continents.

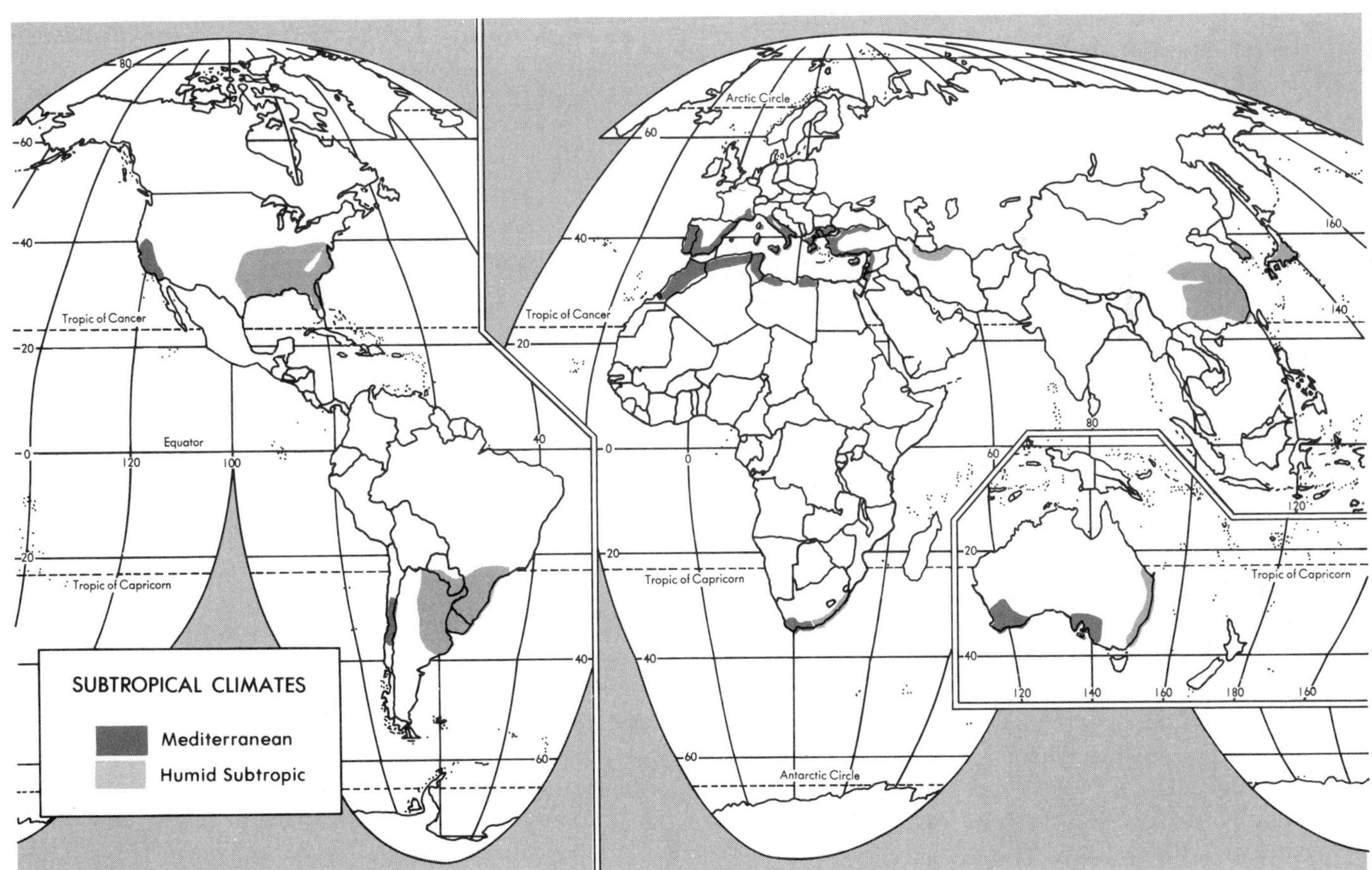

Fig. 10-2. ***Subtropical climates.***

2. The Horse Latitudes, poleward of the Trades, regions of sinking air that warms and dries. They show up especially as permanent cells at sea, as in the Hawaiian High off the North American West Coast.
3. The broad Westerlies poleward of the Horse Latitudes, characterized by their many eastward-moving cyclones following the Polar Front.

As the systems shift with the advance and retreat of the summer sun, they cause the alternating wet and dry seasons.

A reasonably typical area is the California Mediterranean region. In the summer when the sun is overhead near the Tropic of Cancer, the northern fringes of the Tropical Dry climate move northward from the Sonoran Desert to influence California. Offshore the Hawaiian High has also shifted to the north, forcing the storms from the Gulf of Alaska to skirt its poleward edge well north of California. An imported Tropical Dry climate therefore lasts throughout the summer months.

When the overhead sun moves into the Southern Hemisphere, the Hawaiian High tends to follow it south, and the westerly storms from an increasingly active Aleutian Low begin to swing farther south. Not all these storms move across the southern half of California but many of them do, and their low nimbostratus clouds and drizzle rain are typical of winter conditions. Occasionally, cyclones will stagnate (blocked by a combination of mountains and a continental high inland), and heavy precipitation will result. At other times, quite violent storms will develop from maritime polar air detouring far south over subtropical seas and thus becoming increasingly unstable as it is warmed at the bottom. Between storms bright sunny weather prevails, so that a winter in California is not quite the same as a winter in Oregon. Yet frequent cyclones are characteristic, and the full year's rainfall is received during the cold half of the year, often in just a few midwinter months. This pattern is essentially the same on all continents in both hemispheres, and its net result is a Mediterranean winter rainfall pattern on every continent immediately poleward of the tropical deserts.

Mediterranean Temperatures

Because the Mediterranean climate is limited to a narrow coastal zone, the high summer temperatures that should prevail are considerably moderated by oceanic influence. Specifically, the cold current (the same cold current that is found off the coasts of Tropical Dry climates) effectively moderates daytime temperatures. Sea breezes bring cool air over the land, and advection fogs drift ashore from off the current and shield the coastal regions from the direct rays of the sun until almost noon each day.

The high summer temperatures that should prevail in the Mediterranean are considerably moderated by oceanic influence.

Making for even more comfortable living conditions are the cool nights, for summer is the dry season and the accumulated heat of the day radiates off into space very rapidly during cloudless nights. On the average, summer daytime temperatures reach only about 75° to 80° F (24° to 27° C). With their long rainless summers and brilliant sunny afternoons, yet mild daytime temperatures and cool nights, it is small wonder that the Mediterranean regions have long been renowned as resort and tourist areas (Fig. 10.3).

There are exceptions to this pattern. The Central valley of California, insulated by the coast ranges from the cooling effects of the sea, comes close to Tropical Dry summer temperatures. Nights are cool, but summer daytime temperatures average near 100° F (38° C) throughout. Latitude has little effect. For example, Redding in the far north and Fresno in the south display a good deal of similarity. It is distance from the sea that is important. Should the Central valley, then, be classified as Mediterranean climate? In all respects, it is typical except for summer daytime temperatures, and since the only alternative is to establish a separate climatic classification to accommodate it, the valley is usually included with the Mediterranean. It should be remembered that it is an important variation, however.

To a somewhat lesser degree, most of the Old World Mediterranean also deviates in the same way. The Mediterranean Sea lacks a cold current, and, as a landlocked sea, it is warmer than the open ocean. Therefore its cooling influence on the adjacent coastal areas is less effective than elsewhere. Palermo, for instance, in Sicily, is at approximately the same latitude as San Francisco, yet the average temperature for the month of August (both day and night included) is 79° F (36° C), whereas fogbound San Francisco reaches only 60° F (16° C) (Fig. 10-3, Tunis).

Agriculture in the Mediterranean

Short, mild winters and long, warm, sunny summers with virtually a year-round growing season would appear at first to be an ideal agricultural climate. But there is a serious flaw in this agricultural paradise—water. Not only is rainfall inadequate, but it also comes at the wrong time of year. The

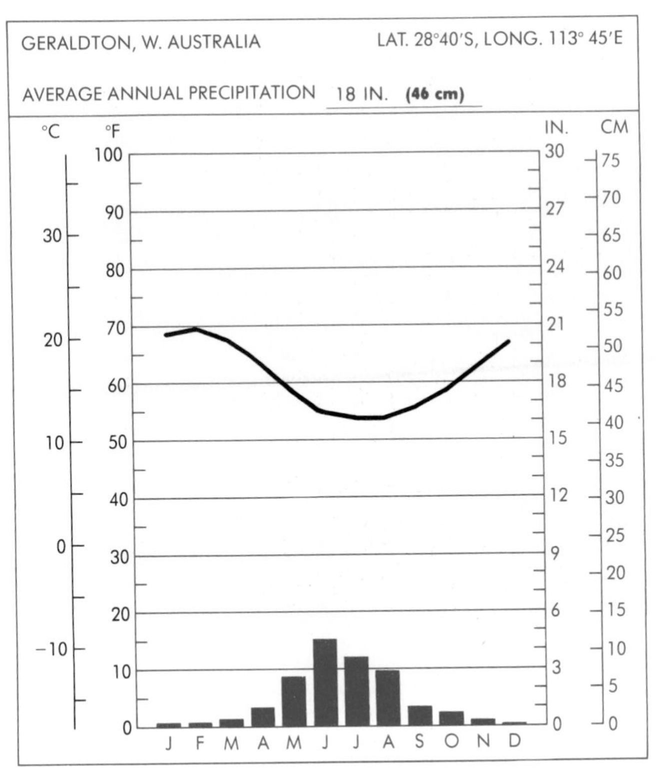

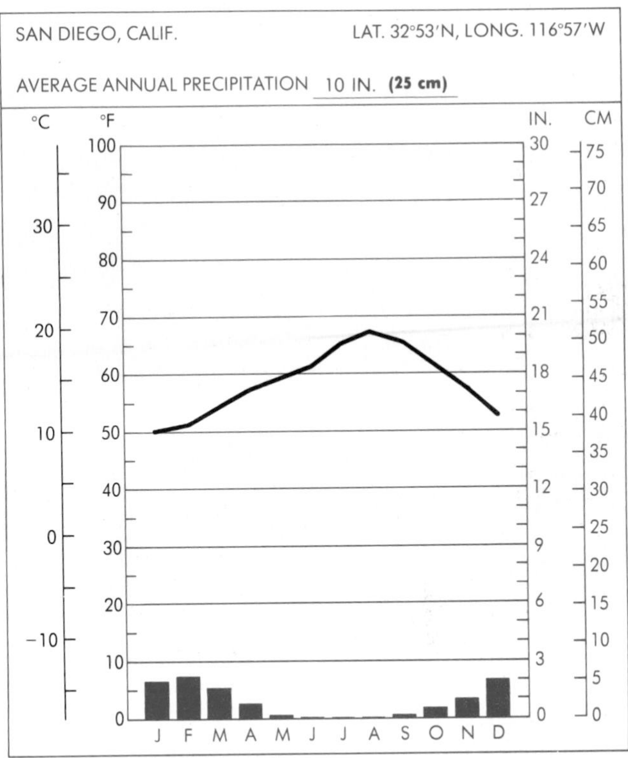

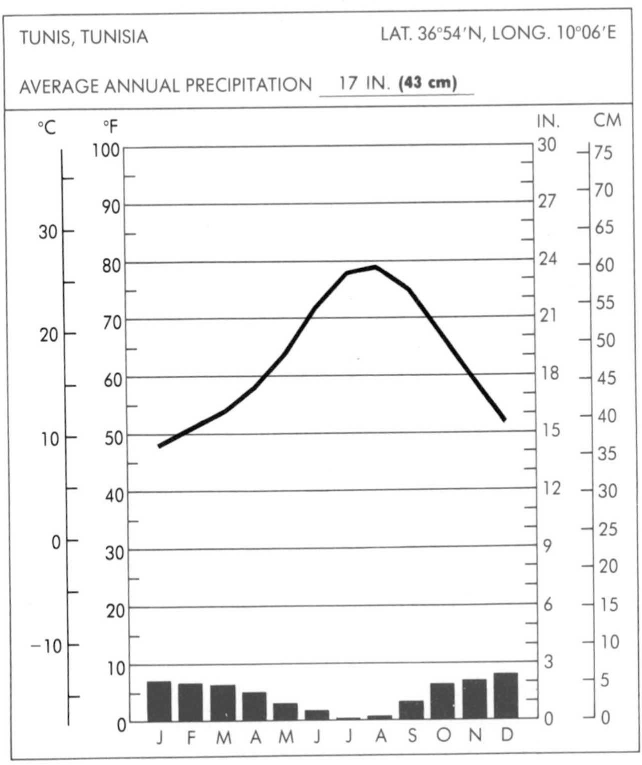

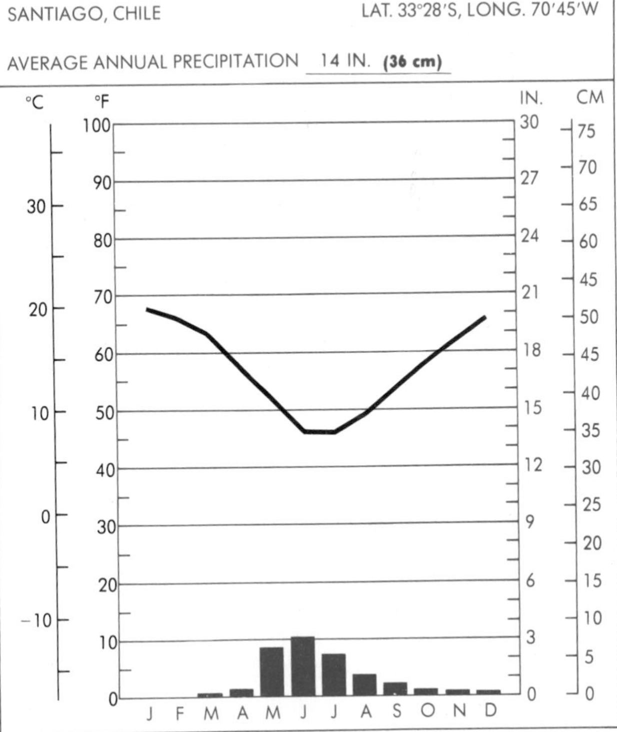

Fig. 10-3. ***Mediterranean stations.***

*Fig. 10-4. **The rolling Tuscan rural country near Siena.** Grapes and wine are the serious business here, although the agricultural scheme involves olives, grains, and vegetables as well. These crops, these methods, and even millions of these people have transplanted themselves very successfully in every Mediterranean locale in the world from South Africa to Chile and California.*

summer growing season is the water-short period, and crops must be limited to those that can withstand prolonged drought. The natural vegetation reflects this condition in its sparseness and the special adaptation of grasses, brush, and occasional trees to survive aridity. Anywhere one travels in the Mediterranean climates, olives, grapes, figs, and wheat are common. These are the standard drought-resistant crops (Fig. 10-4). To a large degree, the rural poverty of the Old World Mediterranean basin stems from a large agricultural peasant population attempting to wrest a living from the land in the face of this climatic defect.

The serious flaw in this agricultural paradise is lack of water.

Given summer moisture, the Mediterranean climate could become highly productive. With irrigation, as in the California Central valley where *exotic streams* (so called because they flow from a moist region into an arid region) are fed by melting Sierra snows, the long growing season and bright summer sun can be great assets. Now the land that had been used only for sparse grazing or low-yielding dry-field crops can produce citrus, cotton, sugar beets, and a seemingly endless range of profitable products. A disadvantage of the Mediterranean climate is that the normal river, reflecting the seasonal rainfall, carries water only during the winter. Yet even here, given sufficient capital, a dam could be erected to save the winter's water in a reservoir for the following summer. The ideal feature, of course, would be a nearby mountain range supporting permanent snow, as in California and Chile. The melting snows of summer can charge exotic streams with abundant flow at exactly the right season. Many would claim that the Mediterranean climate is the world's finest, but without irrigation, it is far from the finest climate for agriculture.

Mediterranean Winters

Winters are moderate in the Mediterranean. Frosts are not unknown, but temperatures seldom drop much below 30° F (−1° C), and then only on infrequent winter nights. Such mildness might seem to be expected in the subtropics, but Northern Hemisphere locations in the interior of the continents at equivalent latitudes experience much colder temperatures because they are invaded by outward-moving continental polar air masses. In North America, the Sierra Nevada forms a higher barrier against such cold air; comparable mountain chains ringing the northern Mediterranean basin are equally effective in blocking out continental air.

SIERRA SNOW

The Sierra Nevada (Snowy Range) stands as a massive barrier along the entire 350 mile length of California's Central valley. There are no natural passes across its 8 to 10,000 foot serrated crestline as the gold-seeking 49'ers and the Donner party (a westbound group of 87 settlers, 40 of whom froze to death while attempting to cross the range in the winter of 1846–1847) discovered long ago. It is also a major obstruction for Pacific storms and oceanic air currents as they attempt to invade the continent from the west. Orographic precipitation is the result. Most of it occurs in the winter during the active storm season, and a large percentage of that is in the form of snow at the higher elevations.

These are not dainty, hesitant, tentative little flurries. The Pacific air flow is relatively warm, always saturated, and extremely persistent, so that when its only alternative is to rise well above the freezing point, snowfall is both inevitable and imposing. A total of 68 inches (152 cm) fell one day in Sequoia National Park, and the *mere average* for a year in little Alpine Country, just south of Lake Tahoe, is 450 inches (1143 cm). If these numbers are impressive, consider the all-time annual record at Alpine County's Tamarack [8000 feet (2438 m)] of 884 inches (3023 cm). That translates into 74 feet (23 m); not many places in the world have ever exceeded such a fall.

So here, poised above the hot and arid Central valley, is a treasure of stored water, glistening in the sunlight and melting at an accelerated rate with the advent of summer. Down the long western flank of the mountains plunge the vibrant rivers, cutting huge canyons as they go to feed the Sacramento in the north and the San Joaquin in the south. This is the dry season's irrigation water, tapped to transform the Mediterranean climate from an agricultural liability into a productive miracle.

Continued successful irrigation depends on ***water control***. This process begins each year as teams of specialists in the mountains sample snow depth to determine total runoff for the ensuing season. Dams at critical points on each river hold or release water on demand. This procedure works nicely most of the time until a radically aberrant year arrives—the winter of 1982–1983 was one of these.

The early Spanish, finding the Central valley to be alternately an endless marsh and a parched desert, retreated to the more benign coast. In 1982–1983 saturation and quasi-swamp conditions arrived prematurely with a record-breaking rainfall season; then came the melting snow threatening the control dams. Disaster, partially averted by a cool early summer, arrived that year in the same liquid form as the usual salvation. Those old Spanish were more clairvoyant than they knew.

Although frosts are not unknown in the Mediterranean, temperatures seldom drop below 30° F (−1° C).

Occasionally, cold air will spill over a low spot in the mountains and flow, often at high velocity, down the Mediterranean slope. The dreaded ***Bora***, at the head of the Adriatic, is a gravity or ***katabatic*** wind of this type originating in the Danube basin. As it loses altitude, it is warmed adiabatically so that when it arrives at the sea, its temperature is a great deal warmer than when it started. It still seems cold relative to the normal temperatures and may be well below freezing, but this phenomenon is rare and winter averages are usually on the order of 50° F (10° C). In the Southern Hemisphere, continental cold air masses in the middle latitudes are unknown, but mean monthly temperatures may still dip below 50° F (10° C) in some Mediterranean climate regions there.

HUMID SUBTROPICS

On the east coast of every continent, almost opposite the Mediterranean regions, is the ***Humid Subtropic*** climate. It, too, is on the margin of the tropics and shares a boundary with a tropical climate—in this case, the Tropical Wet and Dry (Fig. 10-2).

The Humid Subtropic is also on the margin of the tropics and shares a boundary with the Tropical Wet and Dry.

Regions of the Humid Subtropics

In the United States, the Humid Subtropics are located in the southeastern states. Unlike the Mediterranean, this climate is not confined to the seaboard and there are no limiting high mountain ranges. It therefore covers a sizable area, including all the so-called Old South—from central Texas to the Atlantic and from the Gulf to a line roughly along the Ohio River to the Chesapeake Bay. Comparable to this area on the east side of the Eurasian continent is a large part of central China around the Yangtze River, and southern Korea and Japan. The same pattern holds in the Southern Hemisphere where southern Brazil, Paraguay, Uruguay, and the Argentine Pampas make up a large contiguous area. In contrast, in South Africa and Australia, mountain ranges parallel the coast and limit the Humid Subtropics to a narrow strip. The representative area in Africa is very small for this reason and also because Africa scarcely pushes into the subtropics. In Australia the coastal corridor extends from near Brisbane in the north to almost the southern tip of the continent.

Moisture in the Humid Subtropics

The name *Humid Subtropics* points up the major difference between this climate and the Mediterranean—moisture. The annual total is roughly double that of the Mediterranean, averaging 40 inches (102 cm), and there is no dry season.

Here we are dealing with the identical wind and pressure systems that occur on the west coast, but their effect is different. The Trades, for instance, which in the summer contribute to west coast aridity, are onshore in the east, coming off a warm sea and crossing a warm current. Nor do the Trades migrate far enough poleward to affect the entire subtropics. The general flow of air into the summer continental low brings moist air off the sea. In the Far East, we use the term *summer monsoon* to describe this phenomenon. However, all continents, even those in the Southern Hemisphere, have a summer monsoonlike tendency and suck in air from offshore. This air is heated by conduction, and the typical warm-season precipitation comes in the form of frequent afternoon convectional thundershowers. Heat plus moist air equals convectional precipitation, and the Humid Subtropics achieve this equation each summer.

Late summer/early fall precipitation totals are augmented by the typhoons/hurricanes that affect these latitudes except in South America. The coastal regions especially receive moisture from these storms, which are in their general vicinity. So, while the west coast Mediterranean is experiencing a lengthy dry season, the east coast Humid Subtropics are receiving significant rainfall. Often, as in the Far East where the monsoon and typhoon are well developed, the annual rainfall has a definite summer maximum.

Typhoons/hurricanes add to the late summer/early fall precipitation totals in the Humid Subtropics.

Winters in the Humid Subtropics

Winters in the Humid Subtropics are much like those in the Mediterranean. Polar Front cyclonic disturbances swing farther equatorward with the retreat of the overhead sun, and frontal activity gives rise to gray skies and drizzle. Only in the Orient is there a variation on this pattern, for storms spawned in the North Atlantic do not always survive the trip across Asia in the winter. New storms generated along the Polar Front near the upper Yangtze valley somewhat compensate for this lack of moisture, but generally winter precipitation fails to match that of the summer.

In winter, Polar Front cyclonic disturbances swing farther equatorward with the retreat of the overhead sun.

In the Northern Hemisphere some of this winter precipitation is in the form of snow as the great continental cold air masses surge out of the interior. They do not dominate the Humid Subtropics by any means, but cold spells occasionally occur. Where there are no mountain barriers at right angles to the flow of Canadian air in North America, cold can invade all the way to northern Florida and the Gulf for short periods most winters. In addition, more northerly locations such as Virginia and Kentucky receive several minor snowstorms every year. Asiatic Humid Subtropic regions, though protected to a greater degree by high mountains, are subject to a more vigorous winter monsoon pushing Siberian air far out of the interior. With few exceptions, then, periodic snow and cold are encountered everywhere in these regions. Tokyo, Shanghai, and Seoul all have several cold spells each year with snow a regular winter feature.

With few exceptions, periodic snow and cold are encountered everywhere in these regions.

The average temperatures for the winter months, as shown on the climate charts (Fig. 10-5), do not always reflect the few days of freezing weather each month, for they are

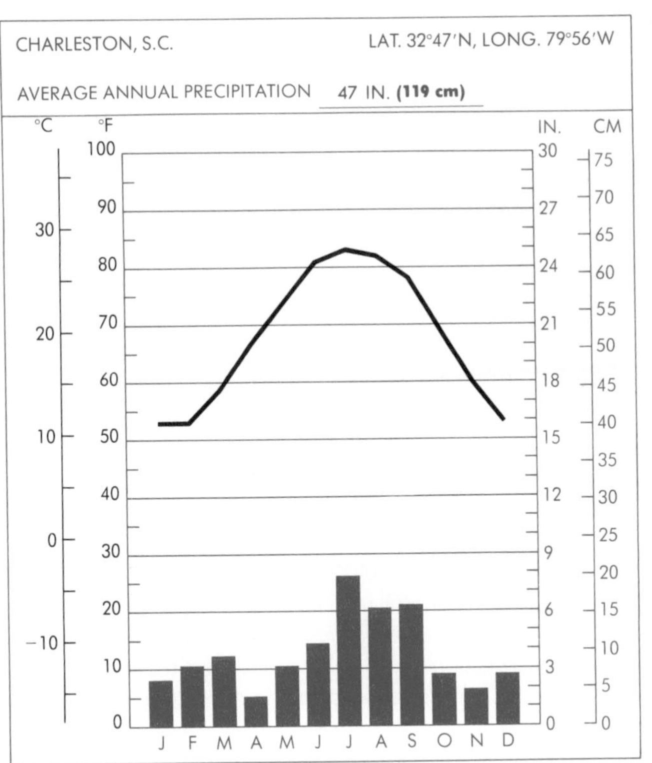

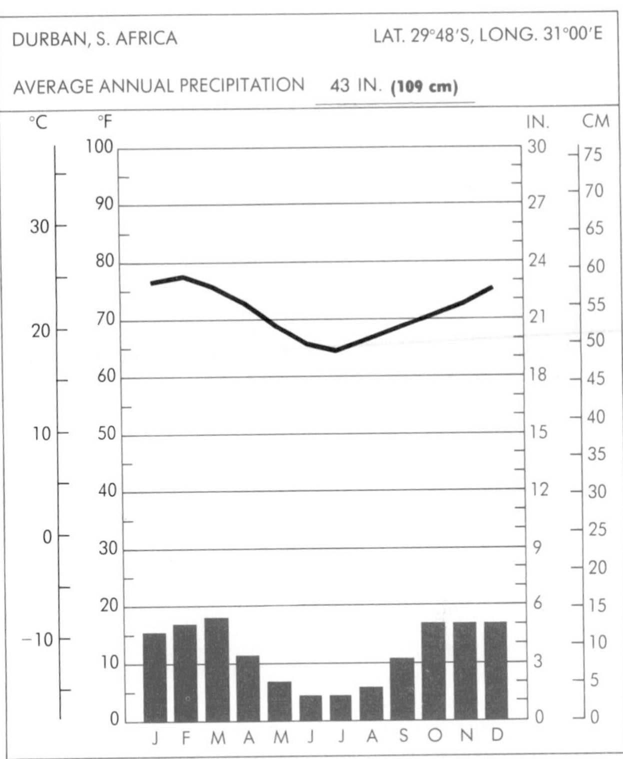

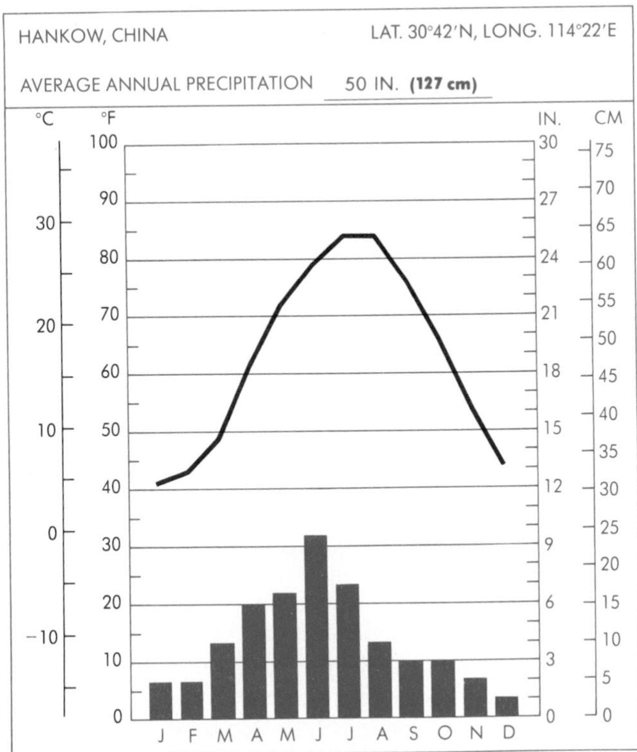

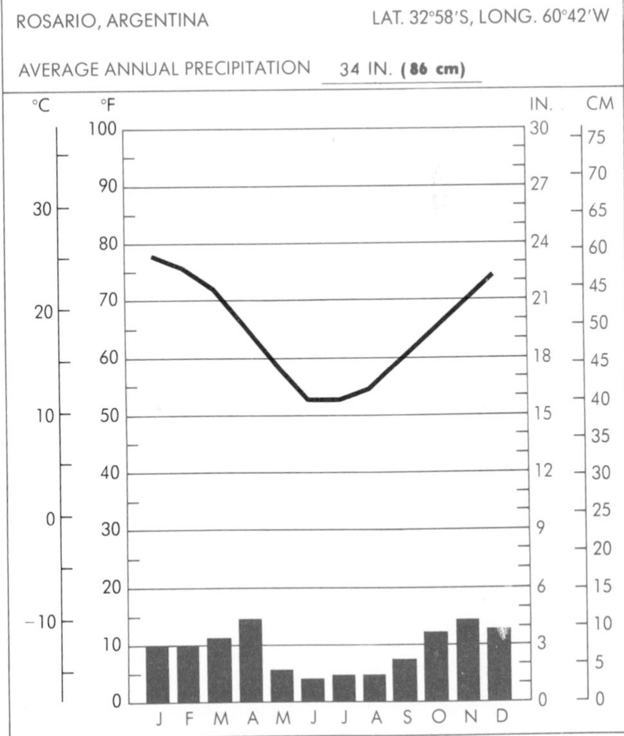

Fig. 10-5. ***Humid Subtropic stations.***

offset by many relatively mild days. When we compare the averages of Northern Hemisphere stations with those of the Southern Hemisphere, where there are no large middle-latitude continents to develop cold air masses, there appears to be no great difference. The cold spells should be recognized, however. An excellent example of their effect in limiting agriculture is the difference between the northernmost limit of frost-touchy, commercially grown citrus in Florida and in southern California. The lake country of central Florida is the heart of the industry in the Humid Subtropics with the northern limit in the area of Jacksonville. In Mediterranean California sizable orchards are common in the San Joaquin valley at least as far north as Fresno. This is 6° to 7° of latitude north of the Florida citrus-growing region, which cannot match the California northward extent because of killing frosts. Yet the coldest month average for both the Mediterranean and the Humid Subtropic regions in general is much the same—in the vicinity of 45° to 50° F (7° to 10° C).

Summers in the Humid Subtropics

Summers are hot in the Humid Subtropics. They are several degrees warmer than the same latitudes on the west coast, for there is no cooling current offshore and no fog. Moreover, the humidity is constantly high because maritime tropical air is dominant, making moderate temperatures feel uncomfortable, even at night. New Orleans, for instance, features four midsummer months averaging close to 80° F (27° C) and almost daily convectional showers. These conditions are comparable to Tropical Wet conditions, and without air conditioning it can be a difficult season. Days are often in the 90s (F) (32° to 38° C), and the hottest month average in the Humid Subtropics is seldom below 75° F (24° C). Florida beaches may be very pleasant as the local land and sea breezes keep the air moving, but only a few miles inland, it becomes excessivley hot and sticky.

With no cooling current offshore and no fog, summers in the Humid Subtropics are hot and constantly humid.

Agriculture in the Humid Subtropics

The Humid Subtropics climate differs in several ways from that of the Mediterranean:

1. Greater total precipitation distributed more evenly throughout the year.
2. Higher summer temperatures.
3. Similar mild winter temperatures, but interspersed with occasional cold spells in the Northern Hemisphere.

While this climate is less comfortable than the Mediterranean, it is basically much more productive agriculturally. A long growing season combined with adequate rainfall in midsummer is the chief advantage, and because the summer precipitation comes from short afternoon showers, days remain bright and sunny. Everywhere in the Humid Subtropics, people are engaged in agriculture.

A long growing season combined with adequate rainfall in midsummer is the chief agricultural advantage.

Intensive rice culture supports a high population throughout the Far East, and farming has been a way of life for generations in the U.S. Cotton Belt. Weaker population pressures have led to more extensive land use in South America and Australia, yet the agricultural potential remains. Where excessive slope or soil inadequacies are a problem, forests often make the land productive and, once cut, regenerate themselves rapidly under the influence of a benign climate. Active reforestation of former cropland is widespread in the United States. The 13 states of the old Confederacy, taken together, produce more wood products today than any other section of the country (Fig. 10-6).

CONCLUSION

Two strikingly different climates occupy similar subtropical latitudes. Mediterranean regions, on the western continental margins, have a split personality when it comes to climate. In the summer, Tropical Dry conditions invade with clear skies, warmth, and drought. Winters bring an invasion of the cool and moist midlatitude west coast conditions complete with frequent cyclonic storms. Agriculture requires summer irrigation, but water is usually available from the melting snowpack in nearby mountains.

Humid Subtropic regions are well watered throughout the year with precipitation from midlatitude cyclones in winter and rain from monsoon-generated convective storms in summer. With the moisture added from passing hurricanes, the result is an excellent climate for agriculture. Both climates have mild temperatures as the latitude would suggest, although the Humid Subtropics are prone to an occasional cold chill from polar outbreaks.

From our discussion of these two transitional climates, we now turn to the midlatitudes where tropical and polar air masses battle for supremacy and the influence of the continental masses is highly significant.

Fig. 10-6. ***Tree farming.*** *Conifers grow rapidly in the Humid Subtropic climate. Here slash pine from a Georgia tree plantation is loaded for transport to the mill as the first step in its conversion to paper.*

KEY TERMS

subtropical climates	***exotic streams***	***katabatic***
Mediterranean	***water control***	***Humid Subtropic***
	Bora	

REVIEW QUESTIONS

1. Why are Mediterranean climates always located on the west coasts of continents and Humid Subtropics on the east coasts?

2. In what way are Mediterranean and Humid Subtropic climates transitional?

3. What causes the distinctive rainfall pattern of the Mediterranean climate?

4. Why do regions with Mediterranean climate *not* have extremely high temperatures?

5. Explain why subtropical regions are so attractive for human habitation and recreation.

6. How has the presence of nearby high mountains influenced agricultural development in regions of Mediterranean climate?

7. Why do regions of Humid Subtropic climate cover more area than regions of Mediterranean climate?

8. Why do the Humid Subtropics have more moisture than does the Mediterranean climate?

9. Why can citrus growers in California plant farther north than citrus growers in Florida?

10. Why are summer temperatures hotter and winters milder in the Humid Subtropics than in the Mediterranean climate?

APPLICATION QUESTIONS

1. Describe how the climographs in Figures 10-3 and 10-5 display the differences between the Humid Subtropics and the Mediterranean climate.

2. Suppose you are a travel agent and you have been assigned to plan two trips for a round-the-world traveler—one with three stops *only* in the Mediterranean climate and one with three stops *only* in the Humid Subtropics. Where and when would you have your traveler go in order to experience a wide range of scenery and local customs in the greatest comfort possible? Write up an itinerary describing the types of conditions your traveler can expect to encounter at each stop.

3. Review your notes on the tropical climates. Which tropical climate is *most like* the Mediterranean? Why? How do they differ? Which tropical climate is *most like* the Humid Subtropics? Why? How do they differ?

CHAPTER 11

MID-LATITUDE CLIMATES

OBJECTIVES

After studying this chapter, you will understand

1. Where the Marine West Coast climate exists, what its characteristics are, and what causes these characteristics.
2. Where the Middle-Latitude Dry climate exists, what its characteristics are, and what causes these characteristics.
3. Where the Humid Continental climate exists, what its characteristics are, and what causes these characteristics.
4. How and why we distinguish between Humid Continental–Long Summer and Humid Continental–Short summer.

Contour planting of corn and alfalfa on this farm in northwestern Illinois retards the erosion of valuable soil during summer thundershowers and spring snowmelt. Farmers and others who live in the middle-latitudes must contend with a variety of climates including soggy marine west coasts, dry rainshadow deserts, or the hot and cold of continental interiors and east coasts.

INTRODUCTION

We now turn to the climates that occupy the broad belt between the subtropics and the high latitudes—the zone of the Westerlies. Here we encounter three climates: (1) Marine West Coast, (2) Middle-Latitude Dry, and (3) Humid Continental.

The basic characteristic of these climates is strong seasonality. The middle-latitude climates are most widely represented in the Northern Hemisphere where the continents are larger than they are at these latitudes in the Southern Hemisphere (Fig. 11-1).

MARINE WEST COAST

The climate of the entire west coast of every continent in the middle latitudes is called ***Marine West Coast.*** It is the third in a series of west coast climates stretching from well within the tropics to the poleward limits of the middle latitudes. It shares a common boundary with the Mediterranean at the margin of the subtropics, which, in turn, merges into the Tropical Dry.

The climate of the entire west coast of every middle-latitude is called Marine West Coast.

Regions of the Marine West Coast

Chile provides an excellent yardstick for illustrating this series. The northern one-third of the country is Tropical Dry, the central one-third Mediterranean, and the southern one-third Marine West Coast. Of all the continents, only Africa misses out, for it does not push into the middle latitudes in either hemisphere. Australia barely does: only Tasmania

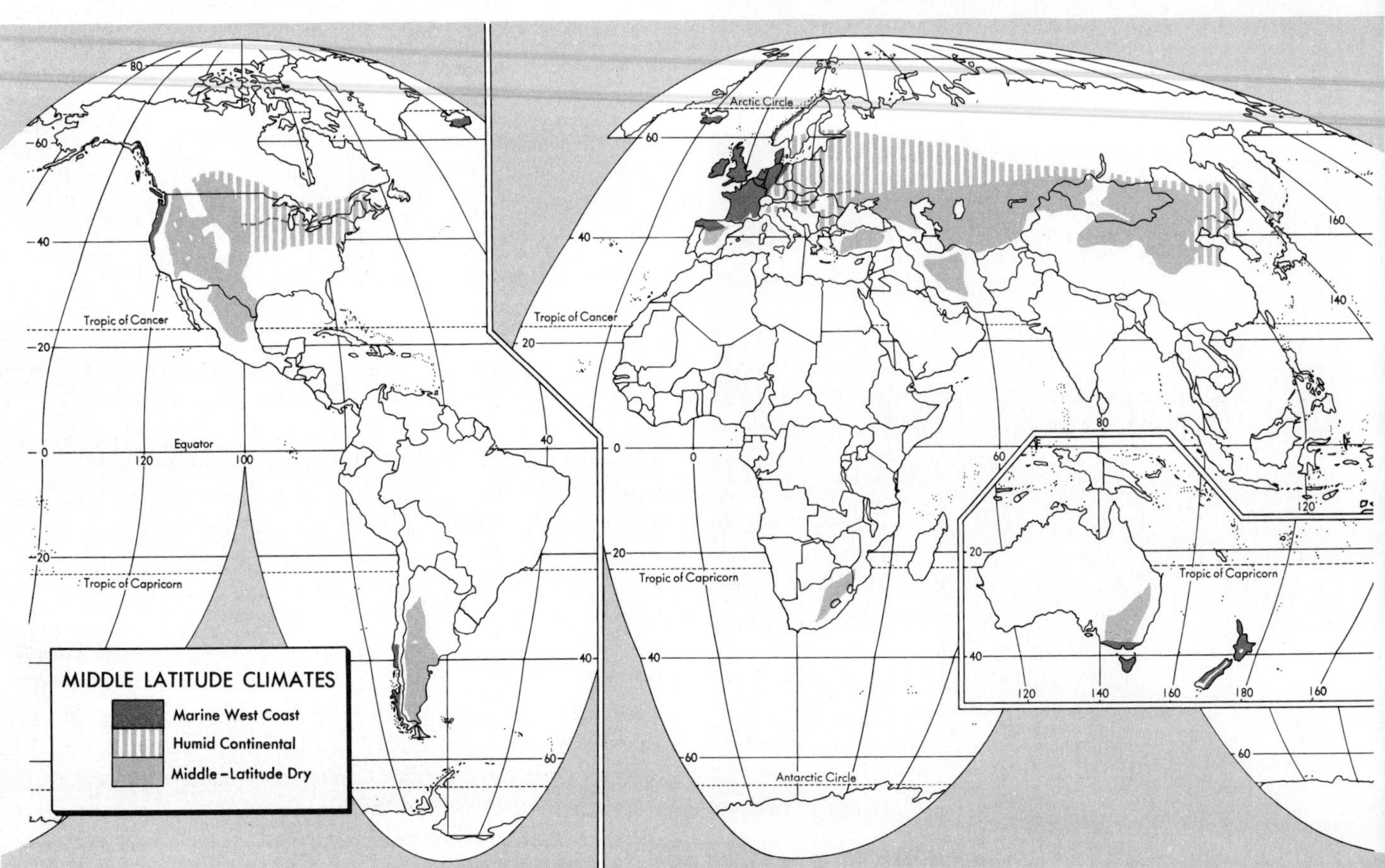

Fig. 11-1. ***Middle-latitude climates.*** *Position of the "crop line," shown by dashes, denotes the transition from long to short summers in the Humid Continental.*

and the southern tip of Victoria in the vicinity of Melbourne are represented. New Zealand, in much the same latitude, can be included with Australia.

The Northern Hemisphere, where the greatest land masses are in the middle and high latitudes, exhibits the most extensive regions of Marine West Coast climate. Along the west coast of North America, this climate extends from northern California to southern Alaska. The belt is narrow, between the fringing mountains and the sea, but a full 20° of latitude are involved. Similarly, in Western Europe the entire coastal zone, from northern Spain to just short of the Arctic Circle in Norway, is classified as Marine West Coast. Included here are northern Spain, western France, the Benelux countries (Belgium, Netherlands, and Luxembourg), the North Sea coast of Germany, Denmark, southern coastal Norway and Iceland, and, of course, Great Britain and Ireland. However, unlike North America, Western Europe has no mountain range closely paralleling the seaboard (except in Norway). The inland climatic boundary is therefore much more difficult to determine. As the word "marine" implies, regions of this climate must be close to the oceans. Where no mountain barrier separates marine from continental influences, as in West Germany and France, we must recognize a zone of subtle transition.

The Northern Hemisphere exhibits the most extensive region of Marine West Coast climate.

Warming Influences in the Marine West Coast

Probably the most striking feature of the Marine West Coast climate is the mildness of the winters at these relatively high latitudes. The average temperature of the coldest month is typically above 32° F (0° C)—often in the low 40s (F) (4° to 7° C). The growing season (the number of days between the last frost in the spring and the first frost in the fall) averages well over 200 days. At such locations as Dingle, Ireland, and Cape Flattery, Washington, it lasts over 300 days. This climate is almost tropical. In the southeastern United States 200 days is regarded as the minimal limit for growing cotton. Therefore, other things being equal, subtropical crops such as cotton could be grown in, say, British Columbia. Other things are not equal, of course, but it is nonetheless remarkable that such mild temperatures should prevail.

Freezing temperatures are not common, but they occur every winter and snow is not unknown. The snow nearly always comes in the form of short-lived flurries, and the fall melts rapidly, especially in immediate coastal areas. Compared with North Dakota or central Siberia at exactly the same latitude, however, the Marine West Coast is a virtual hothouse.

In the Marine West Coast, the average temperature of the coldest month is typically above 32° F.

The oceans are responsible for this moderate climate. These great bodies of water to the west change their temperature only slightly from season to season, and the air masses above them reflect this moderation. Moreover, the prevailing winds in the middle latitudes are from the west, constantly bathing the west coasts with oceanic air. These keep the excessively cold continental air masses of the interior at bay. If reinforced by protective mountain chains, as in North America, or if cold continental polar air is lacking, as in the Southern Hemisphere, the west coasts display very moderate temperatures indeed. Western Europe is the least typical in that Eurasia develops the largest and coldest winter air mass in the world (with the possible exception of Antarctica), and there is no continuous mountain barrier to block its occasional invasion of the west coast. Even here, however, the invasion of cold continental air masses is relatively rare.

Marine West Coast climates are moderate because the temperature of the oceans to the west changes only slightly from season to season.

Another definite warming factor, though probably less important than the general oceanic influence, is the warm current that parallels these west coasts (Map IV). The warm current is now far from the low latitudes and is warm only in relation to the colder waters around it. Nevertheless, it retains some of its residual tropical heat. This heat exaggerates the already moderating influence of the oceanic air masses, adding perhaps 3° to 5° F (2° to 3° C) to the winter temperature average. In the North Atlantic, where the branch of the warm current follows the Norwegian coast into the Arctic Ocean, ports are kept ice free all the way to Murmansk, USSR, and the Marine West Coast climate is pushed northward almost to the Arctic Circle (Fig. 11-2).

In the winter, then, the sea is the source of heat and the continent the source of cold. The moderation of this middle-latitude climate is directly related to its nearness to the sea. Cold-season isotherms depart radically from their normal east/west trend on crossing west coasts, especially in the Northern Hemisphere, and align themselves almost north/

Fig. 11-2. ***Ice-free ports.*** *This photo could be in Alaska, Norway, or Chile. The latitude is 50° to 60° from the equator, yet the little harbor is ice free. There is not much physical space here for building a town at the foot of steep mountains, but there is a valuable resource to exploit. The same warm current that discourages icing causes vertical turbulence and as it invades colder water, supports an abundance of plankton and a numerous fish population.*

south (Fig. 4-20). In this part of the world, latitude is less important as a temperature control than coastal orientation.

Summer temperatures are cool, the hottest month usually averaging in the 60s (F) (6° to 21° C). In this season the imported oceanic air masses cool the region. Although not as striking as the winter warming, this summer cooling is significant when compared with summer temperatures in the interior of the continent.

The general tendency of the sea to maintain much the same temperature the year round is reflected in the limited annual range of the Marine West Coast climate. It is no more than 15° F (8° C) in many places and seldom greater than 20° to 25° F (11° to 14° C)—about the same as the average annual temperature range in the Mediterranean climate. Even some tropical locations have a range greater than that. Such a lack of seasonality is one of the chief features of the Marine West Coast climate and is particularly startling when encountered in the middle latitudes where strong seasonal differences are to be expected (Fig. 11-3).

Precipitation in the Marine West Coast

Despite its pleasant temperatures, the Marine West Coast regions have not rivaled the Mediterranean in attracting tourists and resorts. The reason is their reputation for cloudiness, frequent rain, and a generally dark and gloomy aspect—a reputation that is not wholly unearned. A narrow coastal strip with prevailing winds onshore experiences a good deal of precipitation, particularly if it is backed up by high mountains forcing saturated oceanic air masses aloft.

A narrow coastal strip with prevailing winds onshore experiences considerable precipitation, especially when high mountains force saturated oceanic air masses aloft.

In many places the mountain slopes receive nearly 200 inches (508 cm) each year. Even where there are no mountains, persistent rainfall results from the frontal activity of the many cyclones in the Westerlies. This rainfall is usually in the form of light but constant drizzle, a good deal of low-hanging cloudiness, and mist and fog interspersed with actual rain. The end-product is a great many rainy days each year but less total precipitation than might be expected.

London, for example, enjoys a reputation as a very damp and drippy city. Most days are rainy, the ground is nearly always wet, the air is continuously close to saturation, and umbrellas and raincoats are in order for all seasons. Yet London receives only 25 inches (64 cm) of rain each year. In some parts of the world, this climate would be considered semiarid. Much of London's rainfall is in the form of drizzle or mist, however, and so about 200 rainy days are needed to produce these 25 inches (64 cm). Because of the saturated air and high rate of cloudiness, evaporation is kept to a

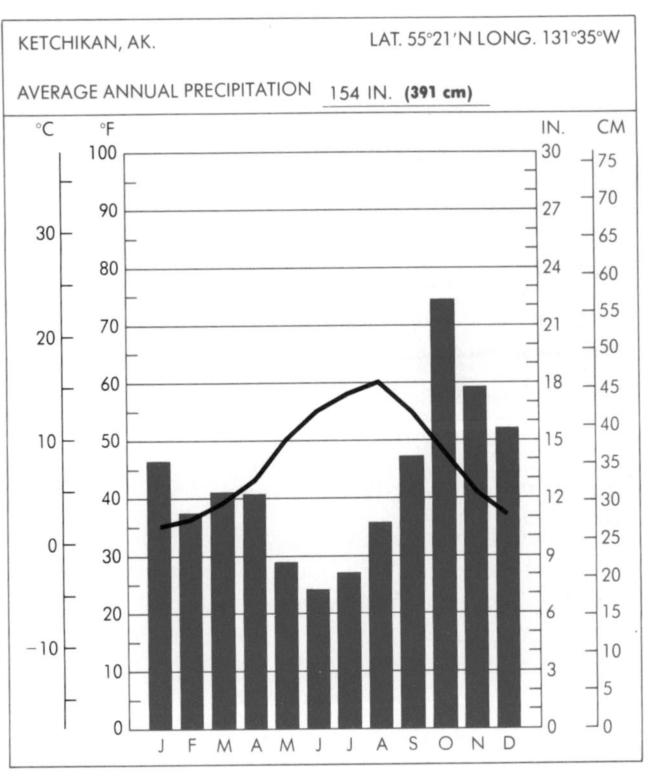

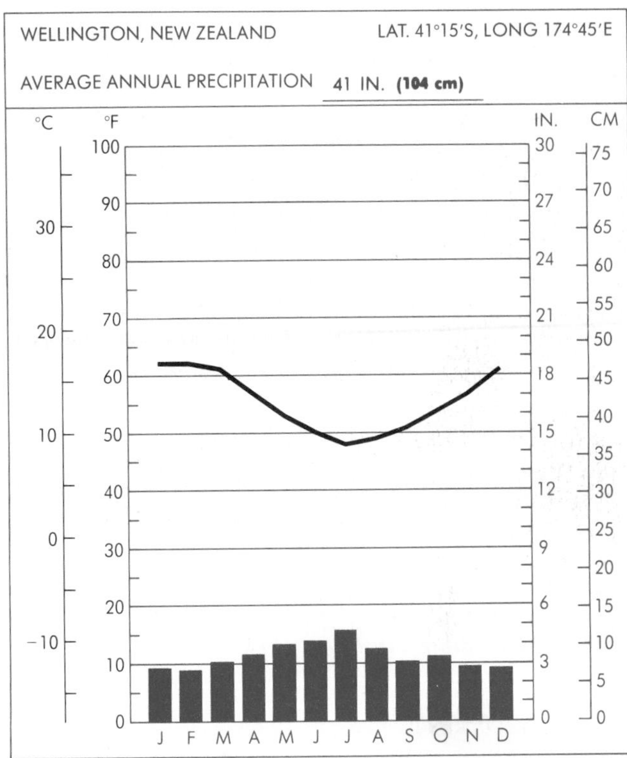

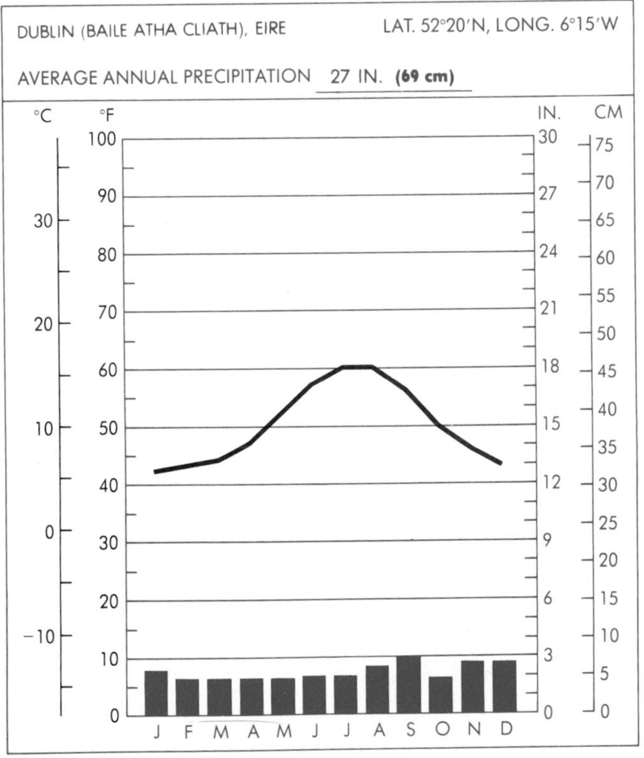

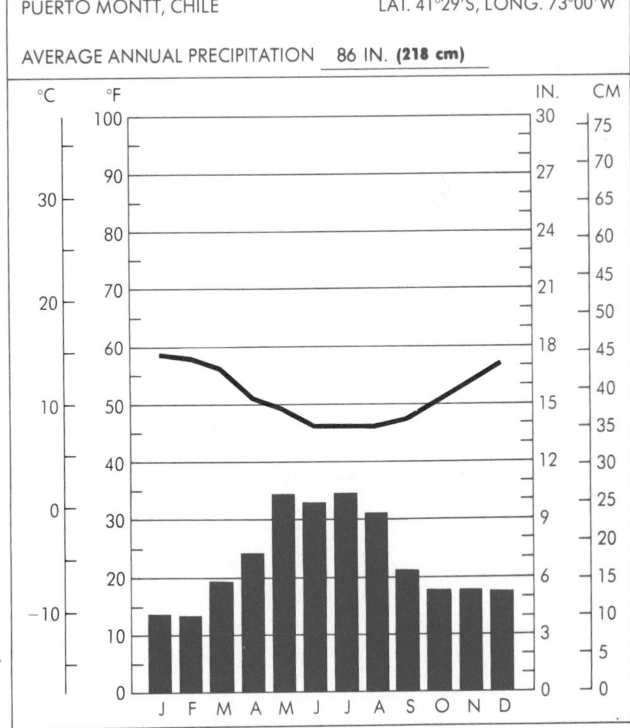

Fig. 11-3. ***Marine West Coast stations.***

minimum and 25 inches (64 cm) is more than adequate for agriculture. The statistics are similar for other rainy Marine West Coast cities. Seattle gets only 33 inches (84 cm) of rain per annum; Portland, 44 (112 cm); and Melbourne, 26 (66 cm).

These totals are scarcely enormous, but all the same life in any of these cities is a wet and often dreary experience, mainly because of excessive cloudiness. All Marine West Coast regions receive well under 50 percent of the possible sunshine each year. Scotland, for example, receives an average of less than one hour of sunlight per day in December and only five and a half hours in June despite the very long summer days. This persistent cloudiness has its bright side: it is responsible for the lessened evaporation. As a result, a little moisture goes a long way and maintains the lush greenness of the landscape and the permanent snowcaps on even fairly low mountain peaks.

All Marine West Coast regions receive well under 50 percent of the possible sunshine each year.

Precipitation totals may vary widely within the Marine West Coast climate, from a low of 23 to 25 inches (58 to 64 cm) to over 200 (508 cm). In terms of human occupation, however, all the large cities and densely populated lowlands are in regions of under 60 inches (152 cm) of rainfall, whereas those areas with more are virtually uninhabited. Part of this situation may simply be that normal drainage cannot cope with over 60 inches (152 cm) and marshes and moors prevail. For a region to receive more than 60 inches (152 cm), it must have steep mountain slopes immediately next to the sea, and such a coast allows little room for humans to find a toehold. The wet coasts of British Columbia, western Scotland, Norway, and southern Chile are all empty or support only tiny fishing villages or lumber operations (Fig. 11-3).

Normally, there is slightly more precipitation in winter. The summers are by no means dry, but there is less cyclonic activity during the warm months. One small area that deviates enough from the average situation to merit mention is the North American coast from northern California to southern British Columbia. Here is a distinct, if limited, summer dry season that lasts from about mid-July to mid-September. It is a beautiful time of year with a great deal more sunshine than in the rest of the year, and daytime temperatures are seldom out of the 70s (F) (21° to 26° C). Although lawns and pastures require some irrigation and forest fires are a problem, generally the dry season is so short that the countryside remains green and the mountain snowcaps glisten without melting away. These conditions exist because the Aleutian Low is weakened and less able to produce vigorous storms, while at the same time the offshore Hawaiian High cell migrates strongly northward, shielding this part of the coast from the few storms that do move eastward. This situation does not last long. By mid-September, the rains begin and, though somewhat sporadic at first, shortly settle into a nine- to ten-month pattern of clouds and drizzle.

The Marine West Coast normally features slightly more precipitation in winter, though the summers are not dry.

Marine West Coast Productivity

Where the land is flat and the rainfall is below 60 inches (152 cm), the Marine West Coast regions have attracted people. Agriculture is possible, though not ideal, the greatest drawback being lack of sunshine rather than temperature. In many parts of the world, grass for grazing has been the best crop.

The greatest drawback to agriculture in the Marine West Coast is lack of sunshine, not temperature.

Not surprisingly, most of our domesticated breeds of grazing animals are of Western European origin and have been introduced successfully into the Pacific Northwest, New Zealand, and Chile—Hereford, Dorset, Shropshire, Percheron, Holstein, and so on. Other crops besides grass are grown; many of them such as oats and potatoes have the special ability to produce well with reduced sunshine.

Forests are the natural vegetation in the Marine West Coast climate regions, and their exploitation is an important industry where they are reasonably accessible (Fig. 11-4). In Western Europe most of the trees are gone, having been cut long ago; in southern Chile and Canada/Alaska, virgin forests remain untouched.

Wood for ships, often rugged terrain, excellent harbors, and nutrient-rich offshore waters conducive to sea life—all have contributed to the growth of the fishing industry in these regions. This maritime heritage is evidenced in extensive merchant fleets and in seaborne foreign trade. Today, trade is intimately tied to manufacturing, and Western Europe, with its profitable mix of mineral resources as well, stands out as one of the world's great industrial and commercial leaders.

Fig. 11-4. ***Cutting ancient forests.*** *Forests are an important resource in the uplands and in the less densely populated parts of the Marine West Coast. In western Washington loggers use modern forestry equipment to handle massive Douglas fir logs cut from trees more than 200 years old. Less than 10% of the original forest in western states remains to be cut, but reseeding of a new crop is the routine followup.*

Fjords

To a degree, the broken coastlines of typical Marine West Coast regions reflect the role of climate in shaping landforms. All these regions are characterized by numerous fjords—steep-sided, glacier-carved inlets—excavated in preexisting river valleys when the climate was cold enough to support ice at sea level (see Chapter 21). Alaska, British Columbia, the Puget Sound, Scotland, Norway, southern Chile, Tasmania, and New Zealand all have fjorded coasts. Thus all these regions, today classified as Marine West Coast, shared a similar but colder climate in the past. Although the tidewater glaciers are gone now, we can read the climatic history in the landforms of carved rock they left behind.

All Marine West Coast regions are characterized by fjords.

MIDDLE-LATITUDE DRY

The ***Middle-Latitude Dry*** climate is so similar to the Tropical Dry that many classifiers lump them together. Furthermore, the areas of Tropical Dry and Middle-Latitude Dry climate are next to each other and merge almost imperceptibly. It is hard to decide exactly where to separate them. Aside from the obvious locational differences between the Tropical Dry and the Middle-Latitude Dry, they are the results of entirely different causes, and there is one major difference between the climates—temperature. We will regard them as separate climates, then, although we must recognize their many similarities.

The one major difference between the Tropical Dry and the Middle-Latitude Dry climates is temperature.

Regions of the Middle-Latitude Dry

Since the middle latitudes are essentially the Westerly belt, any dry climate occurring here must be divorced from the west coast with its onshore humid air masses (Fig. 11-1). The North American representative comes very close to the west coast but is east of the high and continuous Sierra Nevada/Cascade chain. It is therefore in a rainshadow, and the moisture off the sea is precipitated in heavy amounts on the Marine West Coast seaward slope of the mountains. Here is one of nature's sharpest lines. The windward slope receives up to 200 inches (508 cm) of rainfall, whereas such places as Wapato, Washington, immediately in the lee of Mount Rainier, and Owens valley in California, in the rainshadow of Mount Whitney, receives 5 inches (13 cm).

The entire area between the Sierra Nevada/Cascade and the Rockies averages only 10 to 15 inches (25 to 38 cm). This is only half the total dry region, for although the Rockies attract some increased moisture through orographic lifting, in their rainshadow on the high plains it is dry once again. Traditionally, the 100th meridian running through the Great Plains has been regarded as the 20-inch (51-cm) rainfall line (isohyet) in the United States. That is, west of this line, the precipitation is less than 20 inches (51-cm). To the pioneers, here was the beginning of the Great American Desert, and it is so labeled on many old maps. Thus it becomes convenient to designate the 20-inch (51-cm) rainfall line as the eastern margin of the Middle-Latitude Dry climate. (We will explain this choice further in the next section under the topic of aridity.)

A second and even larger Middle-Latitude Dry region exists in interior Eurasia. It is located in the same general latitudes as that of North America but much farther from the west coast. Because Europe has no continuous coastal mountain chain to sharply limit all the moisture to a narrow coastal area, the Marine West Coast in Northern Europe merges into somewhat drier climates farther inland. The decrease in Atlantic moisture with distance eastward is a gradual process. A 20-inch (51-cm) rainfall line therefore lies far back in the interior of the continent.

The western margin of the Middle-Latitude Dry is in the southern Ukraine. From there it stretches eastward into the Caspian basin and Russian Turkestan, interior China north of the Tibetan Plateau, and even the far northeast of China, almost the east coast.

The only part of the Southern Hemisphere with a significant land mass in the middle latitudes is in South America, and even here the area is limited. But Patagonia (the southern half of Argentina) lies in the rainshadow of the Andes, which isolate the Marine West Coast region of southern Chile. Here is a close parallel to the situation in North America where the Middle-Latitude Dry climate closely approaches the west coast.

Aridity

Aridity is, of course, the chief factor that allows us to classify all these regions together. It does not matter whether the moisture arrives in the summer or the winter; the important thing is that there is not enough.

Aridity is the chief factor that allows us to classify all these regions together.

A good deal of the Middle-Latitude Dry is *true desert;* that is, it receives less than 10 inches (25 cm) per annum. The fringing *semiarid* regions, which receive up to 20 inches (51 cm) of precipitation, are also included. One reason why a district with 15 to 20 inches (38 to 51 cm) of rain can be grouped with a region receiving less than 10 inches is the erratic nature of the rainfall (Fig. 11-5). As in the Tropical Dry zones, *the drier the climate, the less reliable the rainfall,* and this situation often means great difficulty in human land use.

Yet all these precipitation characteristics fail to help us separate the Middle-Latitude Dry regions from those of the Tropical Dry, for they are essentially similar. Only the temperature differences can provide us with the needed distinction.

Temperature

Winter temperatures in the heart of the Middle-Latitude Dry regions, especially in the Northern Hemisphere where continental air masses dominate, are well below freezing, and precipitation is in the form of snow. This climate is a far cry from the Tropical Dry where cold-month averages are typically in the 50° to 60° F (10° to 16° C) range. In eastern Montana or Chinese Sinkiang, winter days are frequently far below 0° F (–18°C). The winters make the middle latitudes a different world from the tropics.

Winter temperatures in the heart of the Middle-Latitude Dry regions are well below freezing, and precipitation is in the form of snow.

On the other hand, summers are hot, often excessively so, although they are neither as hot nor as long as in the Tropical Dry. Averages are about 70° F (24° C) for the hottest month. Nights are cool, reflecting the low humidity, and even in the winter, days may be a great deal warmer than the nights (Fig. 11-5).

Temperature, then, particularly in the winter, is the main difference between the Tropical Dry and the Middle-Latitude Dry. Yet, in the zones where these two dry regions merge, there is an extensive area where temperatures range between the extremes of the tropics and the middle latitudes. Northern Mexico, southern New Mexico, Iran, Iraq, eastern Syria, northwestern Argentina, as well as parts of southern Africa and Australia, have dry climates with temperatures that are not quite typical of either Tropical Dry or Middle-Latitude Dry. Winters, for instance, average just above freezing, and summers are in the low 80s F (27° to 29° C). This picture is further complicated by upland plateaus, as in northern Mexico, and Iran; although not high

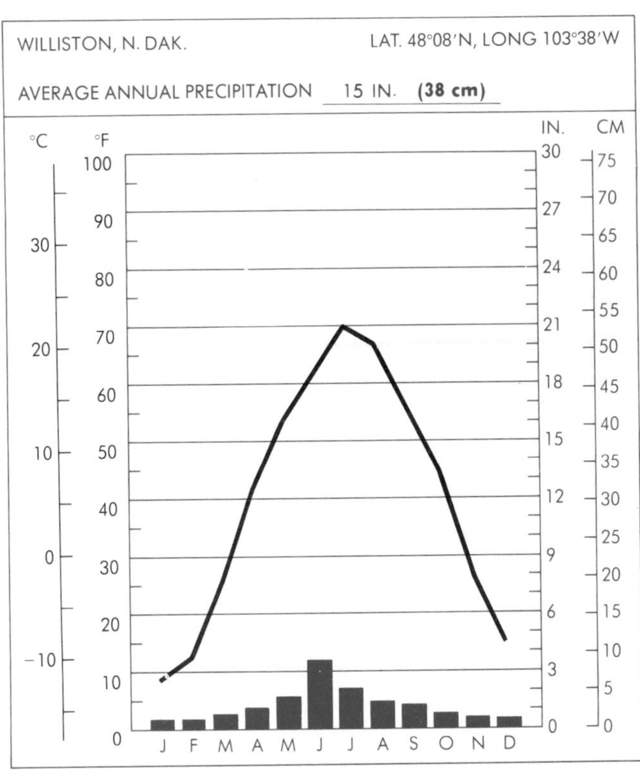

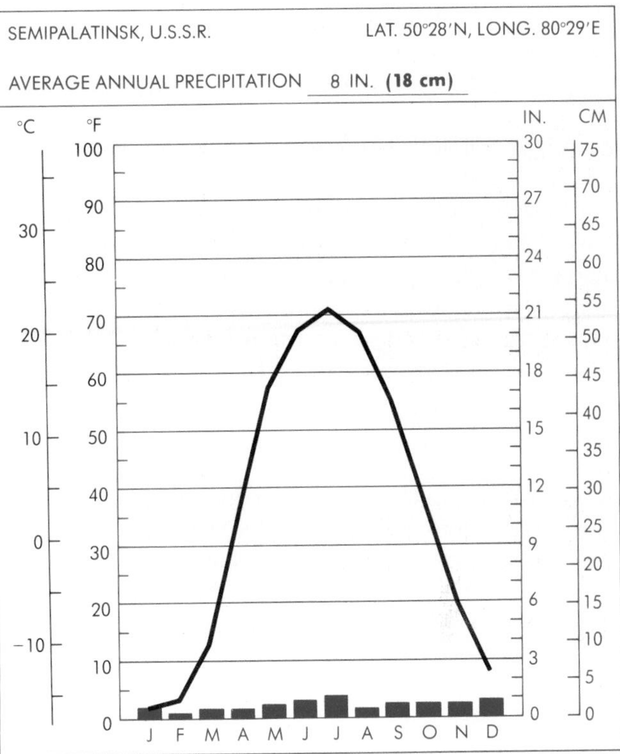

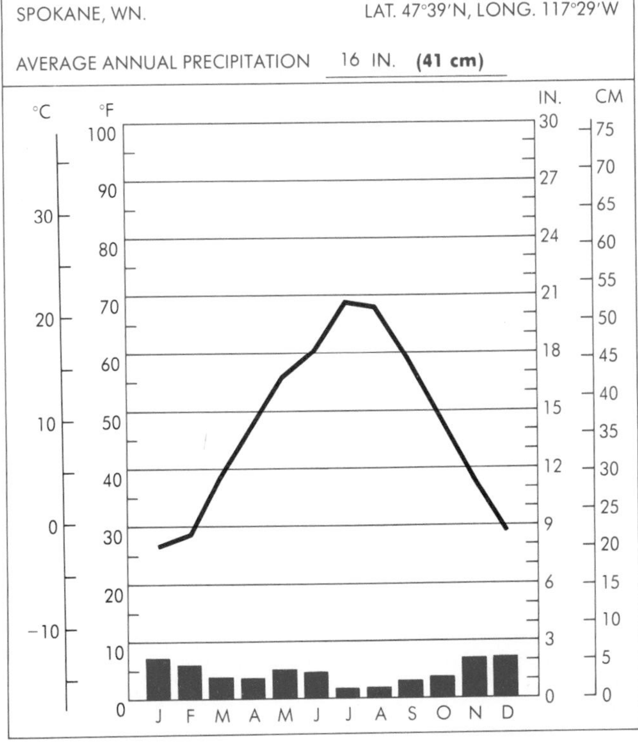

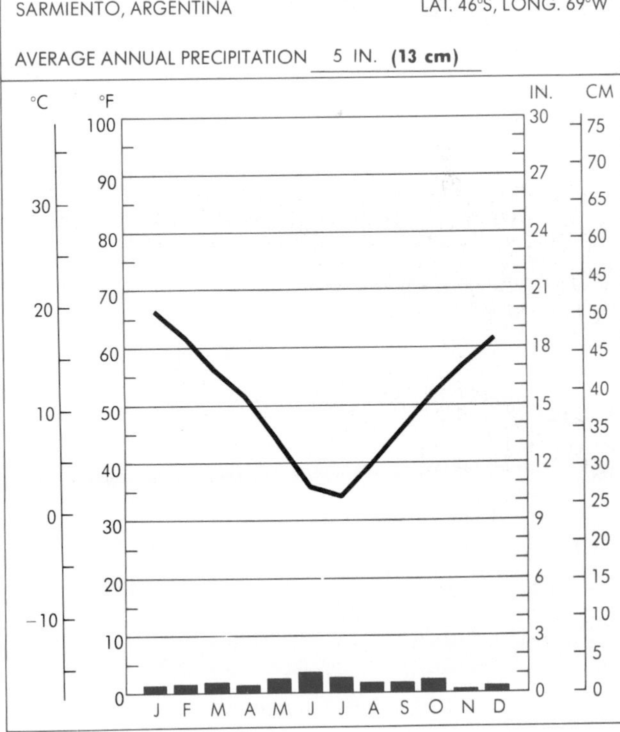

Fig. 11-5. ***Middle-Latitude Dry stations.***

enough to be regarded as mountains, they do slightly affect temperatures.

Classifiers have always had difficulty drawing a separating line through this territory. One solution is to establish an entirely new classification called Subtropical Dry. The problem is that an "almost but not quite" region lacks character of its own to distinguish it from its neighbors. Instead, we simply draw a line at a midway point through this zone of merging. We should recognize, however, that such a line is somewhat arbitrary. Although this border region is often thoroughgoing desert, it is a transitional band between Middle-Latitude Dry and Tropical Dry.

Human Habitation

As in all dry regions, water is the critical factor in allowing human habitation on any significant scale. To a large degree, the geography of the western United States, for instance, is oasis geography (Fig. 11-6). Yakima, Salt Lake City, Reno, Boise, and lower Colorado are all densely populated oases. In between these limited and widely scattered favored spots are extensive unpopulated regions, although here and there rich mineral deposits support some activity despite a lack of water.

To a large degree, the geography of the western United States is oasis geography.

Fig. 11-6. ***Oasis irrigation.*** *Canals bring water from the Wasatch Mountains to this irrigated farm near Salt Lake City, Utah. Farms and cities thrive in arid lands where water is available.*

The semiarid desert margins unlike most comparable regions in the tropics, have been made reasonably productive. Grazing has been traditional, but large-scale grain farming, especially of wheat, has taken over large areas. The reasonably good accessibility to the middle-latitude urban markets is probably the main reason for the greater development of agriculture, but this is far from ideal agricultural land. Both the high plains of North America and the "virgin lands" in the USSR, for example, have suffered from recurrent drought, and they have had a continuous history of crop failure. Some claim that these regions are not agricultural land at all. They believe that these regions should be maintained in permanent grass and that even grazing should be controlled to conform to the rainfall cycles.

HUMID CONTINENTAL

The third middle-latitude climate is the ***Humid Continental,*** which is in much the same latitude as the Marine West Coast and the Middle-Latitude Dry. It exists, however, only in the Northern Hemisphere, for not even South America has enough mass for its full development in the Southern Hemisphere.

The Humid Continental is in much the same latitude as the Marine West Coast and Middle-Latitude Dry, but exists only in the Northern Hemisphere.

Regions of the Humid Continental

In North America virtually the entire northeastern quarter of the United States and adjacent Canada is involved (Fig. 11-1). Its southern boundary is the Humid Subtropic, its western boundary is the 20-inch (51-cm) rainfall line or the Middle-Latitude Dry, and its northern border is essentially the northern limit of agriculture dictated by a shortened growing season. A finger of the Humid Continental pushes far to the west, north of the Middle-Latitude Dry. Decreasing evaporation offsets the slightly lower precipitation, allowing this region to be classified as humid.

In Europe the Humid Continental is found immediately east of the Marine West Coast and extends well into the interior of the USSR. The area is in the shape of a triangle with its broad base to the west. Southern Sweden, both Germanies except the North Sea coast, Czechoslovakia, Hungary, and the Po River valley of northern Italy form this base. To the east, southern Finland, Poland, Romania, and much of European USSR are included. The tip of the triangle is north of the Middle-Latitude Dry region and pushes as far east as Lake Baikal.

Another, smaller triangle-shaped area classified as Humid Continental is in eastern Asia and has its base along the sea. Most of North China east of the Middle-Latitude Dry is included, as are North Korea and the northern half of Japan. The triangle's tip is north of the Middle-Latitude Dry region and follows the Amur valley inland to Lake Baikal.

Precipitation

The word "humid" in connection with this climate means an average precipitation of about 30 inches (76 cm), although it varies from 20 inches (51 cm) at the fringe of the semiarid to over 50 inches (127 cm) near the sea. Nowhere is the rainfall during the growing season too slight for agriculture.

The word "humid" in Humid Continental climate means an average precipitation of about 30 inches (76 cm).

In the winter the bulk of precipitation is of cyclonic origin from the frequent westerly storms, and much of it is snow. Most of the summer rainfall, on the other hand, occurs as convectional showers.

*Fig. 11-7. **Snowstorm.** Spring is still over a month away when the mid-March snowstorm arrives.*

Summer and winter precipitation tends to be approximately equal near the coast (except in eastern Asia), but with increased distance inland, a summer concentration becomes apparent. This is because the dry, cold continental air masses of the interior, which dominate in winter, provide only limited amounts of moisture.

Eastern Asia, even this far north, still reflects the influence of the monsoon, so that strong summer maximums are the rule and winters are almost dry despite occasional winter cyclones. However, Japan, because of its insularity, receives heavy winter precipitation, usually as snow. This pattern of winter cyclonic precipitation and summer convectional showers with no well-defined dry season is similar to the pattern of the Humid Subtropics, but the annual totals are somewhat less and there is a good deal more snow (Fig. 11-7).

Temperature

You will remember that the word "marine" in a climate's title means the sea's moderation of both summer and winter temperatures and a resulting small annual range. The word "continental," on the other hand, means radical cooling in the winter and heating in the summer. Air masses reflecting this large annual range give their character to the temperatures of the Humid Continental climate. Summers are hot and often sticky, as monsoonal influences bring in the maritime tropical air; nights cool off very little. These are the times when Iowa farmers must rationalize that their corn is growing nicely even if conditions are far from ideal for human comfort.

The word "continental" in Humid Continental climate means radical cooling in winter and heating in summer.

It may also be a good deal cooler along the Baltic seafront or in Nova Scotia, and in the far interior the heat is less trying because of drier air despite the high temperatures. Generally, the Humid Continental summers average 75° F (24° C) and are quite warm for their latitudes.

Winters are even more uncomfortable. Cold-month temperatures must average below freezing in order to qualify as Humid Continental. Although near the coast they may not be far below freezing, the humidity creates a penetrating sort of cold. Inland it is drier, but temperatures fall to averages of 10° to 12° F (−12° to −11° C) with a good many days below 0° F (−18° C).

Basically, there are only two pleasant months each year—May and October, the transition periods. Indian summer is a

URBAN CLIMATE

Each year, as we bring more mechanization to world agriculture, the demand for rural labor decreases, small towns lose population, and thousands move to the urban centers. This results in sprawling cities; many of the largest and fastest growing of these cities like London, Moscow, Tokyo, Seattle, New York, Melbourne, and Buenos Aires are in the midlatitudes.

Meteorologists now generally agree that urbanization brings climate modification and the greater the urban area—the more pavement, industry, buildings, and vehicles—the more the climate is modified.

The physical contrasts between urban and rural lands are important. These include

1. surface materials
2. shapes of surfaces
3. heat sources
4. moisture sources
5. air quality

Urban areas, containing more stone, asphalt, and concrete surface materials, have a higher thermal capacity than rural fields and forests. Thus cities heat more slowly but retain heat longer than nearby rural lands.

The three-dimensional character of cities created by multilevel buildings contrasts sharply with the two-dimensional lands beyond the urban fringe. More surface is therefore available for heating, and air movement through the city is slowed greatly by surface friction—wind speed is reduced.

With industry, vehicle use, and domestic heating all concentrated in urban areas, cities are great sources of heat and once again, we see that they will tend to be warmer than rural areas.

The waterproofing of cities—pavements, roofs, drainage ditches—means quick drying after rain, and the general lack of vegetation leaves little opportunity for water that does find a way into the soil to return to the air in transpiration. Thus cities are likely to be drier than rural areas between rainy periods.

Air pollutants from automobile and factory exhaust choke the air over most urban areas and add billions of condensation nuclei for water-droplet formation, thereby increasing the possibility of greater precipitation.

Do these contrasts really lead to climate modification? Numerous studies suggest that they do. Summarizing research results shows the influence of large urban areas on regional climates:

Element (annual means)	*Comparison with Rural Surroundings*
Temperature	1.0° to 1.5° F warmer
Relative humidity	6 percent lower
Dust particles	10 times greater
Cloudiness	5 to 10 percent greater
Wind speed	20 to 30 percent lower
Precipitation	5 to 10 percent greater

Therefore, by spreading urbanism we are actively modifying regional climates and, because carbon dioxide and other greenhouse gases come mainly from our cities, we are probably increasing the potential for global climate change at the same time.

glorious time and can be counted on virtually every October, but spring is liable to be over very quickly. If one wakes up late some May morning, one may miss spring altogether.

Long Summer Versus Short Summer

The size and the relatively great variation within the Humid Continental make it convenient to subdivide the region on the basis of its agricultural potential (Fig. 11-1). This division was first made in the United States because of a very obvious ***crop line*** running east/west across much of the Middle West.

The chief crop used as a climatic indicator was corn. South of a line drawn through southern South Dakota, southern Minnesota, southern Wisconsin, southern Michigan, the Ontario Peninsula of Canada, northern New York, and southern New England, the growing season is long enough for old-fashioned field corn to come to full maturity. (The newer hybrid corns allow a good deal more flexibility, but the climate division was made well before their introduction.) Winter wheat, too, adheres reasonably well to the same northern boundary, for its ability to survive in the field (already sprouted) through the winter is closely related to the severity of the winter temperatures (Figs. 11-8 and 11-9). Spring wheat, planted in the spring, must replace it if

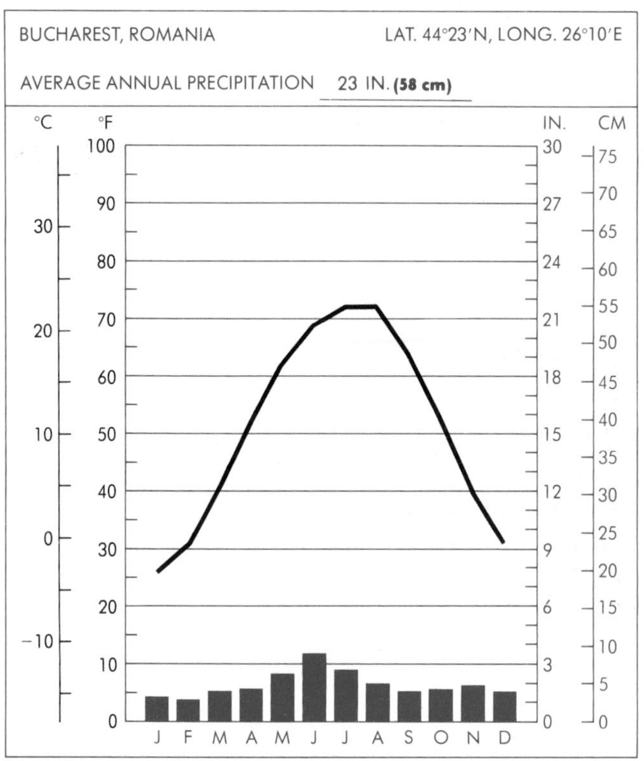

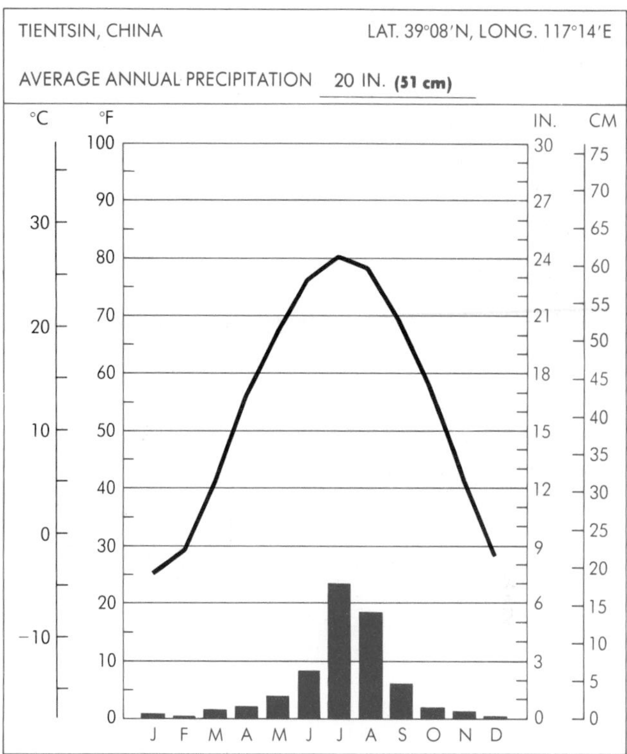

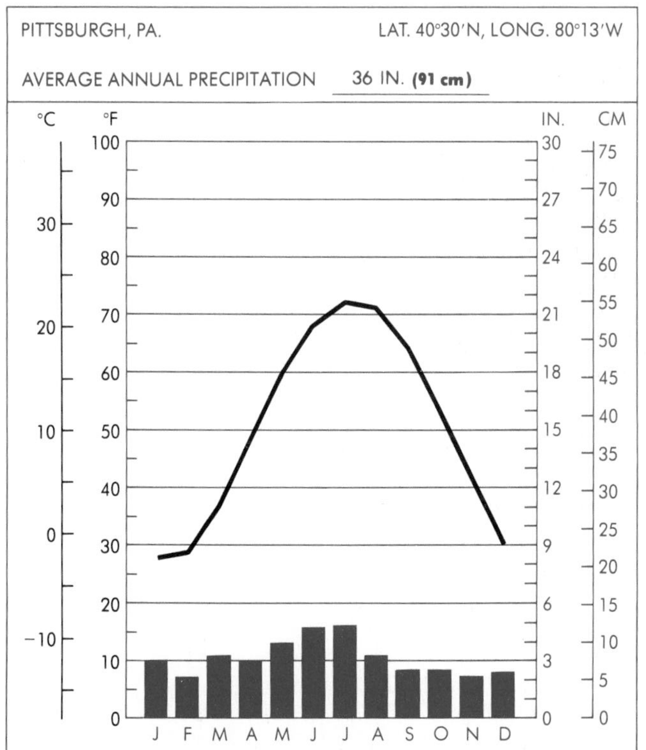

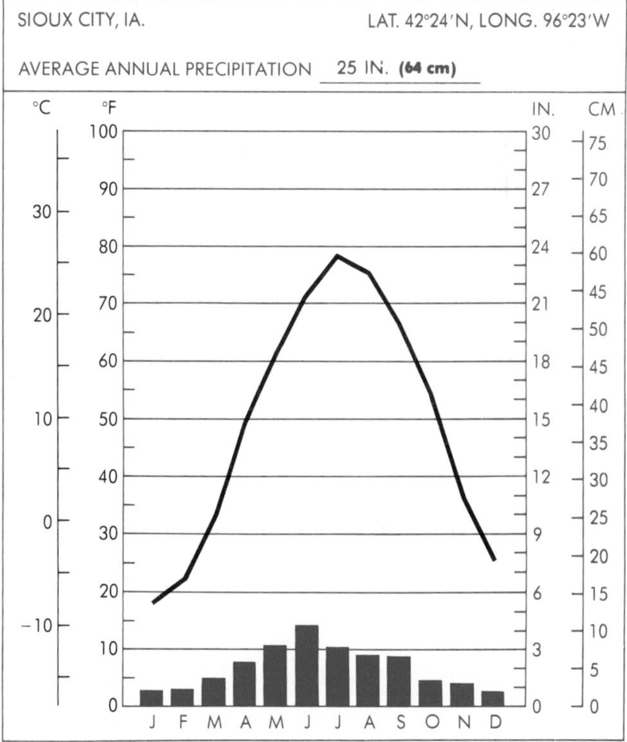

Fig. 11-8. ***Humid Continental–Long Summer stations.***

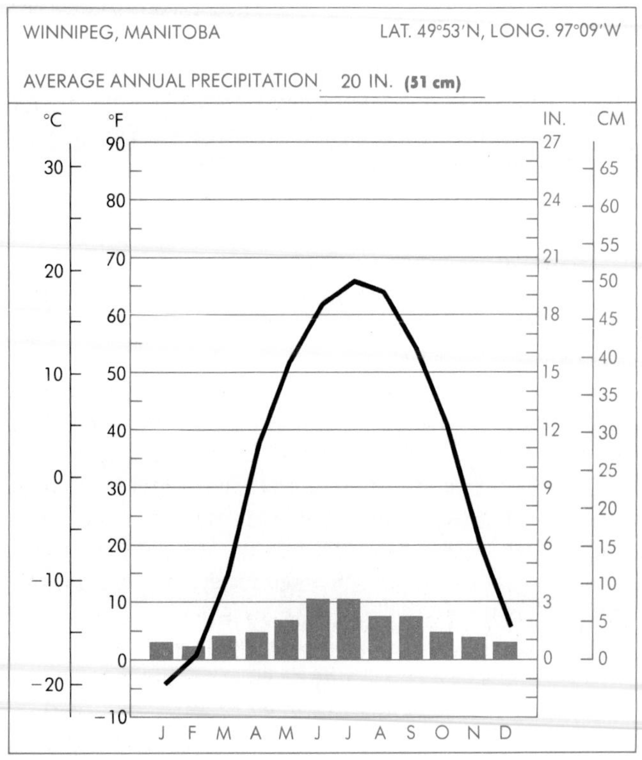

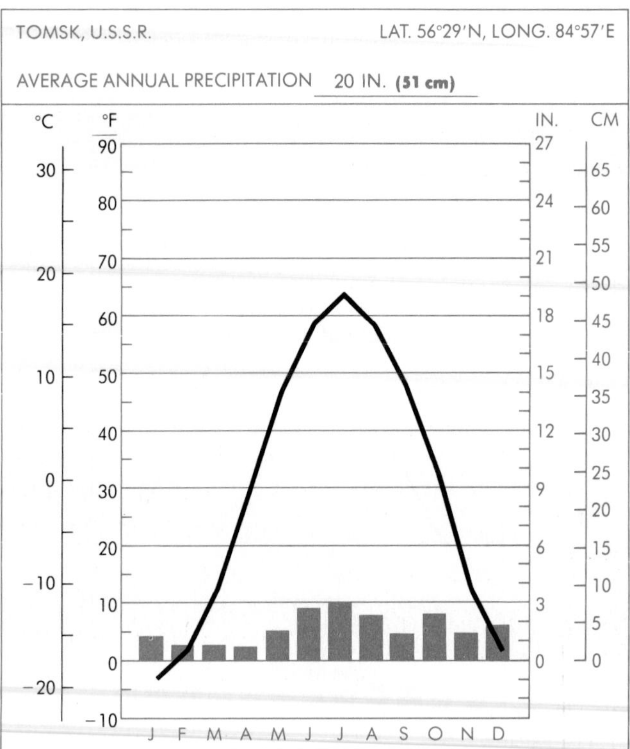

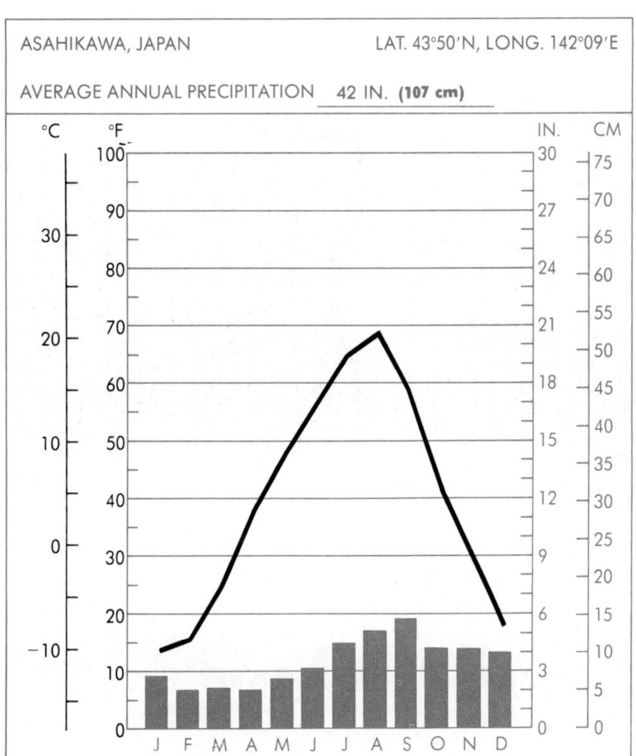

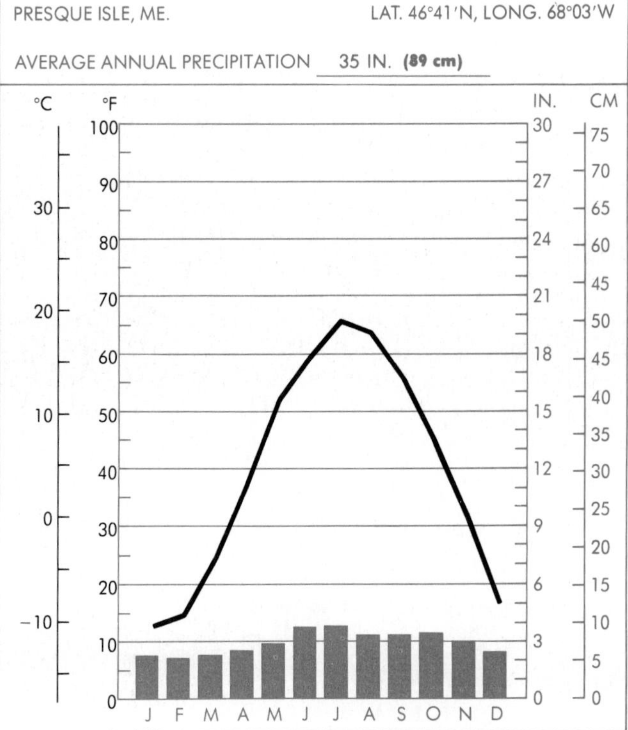

Fig. 11-9. Humid Continental–Short Summer stations.

winters are too cold for survival. The line, then, depends largely on the length of the growing season and, to a degree, on the severity of winter temperatures. We have given the names *Humid Continental–Long Summer* to the region south of the line and *Humid Continental–Short Summer* to that part north of the line.

The Humid Continental–Long Summer and the Humid Continental–Short Summer are the regions south and north, respectively, of the crop line.

Although corn is less popular in Europe, an identical line separates the Danube basin, northern Italy, and the southern Ukraine from the remainder or more northerly portion of the Humid Continental. In eastern Asia, it runs through northern China and North Korea, and between Honshu and Hokkaido in Japan.

Populations in the Humid Continental

The Humid Continental climate is conducive to agriculture, but this factor is not enough to account for the large-scale economic development that has taken place. This development is also closely related to the rise of urban markets and their increasing needs for food (Fig. 11–10). Sea frontage in North America and especially in Japan, Korea, and elsewhere in eastern Asia has encouraged the growth of large cities based on foreign trade. In maritime Europe a similar development has drawn large parts of the adjacent Humid Continental region into its hinterland.

Trade is not the only cause of urbanization; industry based in large part on coal has also played a leading role. At one time in the geologic past nearly the entire middle-latitude region was under the influence of a wet tropical climate, and the accumulation of thick beds of decomposing vegetable material built up in numerous swampy areas. This was embryonic coal, and subsequent pressures eventually formed it into extensive deposits of high-quality fuel.

Not all of today's good coal fields are within the Humid Continental climatic regions, though many are: the Appalachians, East Germany, Poland, the Ukraine, Kuznetsk (Siberia), and northern China. Others are nearby, and this coal availability has led to large-scale manufacturing.

Some maintain that the fluctuating temperatures and humidity of the Humid Continental climate are responsible for a particularly vigorous human population in that area, as opposed to the enervating effects of the tropics. Others con-

Fig. 11-10. ***Satisfying urban markets.*** *Long days of sunshine offset to some degree the short summer in Saskatchewan, Canada. Spring wheat does very well.*

tend that the development within this region has been accomplished despite a thoroughly uncomfortable climate. Whatever the cause, some parts of the Humid Continental, especially in North America and parts of eastern Asia, must be counted among the most prosperous regions in the world.

CONCLUSION

Midlatitude climates occur in the zone of westerly winds. Although these climates have a similar latitudinal position, westerly winds, mountain ranges, and continentality distinguish three very different climatic types.

Marine West Coast, the most windward in location, suffers long bouts with light rain and gray skies. Temperatures are moderate; forests and pasture grasses do well.

Inland, away from marine influence and especially in the rainshadow of major mountain chains, Middle-Latitude Dry grades from true desert to semiarid conditions. Air temperatures show greater annual range but are generally cooler than those of Tropical Dry.

Humid Continental, found only in the Northern Hemisphere and generally east of Middle Latitude Dry, is a snow climate. Winters are cold but summers are warm. Cyclonic storms bring precipitation in the winter, and summer rain comes from convective thunder clouds. Summers decrease in length poleward. With long summers corn and winter wheat do well; with short summers spring wheat is productive. Humid Continental regions are noted for numerous urban/industrial centers.

In the following chapter we conclude our discussion of world climate with a focus on the high latitudes.

KEY TERMS

Marine West Coast
Middle-Latitude Dry
true desert
semiarid
Humid continental
crop line
Humid Continental –Long Summer
Humid Continental –Short Summer

REVIEW QUESTIONS

1. What do all Marine West Coast climatic regions have in common?

2. What is responsible for the mild temperatures of the Marine West Coast climate? Why?

3. Why are Marine West Coast regions so gloomy?

4. What is the difference between Middle-Latitude Dry and Tropical Dry?

5. How do the regions of the Middle-Latitude dry climate differ from one another? Why do we classify them together if they are so different?

6. Why is it correct to say that "the geography of the western United States is oasis geography"?

7. Why does the Humid Continental climate exist only in the Northern Hemisphere?

8. Compare Humid Continental, Humid Subtropical, and Marine West Coast. How are they similar and how are they different?

9. What is the significance of the word "continental" in the Humid Continental climate? What does it help to explain?

10. How are the two divisions of Humid Continental reflected in agricultural crops?

APPLICATION QUESTIONS

1. Using the climographs in Figures 11-3, 11-5, 11-8, and 11-9, distinguish between the midlatitude climates.

2. Imagine you are an independently wealthy traveler. You can go anywhere you want at any time of the year, and you want to visit at least two locations in each of the midlatitude climates this year. Where would you choose to go and *when* should you go there in order to avoid getting rained on, overheated, or chilled on your nature walks? Prepare an itinerary and explain your choices.

3. Do you live in the middle latitudes? If so, how is your area typical of one particular midlatitude climate? Is it atypical or unusual in any way? If you do *not* live in the middle latitudes, explain how your climate differs from the midlatitude climate most closely related to it.

4. If you live in the midlatitude region, check the newspapers for temperature and precipitation data for the past 12 months. Prepare a climograph using these data. Which of the climographs illustrated in this chapter is most similar to yours? Most different? (If you do not live in a midlatitude region, prepare a climograph for your area and compare/contrast it with the climograph most similar to your own and the climograph most different from your own.)

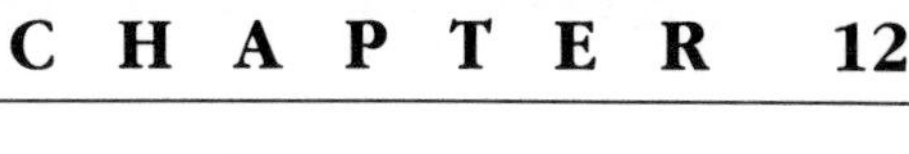

CHAPTER 12

THE HIGH-LATITUDE CLIMATES

OBJECTIVES

After studying this chapter, you will understand

1. Where the Taiga climate occurs, what its conditions are, and what causes these conditions.
2. Where the Tundra climate occurs, what its conditions are, and what causes these conditions.
3. Where the Polar Ice Cap climate occurs, what its conditions are, and what causes these conditions.

The warmth of summer in the Arctic Tundra along Alaska's Brooks Range thaws only the upper few inches of soil—not enough for trees of the Taiga to grow. Bitterly cold and dark winters in the high latitudes are followed by mild summers and nearly 24 hours of sunlight. Under such extremes, only lichens, grasses, and blue berries seem to do well.

INTRODUCTION

To round out our discussion of world climates we now turn to the high latitudes. Here we are essentially beyond the limits of agriculture and practically beyond the limits of human habitation as well. These are largely virgin lands, almost untouched except for a few particularly favored spots. In this sense the high latitudes resemble large parts of the world's dry regions and Tropical Wet areas, and the climate is largely responsible. Long, cold winters are the basic restrictive factor.

Because of their unattractiveness to humans and their general inaccessibility, the high-latitude climates have not been thoroughly investigated. Large areas lack recording stations, and thus ground-based data are limited. Only recently with the aid of weather satellites and interest in mineral extraction has an organized effort been made to understand these climates. Much work needs to be done.

We recognize three high-latitude climates: (1) the Taiga, (2) the Tundra, and (3) the Polar Ice Cap (Fig. 12-1). The first two are largely restricted to the Northern Hemisphere.

TAIGA

The ***Taiga*** climate gets its name from its distinctive type of vegetation—forest made up of a distinctly limited group of conifers. Details of the forest makeup are discussed in Chapter 27, but it is important here to emphasize that only a highly specialized group of plants can survive in a region dominated by the rigorous Taiga climate.

The Taiga climate gets its name from its distinctive vegetation—its special kind of coniferous forests.

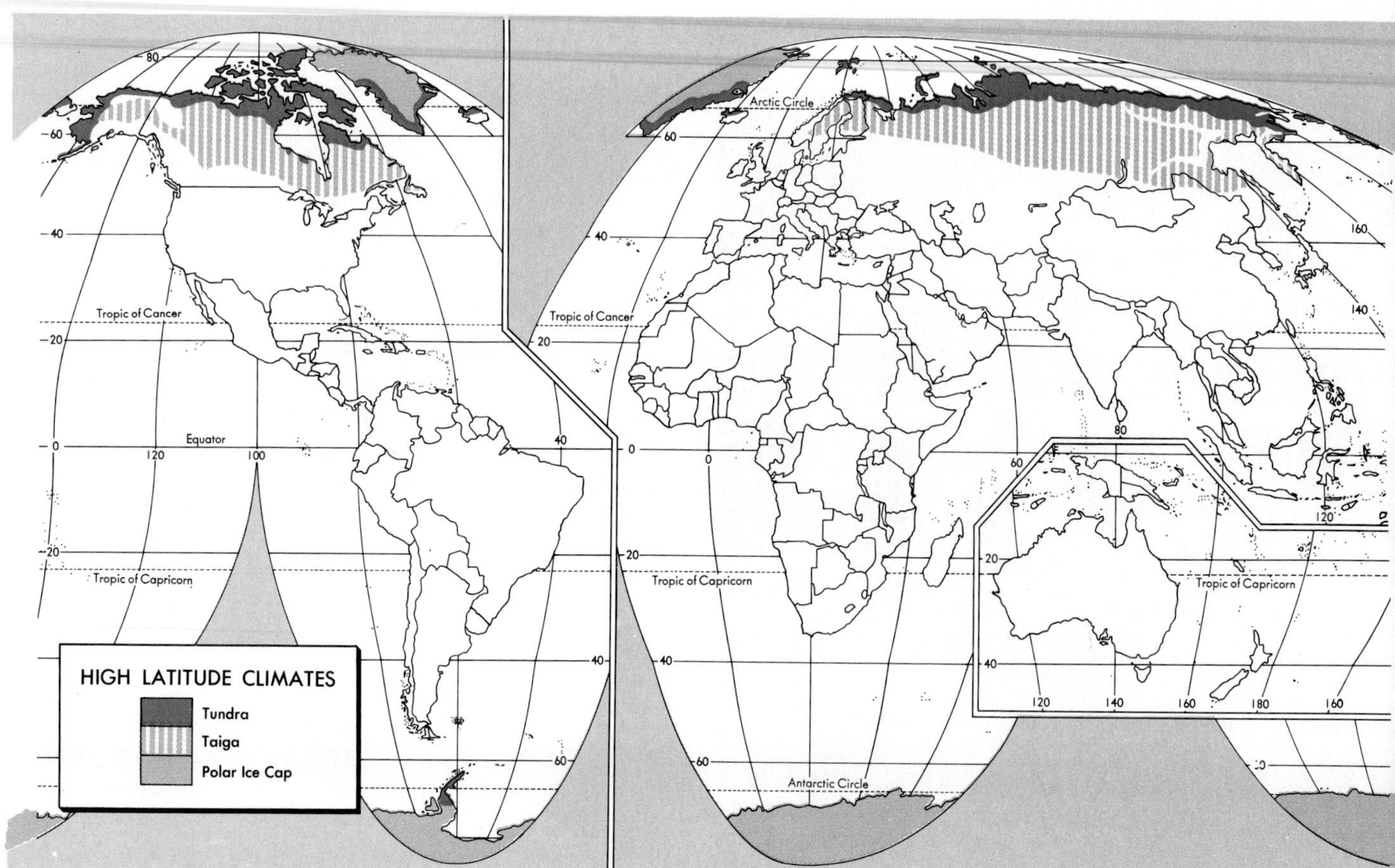

Fig. 12-1. ***High-latitude climates.***

Regions of the Taiga

The North American representative of the Taiga is a long, relatively narrow band extending clear across the continent between roughly 50° and 60° N latitude. Although it closely approaches the west coast in Alaska, the band does not quite reach it, nor does it touch the north coast anywhere except along the margin of deeply indenting Hudson Bay. Only at the shore of the Atlantic, in Labrador, does the Taiga reach the sea. Thus it is a continental climate and exhibits typical continental tendencies.

At very similar latitudes, a comparable belt of Taiga extends across the much larger Eurasian continent, again touching the sea only at its eastern extremity. From northern Sweden and Finland to Kamchatka (USSR) on the Pacific Coast, the Taiga region stretches for over 5000 miles (8050 km) and for most of that length shares a common southern boundary with the northern edge of the Humid Continental climate as it does in North America (Fig. 12-1).

Winter and Summer in the Taiga

The Taiga is dominated by winter. Not only is the winter long (eight to nine months), but for many years the Taiga was thought to be the coldest region on earth. Verkhoyansk, a mining community in eastcentral Siberia, has recorded a minimum temperature of −92° F (−69° C). However, just a few years ago an even colder temperature was recorded at the South Pole [−127° F (−89° C)], and it may be equally cold at the tip of Mount Everest. Nonetheless, winters are excessively cold in the Taiga, averaging far below 0° F (−18° C).

The Taiga winter lasts eight to nine months and was once thought to be the coldest on earth.

In contrast, summers are surprisingly warm despite their short duration. Fairbanks, Alaska, experiences a few daytime temperatures in the 90s F (32° to 38° C) nearly every year. Generally, the hottest month average throughout the Taiga is close to 60° F (16° C), sometimes making a 100° F (56° C) annual temperature range, which is by far the largest in the world (Fig. 12-2).

Precipitation

Because these far northerly continental interiors are dominated by cold high-pressure cells that block the intrusion of marine air, very little moisture is available throughout much of the year. Snow falls in limited quantities, but it is dry and powdery. However, because no melting occurs for months at a time, the entire winter's snowfall is visible all at once, a situation that does not prevail in many areas where snowfall is much heavier. In addition, dry snow is rapidly picked up by and whirled around by high winds, so that blinding blizzards may actually involve very little new snow. Drifting against obstructions of any kind in the path of the wind is also common. The snowfall is not excessive, but the individual who is forced to dig through a 9-foot (2 m) drift to get out of the house will not easily be convinced of the low accumulation. Actually, when the total annual fall of this powdery snow is melted down and measured in inches of moisture, it amounts to very little.

Because the far northerly continental interiors are dominated by cold high-pressure cells, little moisture is available.

By far, the greatest quantity of moisture comes in the form of rain from widely scattered summer convectional showers. Summer precipitation averages 5 to 15 inches (13 to 38 cm), which would indicate desert areas in most parts of the world, yet this precipitation supports forests in the Taiga. The greatly decreased evaporation at these high latitudes as well as the fact that the bulk of rain is concentrated during the short growing season allows tree growth (Fig. 12-3).

Taiga Populations

Who lives in such a climate? Not very many people do; it is populated only thinly and sporadically. The agriculturalist is conspicuously absent, but the miner is there, attracted to iron in Sweden, gold in Siberia, and uranium in Canada. Although the trees are not large and regenerate themselves very slowly, the forests can be exploited. Thus far, forestry is profitable in only a few places—chiefly along the populated fringes as in Scandinavia and Quebec. One indication of the remoteness of the Taiga is that it is the last stronghold of the fur trappers; cold winters produce fine pelts. If animals still abound, then, there cannot be many people.

TUNDRA

Like the Taiga, the ***Tundra*** is named for the specialized plants that grow and flourish under the restrictive influences of this high-latitude climate. Trees, for instance, are absent,

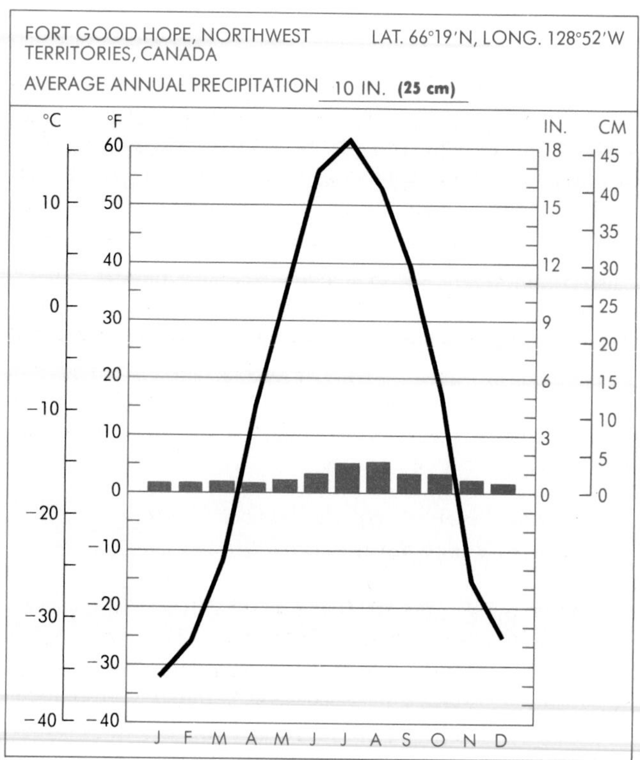

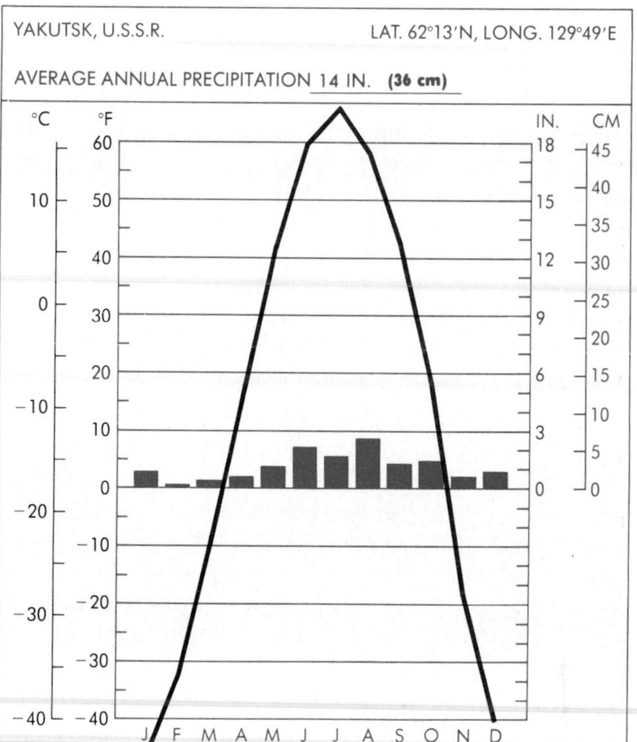

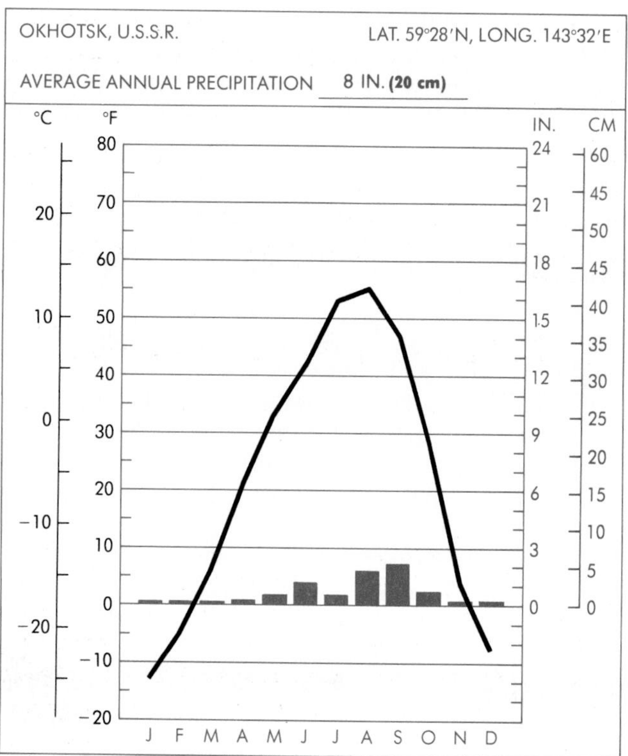

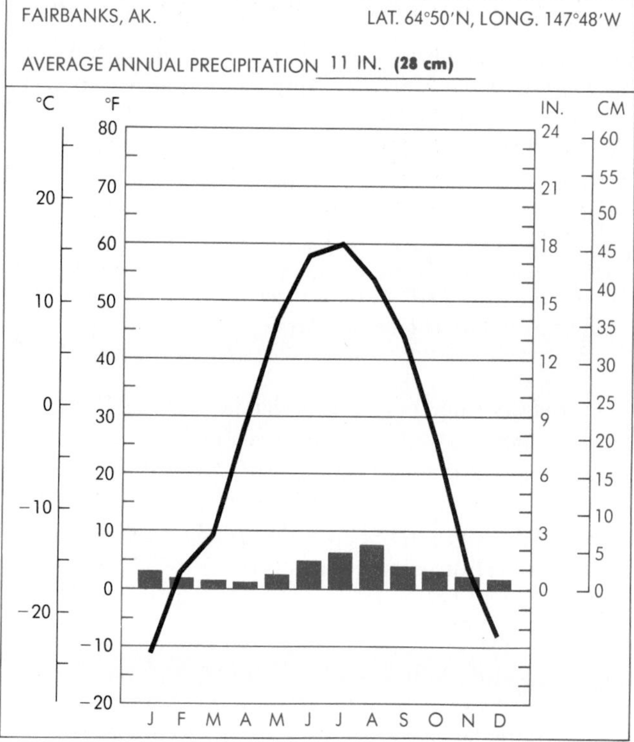

Fig. 12-2. Taiga stations.

Fig. 12-3. ***Logging in Sweden.*** *Spring comes late in Sweden, but when the ice finally goes out of the rivers an avalanche of saw logs, poles, and pulping logs, accumulated during the winter, moves downstream to the mills. To get an idea of the size of these logs, check against the man and dory at the lower right. Compare them with the timber being taken out of the Washington forest (Fig. 11-4).*

making the boundary between Taiga and Tundra the poleward limit of trees. Only certain sedges, grasses, wild flowers, low brush, and primitive mosses and lichens are represented. Since these are all shallow rooted, the frozen subsoil is not a hindrance to growth. These plants can remain dormant during a long cold season, and they have the ability to mature and reproduce themselves very rapidly once the short, cool summer commences.

Only certain sedges, grasses, wild flowers, low brush, and primitive mosses and lichens are represented in Tundra regions—there are no trees.

Regions of the Tundra

The Tundra is a coastal climate, never extending very far inland (Fig. 12-1). It involves the entire north coast of North America, including the Canadian Arctic islands as well as the ice-free shores of southern Greenland. In Alaska it is also found facing westward on the Bering Sea, and although the Aleutians and Alaskan Peninsula are not highly representative, they are usually classified as Tundra since they lack trees. The Alaskan south coast, however, in the vicinity of Anchorage and Cordova, is neither Tundra nor Taiga and is probably best included as a cold subset of the Marine West Coast. The Tundra also merges with the Marine West Coast in northern Norway, and from there the Tundra fringes the entire north coast of Eurasia.

Marine Characteristics

Because of this coastal orientation, the Tundra displays certain marine characteristics. We have so far thought of "marine" climates as regions of mild temperatures as opposed to harsher "continental" conditions. Of course, the Tundra temperatures are far from benign. However, we must remember the latitudes involved here. When compared with the somewhat lower latitude but continental Taiga, the Tundra really has quite moderate temperatures. The coldest winter month averages only a little below 0° F (−18° C), even though offshore waters are frozen much of the time. This may seem cold, yet the Taiga winters are usually a good deal colder. Summers are comparably cool, reaching a warm-month average of about 40° F (4° C), approximately 20° F (11° C) less extreme than interior locations (Fig. 12-4).

Compared with the Taiga, the Tundra has quite moderate temperatures.

Permafrost

Such an annual temperature range results in a condition typical of the Tundra (but also present in scattered locations in the Taiga) called ***permafrost,*** a permanently frozen subsoil. The low, dense vegetative cover effectively insulates the soil, and, together with the short, cool summers, there is simply not enough heat to allow thawing below a surface foot or more. This is not a short-run phenomenon. The almost perfectly preserved remains of long extinct wooly mammoths and saber-toothed cats have been discovered

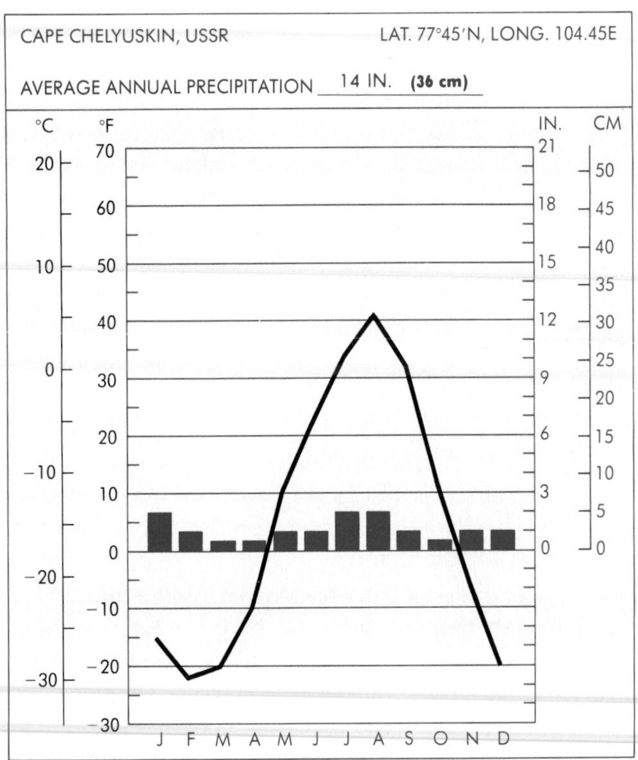

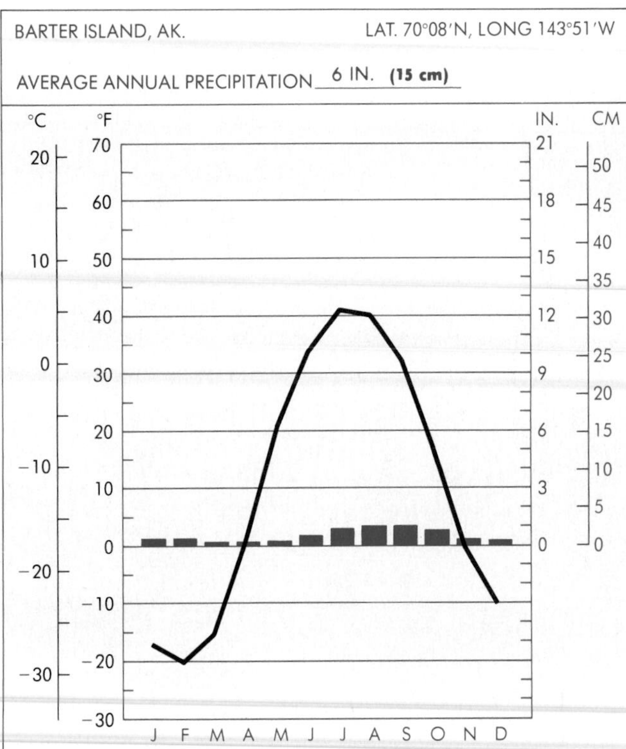

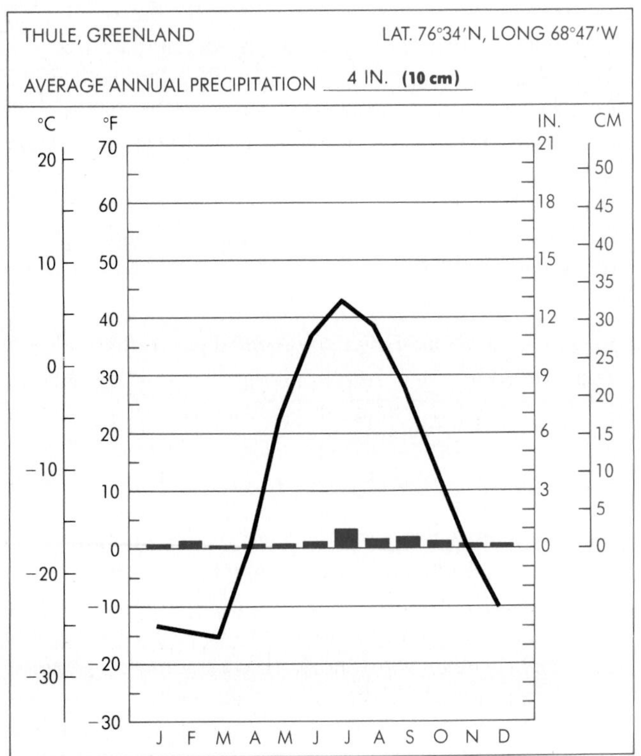

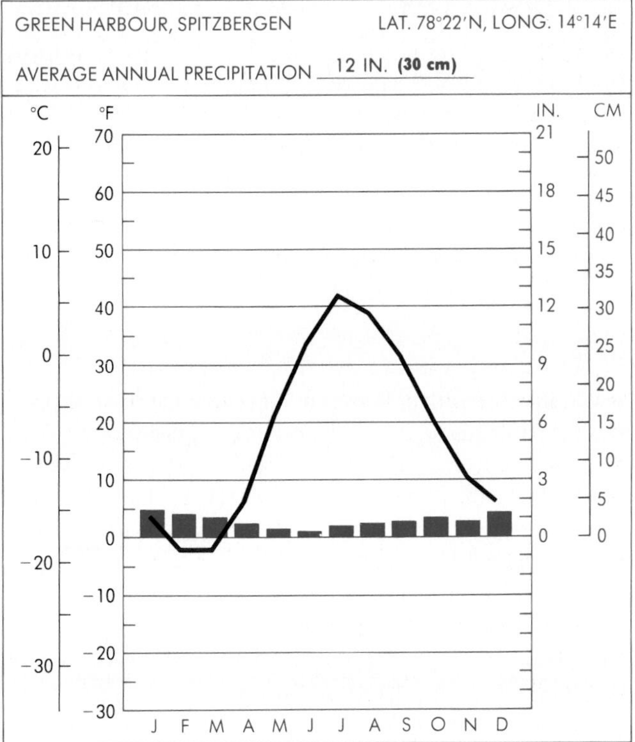

Fig. 12-4. Tundra stations.

Fig. 12-5. ***Building on permafrost.*** *Snaking for hundreds of miles across the tundra, the Alaska pipeline has a complete insulation system consisting of steel-jacketed fiberglass panels, weatherproof expansion joints, and special foam insulated modules at support and anchor assemblies. Without such insulation, absorbed summer solar heat and heat from the oil in the pipe would melt the permafrost causing the pipeline to sink below the surface.*

imbedded in the permafrost deep freeze. We must regard it as fossil ice that has *never* thawed once the climate cooled to its present extreme. If the insulating layer of surface vegetation is disturbed, the upper part of the permafrost will thaw to greater depth during the summer. Because permafrost becomes unstable when it thaws, construction projects must be specially designed to insulate the ground from building and other heat sources (Fig. 12-5).

A condition typical of the Tundra and in scattered locations in the Taiga is permafrost.

Permafrost near the surface also undoubtedly determines the poleward margin of trees. Further south in the Taiga, winters are even colder than in the Tundra and the ground is solid ice. The forest is dormant over the winter, however, and Taiga summers are long and warm enough for the frozen earth to thaw to a depth of several feet. This condition permits reasonable drainage of the soil and sufficient root space for trees to manage despite severe winters. Depth of summer ice is critical.

The Tundra usually has poorly developed surface drainage on the flat-to-rolling coastal terrain. The solid ice inhibits any loss of moisture vertically, and the soil becomes a soupy mud. Alternating shallow ponds and quacking bogs, where tangled roots and matted vegetation "float" on the mud, make the summer tundra landscape a difficult one to cross. Water is everywhere despite a low receipt of annual moisture amounting to only 5 to 15 inches (13 to 38 cm). As in the Taiga, most of this moisture comes as summer rain, some as convectional showers, but some from arctic front storms that are also probably responsible for the common winter blizzards.

In the summer, water is everywhere in the Tundra despite the low annual moisture.

Tundra Populations

Obviously, this habitat is a difficult one for humans, and traditionally the Lapp reindeer herders of the Scandinavian countries and scattered groups of Eskimos, living chiefly off the sea, have been the only inhabitants. Today, a few scientific, military, and big oil company installations are intruding into the Arctic solitude and are attempting to solve the problems of modern living under unique and adverse conditions.

POLAR ICE CAPS

The ***Polar Ice Caps*** are found in Antarctica and Greenland, the only high-latitude land areas permanently covered by masses of ice large enough to create a distinctive climate

THE LAPPS

The ***Lapps*** of far Northern Europe have very successfully occupied a high-latitude Taiga/Tundra environment for almost 2000 years. Their numbers are not large (currently about 41,500), and they are spread thinly along the northern periphery of four different countries: 25,000 in Norway, 10,000 in Sweden, 4500 in Finland, and 2000 in the USSR.

The Lapps are slender, dark haired, and short statured among their Scandinavian neighbors, and speak a Finno/Ugric language (similar to Finnish and Hungarian). Their anthropological background remains obscure, but whatever their origins, they have moved into what for most would have been a very difficult environment. What is more, they have managed to preserve their race, their language, and elements of their culture on the extreme northern fringe of European civilization.

The key to the Lapps' survival over the centuries has been the reindeer. Although individually owned, reindeer are communally herded in a tightly organized nomadic husbandry system that involves an annual migration of up to 300 miles (480 km) from the summer mountain pastures to the coast and back. Historically, Lapp culture and reindeer-herding nomadism have been inseparable.

However, life is beginning to change for the Lapps. For those who value the old ways, modernism has raised its "ugly head"; for those who equate modernism with progress, the "quality of life" has improved immeasurably. The reindeer is still herded, but its range is partially fenced, snowmobiles aid in animal control, national boundary restrictions now limit former unregimented wanderings, and only the herders migrate with their animals, the women and children remaining in new, permanent dwellings.

The old self-sufficiency is threatened by an encroaching European money economy, and not all of the Lapps' money requirements to satisfy their current life-style can be met by reindeer herding or selling handicrafts to tourists. Many are being forced into wage labor in mines, construction gangs, or commercial fisheries. For most Lapps, however, life is easier than in the "good old days." Large, modern, totally planned towns have brought twentieth-century services and facilities to the very heart of Lapland.

Although outnumbered in their own land, like the American Indians and Canadian Eskimos, the Lapps are not a dying race. Their numbers have actually increased in the past few centuries. It is their traditional way of life that faces change, if not possible extinction.

(Fig. 12-1). The Arctic Ocean, too, in the immediate polar region, is frozen year round and undoubtedly displays a similar climate, but since we have agreed not to classify oceanic climates, it is not considered here. By the same token, we have not classified highland climates, and much of interior Greenland and Antarctica is well over 5000 feet (1500 m). In the interest of consistency, then, we should consider only the relatively low parts of these ice-covered lands, but with so little specific data available it has been usual simply to make a few generalities and to ignore the variations caused by elevation. Gradually, climatic records are being accumulated, and eventually we will be able to present greater detail.

The Polar Ice Caps are found in Antarctica and Greenland.

Temperatures on the Polar Ice Caps

We know that the Polar Ice Caps are very cold (Fig. 12-6). The old Little America station established by Admiral Byrd on shelf ice at the Antarctic continental margin has recorded a coldest month average of −34° F (−38° C). But winter temperatures routinely drop well below −100° F (−73° C) at the Soviets' Vostok Station near the South Pole. Probably a winter average somewhere between the Little America and the Vostok figures would be about right—colder than the Tundra and possibly colder than the Taiga.

Of greater significance than the winter cold is the fact that

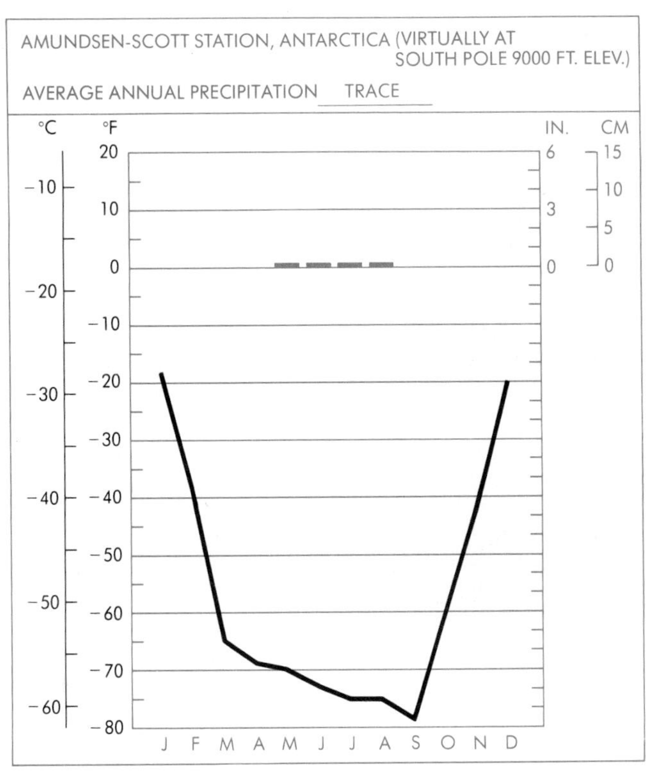

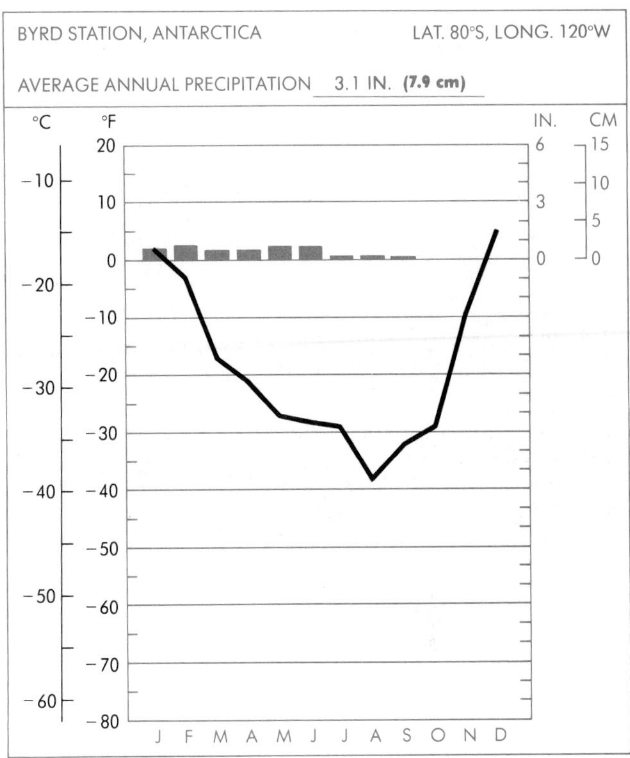

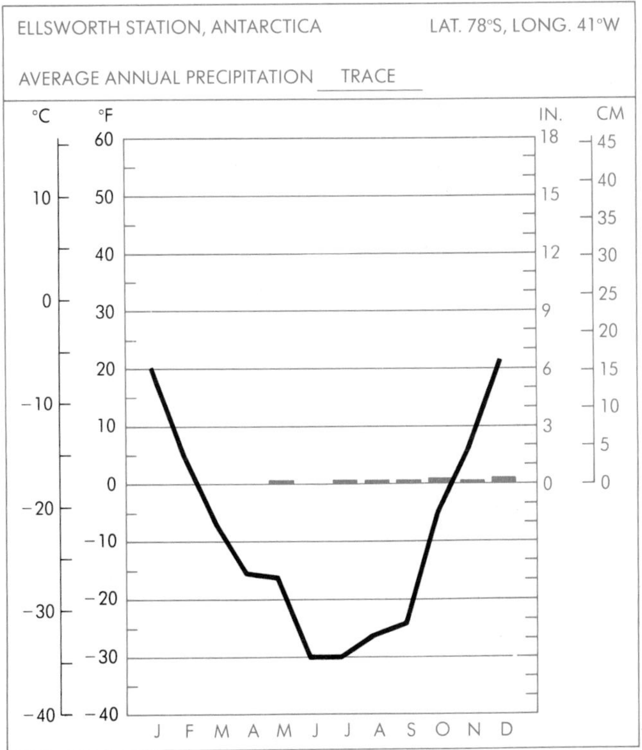

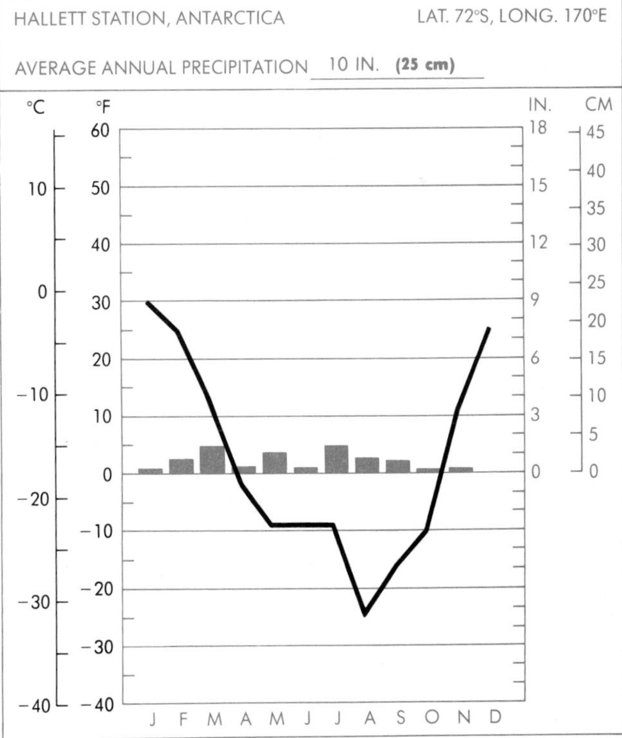

Fig. 12-6. ***Polar Ice Cap stations.***

Fig. 12-7. ***Antarctica.*** *There are permanent settlements these days in several parts of Antarctica. One is the U.S. Navy base at McMurdo Sound. These little communities are drawn here through scientific curiosity and the desire to establish a political presence rather than by Antarctica's climatic comforts.*

the warmest month averages do not reach as high as 32° F (0° C). Even at relatively mild Little America, the summer average is only 23° F (−5° C). This means minimal thawing and a maintenance of the ice mass, despite low precipitation, and only slight possibilities of vegetation even if soil were exposed. In fact, certain lichen species have been encountered on rocky outcroppings, so even here vegetation is not absolutely lacking.

The warmest month averages of the Polar Ice Caps do not reach 32° F (0° C).

Precipitation

Precipitation has not been accurately measured, but it is unlikely that the cold air masses can give off more than 10 inches (25 cm) per annum, and this is in the form of dry snow and minute ice crystals. With a constant loss of ice as great continental glaciers move out toward the sea, discharging icebergs at their periphery, sufficient replacement through precipitation is needed to sustain them. Extreme wind and blizzard conditions are more frequent here than in the Tundra, because accelerated air drainage causes high-velocity winds to blow down from the elevated interiors (Fig. 12-7).

The cold air masses in the Polar Ice Caps are likely to give off no more than 10 inches of precipitation annually, the form being dry snow and minute ice crystals.

CONCLUSION

Of the earth's major climates those of the high latitudes are the most severe, though there is some variability here. Taiga shows the influence of continentality with very cold winters, colder than the sea-moderated Tundra, and, perhaps surprisingly, the Taiga may warm to above 90° F (32° C) on occasion. Because evaporation into frigid air is minimal, low precipitation amounts in these climates effectively maintain moist conditions. Even though the high-latitude climates present a barrier to human activity, these regions are of increasing interest owing to the abundance of mineral resources.

Figure 12-8 is a summary of all the climates and their controls presented in Chapters 9 through 12. Although there is much variation between climatic types, more than a little regularity characterizes their overall distribution. We know that controls are central to understanding climates and their distributional pattern. Latitude, for example, exerts the greatest control, but marine and continental effects are important, as are prevailing winds, pressure patterns, mountain barriers, ocean currents, and storm tracks. Thus the apparent climatic chaos initially seen in Map 1 may now be viewed as remarkably regular, with climates responding in a

Climate	*Location*	*Characteristics*	*Controls*
Tropical Wet	Doldrum belt and trade wind coasts	No seasons, always warm; rainfall: 65″ to 400″+ (min. 2.4″ per month), convectional, orographic, weak equatorial lows	Low latitude, low pressure (ascending air), mountain barriers (windward slopes, onshore Trades)
Tropical Dry	Tropical west coasts	Hot (hottest on earth) and dry; rainfall: less than 20″ (variable), convectional	Low latitude (Tropics), mountain barriers (leeward slopes, offshore Trades), subtropical high-pressure cells (descending air), cold ocean current
Tropical Wet and Dry	Transition between Doldrum belt and Tropical Dry; monsoon coasts	Always warm; pronounced wet and dry seasons, rainfall more than 20″, orographic, convectional	Low latitude, wind reversal (monsoons), annual N-S shift of wind and pressure patterns (offshore wind, descending air dry season; onshore wind, ascending air wet season)
Mediterranean	Subtropical west coasts	Mild temperatures (summer and winter); winter rain, summer drought, 15″ to 20″ average per year, frontal, orographic	Subtropical latitude, annual N-S shift in wind and pressure patterns (onshore Westerlies and midlatitude cyclones in winter, descending air from subtropical high in summer), cold ocean current, coastal location
Humid Subtropic	Subtropical east coasts	Warm, humid summers; mild winters with cold spells; moist all year, rainfall: 40″ per year is average, midlatitude cyclones, convectional, tropical storms	Subtropical latitude, annual N-S shift in wind and pressure patterns (Westerlies midlatitude cyclones in winter, onshore Trades from subtropical high pressure in summer), warm ocean current
Marine West Coast	Middle-latitude west coasts	Mild winters, cool summers; cloudiness, frequent light rain and drizzle, precip. 20″ to 60″ (200″+ on some coastal mts.), frontal, orographic	Middle latitude, onshore Westerlies, traveling midlatitude cyclones and anticyclones, warm current
Middle-Latitude Dry	Leeward (eastern) side of mountain chains in the midlatitudes	Similar to Tropical Dry with hot summers but with cooler winters and snow; arid: precip. 10″ or less (true desert), or 10″ to 20″ (semiarid), highly variable, frontal, convectional	Middle latitude, Westerlies, mountain barriers (rainshadow, descending air)
Humid Continental	Continental interior and east coasts of Northern Hemisphere middle latitudes	Hot, moist summers; cold winters with snow; no dry season; precip. 20″ to 50″+, frontal in winter, convectional in summer	Middle latitude, large land mass, traveling midlatitude cyclones and anticyclones, monsoons
Taiga	High-latitude continental interiors of the Northern Hemisphere	Long, cold winters; short, warm summers; persistent winter snow cover with drifts; precip. 5″ to 15″ summer max., frontal in winter, convectional in summer	High latitude, large land mass (continental interior), storms (arctic and polar front)
Tundra	High-latitude coasts in the Northern Hemisphere	Long, cold winters (but warmer than Taiga); short, cool summers; precip. 5″ to 15″, frontal in winter, convectional in summer	High latitude, coastal location, storms (arctic front)
Polar Ice Caps	Antarctica and Greenland	Severely cold winters, cold summers, precip. 10″ or less	High latitude, large land mass, frozen seas

Fig. 12-8. ***Summary of climates and controls.***

more or less predictable way to a complex of controls that vary over the earth's surface.

With an understanding of atmosphere and associated weather and climate now firmly within our grasp, we can go on to consider the hydrosphere. As we proceed, note that the two "spheres" are closely intertwined. Indeed, with each major shift in focus from atmosphere to hydrosphere, lithosphere, and biosphere, we bring with us what we already know and use it as a basis for furthering our overall understanding of the earth system.

KEY TERMS

Taiga
Tundra
permafrost
Lapps
Polar Ice Caps

REVIEW QUESTIONS

1. Even though the Tundra climate occupies the more poleward position, why is it considered milder than the Taiga climate?

2. Why are the high-latitude climates home to such a low percentage of the world population?

3. Why are names for vegetation—taiga and tundra—appropriate for high-latitude climates?

4. Why are there trees in the Taiga but not in the Tundra if the Tundra is in fact the warmer climate?

5. Why does there seem to be more snowfall in the Taiga than there really is?

6. How does the Tundra compare with other marine climates? How does it differ?

7. How does the Taiga compare with other marine climates? How does it differ?

8. Why are the high-latitude climates considered moist even though they receive less than 15 inches (38 cm) of precipitation per year?

9. Why is there so little vegetation in the Polar Ice Cap climate?

10. How has life changed for the Lapps since the beginning of the twentieth century?

APPLICATION QUESTIONS

1. Using the climographs in Figures 12-2, 12-4, and 12-6, distinguish between the high-latitude climates.

2. Not many people live in Antarctica, but a lot of people seem to like to visit it. In fact, some companies specialize in Antarctic expeditions. Go to the library and find out what is involved in a modern "vacation" to the Antarctic and plan an imaginary trip. Be sure to detail when you would go, what means of transport you would use, what you would do when you got there, what possible dangers you would face, and what you would need to wear.

3. We know that the Lapps survive in the Taiga and Tundra because of the reindeer, but how do the reindeer survive? Find out how the reindeer are specially adapted to their harsh environment.

4. Research and describe one environmental danger threatening the Taiga, Tundra, or Polar Ice Cap regions. What should be done about it?

PART THREE

EARTH'S WATER

CHAPTER 13

GLOBAL WATER DISTRIBUTION AND CIRCULATION

OBJECTIVES

After studying this chapter, you will understand

1. How water moves through the hydrologic cycle.
2. How and why water behaves as it does below the surface.
3. The difference between porosity and permeability.
4. The importance of the water table.
5. The behavior of various forms of groundwater.
6. How the various forms of surface water behave.
7. Why surface and deep-water oceanic currents exist and what patterns they follow.
8. Why oceanic tides exist and what patterns they follow.

Water from sea surface evaporation moves landward with the wind, precipitates as rain or dew, fills the streams and groundwater, and flows again to the sea in an endless cycle. In this aerial view of Honolulu and Pearl Harbor on the Hawaiian island of Oahu, clouds are building over the Koolau Range soon to release their stored water as part of the cycle.

INTRODUCTION

Earth is called the "blue planet" because oceans, lakes, and clouds dominate the view from space. Our home is unique in the solar system with its abundance of water, which affects every aspect of the earth's system. Water is a major component of plants and animals, a transporter of dissolved minerals, and a raw material in photosynthesis. Running water with its erosive power helps to wear away mountains and fill valleys with sediment. Breaking waves sculpt the shore into myriad forms, and lakes, ponds, marshes, and swamps dot the countryside.

In Part Two we saw that water is a major element in the dynamics of weather and in patterns of global climate. Now, in Part Three, we build on this knowledge with a direct focus on the hydrosphere. We begin with a general exploration of the hydrologic cycle; then we emphasize its various components such as soil water, groundwater, surface water, and water in the ocean. In Chapter 14 we extend the discussion to water resources and problems in water use.

THE HYDROLOGIC CYCLE

There is only so much moisture on the earth. None is lost or gained from outer space, but at any given moment some of it is in the sea, some on the land in the form of water or ice, some in the crust itself as soil and groundwater, some as suspended water droplets and gaseous water vapor in the atmosphere. The cycle by which water changes in both form and place is called the ***hydrologic cycle*** (Fig. 13-1).

All the world's water comes from the oceans, and most of it ultimately returns to this source. Evaporation from the sea provides the water vapor for rain and snow. If it condenses and precipitates over the oceans, then it returns immediately. If, instead, water vapor is carried over the land, a number of things may happen before it finds its way back. It may fall as snow and remain on the land until the spring melt before flowing back to the sea via rivers. Or it may become part of a semipermanent ice mass and be detained many hundreds or even thousands of years.

All the world's water comes from the oceans, and most of it ultimately returns to this source.

After condensing, much water vapor will fall as rain, some flowing off immediately across the surface, some remaining in lakes and swamps, some evaporating directly or through plants to be precipitated again, and some soaking the soil on its way deep into the ground to move gradually downslope to the sea or to appear at a lower level as surface springs and seepages.

This descriptive model is not absolutely perfect, although given time it comes very close. The lag in time before moisture returns to the ocean can be so long that water or ice held in lakes or glaciers may appear to be permanently lost to the dynamics of circulation. However, ice both melts and

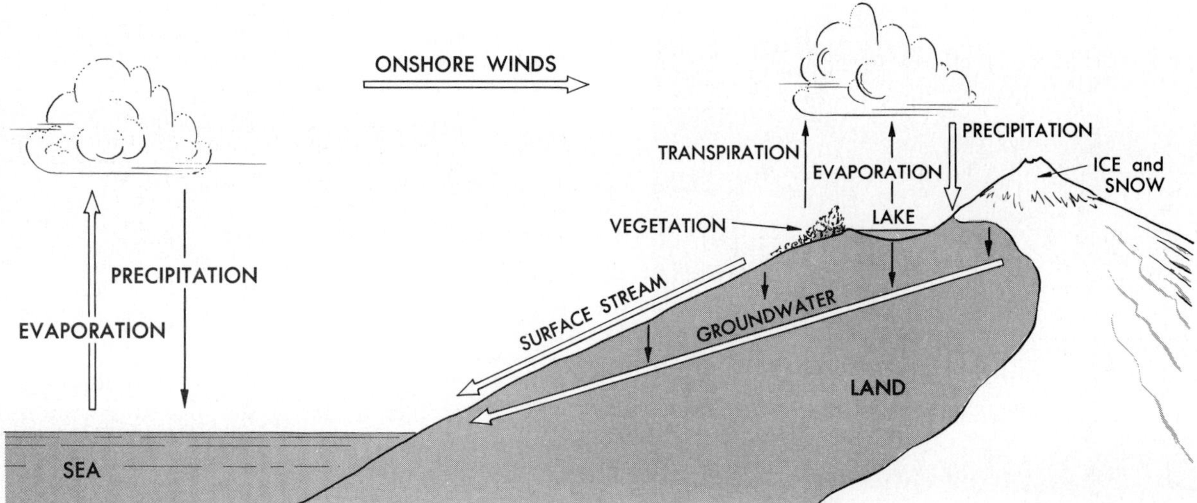

Fig. 13-1. ***The hydrologic cycle.*** *The basic cycle involves four elements: (1) evaporation of moisture from the sea, (2) movement of water vapor from over sea to over land, (3) precipitation of rain or snow onto the land, and (4) the return of this precipitation via gravity drainage.*

the surface or even on it after a particularly heavy rain. Normally, however, it is at least a few feet down. In desert regions the water table may be hundreds of feet below the surface.

In an area of deep, homogeneous, permeable material, the groundwater table is the top of the saturated zone.

Under usual conditions a water table will roughly follow the contours of the terrain that it underlies, rising under hills and declining under valleys (Fig. 13-4). This characteristic may seem somewhat peculiar because we think of water as having a level surface and we assume that the water table should too. In fact, percolation from above raises the water table in higher areas, and seepage into valley streams lowers it.

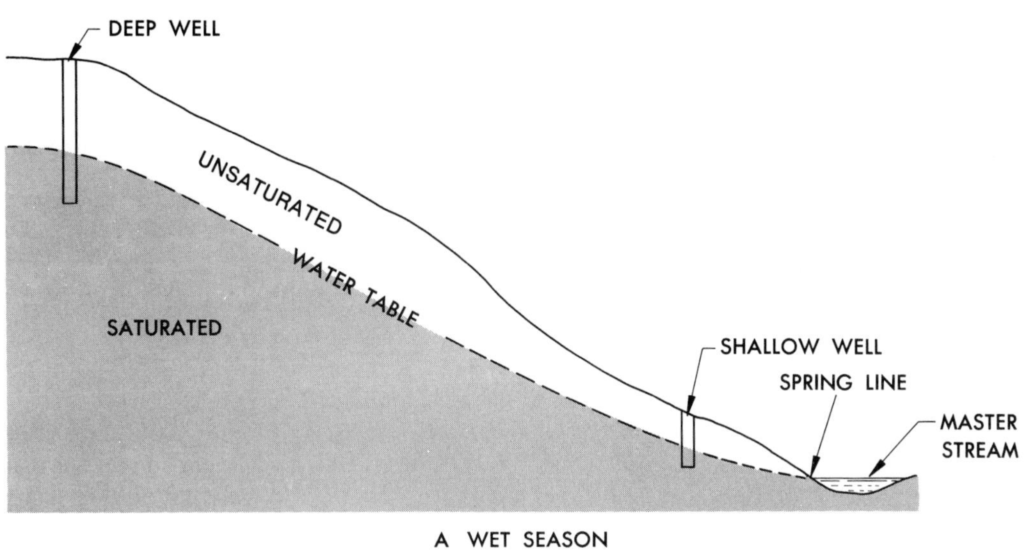

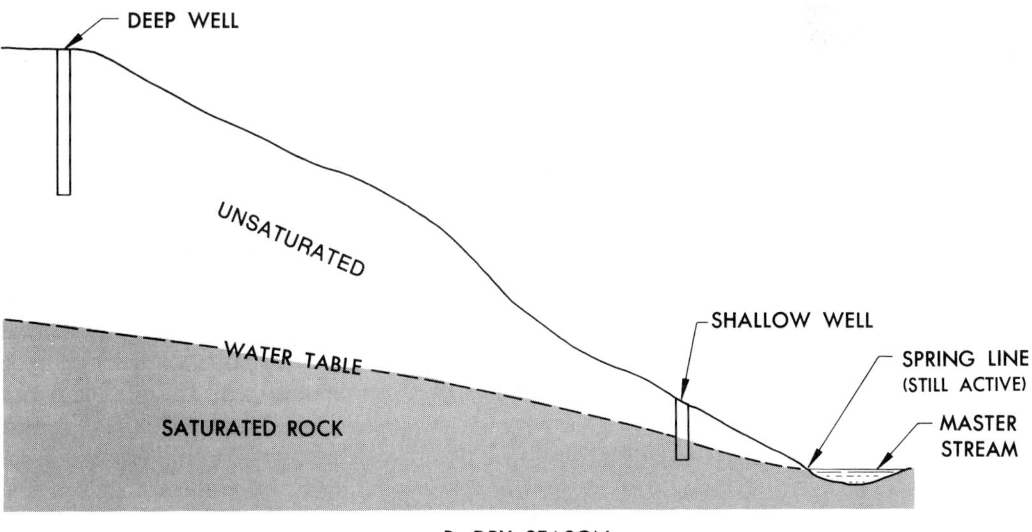

Fig. 13-4. ***The groundwater table.*** *(A) The water table tends to follow the surface contour. (B) During an extended dry period, it flattens out considerably, though still intersecting the surface at the spring line. As a consequence, the well at the top of the hill, despite its depth, is in a worse position than the shallow one on the lower slope.*

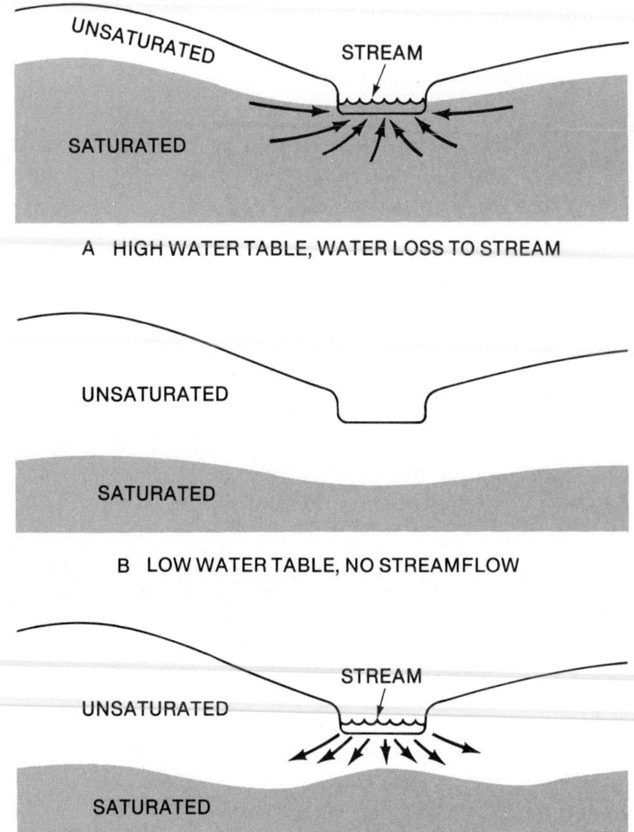

Fig. 13-5. ***Groundwater and surface streamflow.*** *(A) Where groundwater is high, as in moist areas, flow from the saturated zone increases streamflow. (B) With a low water table and no runoff to support streamflow, stream courses are dry. This is typical of arid regions most of the year and of humid regions after severe drought. (C) When stream channels carry runoff, as they do after an intense desert thundershower, seepage from the stream bed refills the groundwater.*

Surface streams usually carry runoff as well as groundwater. During long periods of drought, when there is no runoff, groundwater alone maintains river flow. Obviously, if the drought is long enough to allow the water table to sink below the level of the river bed, stream flow will cease (Fig. 13-5). This helps to explain why stream courses in arid regions do not carry flows year round unless fed by exotic streams from nearby wet areas. When the general level of the water table is well below the surface, only runoff can maintain stream flow. Such stream flow usually lasts only a short time, but there is an interesting twist: water in streams of deserts and seasonally dry regions seeps into rather than out of the groundwater. Consequently, water in channels under these conditions serves to replenish groundwater supplies.

During long periods of drought, when there is no runoff, groundwater alone maintains river flow.

Aquifers

Aquifer is the general term for a permeable rock mass or sediment layer that can hold, transmit, and supply water. Thus far we have been discussing underground layers with relatively homogeneous permeability, but many different strata may be present under a given region, each with its own characteristics. An important example is a permeable layer, such as sandstone, confined between impermeable strata and tilted so that it intersects the surface (Fig. 13-6). Water entering the exposed sandstone outcrop percolates along it as if in a pipe. Normal groundwater will be limited to the zone above this ***confined aquifer,*** for the impermeable layer immediately atop the sandstone stops its downward movement and also acts as a barrier between the water in the aquifer and the groundwater.

Water entering the permeable sandstone, confined between impermeable strata and tilted so that it intersects the surface, percolates along it as if in a pipe.

We will consider the topic of groundwater resources and well drilling in the next chapter, but a word about wells bored into confined aquifers is appropriate here. Shallow wells tap only the local groundwater that depends on seasonal rainfall, and in an arid region it is easily exhausted. Wells tapping the deeper confined aquifer have a more continuous supply, especially if the intake is in a better watered area. In addition, since the permeable surface outcrop is often the highest point in the aquifer, wells drilled to this lower level are usually ***artesian.*** That is, as long as the outlet is lower than the intake, the hydrostatic pressure within the closed system will make water rise in the well. Some artesian wells flow; others merely rise above the level of the general water table. Either type is called *artesian* (from the first recognized well of this type in the French province of Artois).

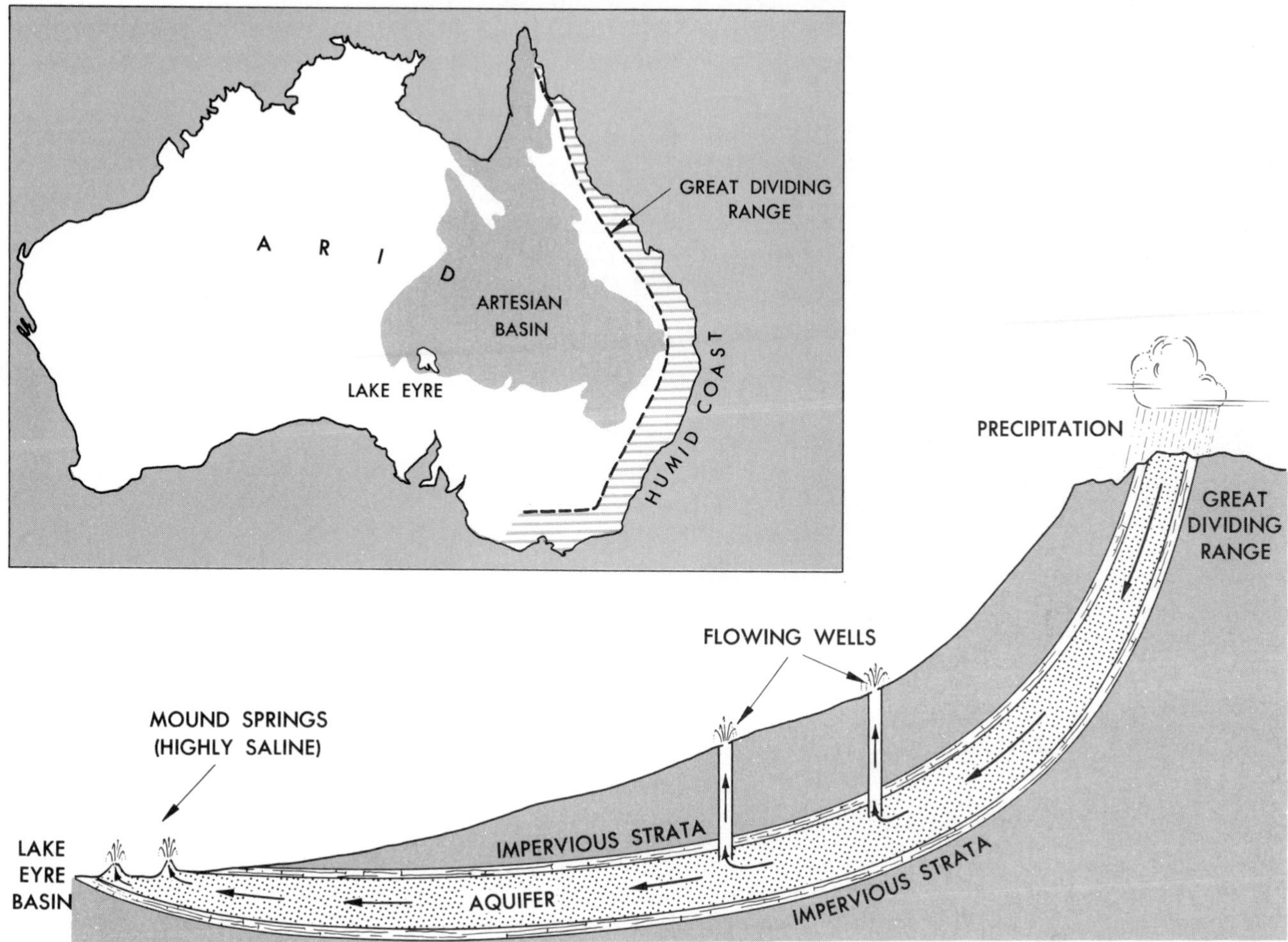

Fig. 13-6. ***Generalized cross-section of Australia's Artesian basin.*** *Shown is the typical configuration of an artesian system. Part of the heavy rainfall along the Great Dividing Range finds its way into the exposed aquifer that carries it downslope under the semiarid interior. It finally returns to the surface in the greater Lake Eyre basin below sea level. Because the outlet of the wells and the mound springs are below the level of intake in the mountains, they are artesian. Salinity in this particular example increases as the groundwater moves along the aquifer. Similar artesian systems are found in other parts of the world; in North America, the Dakota Sandstone, which underlies the northern Great Plains, has its intake in the Rockies.*

Hot Groundwater

The ***heat gradient*** of the earth's crust, that is, the increase of heat with increase of depth, may affect groundwater moving along a deeply buried aquifer. Theoretically, there is a depth limit to which rock can contain water. At several miles down, the overlying strata exert so much pressure that the pore space in the rock becomes too small to accommodate a liquid. Water may be present, but it is chemically combined in the rocks. Above this limit, enough heat still exists to raise the temperature of moving groundwater. Water from such deep sources is discharged in the form of a hot spring.

Water from deep sources is discharged in the form of a hot spring.

THOUSAND SPRINGS

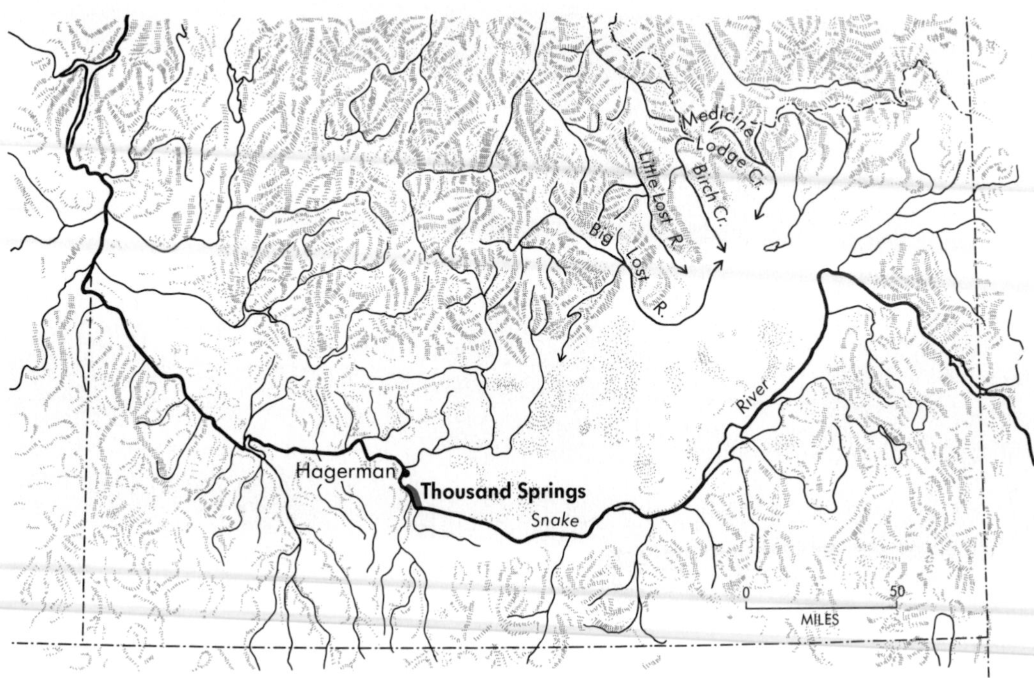

If we simply measure total volume of water delivered per hour, we find that perhaps the most remarkable series of springs anywhere in the world occur in the Snake River canyon of southern Idaho. Here for 10 miles (16 km) upstream from the town of Hagerman literally thousands of vigorously flowing springs issue from a porous stratum exposed along the north wall of the canyon. No one is absolutely certain of the origin of this huge volume of water, but two pieces of evidence point to a likely source: (1) no tributaries of any kind enter the Snake from the north for a 200-mile (322-km) stretch short of Hagerman, despite the high Rockies and continental divide not too far off; and (2) several streams drain southward from the Rockies, notably the sizable Big and Little Lost Rivers, which disappears abruptly into porous lava beds.

If this is the source of the Snake River springs, then we must picture an aquifer, probably a buried *vesicular* lava sheet connecting the Lost River sinks with canyon wall flows. Vesicular lava, when molten, is highly charged with gases. As it cools and congeals, the tiny apertures created by the escaping gases result in an exceptionally permeable layer. The underground water does not percolate sedately through this kind of aquifer but flows virtually unimpeded as in a pipe.

This type of spring is not rare, but much more common are the hot springs, geysers, and steam vents associated with regions of recently active vulcanism. Here the heat of the earth's interior has been brought well up into the crust where it is easily encountered by even very shallow groundwater. Such well-known tourist attractions as Yellowstone Park, the Rotorua district of New Zealand, and Iceland are fantastic collections of every conceivable type of thermal phenomenon, and each is underlain by hot volcanic rocks (Fig. 13-7). To be sure, cooling magma emits hot water, steam, and other gases, so not all this activity is a result of heated groundwater. But surface waters percolating downward to contact the hot rocks below are without doubt important elements.

These hot springs may achieve very high temperatures indeed; some may even reach the boiling point. If the heat is even greater, then the water becomes steam and issues through vents in the rock as jets. We use the term ***fumarole***

Fig. 13-7. ***Little mud volcanoes.*** *When a hot spring is fouled by viscous mud (much of it a result of chemical decomposition of rock under the influence of hot water), it finds its way to the surface via these little mud volcanoes. They swell and then collapse one after the other with "plopping" sounds. Yellow-stone mud volcanoes are sited directly over a thermal plume that brings heat from the mantle close to the surface in the middle of the North American plate. The entire geologic history of the Park region has been one of active vulcanism—two calderas (collapsed volcanic cones) one within the other, massive lava flows, and periodic explosions. Little mud volcanoes and related thermal phenomena are signals that the entire district is far from dormant.*

(or ***solfatara***) if the pure steam is contaminated with magmatic gases and emits an odor. ***Geysers,*** which are merely thermal springs that erupt at intervals, are of this same origin. They require a long tube reaching from the surface to the heated rocks below. As the tube fills, the superheated water at the bottom is under too much pressure from that above to turn into steam. When all the water reaches the boiling point and some bubbles out at the top, pressure is released and the water at the bottom flashes into steam causing an eruption. The great volume thrown out by many geysers indicates that an underground reservoir must be connected with the tube.

SURFACE WATER

Water on the earth's land surface takes many forms—from ponds and rivers to lakes, swamps, marshes, and reservoirs. The water contained can be saline or fresh, but all such features are intimately connected with the hydrologic cycle. Moreover, they are of direct concern to us not only as a source of water, but also as food, transportation, and recreation.

Runoff and Stream Flow

When the rate of rainfall or snowmelt exceeds the soil infiltration rate—a common occurrence in humid regions after soil water is fully replenished—water flows along the surface as ***runoff.*** At first it may collect behind small dams of leaves or twigs that hold it back temporarily, but as it builds up, it begins to move as ***overland flow.*** Such flow is not easy to see, although, if you look closely at the surface of a grassy hillside during a heavy rainshower, you may notice a thin sheet of water moving through the stems to lower elevations. Overland flow gives way to more organized movement through channels, called ***stream flow***—first as small rivulets, then small tributary streams, and finally major rivers.

When the rate of rainfall or snowmelt exceeds the soil infiltration rate, water flows along the surface as runoff.

Runoff itself does not provide all the water for most streams. As we have seen, flowing groundwater maintains stream flow between periods of heavy runoff. The total area drained by a major river and its tributaries carrying runoff

Fig. 13-8. ***Drainage basin of the Ialomita River, Romania.*** *With headwaters in the Transylvanian Alps, the Ialomita River flows into the Danube River east of Ploiesti. Its drainage basin area is 3340 square miles (8500 sq km). Note the typical treelike pattern of the drainage system.*

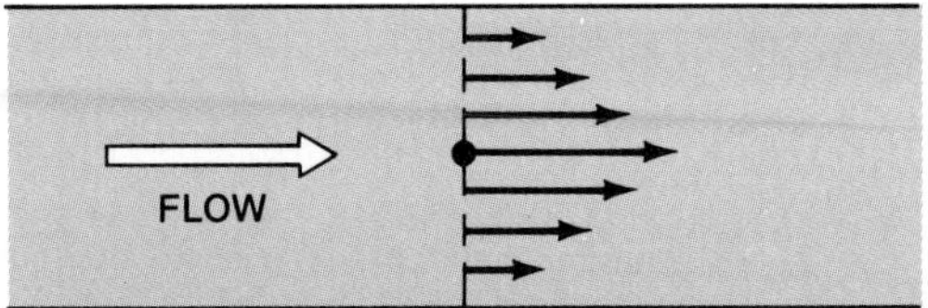

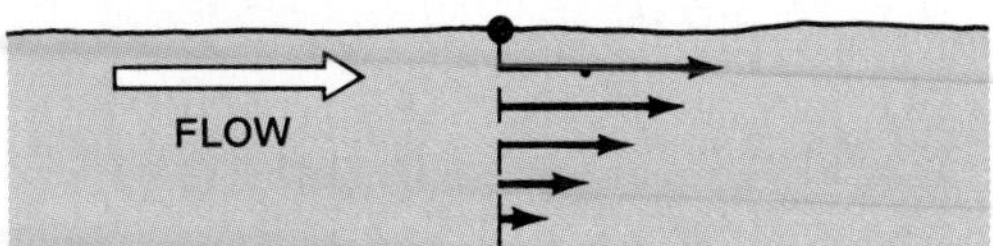

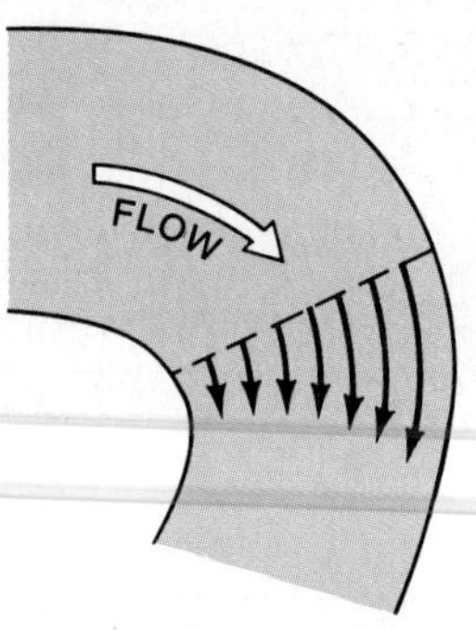

Fig. 13-9. ***Idealized views of streamflow.*** *Arrows show water movement in one second. When the channel is straight (A), maximum surface flow is near the center of the stream but decreases with depth (B). The highest velocities in a curved reach (C) are pushed to the outside of the curve. Thus streamflow is strongly influenced by friction and the curve of the channel. Most streams have far more complicated flow patterns owing to channel irregularities caused by rock outcrops, boulders, and deposited materials.*

and groundwater contributions is its *drainage basin* (Fig. 13-8). The size of the drainage basin depends on topography and water supply, and can be very large. The Amazon basin of South America, for example, is over 2,300,000 square miles (6,100,000 sq km).

The total area drained by a major river and its tributaries carrying runoff and groundwater contributions is its drainage basin.

How does water flow down stream channels? Most people are great water watchers, so you have probably observed some of its characteristic movements:

1. Not all the water in a particular length (or *reach*) of a stream moves at the same rate (Fig. 13-9). Friction along the sides and bottom of the channel slow velocity. Thus water moves most rapidly at the center of the stream near the surface—*if the stream is following a straight course.*
2. Along curved sections the highest speeds occur on the outside of the bends and the lowest speeds on the inside of curves.

The importance of these relative velocities will become evident when we consider stream erosion in Chapter 20, but here our concern is with *discharge,* the amount of water flowing downstream. The U.S. Geological Survey keeps gauging stations on all major rivers in the country to estimate water resources. Survey hydrologists use a *current meter* resembling a cup anemometer (Fig. 13-10) to measure velocity at numerous positions across the stream. A single measurement, as we now know, would not be as representative of the whole cross-section. Instead, we need to average many measurements to arrive at average velocity. We then multiply average velocity by the cross-sectional area. This lets us estimate stream discharge, which we usually express in cubic feet per second (or cubic meters per second).

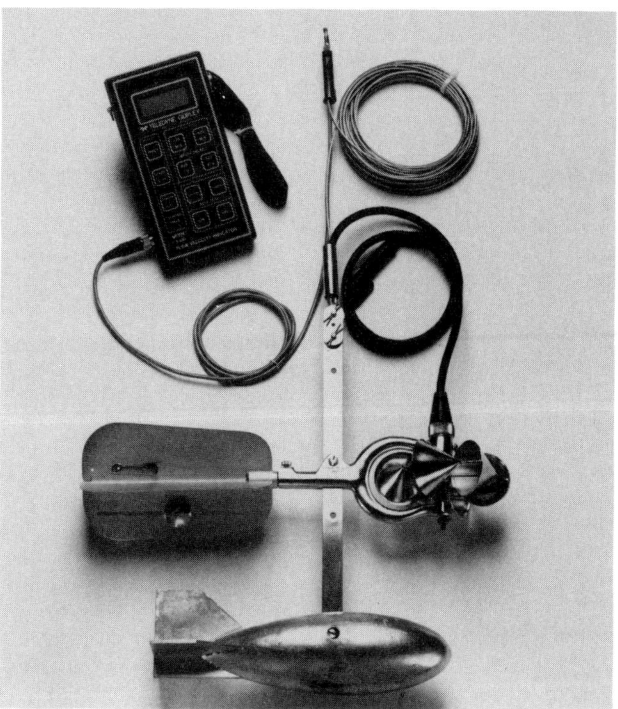

Fig. 13-10. ***Water current meter.*** *Like an anemometer used to measure air flow, the water current meter uses cups that respond to water movement. The device, suspended from an overhead cable and kept vertical by a finned weight, is placed at numerous positions in a stream. As the cups whirl, velocity is displayed electronically.*

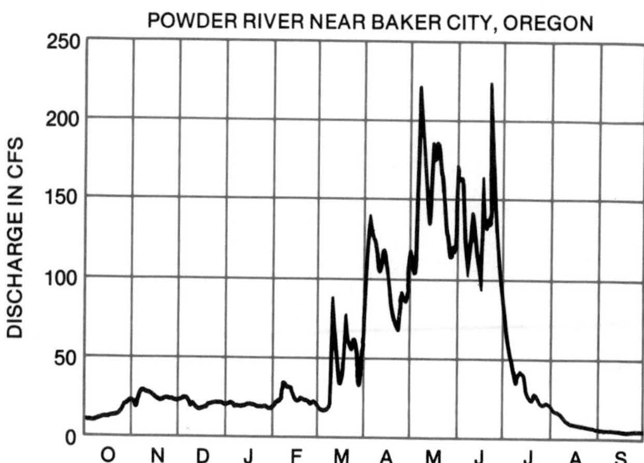

Fig. 13-11. ***A hydrograph in eastern Oregon.*** *Based on daily measurements of stream discharge in cubic feet per second (CFS), a hydrograph illustrates streamflow characteristics over time. This year-long hydrograph of the Powder River near the town of Baker City in eastern Oregon shows that discharge is greatest between March and July when snow is melting and least during the dry summer months. Low flows are maintained by flowing groundwater.*

A plot of discharge fluctuations over time is called a ***hydrograph,*** which at a glance can show the effects of heavy rain and runoff on stream flow, the timing of floods, and even the portion of river flow that is maintained by groundwater seepage (Fig. 13-11).

A hydrograph shows the effects of heavy rain and runoff on stream flow and the timing of floods.

Ponds, Lakes, Marshes, and Swamps

Flowing streams are important links in the hydrologic cycle, but, taken together, water bodies occupying topographic depressions contain a much larger quantity of water. Lakes alone contain well over 90 percent of land surface water, whereas flowing streams carry a scant 1 percent at any given time. ***Ponds*** are small lakes. They are often shallow enough for aquatic plants to grow on the sunlit bottom. Many ponds in arid areas and seasonally dry regions have water only during wet months. The ***vernal ponds*** of California and the Pacific Northwest, for example, contain water only in the spring but serve as the focus of an early season flurry of plant growth and animal life (Fig. 13-12).

Lakes are larger water bodies surrounded by land and may be filled with fresh or salt water. Evaporation from the broad surface of lakes not only returns water to the atmosphere but also leaves dissolved salts behind (Fig. 13-13). Thus, if excess salts are not reduced by the diluting effects of rainfall and fresh water inflow, and are not flushed out by outlet streams, water becomes too saline to drink or use for irrigation. Some of the world's largest lakes are saline. They are so large that they even carry the name of seas: Caspian Sea, Aral Sea, Dead Sea, Salton Sea. These and others like the Great Salt Lake in Utah, Lake Eyre in South Australia, and Lop Nor in China's Tarim basin have too little precipitation and runoff under Tropical Dry and Middle-Latitude Dry climates to compensate for evaporation.

Evaporation from the broad surface of lakes returns water to the atmosphere and leaves dissolved salts behind.

In contrast, humid regions support fresh water lakes. Many, like the Great Lakes of North America and lakes

Fig. 13-12. ***Vernal pond in southern Oregon.*** *During a short period each spring, this pond and many others like it play host to numerous aquatic plants, insects, and frogs.*

Fig. 13-13. ***Freah water lake.*** *This mountain lake with several inlet streams and an outlet is typical of lakes in moist environments. Even though evaporation at the surface leaves dissolved salts behind, salts are continually diluted at the inlet and removed at the outlet. Lakes in arid regions, in contrast, often do not have an outlet stream, and salts left behind by evaporation result in highly saline, less usable lake water.*

Vänern and Vättern of Sweden, fill valleys and other depressions carved by glaciers. Indeed, a glance at a world map reveals thousands of lakes in the glaciated zone across Northern Europe and North America and on the flanks of glaciated mountain ranges. Rift valleys and other structural basins may also hold lakes, like the chain of lakes in the African Rift.

Geologically speaking, lakes are short-lived. Climate change may rob a lake of its water supply, leaving only glistening salts on the old lake bed. Outlet rivers may cut away the exit so deeply that the lake water can no longer be impounded. Inflowing streams may deposit so much sediment that it fills the basin. Even the remains of fish and aquatic plants may accumulate to the point that open water cannot be retained.

Lakes are short-lived because of climate change, sediment deposits from inflowing streams, and the accumulation of fish and aquatic plants.

Advanced stages of lake filling are often marked by the presence of ***marshes*** or ***swamps***. Marshes, unlike swamps, contain mostly nonwoody plants such as grasses and sedges. Trees are more characteristic of swamps. Both marshes and swamps are classed as "wetlands" but are not limited to old lake sites; they occur whenever drainage is poor and water saturates the soil for long periods. Other typical sites for wetlands include deltas, flood plains, and wet coastal areas.

It is not surprising, then, that few lakes exist more than a few thousand years, but, as we will see in Part Four (Landforms), the earth's topography is anything but fixed. Climates change too. The coming and going of lakes is likely to persist well into the future.

OCEANS

The world's oceans dominate the hydrologic cycle. This vast reservoir covers well over two-thirds (71 percent) of the earth and contains more than 97 percent of all the water on the planet. Most rain falls directly to the ocean, and over 90 percent of precipitation that falls on the land comes from ocean evaporation. In a way the ocean is much like a large lake with countless inflowing streams but no outlet except evaporation. We would expect high salinity. Indeed, average salinity is around 3.4 percent, and the most abundant salt is the sodium chloride we use at the table. We would also expect the ocean to become even more salty with time, and yet it does not. One of the great mysteries still to be unraveled is the fate of excess salt.

Most rain falls directly to the ocean, and over 90 percent of precipitation that falls on the land comes from ocean evaporation.

Within the seemingly constant and random motion of the waters of the sea there is a fundamental order. Two basically predictable mobility patterns are (1) surface and deep-water currents and (2) tides. In the remainder of this chapter, we will examine these oceanic patterns.

Surface Currents

Some combination of prevailing winds, temperature variations, salinity differences, and bottom configuration sets the surface currents in motion. Once they begin to move, an additional factor comes into play—the deflective force of Coriolis. The result is circular hemispheric currents that always flow clockwise in the north and counterclockwise in the south. Along east coasts are "warm" currents flowing poleward. Along west coasts (at least in the tropics and subtropics) "cold" currents move equatorward.

Along east coasts "warm" currents flow poleward, and along west coasts "cold" currents move equatorward.

Warm and cold are relative terms, so that a current flowing poleward from the tropics into higher altitudes is warmer than the water it is flowing through. That makes it a warm current. Once that same current has crossed the high middle latitude and has moved equatorward to tropical waters, it is called cold, for a large part of its original heat has been dissipated and it is now colder than the water through which it is flowing. This is the standard pattern, regardless of ocean basin or hemisphere (Fig. 13-14 and Map IV).

ATLANTIC CURRENTS The most powerful of the poleward-flowing warm currents is the *Gulf Stream,* a veritable river in the sea, carrying huge volumes of indigo tropical water far to the north at speeds of up to 3 knots. Because of its distinctive color, the current is readily visible to the observer crossing it off the east coast of the United States. The balmy temperatures and occasional cumulus clouds and storms associated with it are reminiscent of the tropics.

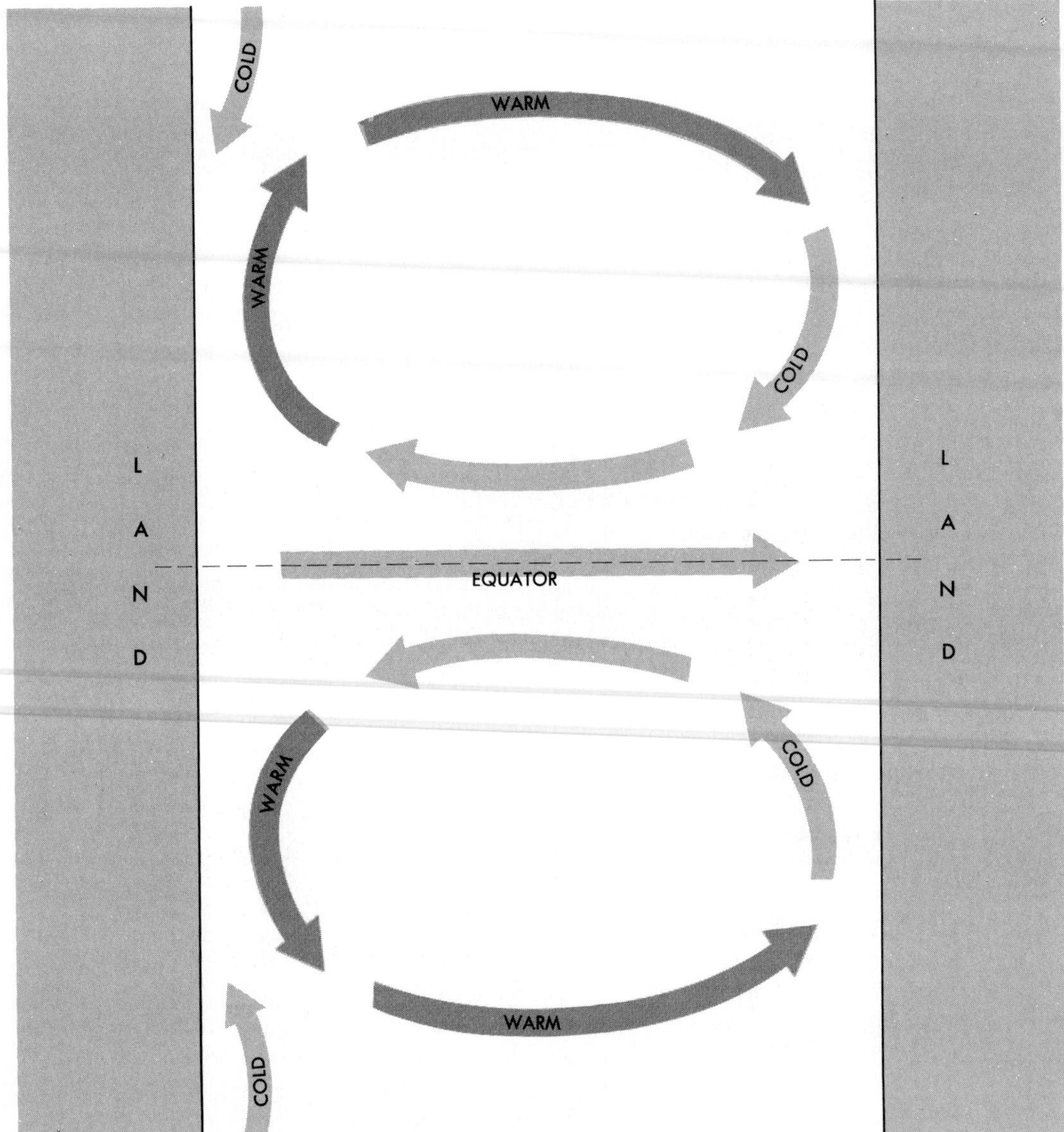

Fig. 13-14. ***Basic ocean current circulation.*** *This standard pattern is recognizable in every major ocean basin.*

Even this great current must respond to the influence of the earth's rotation. As it moves past Cape Hatteras, North Carolina, the Gulf Stream veers away from the coast toward the northeast, pulling in behind it a cold current from the Arctic called the *Labrador Current.* As the brilliant blue tropical waters and the icy green current from the Arctic merge over the Outer Banks off Newfoundland, the cold runs under the warm. Virtually permanent fog occurs here, as warm air is abruptly cooled and the roiling waters mix the rich nutrients ideal for plankton and fish life.

With the merger of the blue tropical water and the icy green current from the Arctic over the Outer Banks off Newfoundland, the cold current runs under the warm.

Now the Gulf Stream tends eastward, pushed along by the Westerlies. In the process it loses some of its vigor as it widens out, and it spills minor eddy currents both north and south as it goes. Upon reaching the coast of Western Europe in the vicinity of the British Isles, it once again tends to curve

THE GULF STREAM

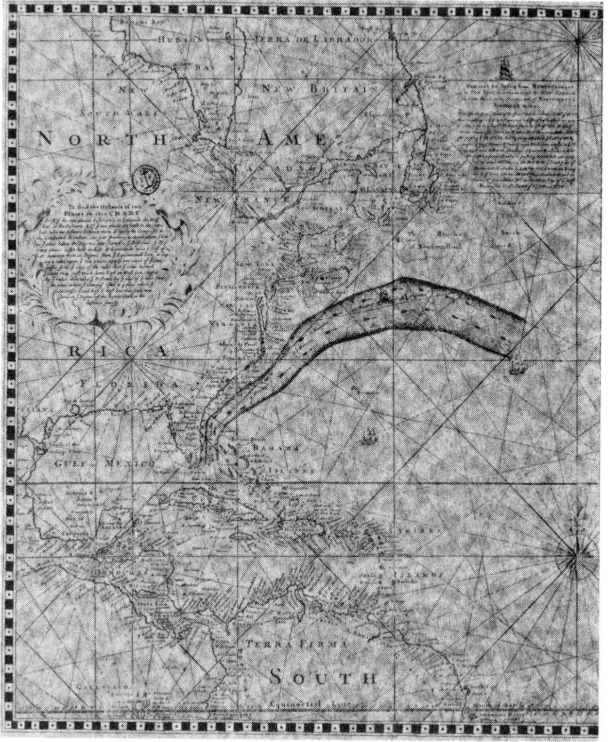

Franklin's map of the Gulf Stream.

In 1769 Benjamin Franklin (1706–1790), then Postmaster General, ordered the first chart of the Gulf Stream to be drawn as an aid for British mail ships. New England captains had long known something about the great eastward-flowing current through their close contact with fishermen working the outer banks. They would use it to their advantage heading east and carefully shun it on the westward passage. Englishmen, "too wise to be counselled by simple American fishermen," consistently took two weeks longer on the Liverpool-to-Boston run than did their American counterparts. They learned reluctantly.

Once that practical problem was cleared up, there still remained a tremendous shortage of knowledge about the Gulf Stream. Organized oceanographic investigation had never really been attempted. Yet there was an untapped source of information on ocean and wind currents that no one had considered seriously: ship's logs. How about collecting all those data? Young Lieutenant Matthew Fontaine Maury (1806–1873) became an early champion of this approach, reasoning that a large number of ship's logs, taken together, would present a significant overall picture of ocean conditions.

Maury was a scientist and curious scholar disguised in a naval uniform. Throughout his long career he quested for and championed the accumulation of knowledge in all things nautical. Accordingly, in 1842 when he was put in charge of the newly organized Depot of Charts and Instruments (later to become the U.S. Naval Observatory and Hydrographic Office), he quickly began collecting ship's logs. From that project emerged a document called *Wind Currents and Charts,* adopted internationally and frequently brought up to date by new observations.

Through the years such a great store of data on the oceans accrued that Maury felt he should make this material available in a book—*The Physical Geography of the Sea,* published in 1855. It eventually ran to 19 editions and was translated into five foreign languages. One large chapter dealt with the Gulf Stream. It began:

> There is a river in the ocean. In severest droughts it never fails, and in the mightiest flood it never overflows. Its banks and bottom are of cold water while its current is warm. The Gulf of Mexico is its fountain and its mouth is the Arctic Ocean. It is the Gulf Stream. There is in the world no other such majestic flow of waters. Its current is more rapid than the Mississippi or the Amazon and its volume is a thousand times greater.

(Actually its volume is three thousand times greater than that of the Mississippi.)

Maury has been called the Father of the Naval Academy, and today a building at the Academy bears his name. In addition, still at the top of every chart issued by the U.S. Hydrographic Office is the notation "founded upon the researches made and the data collected by Lt. M. F. Maury, U.S.N."

---

to the right. The bulk of this *North Pacific Drift* flows southward much cooler and more subdued. A lesser branch, split off by Ireland, swings northward, hugging the Norwegian coast past North Cape and eventually fading out along the Russian north coast. Cool as it has become by now, this branch warms these high latitudes and is responsible for ice-free harbors far north of the Arctic Circle.

The main current, cool now as it moves to the south and into tropical water off the coast of Portugal and Morocco, is called the *Canaries Current.* Its generally low temperature is reinforced by the upwelling of cold, nutrient-rich waters inshore as it pulls to the right. Once again conditions are highly favorable for oceanic microorganisms, which are basic fish food (Fig. 13-15). As the southerly reach is con-

Fig. 13-15. ***A fishing community in Japan.*** *Every modest cove in Japan, it seems, supports a little fishing community. The broad* Kuro Shio (Japanese Current) *sweeping along their coast encounters the cold southward probing* Oya Shio (Okhotsk Current) *just offshore, and the mingling of the waters encourages plentiful marine life.*

cluded, the Canaries Current becomes tropical and, warming rapidly, continues clockwise to flow due west north of the equator. Here it is termed the *North Atlantic Equatorial Current.*

At the center of this great North Atlantic whirlpool is a region unlike anything else in all the world; this is the *Sargasso Sea.* Each ocean has the same circular pattern of currents, and in every case its center is in the permanent high-pressure cell of the generally calm Horse Latitudes. Only the North Atlantic combines this quiet, currentless, windless sea with the great accumulation of sargassum, a brown algae also known as Gulfweed. Seemingly endless miles of ocean contain the algae brought in originally by the Gulf Stream from the West Indies where storms have torn it loose from its rocky coastal habitat. Through countless thousands of years, the sargassum has evolved the ability to maintain and reproduce itself in midocean.

The center of each ocean is in the permanent high-pressure cells of the generally calm Horse Latitudes.

The old legend of the Sargasso Sea as a graveyard for ships has a small basis in fact. Given enough time, any floating debris is likely to find its way into the heart of the spiraling current circulation. The Bermuda High is a permanent area of calms that makes sailing difficult. But the sargassum is scarcely thick enough to entrap any seaworthy craft, and the facts have been greatly exaggerated through the years.

The South Atlantic contains a mirror image of the North Atlantic current circulation. The poleward-flowing *Brazilian Current,* following the South American coast, is warm; the broad and still warm *West Wind Drift* flows to the east; the large and vigorous *Benguela Current,* with its associated upwelling, draws Antarctic water equatorward off of western Africa; and finally, the westward-moving *South Atlantic Equatorial Current* completes the circulation. Even the cold Labrador current finds a counterpart in the *Falkland Current* of the South Atlantic.

The South Atlantic contains a mirror image of the North Atlantic current circulation.

Between the two equatorial currents running to the west is an apparently out-of-place *Equatorial Countercurrent* that follows an opposite course. This is caused by surface water piling up on the west side of the Atlantic by the constantly blowing Trades and the easterly Equatorial Currents. The water level at the Atlantic end of the Panama Canal is several feet higher than that at the Pacific end—even though sea level is supposed to be the same the world over. However, the winds are onshore on the east side of the isthmus and offshore on the west. Here is the phenomenon that causes the Equatorial Countercurrent. As the major warm currents of each hemisphere curve away to flow poleward taking some of this water surplus with them, the remainder flows "downhill" to the east in the windless Doldrums. This same situation supposedly causes the high current velocities of the *Florida Current* and other elements contributing to the

Gulf Stream as they flow east to northeast through the restricted channels of the West Indies.

PACIFIC CURRENTS The Pacific currents display the same basic pattern as the Atlantic. The *North Pacific Equatorial Current* curves northward off the coast of Southeast Asia and becomes the *Japanese Current* (*Kuro Shio* or *Black Current* because of its dark blue color). This warm flow in turn curves right when it reaches central Japan to become the *North Pacific Drift*. In so doing, it draws in behind it a southward-probing finger of cold water from the Arctic called the *Okhotsk Current (Oya Shio)*. Here again, the cold current underruns the warm, developing a region of fog and often turbulent seas. Unlike the North Atlantic, the Aleutian chain and Gulf of Alaska do not allow more than a very minor part of the North Pacific Drift to invade Arctic seas. Essentially the entire current turns south to parallel the North American coast and become the cold *California Current*.

The Pacific currents display the same basic pattern as the Atlantic.

In the eastern Pacific the *South Equatorial Current* is a powerful stream as is the *Equatorial Countercurrent* to the north of it. Although the countercurrent maintains itself across the entire ocean basin, the South Equatorial Current runs into difficulties as it progresses westward into the island world of the east Indies and Australia. The general tendency is to curve south, but each island group and narrow strait forces adjustments and deflection of some part of the stream. In the absence of a continental mass to force poleward, it tends to become badly disorganized. One fraction, called the *East Australia Current,* moves along the coast of the southern continent and is eventually picked up by the Westerlies and swung east into the West Wind Drift.

On reaching the far tip of South America, the current turns northward, pulling with it large quantities of Antarctic water. As a result, the world's greatest cold current comes into being: the *Humboldt Current,* which maintains its low temperature through massive upwelling. It flows along the coast of South America to the equator. The penguin, classic symbol of Antarctica, has no business way up north at the equator, but there he is portly and dignified, on the Galapagos beaches, nourished by cold, imported Antarctic waters. Here again, but on a larger scale than in other oceans, the upwelling of nutrient-laden cold water stimulates the active growth of plankton and produces a great concentration of fish. However, as we have seen in Chapter 9, the arrival of El Niño can change all this.

INDIAN OCEAN CURRENTS The Indian Ocean is somewhat different from the others. South of the equator a reasonably typical circular pattern exists, but the limited sea of the Northern Hemisphere is so strongly affected by the seasonal monsoon winds that no permanent currents exist and the surface waters tend to move in opposite directions in winter and summer.

Deep Sea Currents and Waves

All the currents we have described to this point are surface currents, propelled in large part by winds. On occasion over continental shelves, they scour the bottom, but at most they represent only the movement of surface waters to a depth of a few hundred fathoms. What about the great mass of the sea? Does it move in any coordinated, predictable manner?

We know little about the ocean depths, but in theory the warm and thus lighter waters of the tropics should drift poleward at the surface, whereas the cold Arctic and Antarctic waters subside and move equatorward along the bottom. Experimental testings beneath the Gulf Stream have indicated that a current of cold water does indeed move in opposition to that at the surface, but no widespread, definable pattern has yet been found. In addition, deep sea photographs have shown well-defined ripples in the bottom sediments, very similar to those formed by waves near the shore. This evidence suggests the presence of waves or current in the ocean depths.

A current of cold water moves in opposition to that at the surface, but no definable pattern has yet been found.

The tides, of course, activate all the ocean waters in their daily rhythms, as do the sporadic *tsunamis*. There is also some cause to believe that deep waves, quite unlike the normal surface waves, occur widely. At the moment, however, we have only vague hints about deep sea movement, and oceanographers will require years of work before they can solve the riddles.

Tides

BASIC CAUSES Theoretically, the oceanic tides move in a simple and highly predictable way. We know their basic cause—the gravitational attraction of the moon and sun. We

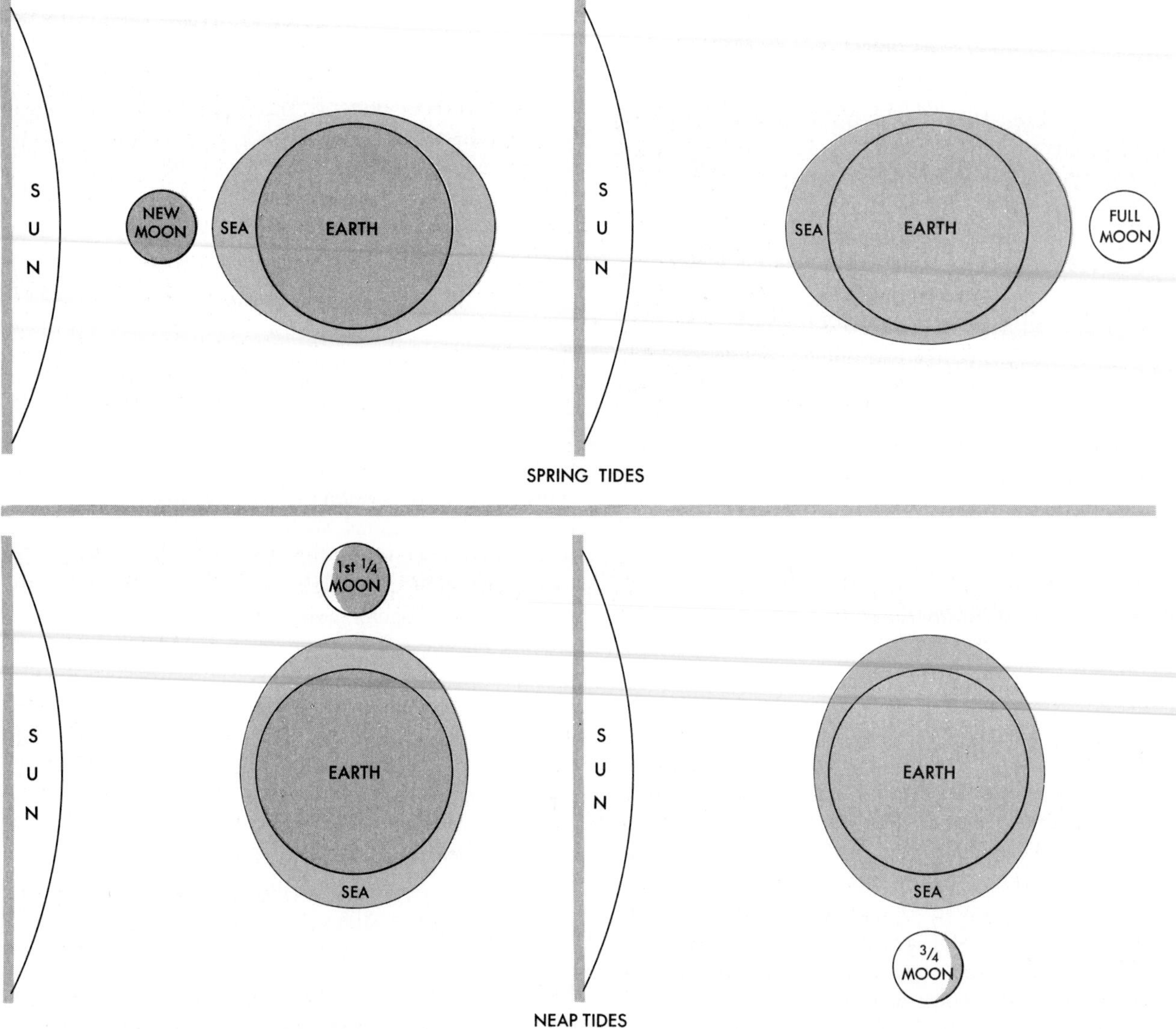

Fig. 13-16. ***Idealized relationship of tides to phases of the moon.*** *The basic cause of tides is the same everywhere, although countless variations result from strictly local factors.*

also know that the moon, though many times smaller than the sun, is so much nearer the earth that it exerts roughly double the force of the sun. Even the ancients were somewhat aware of the relationship between the moon and the tides, for in most places on the earth the time of high and low tide progresses by 50 minutes each day as the moon rises 50 minutes later. In addition, the height of each tide varies regularly each month as the moon waxes and wanes (Fig. 13-16).

When the moon is new (dark), it is between the sun and the earth, and the combined forces of both act to pull the mobile waters away from the earth into a bulge. At the same time, a similar oceanic bulge appears exactly opposite as a result of the moon pulling the earth's solid mass away from the fluid water. Now as the earth rotates on its axis, every coastline is exposed to each bulge 12 hours apart.

When the moon is new, the combined forces of the sun and earth pull the mobile water away from the earth into a bulge, at the same time that a similar oceanic bulge appears exactly opposite.

Fig. 13-17. ***A 20-foot (6-m) tide in New Brunswick.*** *Did all those people drown?*

A roughly comparable situation occurs once again during the full moon when the earth, moon, and sun are all in line, but this time the earth is between moon and sun. Two sizable oceanic bulges once more develop, and each location experiences two high tides. The highest tides of each month, a result of moon and sun acting in concert (new moon and full moon), are called ***spring tides.*** (This is an unfortunate term, for they have nothing to do with season.)

Subsequent tides exhibit the same sequence of daily occurence. The high tides become progressively lower until at the quarter phases of the moon, the relative positions of the sun and moon are at right angles and their forces oppose one another. Now the tidal bulges are much less exaggerated, and the mildest high tides, or ***neap tides,*** occur. So each day there are two high tides and two low tides six hours apart, but the greatest of these are at the new and full moon every month and the least are at the quarters.

The greatest tides are at the new and full moon every month, and the least are at the quarters.

VARIATIONS Although the attractive force of moon and sun is most certainly the basic cause of tidal movements, the actual character of the tides at any given place varies widely. Puget Sound has 12-foot (4-m) spring tides, Tahiti has only 1-foot (0.3 m), and parts of Korea, Alaska, and the Siberian Pacific Coast have 30- to 40-foot (9 to 12 m) tides. Yet all face out on the same ocean. The tides at opposite ends of the Panama Canal vary over 15 feet (5 m). Not even the typical cycle of two high and two low tides advancing 50 minutes each day is universally the same. The Gulf of Mexico has only one rise and fall; the Pacific Coast of the United States has two but of differing heights; and Tahiti does not have the 50-minute advance each day of its mild but otherwise normal tides (Fig. 13-17). These great variations, sometimes occurring only a few miles from each other, are the result of both bottom topography and local coastline configuration. A number of locations have such large tidal ranges that they are considered suitable sites for harnessing the kinetic energy of the tides to convert to electric energy in much the same manner as hydroelectric dams on rivers (Fig. 13-18).

The great variations in tide patterns are the result of bottom topography and local coastline configuration.

In inshore waters, tidal currents may reach high velocities, and shipping is advised to avoid certain restricted channels during flooding or ebbing tides. The Norwegian Maelstrom in the Lofoten Islands creates devastating whirlpools. Much-used Seymour Narrows between Vancouver Island and the mainland, a part of the Inside Passage to Alaska, has proved to be so dangerous (15-knot currents) that demolition of shoal rocks has been carried out in an attempt to tame its fury. Areas along a broken coastline, where retreating tides from opposite shores meet in frothing eddies, are both frequent and dangerous (Fig. 13-19).

Many river ports, such as Calcutta on the Hooghly, have ***tidal bores***—a virtual wall of water from a few inches to a foot or more high rushing up the estuary (river mouth) in the flood tide. Here and there a bore becomes so exaggerated that all navigation must be suspended temporarily. To produce a bore, a river mouth must experience not only a

Fig. 13-18. ***Tidal power plant, Saint Malo, France.*** *The world's largest tidal power plant is at the mouth of the Rance River Estuary at Saint Malo.*

Fig. 13-19. ***The perils of narrow channels.*** *Narrow channels not only confine and magnify normal tidal currents, but also frequently spawn a dangerous churning. Small craft are advised to stay clear of such sites when the tides are turning.*

high tidal range, but also an obstruction such as a bar or spit that holds the rising water back until finally it is overcome with almost explosive force. Then if the estuary is funnel-shaped, as is the case at Hangchow Bay, China, where perhaps the world's best known tidal bore occurs, the height of the advancing wave builds up rapidly. Most rivers discharging into the sea do not experience a bore, but nearly all are troubled by tides backing up normal stream flow, causing upstream flooding and allowing saline water to invade irrigation outlets and the like.

CONCLUSION

In the hydrologic cycle, water circulates between the hydrosphere, atmosphere, lithosphere, and biosphere. Its pathways can be quick and simple, with evaporation from the sea and return directly as rainfall; or they can be long and complex, with semipermanent loss in ice sheets and long-term storage in groundwater. The hydrologic cycle not only contains all the possibilities, but we can see now that many factors of physical geography affect water movement. For example, solar heating causes evaporation and drives the wind and pressure patterns that propel the great ocean currents. Gravity pulls water into the soil, sends rivers to the sea, and causes the tidal bulges. Earth rotation sets in motion the daily ebb and flood of the tides and gives the Coriolis force the power to bend ocean currents. Even topography, soils, rock types, and life itself affect the movement of water through the cycle.

With the hydrologic cycle, its myriad pathways, and its intimate relationship to the rest of the earth system in mind, we can now discuss, in the next chapter, water resources and problems in water use. Later, in Parts Four and Five, our knowledge of weather and climate and hydrology will help us to understand landforms, soils, and vegetation.

KEY TERMS

hydrologic cycle
infiltration
soil-water zone
percolation
groundwater zone
transpiration
porosity
permeability
groundwater table (water table)
aquifer
confined aquifer
artesian
heat gradient
fumarole
solfatara
geysers
runoff
overland flow
stream flow
drainage basin
discharge
current meter
hydrograph
ponds
vernal ponds
lakes
marshes
swamps
spring tides
neap tides
tidal bores

REVIEW QUESTIONS

1. Start with evaporation from the ocean surface and describe the various pathways water may take on its return to the sea.

2. How does the depth of the water table affect stream flow?

3. Distinguish between porosity and permeability.

4. What is an artesian groundwater basin? How can a well bored into an artesian aquifer yield free-flowing water?

5. How are geysers and fumaroles similar? How are they different?

6. Why is it necessary to understand the characteristics of streamflow to prepare a hydrograph? Of what value is a hydrograph?

7. Why do lakes tend to disappear after a few thousand years?

8. Describe and name the currents in the Atlantic, Pacific, and Indian oceans. How is the pattern of ocean currents related to the global wind and pressure system?

9. Account for the overall regularity of ocean tides.

10. Under what conditions are tidal bores generated?

APPLICATION QUESTIONS

1. Design an experiment that will allow you to demonstrate the different rates at which soil absorbs rainwater.

2. Find out about the water table in your area. Does groundwater play an important role in providing drinking water for your community? How is it tapped? Is it in danger of contamination? Explain how and suggest ways of protecting it.

3. Describe the largest and the smallest bodies of surface water in your area. How are they alike and how are they different?

4. Is there a stream gauging station in your area? Find out where it is located and why its work is important to your community. Construct a hydrograph with the data available.

5. Compare and contrast the surface current patterns in the Atlantic and Pacific, both Northern and Southern hemispheres. Since both are influenced by the Coriolis force, what accounts for their differences?

6. Look in a good world atlas and find five places in the world that exhibit variations on the basic tidal pattern. Describe them and see if you can account for their characteristics.

CHAPTER 14

THE EARTH'S FRESH WATER

OBJECTIVES

After studying this chapter, you will understand

1. How the hydrosphere is tapped to provide the necessary quantity and quality of fresh water when and where it is needed.
2. How electrical power is generated by streams and geothermal resources.
3. Water use problems of depletion and pollution.

The principal use of fresh water—irrigated agriculture.

INTRODUCTION

The next time you use water in your home for cooking; watering your lawn, garden, or plants; bathing; or disposing of wastes, think about where that water comes from. Trace it back to its origin. How was it made available to you? Think about how much you pay for it now and then about what you would do if it became short in supply or even unavailable. How high a price would you be willing to pay to continue receiving water in your home as you do now?

In the next few pages we will explore certain aspects of the management and use of water, the most precious component of the earth's physical system. This will help us reach a better understanding and appreciation of the hydrosphere.

FRESH WATER RESOURCES

Ninety-seven percent of the earth's water is in the salty oceans. Much of the remaining nonsaline water is locked up on the continents in ice caps, glaciers, more-or-less permanent snow fields, and inaccessible groundwater. Therefore, only a relatively small amount of water in the hydrosphere is potentially suitable for direct human use. There is no global shortage of fresh water, but its global distribution is uneven. Variation in climate, geology, and political jurisdictions may limit the supply of water in a given location. If continental precipitation were evenly distributed over land surfaces, all land would receive nearly 33 inches (83 cm) of water each year. If subsurface geology, landforms, and soils were globally uniform, it would be a simple matter to locate groundwater sources. However, such uniformity does not exist, and humans through the millennia have devised numerous methods to transport, store, and manage water from streams, lakes, and wells for drinking, agriculture, industry, power generation, waste removal, and navigation.

There is no global shortage of fresh water, but its global distribution is uneven.

Reservoirs and Groundwater

Capturing and storing water from stream runoff or collecting it directly from rainfall is a practical and ancient endeavor. In areas where the natural supply of precipitation is insufficient to sustain human population or maintain irrigated agriculture, the acquisition, storage, conservation, and intensive management of water resources are critical. Placing dams across rivers to form reservoirs has been a traditional method to keep stream runoff from removing needed water to the ocean. Digging wells into unconsolidated rock to reach groundwater and drilling into hard bedrock to reach confined aquifers for a water source are common ways to establish a more reliable water supply in arid climates or climates with a distinct dry season. In a number of locations around the world, confined aquifers are tapped or naturally bring water to the surface (see Chapter 13).

The Santa Clara Valley in California was once an agricultural region dominated by irrigated orchards and specialty crops, but it has become urbanized over the past 30 years. This change has essentially removed its agricultural base but not its need for water. Most of the valley, which is located at the south end of San Francisco Bay, receives variable annual precipitation, usually amounting to below 20 inches. A Mediterranean climate prevails, with all but a scant amount of rainfall occurring in the winter season. In the valley's agricultural days it was essential that crops have water available through irrigation in summer. Therefore the inhabitants established numerous wells in the valley and a few reservoirs in the surrounding hills to deliver water when needed. Over time, as water was withdrawn from below the valley surface, the water table receded and wells were made deeper in order to draw water from the greater depth. Figure 14-1 shows an overall trend of increasing well depth for most of this century.

REPLENISHING GROUNDWATER SUPPLIES Groundwater levels can go both ways: water can be taken out of the ground, and water can be put into the ground. It is not necessary to wait around for nature to recharge the supply. In Santa Clara Valley surface water that normally flowed into San Francisco Bay during winter could be captured in reservoir systems specifically designed to allow percolation into the natural underground reservoir in the unconsolidated rock material below the valley floor. Figure 14-1 reflects the introduction of reservoirs and *percolation ponds* to recharge groundwater during the 1930s and 1940s. After World War II, the water demands of an increasing population in the valley contributed to another significant lowering of the water table and water had to be withdrawn at still greater depths. Since the 1960s water supplies in Santa Clara Valley have been augmented with water from California's Central Valley. Thus the increase in the water table during the 1970s and 1980s might not have been as dramatic if additional water sources had not been found.

Because most of the rain in Santa Clara Valley (and most of California) arrives during the winter months, the long-empty stream channels, frequently grown up with brush and weeds during the dry season, simply cannot handle the runoff when the rains finally come. The result is serious

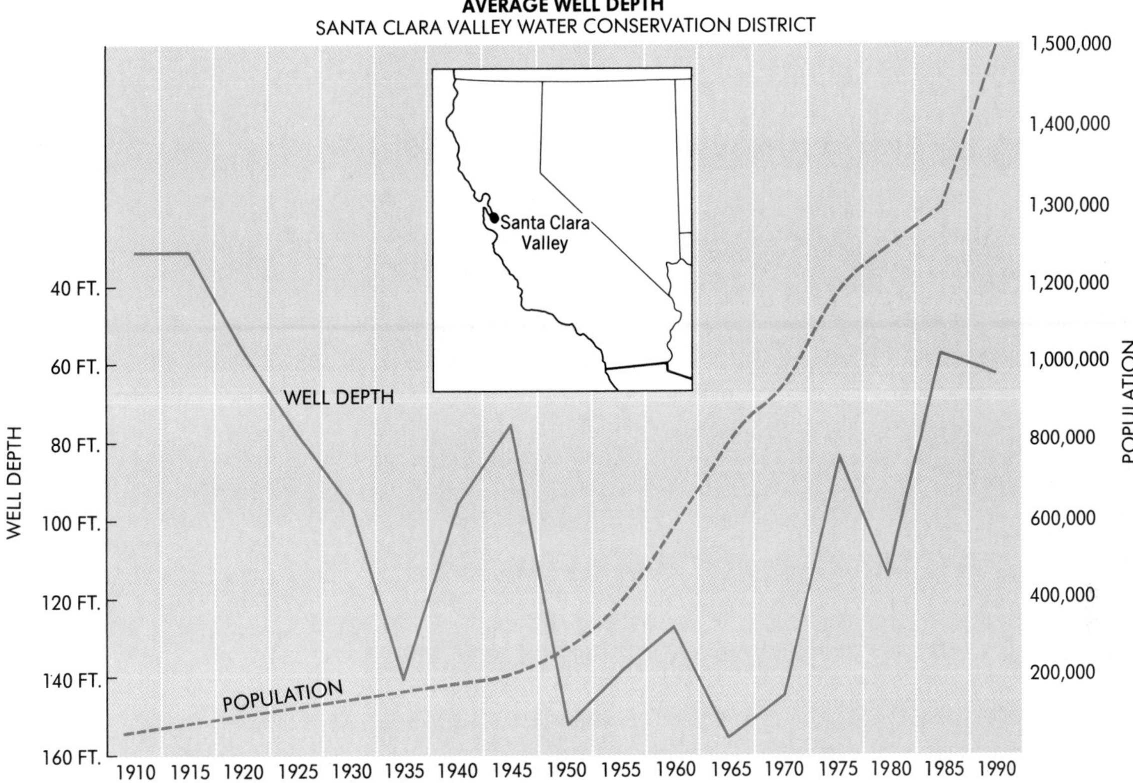

Fig. 14-1. ***Well depth in Santa Clara Valley.*** *A general trend toward increased depth is apparent. The fluctuations are, to a degree, a result of unreliable rainfall, but of greater significance has been the construction of reservoirs to recharge underground strata, as in the late 1940s. The sharp upturn in 1965–1975 reflects the introduction of Central Valley water.*

flooding as well as the loss of valuable water that would have been welcomed the summer before (Fig. 14-2).

Dams and reservoir systems constructed to impound a seasonal surplus of water and to enhance groundwater storage (Fig. 14-3) have many positive effects on water-related problems.

1. They control flooding.
2. They conserve water for use when it is needed.
3. They provide lakes for recreation.
4. They help recharge the groundwater supply by allowing water from holding basins to seep into the ground.

PROBLEMS OF EXCESSIVE GROUNDWATER REMOVAL Recharging the groundwater supply is important not only because underground water must be replenished, but also because the subsurface storage capacity may become reduced. When water is withdrawn from a valley underlain with deep deposits of unconsolidated rock, the water-holding capacity of the material is permanently reduced because the porous rock is compressed and compacted as large reserves of water are removed. Continual resupply is therefore essential to maintain an effective groundwater reservoir, particularly in urban areas and in areas of intensive irrigation. ***Land subsidence,*** the lowering of ground surfaces as a result of excessive groundwater extraction and subsequent compaction, has reduced the recharge capacity of aquifers in many areas. For example, land in the Tokyo area has sunk by as much as 16 feet (4.9 m) in the last 75 years. Similar subsidence from groundwater withdrawal has occurred in Mexico City, the San Joaquin Valley (California), and Phoenix, Arizona. Measures have been taken to stop the loss of subsurface storage, mainly through the use of artificial recharge systems that at least attempt to replace water as quickly as it is withdrawn.

When water is withdrawn from a valley underlain with deep deposits of unconsolidated rock, the water-holding capacity of the material is permanently reduced.

Fig. 14-2. ***California winter floods.*** *Marin County, California, has too much of a good thing at the wrong season. Although people in some areas of California experienced water shortages during drought years in the late 1970s and late 1980s, in many other years winter rains bring floods.*

Fig. 14-3. ***Water storage.*** *Anderson Reservoir in the hilly country bounding Santa Clara Valley of California (dam in left foreground). The recreational feature of the artificial lake is merely one of several in this type of multipurpose development.*

In some cases the groundwater being withdrawn from its subsurface reservoir is ***fossil water***—water that has taken thousands of years or more to build up in the aquifer. Clearly, it takes far more time to replenish such supplies, either naturally or artificially, than it does to extract them. As a result, the resource becomes severely depleted.

Another problem associated with excessive groundwater extraction is that it can allow seawater to intrude into

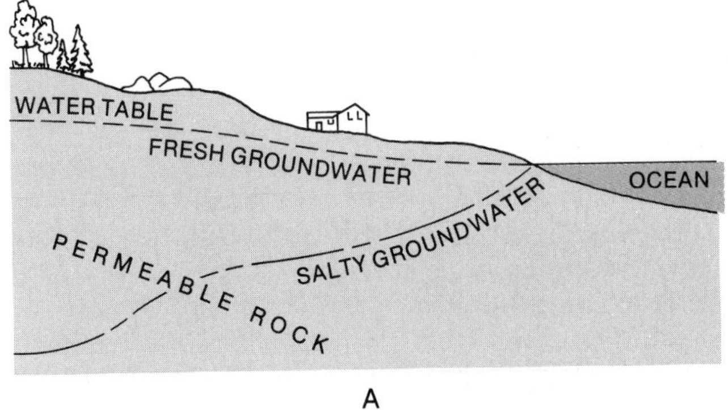

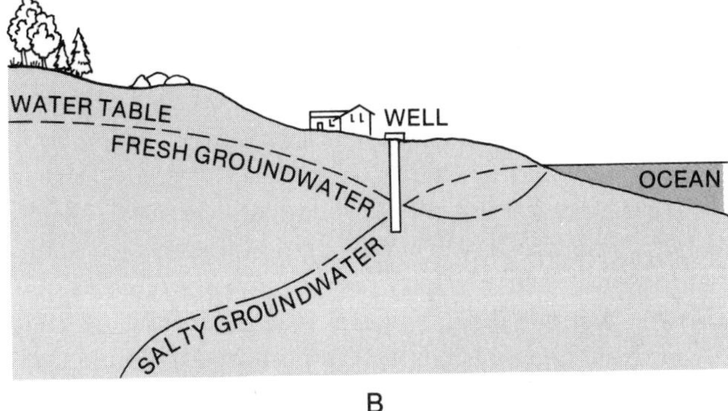

Fig. 14-4. ***Salt water intrusion.*** *(A) Along a coast the denser salt water from the ocean wedges below fresh groundwater in permeable rock. (B) When a well is installed and water is withdrawn too rapidly over time, the salty water can replace fresh water near the bottom of the well. The well has been "salted out."*

permeable groundwater strata along coasts. In humid coastal regions the water table is normally above sea level, and the greater relative density of seawater keeps the salty ocean-derived water wedged well below the fresh groundwater (Fig. 14-4A). However, if a well is put in and too much water is removed, hydraulic pressure is diminished and salty seawater can enter the fresh water zone (Fig. 14-4B). Thus the fresh water well will be ***salted out,*** which will often end in a local disaster unless a certain minimum groundwater table can be maintained.

SURFACE WATER/GROUNDWATER RELATIONSHIPS Generally, it is easy to see the relationship between the level of a lake, pond, swamp, or stream and the groundwater supply. Since much of the surface water feeds the underground strata, a lowered water table increases withdrawal from surface water bodies through percolation, thus lowering their levels. Because of their location at valley bottoms or in local terrain depressions, most of these surface water features also receive groundwater input from upslope. Thus each part of the system sustains the other. Local and regional planners generally recommend farm ponds, reservoirs, and flooded percolation basins as effective aids in stabilizing streamflow and well and lake levels.

EVAPORATION When water is kept in large reservoirs and little provision is made for directing it into an available aquifer, evaporation can be a problem. One advantage of storing water underground rather than on the surface is that this method almost completely negates losses resulting from evaporation. Subsurface storage is not always possible, however.

The places in the world that have the most need of water storage have long, hot summers. Under these conditions the effects of evaporation can be enormous. No two reservoirs have precisely the same rate of loss to evaporation; variations depend on such factors as climate, dimensions of water surface exposure, and percolation rates. Experiments in the northern Negev region of Israel have indicated losses in excess of 75 percent from evaporation alone. Obviously, any method that will cut down on such a loss rate must receive serious consideration. Some interesting work with the use of nontoxic chemical films to inhibit evaporation appears to hold real promise for both reservoirs and percolation ponds, especially those that are not used for recreation. Water skiing and fishing are not wholly compatible with an effective film cover.

Water Quality

Today we are also faced with threats to the quality of groundwater such as leakage from septic systems, gasoline storage tanks, hazardous waste sites, sewer lines, and mining operations. Pesticides and nitrates from agricultural fertilizers have also entered groundwater supplies. In parts of Nebraska, where water pumped from the Ogallala aquifer has reentered the ground after being used for irrigating crops, the water has become undrinkable and too saline for many crops.

Salts in groundwater and soil are even more serious problems in desert areas that have been made agriculturally productive through elaborate and expensive canal and drainage systems. High temperatures rapidly evaporate surface water on irrigated fields, leaving behind various salts that eventually make the soil so saline that crops can no longer tolerate it. This process is known as salinization (Fig.

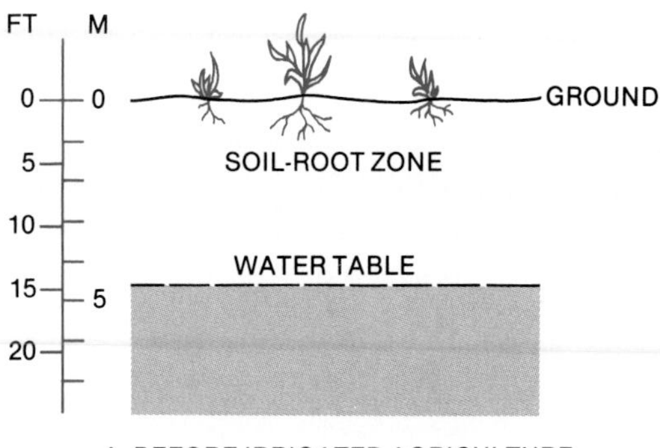

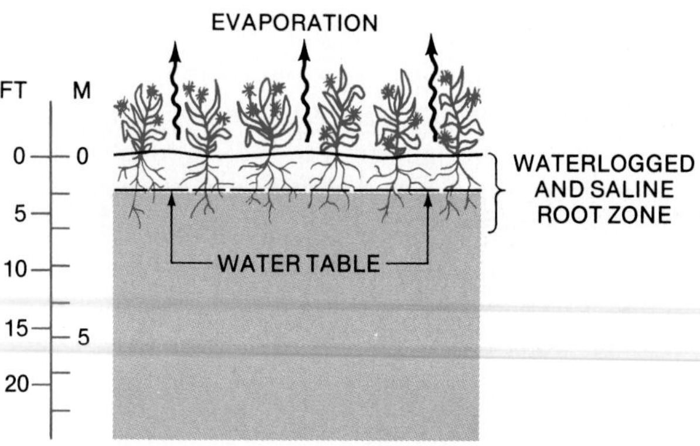

Fig. 14-5. ***Salinization–waterlogging.*** *This simple diagram shows the salinization–waterlogging process in poorly drained portions of hot, semiarid, and arid regions. (A) Near-surface water table in lower parts of valleys and basins. (B) Water table rises with introduction of irrigation water for crops; summer heat evaporates water, leaving behind salty surface crust and saline soil as root zone becomes waterlogged.*

14-5). (For additional discussion of salinization, see Chapter 24.) In desert areas the addition of irrigation water over time often raises the water table. The soils become saturated as excess water is unable to drain away from the root zone. This ***waterlogging*** restricts plant growth and lowers crop yields. In some cases the negative effects of salinization and waterlogging can be controlled by more efficient application of water and the installation of drains to take away excess water.

Even if water can be successfully drained out of agricultural areas where soils are saline and waterlogged, what is to be done with the drainage? This water sometimes contains not only high levels of salt, but also toxic levels of naturally occurring elements or compounds such as cadmium, chromium, arsenic, or selenium. In California, for example, water from streams in the northern part of the state, where water is relatively plentiful, has been transferred to the semiarid southern portion of the Central Valley for irrigation. Extensive corporate farms and ranches have blossomed over the past 30 years. As more and more water was pumped onto the land, however, waterlogging and salinization became problems; crop yields declined as soils held the water and accumulated evaporative salts. In some areas naturally occurring selenium became part of the soil-water solution. When present in high concentrations, however, selenium affects wildlife production and can be toxic to many forms of life. The potential danger of the selenium went unrecognized as an extensive drainage system was installed to remove excess water from agricultural land in the southern end of the Central Valley. In 1985, before the principal drainage canal reached its proposed destination in an arm of San Francisco Bay, its construction was finally stopped in response to concerns about the environmental impact on the bay.

The drain water was then diverted into Kesterson Marsh, a national wildlife refuge in the Central Valley located in the Pacific Flyway, the migratory path for millions of ducks and geese in western North America (Fig. 14-6). This wetland had already been designated a repository for farm wastewater from other parts of the valley. Soon after the canal's water began emptying into the marsh, the concentration of selenium began causing death and deformities in certain waterfowl species. The marsh became an environmental hazard, and in the late 1980s a project to clean up the marsh was begun at a cost of millions of dollars.

Meanwhile, waterlogging, salinization, and the buildup of pesticides continue to be problems in the agricultural areas to the south. The Kesterson Marsh is a good geographic example of how the earth's physical system is interconnected. When the immediate need for water is relieved by transfer from a region of surplus, what becomes of this water is influenced by local climate and topography, and there is a tremendous impact on the water table, drainage, soils, and eventually agricultural yields and wildlife habitats.

Aquifer Resources

The water in confined aquifers is another major source of fresh water in many parts of the world. One of the largest artesian systems in the world underlies much of interior Queensland and northern New South Wales in Australia (see Fig. 13-6). There a confined and highly permeable gravel bed receives water at its highest point in the humid Great Dividing Range and carries it for almost 1000 miles (1609 km) to the west beneath an arid-to-semiarid country

Fig. 14-6. ***Kesterson Marsh.*** *A waterfowl habitat contaminated by irrigation drain water.*

where water is at a premium. Many hundreds of wells tap this aquifer, supplying life-supporting water to what would otherwise be a desolate region.

Inevitably, however, too much water has been withdrawn, and most wells that at one time flowed freely at the surface must now be pumped. In addition, the quality of the water deteriorates with the distance it moves through the aquifer so that only in the eastern half of the basin is its salinity low enough to be reasonably useful. Even here the water is too brackish for general irrigation, and so it is used primarily for watering stock.

Although the Lake Eyre Basin is near the heart of the Australian continent, it is below sea level. This is where the lower end of the aquifer of the Great Artesian Basin reaches the surface. The highly saline water flows out here as artesian springs, evaporating almost as rapidly as it appears and creating salt deposits at each outlet. These are called ***mound springs,*** building up to a foot or two in height. They are the result of a unique combination of conditions: high salinity, hydrostatic pressure, and a hot, dry climate (see Fig. 13-6).

Similar artesian systems are found in other parts of the world, although none is as large as that in Australia. The Dakota sandstone, underlying the northern Great Plains in the United States, has its intake in the Rocky Mountains and is a major source of water for an extensive semiarid region to the east of the Rockies. North of the Po River in northern Italy is a line of flowing springs called the Fontanili. Fed by melting snows in the Alps, the underground water moves down through an aquifer of coarse glacial and old stream deposits. Many of the north Saharan oases exist because aquifers transport Atlas Mountain waters far out into the desert to reappear as springs.

Water Transfer

How do we attempt to solve the problem of supplying sufficient water of acceptable quality at the right place at the right time? Unquestionably, problems of supply exist in a world where the human population is increasing by over 80 million people a year (late 1980s). The demand for drinking water alone is an enormous one, to say nothing of the demand for water in large urban environments where flush toilets, green lawns, and water-swept driveways and sidewalks are taken for granted. For the world as a whole, domestic and recreational needs for water account for only 6 percent of the total fresh water used. Irrigated agriculture, with 73 percent of the total, consumes by far the greatest proportion of fresh water. Industry uses the remaining 21 percent. Of course, these percentages will vary widely from region to region, depending on the area's level of industrial and agricultural development and population density. In highly industrialized countries, for example, up to 80 percent of fresh water available may go to industry.

The principle of conserving water for use during the dry season brings us to the concept of transporting water from surplus to deficit regions.

AUGMENTING THE VOLGA

The Volga at Volgograd (Stalingrad)

There was a day in the distant past when the "Mother Volga" flowed freely to the Caspian Sea, gathering the waters from its tributaries as it went and delivering a huge quantity of fresh water at its mouth. But the modernization of Russia has placed a major emphasis on the use of the river as a reliable navigation artery, as a generator of hydropower, and as a seemingly endless source of industrial and agricultural water. The major restriction on the Volga's free flow has been a series of dams that have transformed a majestic rolling stream into a series of quiet lakes, and ever increasing water withdrawal has resulted in the "Caspian Problem."

During the past 45 years the level of the Caspian Sea has gone down dramatically (over 10 feet (3.0 m) and continuing to drop), diminishing fisheries, shoaling harbors, and increasing salinity. If this trend can be alleviated, or even balanced at today's level, it will be a major accomplishment. Is it possible to inject more water into the Volga system to solve the Caspian Problem, increase power production, and supply additional water to irrigate the arid wastes? Soviet engineers and planners began asking just this question as far back as 1932, and an intriguing plan has gradually evolved.

At about 61° N, just west of the Ural Mountains, the headwaters of three rivers are only a few miles apart. Two of these, the Pechora and Vychegda, flow north to the Barents Sea discharging quantities of unused water. The third, the Kama, is a major tributary to the Volga, flowing into it at a big bend near Kazan about 600 miles (966 km) upstream from the Caspian Sea. The proposed project would dam all three rivers near their headwaters to form a reservoir at the low interstream divide. A large part of this impounded water would then be released into Kama drainage to find its way directly to the Volga—in essence reversing the flow of the Pechora and Vychegda. In the process of adding water to the Volga, the new dams would produce power for the far north, and the Kama and its reservoir would be opened up to extensive navigation.

Critics of the project have been quick to point out the loss of agricultural, pastural, and forest land through flooding. The huge reservoir would cover 6000 square miles (15,500 km^2), which is almost as much as Lake Ladoga, European Russia's largest fresh water body. Probably 60,000 people would have to be resettled. Migratory fish running upstream from the Arctic Ocean would find the depleted and dammed rivers almost impossible to negotiate, although this loss could be more than offset by planting millions of other useful fish in the reservoir.

Are such efforts at reengineering nature really worth the price (estimated at something in excess of 1 billion rubles)? Even if money costs balance out through gains in agriculture and power generation, and so on, how do we measure the costs of relocated people and damage to fragile environments? Engineers tend to ask the question, "*can* it be done?" Geographers and planners ask, "*should* it be done?" At this point the answer to the first question is yes. Soviet officials are still mulling over the second.

We have already considered the principle of conserving wet-season surplus water in reservoirs and underground holding basins for use during the dry season. This same kind of thinking, applied on a regional basis, leads to the concept of transporting water from regions of surplus to regions of deficit. The pipe was invented to take water out of the river and to bring it into the house, or to move it into the next county, or to transfer it from the wet side of the Rockies in Colorado *under* the continental divide to the dry side, or to impel it 300 miles (483 km) out into the Australian desert to the gold mines of Kalgoorlie, or to transport it 1000 miles (1609 km) from northern California to the southern part of the state.

Why not transport Columbia River water to the arid southwestern United States, or Mackenzie River water of Arctic Canada to America's Great Basin, or the northward

Fig. 14-7. ***Water transfer.*** *Water from northern California is diverted to the southern part of the state through the California Aqueduct. This water is subject to many uses along the way. Here on the west side of the San Joaquin Valley desert has abruptly turned agricultural green; in metropolitan Los Angeles and San Diego growing needs for domestic and industrial water are met by the imported water. Today Californians face issues of cost for additional water transfer, continued growth and demand, and conservation and quality of water.*

drainage of the Ob River in western Siberia south to the warmer, arid region around the Aral Sea in Soviet Central Asia or to the Kama–Volga River project west of the Urals? It seems that the only limit is need coupled with engineering technology and cost feasibility.

People in the northwestern United States, however, are unlikely to react with great enthusiasm to the redirection of Columbia River water, nor will the Canadians be eager to syphon off Mackenzie water to irrigate cotton in California. The skyrocketing demand for water in southern California and Arizona in recent decades has indeed resulted in acquisition and reallocation of water from northern California (Fig. 14-7) and the Colorado River. Diversion and reallocation have pleased neither northern Californians nor Mexicans, who protested the salty trickle of Colorado River water that entered their country at the U.S. border.

The problem of water transfer centers on agreeing on a point where local and regional interests and human aesthetic values become as important as the progressive use of resources. We must and will use our resources, but perhaps when we begin to substitute the word "rational" for the word "progressive" (and then define rational to the satisfaction of one and all), we will reach an appropriate balance in water use.

The problems involved in water transfer have led us to emphasize successful use of the water we now have through recycling.

Recycling

The problems involved in water transfer have led us to emphasize successful use of the water we now have through *recycling.* Water is never used up. Whether we drink it, flush it, wash or irrigate with it, the water shows up sooner or later somewhat changed. We call it polluted. The trick is to pollute water less and to clean it up when it does become polluted.

The cleaning operation can have two or even three benefits:

1. If we can eliminate all the harmful effects of pollution or contamination such as water-borne diseases, destroyed and weakened wildlife, odors, and the like, it is well worth the cost involved.
2. The byproducts of the cleanup process may aid in cutting down that cost, as Milwaukee discovered many years ago when its sewage treatment facility yielded a profitable commercial fertilizer.
3. Clean water can be used again.

If drinking recycled sewage water offends your sensibilities, be assured that you have probably been doing exactly that for years. Most towns on rivers draw their drinking water from upstream and discharge their sewage downstream. If yours was the only town on the river you might feel better about your water, but there are usually people upstream using water the same way you do, and a little filtration and a dash of chlorine is your only protection. Actually, carefully monitored water recycling, under an enforced code, would ensure a much higher quality water than most of us now drink.

WATER FOR POWER

Petroleum, natural gas, and coal are primary sources of the energy we use. They are all fossil fuels, extracted from the earth and typically burned to produce heat, steam, and, ultimately, electricity. We also use nuclear power to produce electricity. However, water in nature can be harnessed directly to give us electrical energy—for example, when it is moving down streams under the influence of gravity or when it is heated naturally in the earth's interior and captured as steam to turn turbine generators.

Hydropower

Everyone understands that water runs downhill because of gravity. Happily, we have figured out how to convert this energy of motion into a form of energy that is more versatile in application—electricity. The energy produced from falling water at hydroelectric dams on streams accounts for a small proportion of the total primary energy produced from major sources, approximately 4 to 14 percent depending on the region. At present the largest hydroelectric plants are at Grand Coulee on the Columbia River in Washington State and at Krasnoyarsk on the Yenisey River in the Soviet Union.

Water in nature can be harnessed directly to give us electrical energy.

Significant potential for hydroelectric power generation exists, especially in developing countries, and electric output is expected to increase rapidly over the next 30 years. For example, the Itaipú hydroelectric complex near completion on the Piraná River along the Brazil–Paraguay border will have the world's largest production capacity (Fig. 14-8). To provide much needed energy for its people, the government of India has developed hydroelectric dams on rivers flowing from the foothills of the Himalayas. In Africa, several large dams that include hydroelectric-generating plants are planned for the major rivers of the continent.

In the past two decades advances in technology have resulted in "mini" and "micro" hydropower developments on smaller streams that produce energy for local consumption. These smaller projects are believed to have a less negative impact on the environment than larger, regional scale projects. Nevertheless, large dams will continue to be built, but social, political, and environmental problems may limit

Fig. 14-8. ***Itaipú power-generating complex.***

construction. Many large hydroelectric projects proposed in recent years have created controversy. For instance, the governments of Austria, Czechoslovakia, and Hungary are cooperating to construct two major dams, primarily for electric power generation, on the Danube River between Vienna and Budapest. The flooding produced by these dams along a 120-mile segment of the Danube will destroy a large wetland forest, contaminate groundwater and tributaries with polluted mainstream Danube water, and endanger plant and animal species unique to the area. Despite the protests raised by citizens concerned about the detrimental environmental impact, the project has begun and completion is expected by 1996.

Large-scale hydroelectric power-generating projects can create problems. Here are some of them:

1. Agriculture and hunting land may be lost to large reservoirs; dam construction proposals in Africa raise concerns about inundation of land used for crops and hunting.
2. Water reservoirs complicate the control of parasitic diseases such as schistosomiasis; the Aswan High Dam on the Nile River has enhanced the downstream habitat of the snail carrying the parasite that causes the disease in humans.
3. Dams constructed on silt-laden rivers trap sediment and eventually lessen power-generating capacity, since the supply of falling water is diminished. The Aswan High Dam, for example, captures silt and also controls floods, thereby cutting off the natural supply of silt and its nutrients to the cultivated floodplain of the lower Nile. With less deposition in the Nile delta, the waves and currents of the Mediterranean cause the shoreline to recede and valuable land is lost.
4. People may be displaced from towns and villages in the stream valley that becomes flooded by the dam's reservoir.
5. Flooding may destroy unique and valuable natural habitats and recreation sites.

Despite the possible negative effects, hydroelectric power generation offers some advantages over other methods of power generation and a number of peripheral benefits as well. Among these advantages and benefits are the following.

1. Water viewed as a renewable resource.
2. No problems with air pollution and radiation leaks.
3. Little effect on water quality.
4. A reliable water supply provided by reservoirs.
5. Improved navigation.
6. Enhanced fisheries.
7. Increased recreation opportunities.

Geothermal Power

Violent forces such as geysers and steam jets are not merely spectacular scenery; they have occasionally been put to productive use. A pioneer venture in the volcanic sub-Apennines of Italy was producing practical electricity from natural steam as early as 1913, and Luther Burbank was involved in drilling shallow steam wells at The Geysers near Santa Rosa, California, in the 1920s. These two sites, under new management and much expanded, are among the world's largest geothermal projects today. Other important projects are operating in New Zealand, Japan, Hungary, Iceland, Chile, Mexico, and the Philippines, and experimental work is being done in many other nations.

The earth's interior heat is an excellent source of energy, but with the technology currently available it is difficult and expensive to extract. To be economical at present, pockets of molten rock must be within 10,000 feet (3049 m) of the surface. However, with improved equipment and technology we expect to be able to tap expanded sources of useful heat at a depth of up to 16,000 feet (4878 m) in the first decade of the twenty-first century.

Electrical power is derived from a geothermal source much as it is derived from conventional coal- or oil-fired thermal plants.

Electrical power is derived from a geothermal source in much the same way as it is derived from conventional coal- or oil-fired thermal plants. Coal burned under a boiler, for example, produces steam, which is directed to the turbine blades of an electrical generator. In the geothermal plant nature supplies the heat and the boiler as well. It therefore offers a huge theoretical advantage in reduced cost *and* air pollution (Fig. 14-9). Recent experiments in the United States, Japan, the United Kingdom, and France indicate some success with the ***hot, dry rock system*** of producing steam. Surface water is pumped to great depths into fissures in hot rock, fissures where water is not naturally present. Then the heated water is pumped back under pressure to the surface where it is released as steam for generating electricity.

Natural steam and hot water, however, are always to some degree contaminated by dissolved salts and gases. Consequently, corrosion and scaling of equipment results. For example, an immense underground hot water reservoir is located near California's Salton Sea, but until recently its use has been impeded by an equally immense salt content.

Fig. 14-9. ***Harnessing geothermal activity.*** *Groundwater flashing into steam as it contacts hot subterranean rocks has the potential for cheap electric power production, for it does not need fuel and boilers to activate generator turbines. Here in New Zealand the geothermal plant is located at riverside for ease of condensing, and the steam is brought to it by a group of huge pipes. Steam plumes mark well-head sites.*

Furthermore, it is difficult to deal with hot wastewater and brine discharges.

We probably should not regard geothermal power as a universal substitute for traditional energy sources, any more than we should regard solar or wind-generated power as substitutes. In terms of practical development, however, the technology is much more advanced today than it was just a few decades ago. In addition, geothermal activity is not present everywhere. In this instance geography must be balanced against distance to places requiring the energy. Given the right geographic location and combination of physical and economic attributes, geothermal sources can add significantly to our total energy package by providing cheap and efficient power.

CONCLUSION

The resources of the hydrosphere are critical to the continuation of life on this planet. Whether we use water for drinking, irrigation, industrial cooling, generating electricity, or recreation, we must consider it a precious resource and not take it for granted. Knowledge of the way in which water cycles through the earth's physical system, and how it precipitates, runs off the land, is stored, and evaporates and then condenses to form rain clouds once again is valuable in itself. Even so, knowledge as well as appreciation of how we depend on and use water resources, either on the land or in the oceans, is also important. These resources require careful stewardship and vigilance. As reservoirs, percolation ponds, wells, aqueducts, canals, power plants, and productive irrigated farmland all remind us, we cannot afford to be complacent toward water.

KEY TERMS

percolation ponds
land subsidence
fossil water
salted out
waterlogging
mound springs
recycling
hot, dry rock system

REVIEW QUESTIONS

1. How do we store and divert surface water in streams to replenish groundwater supplies?

2. What are the benefits of dam and reservoir projects *other than* water storage?

3. How can groundwater wells in coastal zones become "salted out?"

4. How are the quantity and quality of water in the Great Artesian basin of eastern Australia affected by topography and climate?

5. Describe an actual example of a major water transfer project that takes water from one river drainage basin and transports it to another.

6. What problems may occur when a dam is constructed on a river? What benefits may occur?

7. How is electricity generated in geothermal plants?

APPLICATION QUESTIONS

1. Find out where the source of water is for your community. Is there more than one source? Does it come from groundwater, a stream, or a distant river basin by aqueduct?

2. Make a list of all recreation activities you can think of that are associated with fresh water. What quantities of water do they rely on? Do any of these activities alter the quality of the water used?

3. Telephone or write your local utility company and ask how the electric power you use in your home is generated. Is the electricity brought into your area from a great distance?

4. Check your library for recent magazine articles about fresh water and environmental concerns. What issues are being debated?

ST. JAMES

CHAPTER 15

OCEANIC RESOURCES

OBJECTIVES

After studying this chapter, you will understand

1. Why the continental shelves are the most important part of ocean bottom topography.
2. How minerals and food are extracted from the oceans.
3. The problems of desalinizing ocean water as a method of providing fresh water.
4. How ocean tides are used for energy production.

Cannery Row at Monterey, California, in the late 1930s. The sardine canning industry thrived in Monterey for about 30 years until the fishery off the California coast was significantly reduced by overfishing and by changes in ecological conditions.

INTRODUCTION

If you have been to the seashore, you have no doubt looked out over the ocean and been impressed by its vastness and fascinated by the energy of waves breaking on the shore or crashing against a cliff face. Have you ever thought about what the ocean provides beyond this aesthetic experience? How do people use oceans? Ocean waters cover over 70 percent of the earth's surface, but the resources of the oceans, the food and minerals they may provide, are geographically variable. Thus the actual and potential extraction of food and minerals from the seas and ocean bottoms is not the same everywhere. For example, some ocean regions may have an abundance of fish and others may not. Now that we have begun to understand the importance of ocean resources, we must also begin to understand their geographic complexity and the need for conserving and protecting the ocean's abundant life and minerals.

In the previous chapter, we focused on the management and use of water resources on land. Now it is time to turn to the great salt oceans of the planet, which cover approximately 71 percent of the surface. The major ocean basins are without question large and deep. Average depths run in the neighborhood of 13,000 to 14,000 feet (3960 to 4270 m), or almost 3 miles (5 km). The basins hold about 326 million miles^3 (1.36 billion km^3), or about 97 percent of the earth's water. The general physical configuration of the ocean bottoms is shown in Figure 15-1.

OCEAN BOTTOM TOPOGRAPHY

At present, the most important part of the ocean, as far as resources are concerned, is that which overlies the submerged margins of the continents. In some cases this ***continental shelf*** extends for many miles offshore; elsewhere it may be very narrow. The continental shelf eventually falls away sharply to greater depths forming one of the most imposing slopes in the world (although few of us have seen it). Varying in relief from place to place, the average height of this underwater continental slope is about 12,000 feet (3658 m), with a maximum height of as much as 30,000 feet (9146 m).

The most important resources in the ocean are found overlying the submerged margins of the continents.

At the base of the continental slope is a broader, gentler slope, the continental rise, which is made up mostly of accumulated sediment and stretches between the continental slope and the deeper ocean bottom.

Although the deeper parts of the ocean that extend beyond this rise have some irregular topography, in many areas the bottom forms an extensive ***abyssal plain.*** Volcanic islands and midoceanic ridges associated with plate tectonics (Chapter 17) complete the picture of ocean floor topography.

The Continental Shelf

Where it exists, the continental shelf is considered a legitimate part of the continent—just a temporarily flooded part (Fig. 15-2). Even a minor lowering of sea level would expose vast tracts of this sea bottom, as it has undoubtedly done in the past. Dogger Bank, a particularly shallow fishing ground in the North Sea, is part of the continental shelf today. Yet trawlers have brought up bits of wood and artifacts indicat-

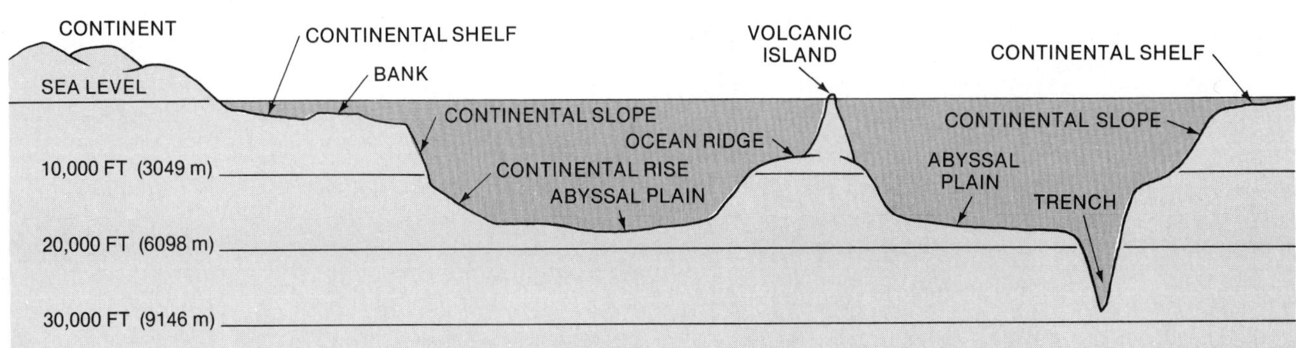

Fig. 15-1. ***Generalized profile of the ocean bottoms (vertical scale greatly exaggerated).***

ing that it was recently forested and inhabited land. Off the United States' Atlantic Coast mammoth teeth, giant sloth bones, and bits of peat containing grasses and twigs have been dredged up from a bottom now 50 fathoms (91 m) below the sea.

The actual breakoff point between the continent and the ocean depths is at the outer edge of the continental shelf, wherever it might be. Oceanographers used the arbitrarily selected 100-fathom (183 m) line as the critical point for many years, but increasing knowledge of undersea topography has complicated our view of the region. In 1953 the International Committee on the Nomenclature of Ocean Bottom Features redefined the continental shelf as "the zone around the continents, extending from low-water line to the depth at which there is a marked increase of slope to greater depth." In other words, shelves vary in depth; we know where a shelf begins by the appearance of a sharp break in topography.

If every continental shelf were exposed and only the true ocean basins contained the seas, the land surface of the earth would measure about 35 percent and the ocean surface would recede to 65 percent.

LIFE ON THE SHELF The continental shelf is the part of the sea bottom that is most like land—life abounds. The critical element is light, which penetrates in sufficient quantities to support photosynthesis. In this softly lighted blue-green world, vegetation growth is plentiful. It is characterized not merely by seaweed anchored to the bottom rocks (Fig. 15-3), but also by myriad swarms of ***phytoplankton***—those minute forms of floating plant life that form the base of the marine food chain.

The continental shelf is the part of the sea bottom that is most like land—life abounds.

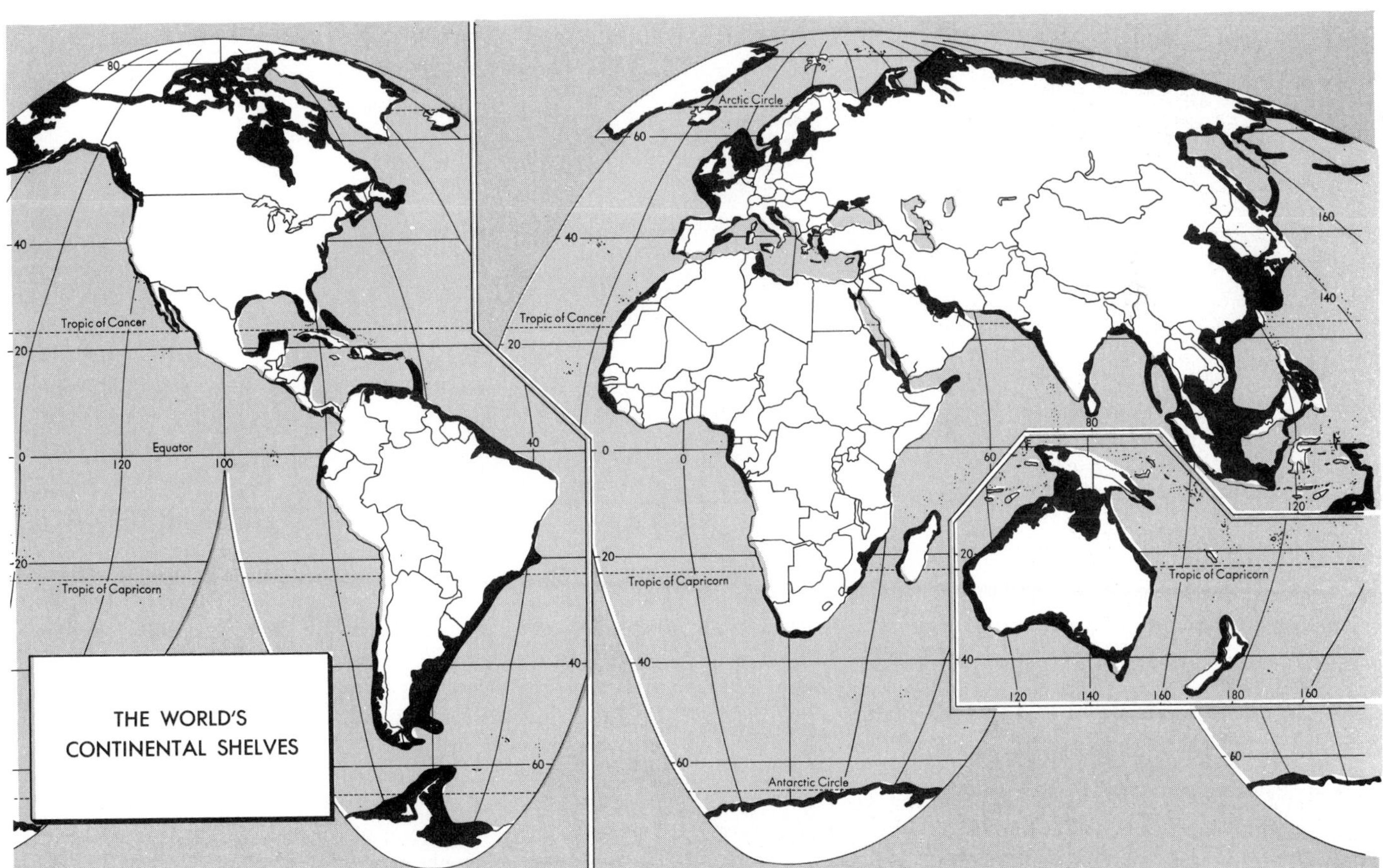

Fig. 15-2. The world's continental shelves. The great ocean basins lie beyond the margin of these shelves.

SARDINES AND MONTEREY CANYON

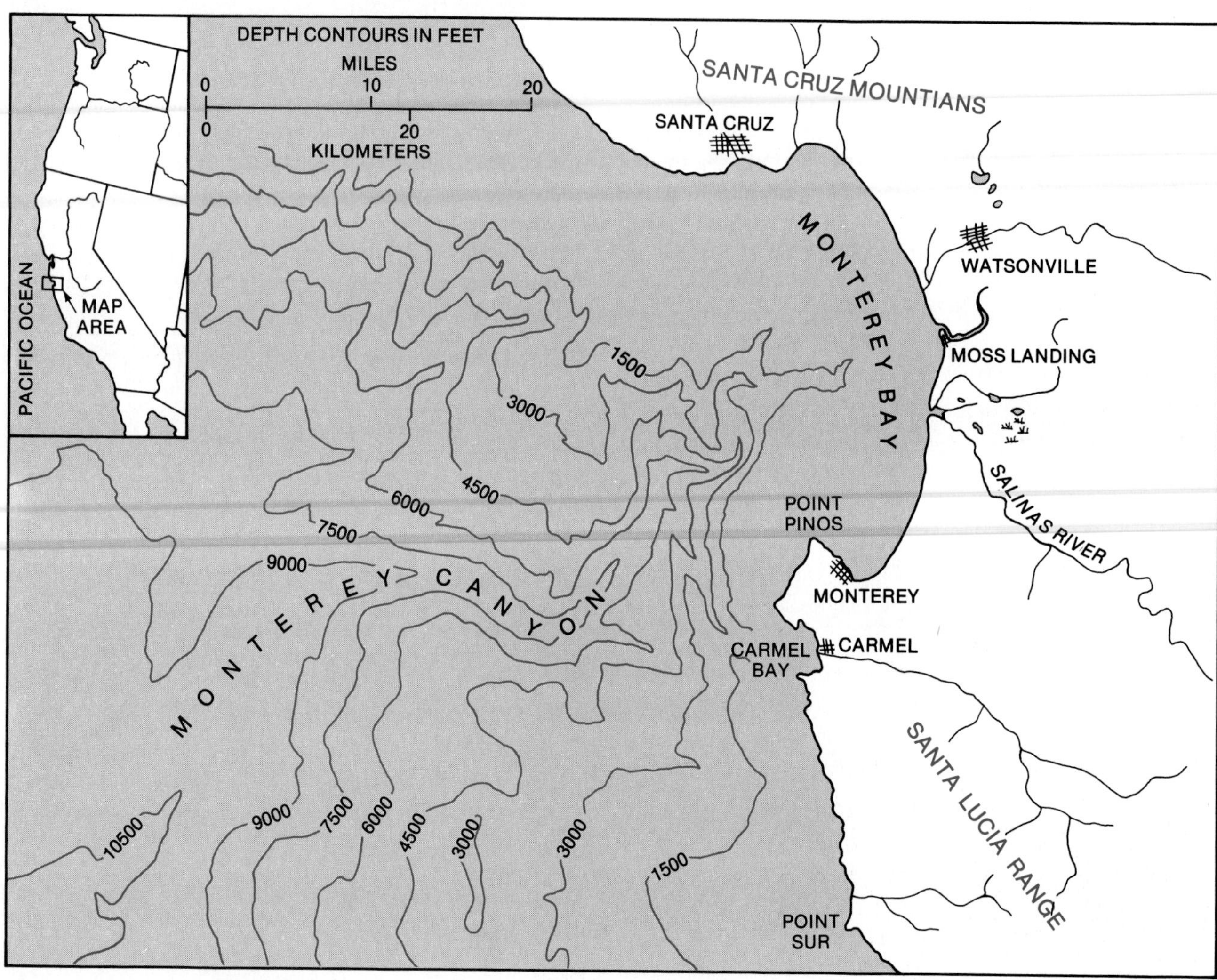

Monterey Canyon is not a land feature. It is a steep-walled submarine canyon that begins very close to shore in Monterey Bay on the central coast of California. From here it sweeps to the southwest below the level of the continental shelf eventually to achieve a depth of 6500 feet (2000 m) at 25 miles (40 km) from shore before continuing its winding course at a gentler gradient into the depths of the Pacific Ocean.

This deep canyon has a significant influence on the environment and resources of Monterey Bay and adjacent ocean. At these middle latitudes along the Pacific Coast of North America the ocean near shore is subject to ***upwelling***. That is, deeper, colder water replaces the sun-warmed surface water as it is pushed aside by prevailing winds. This cold, upwelled water brings with it oxygen and nutrients to replenish ecosystems that produce the plankton on which fish and sea mammals depend. The presence of the Monterey Canyon enhances the fishery off this coast, since it is a local source of very deep, cold, and nutrient-rich water. Here schools of fish congregate to feed and spawn.

The ocean floor topography and upwelling account, in part, for the great Pacific sardine (*Sardinops caerulea*) fishery, which was so productive here in the first half of the twentieth century. Monterey, once the capital of early California, emerged as a commercial fishing port in the first decade of this century, supported by one of the largest fisheries in the Western Hemisphere. The national and international demand for food during World War I firmly established the Monterey sardine canning industry. At its peak years in the late 1930s, the local sardine catch exceeded 1 billion pounds or 500,000 tons (454,000 metric tons) annually.

By 1940 most of the sardines were being processed in "reduction" plants that converted whole fish to oil for industry and dry fish meal for livestock feed. This was a more profitable use of the resource than was canning sardines for human consumption. By the 1950s the "silvery billions" described by John Steinbeck in his colorful novel *Cannery Row,* set along the Monterey waterfront in the 1930s, had declined to a level where catches could no longer support the fish-processing industry at Monterey. Cannery Row now attracts tourists who come to see architectural remnants, to savor hints of the once pungent past, and to treat themselves to the delights of the world-class marine exhibits at the Monterey Bay Aquarium.

Overfishing would be the simple explanation for the decline of the sardine population off Monterey, but that is only part of the picture. An additional reason is that the Pacific sardine shares these waters with the northern anchovy (*Engraulis mordax*). "Stock switching," the replacement of one species with another in a given fishery, was triggered by environmental as well as human factors in the Monterey fishery. The anchovy, a commercially less desirable species, was able to take over the sardine's ecological niche, a niche the sardine can only regain slowly.

The replacement was hastened when the sardines were not able to reproduce successfully during some of the high harvest years. This low reproduction was the result both of taking nearly all adult sardines in some years and of the warming of the regional waters during this period when a slight climate shift took place. This shift upset the timing of the seasonal sardine hatch, which made the young more subject to predation by other species. Hence there were drastically fewer sardines but not necessarily fewer fish.

Research into the fossil record in the marine sediments of Monterey Canyon and vicinity indicates that even without human impact the sardine population has changed significantly through time. The probable cause was a variation in climate that affected the intensity of upwelling and the temperature of the ocean water. Although fish populations along the Monterey coast naturally shift in species composition, here is one example of an ocean resource that could have benefited from more prudent exploitation and the application of ecological knowledge to improve long-term commercial management.

Fig. 15-3. Ocean harvest. Edible seaweed of many kinds are harvested from the continental shelf in Japan. This green vegetation, which responds to the availability of light, heat, and nutrition, is only one link in the rich and complex marine biosphere.

Phytoplankton are eaten by larger *zooplankton*—small animal life forms that are one notch up the food chain. Zooplankton are eaten in turn by small fish and crustacea, and they in their turn by still larger predators: "little fish have bigger fish lurking near to bite 'em, and they in turn have bigger fish, and on *ad infinitum*" (after a ryhme by Jonathan Swift, 1733).

The nutrients needed to sustain this frenzied swarm are washed down by continental rivers, imported by sweeping ocean currents, and delivered from the deep by coastal upwelling. In addition, there is the constant soft precipitation of the organic remains of plankton, which expire after a life span of a few days or even hours, to complete the regenerative cycle of life. One season's residue of plankton nourishes the next generation of life on the shelf. Many of the world's great fisheries are in the seas above the continental shelf, especially those based on *demersal*, or bottom-dwelling, species of fish such as cod or halibut.

SHELF TOPOGRAPHY It would seem that the continental shelves should be smooth surfaced because of wave action and continuous sedimentation, but this is not necessarily the case. The floor is frequently irregular, in particular where continental glaciers have run far out to sea and not only scarred and gouged the surface, but also dumped huge quantities of debris that stand up today as ***banks***. Banks refer to sea bottoms, regardless of their origin, that are broad, shallow, and usually only 200 to 300 feet (61 to 91 m) below the ocean surface. They typically have well-established fisheries because of the excellent habitat for bottom species.

Much more physically dramatic than banks or minor irregularities are the widespread canyons that score the continental shelves to great depth. Usually beginning close to shore near river mouths, these deep ***submarine canyons*** are incised dramatically into the shelf. Some of them reach a depth of a mile (1.6 km) or more below sea level and are frequently compared to the Grand Canyon of the Colorado River (Fig. 15-4). The deepest appears to be Great Bahama Canyon with a maximum wall height of over 2 miles (3 km), but a Russian survey of Zhemchug Canyon in the Bering Sea indicates it is by far the largest in volume.

We do not completely understand how submarine can-

Submarine canyons are important features and sometimes contribute directly to the local abundance of sea life on the continental margins.

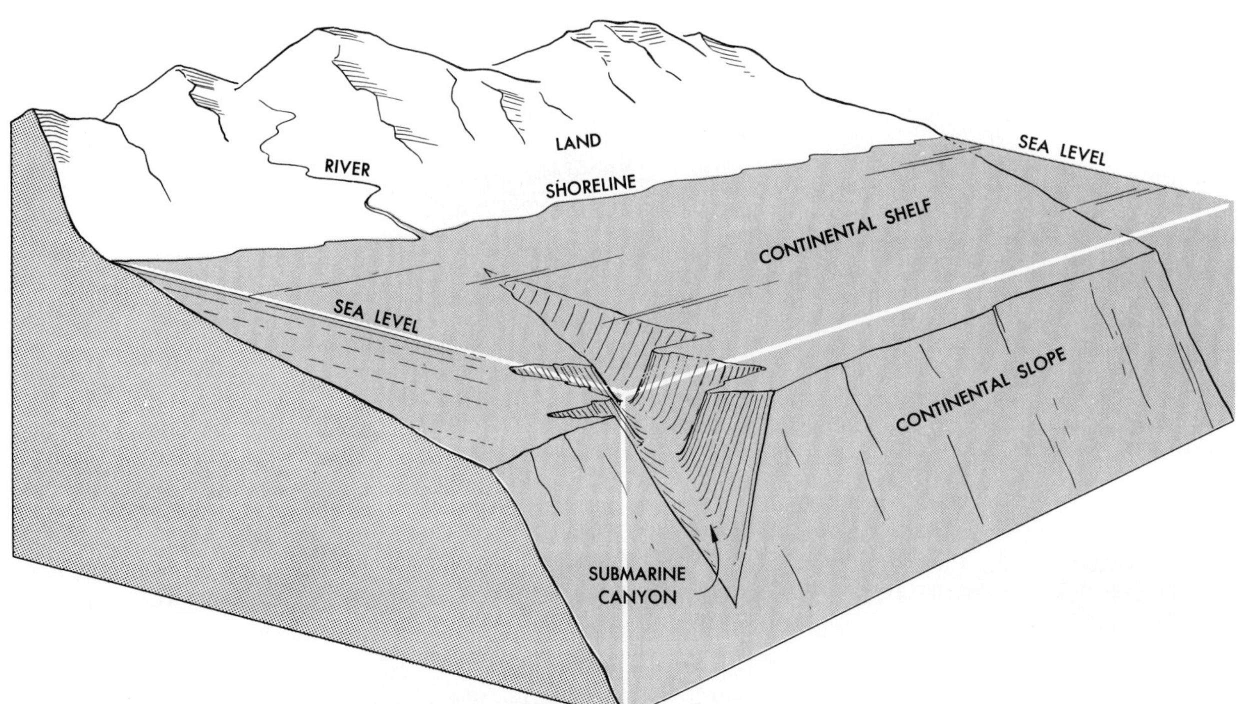

*Fig. 15-4. **Diagrammatic view of a submarine canyon.** These canyons may reach depths of over 1 mile (1.6 km) and are often situated immediately opposite the mouth of a major river.*

yons are formed and maintained. They may be the result of river erosion when the surface was above sea level; they may have been caused by ***turbidity currents,*** those dense, heavy flows of water, mud, and sand that cut the continental shelf; or they may be ancient canyons initially carved when rifting uplifted and split apart continents to form new oceans (see Chapter 17). The ancient canyons could then have subsided as the oceans became larger, and canyon depths could be maintained by turbidity currents fed by sediment carried by rivers and shore currents. Whatever the origin of submarine canyons, they are important features and sometimes contribute directly to the local abundance of sea life on the continental margin.

MINERALS

All the known elements of the earth's crust are presumably represented in the salts of the sea, for the erosion of the land is largely responsible for the ocean's salinity and mineralization. No sooner is a land mass raised above sea level than it is attacked by rivers, winds, waves, and ice, and the entire surface with all its mineral constituents is whittled away to be returned to the sea (Table 15-1). But there are many problems involved in recovering selected minerals for human use. In some instances the problem is a chemical one and in others an economic one. At present, relatively few salts are removed directly from seawater in commercial quantities. The age-old evaporation of seawater by the sun still furnishes sodium chloride to many coastal regions, and more recently magnesium and bromine have been successfully isolated on a large scale at relatively low cost.

On numerous occasions in the geologic past, the oceans intruded into shallow embayments far inland where their

All the known elements of the earth's crusts are presumably represented in the salts of the sea.

Table 15-1 Seawater Concentration of Elements

Element	*Tons per Cubic Mile*	*Metric Tons per Cubic Kilometer*	*Element*	*Tons per Cubic Mile*	*Metric Tons per Cubic Kilometer*
Chlorine	89,500,000	12,137,853.95	Nickel	9	1.22
Sodium	49,500,000	6,713,114.75	Vanadium	9	1.22
Magnesium	6,400,000	867,958.27	Manganese	9	1.22
Sulphur	4,200,000	568,597.62	Titanium	5	0.68
Calcium	1,900,000	257,675.11	Antimony	2	0.27
Potassium	1,800,000	244,113.26	Cobalt	2	0.27
Bromine	306,000	41,499.25	Cesium	2	0.27
Carbon	132,000	17,901.64	Cerium	2	0.27
Strontium	38,000	5,153.50	Yttrium	1	0.14
Boron	23,000	3,119.23	Silver	1	0.14
Silicon	14,000	1,898.66	Lanthanum	1	0.14
Fluorine	6,100	827.27	Krypton	1	0.14
Argon	2,800	379.73	Neon	0.5	0.07
Nitrogen	2,400	325.48	Cadmium	0.5	0.07
Lithium	800	108.49	Tungsten	0.5	0.07
Rubidium	570	77.09	Xenon	0.5	0.07
Phosphorus	330	44.63	Germanium	0.3	0.04
Iodine	280	37.87	Chromium	0.2	0.03
Barium	140	·18.83	Thorium	0.2	0.03
Indium	94	12.75	Scandium	0.2	0.03
Zinc	47	6.37	Lead	0.1	0.01
Iron	47	6.37	Mercury	0.1	0.01
Aluminum	47	6.37	Gallium	0.1	0.01
Molybdenum	47	6.37	Bismuth	0.1	0.01
Selenium	19	2.58	Niobium	0.05	0.007
Tin	14	1.90	Thallium	0.05	0.007
Copper	14	1.90	Helium	0.03	0.004
Arsenic	14	1.90	Gold	0.02	0.003
Uranium	14	1.90			

Fig. 15-5. ***Salt mines.*** *Early nineteenth-century Midwestern pioneers depended on "salt licks" for their domestic salt. They took the salty water from naturally occurring springs and laboriously boiled it down to its essential crystals. The licks signaled thick strata of underground salt left behind by evaporating Devonian seas. Today they are being tapped by deep mines. Almost 0.25 of a mile (0.4 km) below Detroit are these massive caverns hewn out of solid salt and crossed by giant trailer trucks.*

waters were trapped and slowly evaporated away, depositing their salts in thick beds. These deposits may be mined today, sometimes at great depth, by sinking shafts or forcing hot water into them to form brines that can be piped to the surface. Some of the largest underground caverns ever made by humans underlie the city of Detroit where thick salt strata have been mined for many years. They date back to Silurian times when the tropical temperatures evaporated a sea that overlay much of the eastern United States. A further aid to mining efforts is the fact that salts precipitate selectively; they therefore concentrate in various differing and reasonably pure layers (Fig. 15-5).

On other occasions, portions of ancient seas have been trapped in underground cavities or porous rock, and there concentrated brines are drawn off today through wells for processing. The Dead Sea, so saline that its waters sustain no life, is the remnant of a former larger sea. In this case the briney water lies at the surface and is therefore easy to "mine" for its minerals.

Living Intermediaries

Where humans cannot cope with the chemical problems of isolating certain salts, plant and animal life in the sea may act as intermediaries, selecting out and concentrating a particularly valuable mineral. If large numbers of their remains are deposited along with the sediments of old evaporating seas, they may form a rich stratum that can be recovered by mining. Our familiar commercial petroleum is also of organic origin, formed as a result of the precipitation of dead plants and animals to the bottom of shallow seas. We do not fully understand the chemistry of decomposition and the way heat developed by subsequent rock deformation forms a liquid hydrocarbon. We do know, however, that without the sea and the life that inhabits it we would have no oil, for it cannot be made economically in the laboratory.

Petroleum

Petroleum is by far the most valuable mineral drawn from the sea today. It occupies a slightly different category from oceanic salts and minerals extracted directly from seawater. Oil is sucked up from the porous rocks in the continental shelf in much the same manner as an oil well does on land. The fact that the shelf is moderately flooded leads to the use of floating platforms as a work base (Fig. 15-6). These huge platforms, as well as the recovery systems associated with them, are continually monitored and being improved, but they can pose a risk to workers and to the ocean environment when damage occurs from storms and accidents. At many locations around the world—the Gulf of Mexico, Indonesia, California, Alaska, Australia, and New Zealand—offshore reserves are exploited, and many other areas are being explored for economically recoverable oil and minerals.

Methods of Exploitation

Most resources hidden away in submarine strata can be removed only by old-fashioned hardrock methods, difficult as it is to imagine. ***Adits*** (horizontal shafts) do extend from mine heads on the shore out under the sea in a number of places, but again we are involved only with the limited continental shelf. To tap mineral deposits locked in a rock bond

Fig. 15-6. ***Floating platform.*** *An offshore oil rig in the Gulf of Mexico.*

below 10,000 to 15,000 feet (3000 to 4600 m) of seawater will surely require some wholly new concept in mining technology.

There is another method: a simple sweeping, or vacuuming, of loose materials from the sea floor. Most people don't think of sand and gravel as minerals, but huge quantities of them are required. They may often be acquired more cheaply offshore, as byproducts of deep-channel drilling, than on land. Shell, a source of basic lime, is dredged up in a similar fashion. Diamonds, lying about mixed with gravel on the continental shelf, are being retrieved off the coast of Namibia by giant vacuum hoses.

FISHERIES

The present worldwide oceanic fisheries harvest[1] measures out at roughly 110 million tons (100 million metric tons) per annum. Although it is increasing each year, the rate of increase has slowed since the 1970s possibly because of overfishing of some species, increased pollution, higher cost of fuel, or outdated processing plants and equipment.

[1]This harvest includes species of fish, sea mammals, crustaceans, mollusks, sponges, and algae.

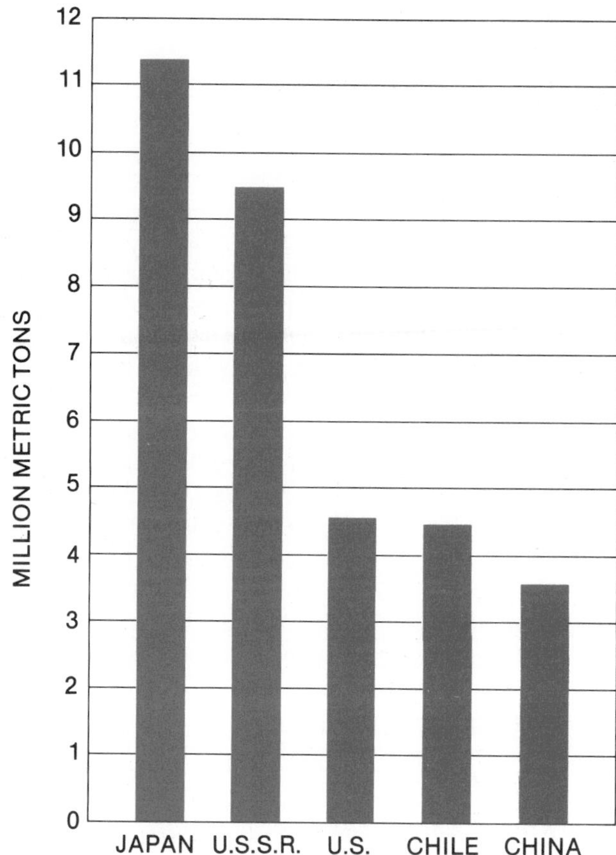

Fig. 15-7. ***Average annual catch of major fishing nations, 1983–1985.*** *(Source: World Resources Institute, 1989)*

Over 90 percent of the total catch is fish, about half of which are consumed directly and the other half converted into fish meal, mostly for livestock feed. The top five fishing nations take about one-third of the annual catch (Fig. 15-7).

The total fish landed each year by all nations is a small fraction of the ocean's entire annual biological production, but only a few species are considered worthy of the chase because of marketing and preference considerations. Only about 20 species, including herring, cod, mackerel, and redfish, are taken in yearly amounts of over 110,000 tons (100,000 metric tons). As a consequence, many fisheries have been severely depleted or too poorly managed to sustain yields.

Some estimates of maximum sustainable yield for the selective commercial catch are around 300 million tons (272 million metric tons) per year. A more conservative United Nations Food and Agriculture Organization estimate places maximum sustainable yield at 100 million tons (91 million metric tons), a figure apparently surpassed in the late 1980s. In recent years overexploitation and natural environmental factors have resulted in declines of North Atlantic cod and herring, California sardines, and Peruvian anchovies.

Fishing Nations

The two most heavily fished regions in the world are the North Atlantic and the North Pacific oceans. Favorable ocean currents, bottom topography, and human population combine to make these fisheries among the most valuable in the world. Countries such as Norway, Denmark, the United Kingdom, and the Soviet Union have traditionally sent their fishing fleets out into the Atlantic in search of food. In the North Pacific, Japan, China, and the Republic of Korea dominate the fishing industry. Japan is the number one fish-catching nation in the world and takes over half the annual North Pacific catch. In the tropics and the Southern Hemisphere the important fishing nations are Peru, Thailand, Indonesia, and India. Among developing countries the share of annual harvest is increasing markedly as these countries strive to supply more protein for their citizens' diets and to increase fish exports to bring in badly needed cash.

The two most heavily fished regions in the world are the North Atlantic and the North Pacific oceans.

Fig. 15-8. ***The international fishing industry.*** *In some societies fish is the major source of protein in the national diet. Even though the number of vessels in the fishing fleet and the size of the catch do not loom large in international fisheries statistics, the industry is of extreme local significance.*

Fisheries Management

Marine fisheries may be viewed as a renewable resource that must be managed wisely. As such ocean fisheries should be exploited to the point of maximum yield within the limits of maintaining or improving the resource while protecting the complex interrelationships of ocean ecology. This objective sounds easy until we try to put it into practice. Although we have learned something of what makes the oceans tick, major research still lies ahead.

In addition, the fishing industry has always suffered from a gap between "ecological" management and "commercial" management of the resource. Although people may know quite a lot about a particular species' life cycle, reproductive capacity, and environmental requirements, they may all but ignore this knowledge when exploiting that species.

To effect a long-range, international conservation plan, all concerned nations must agree to rules to manage the fisheries. Such an agreement will be very difficult to achieve given past performance. Traditionally, the fourteen major fishing nations (not to mention several dozen lesser nations where fish may be the "staff of life") have been less than enthusiastic about agreeing on and enforcing international rules (Fig. 15-8).

DESALINIZATION

People have always been intrigued with the idea of using the sea as a source of drinking water, especially inhabitants of the Mediterranean and Tropical Dry climates where rainfall is limited or nonexistent and yet the ocean is at their front door. In total there are some 18,500 miles (30,000 km) of such dry seacoast where the only barrier to an unlimited supply of fresh water is a modest salt content. Surely, it is said, there must be a simple way to solve this little problem. Clearly, the solution is not simple at all; otherwise these extensive regions would be blooming instead of continuing to suffer water shortage.

Many nations are looking critically at the sea for fresh water.

Many nations are looking critically at the sea for fresh water for several reasons. For example, many underground water sources are exhausted, surface supplies are inadequate, and even traditionally pure and sweet rainwater must pass through an increasingly contaminated atmosphere and lithosphere. The pollution of many water sources is also a problem. The United States currently uses over 600 billion gallons (2262 billion liters) of water per day (40 percent going to irrigated agriculture). This figure continues to increase dramatically, even though the nation's sources essentially remain the same. Under the pressure of need they are perhaps being used a little more efficiently, but at the same time their quality is deteriorating and the cost of delivery is increasing. Conservation can help. We can live with fewer green lawns, less frequent showers and toilet flushings, but these are not cures for an escalating problem that will not go away.

The United States Bureau of Reclamation, Department of the Interior, and similar bureaus in other countries are beginning to make a systematic attack on the seawater desalinization problem (as well as brackish groundwater and polluted surface sources). For the first time, they are in a position to use sophisticated technologies backed by national treasuries. The effort is truly international. With high inflation, however, especially in developing countries, governments pressed to establish spending priorities have tended to scale down research and development funds for a whole array of promising ventures, desalinization among them.

Fresh Water Extraction

Fresh water is extracted from seawater (1) by distillation processes, and (2) by ***reverse osmosis***. The first method is more common and has a number of variations, but generally distillation processes heat salt water until it vaporizes; cool the water vapor, allowing it to condense; and collect the fresh water.

Reverse osmosis, a newer process, uses plastic membranes that allow water molecules to pass through under pressure but block undesirable dissolved compounds. Although the reverse osmosis technique is less expensive to install than a conventional distillation plant, it produces water that is only 95 to 99 percent free of salts.

The Yuma, Arizona, Desalting Plant on the Colorado River uses the reverse osmosis technique. The plant's supply is not ocean water, but the salty water of the lower Colorado River basin. The river in its lower reaches has become salinated from the drainage water of intense irrigation in the deserts of both Arizona and southern California. It was constructed after an agreement between the United States and Mexico obligated the United States to process unusable river water before it crossed the border into Mexico.

No Universal Solution

No single desalinization process can be applied universally. The types of fuel available, the size of plant, the desirability of processing slightly brackish groundwater versus seawater—all these are elements to consider. For instance, the tiny oil-rich nation of Kuwait on the shore of the Persian Gulf has one of the highest per capita incomes in the world. It meets all its fresh water requirements by converting seawater to fresh, but since the population is so small (about 2 million) and cost is no object, we can scarcely point to the Kuwait experience as typical. In most places, the cost per gallon of fresh water gained from a desalinization process is so high that almost any other alternative is less expensive.

ENERGY FROM TIDES

In the past few years, as the cost of producing electricity from fossil fuel has risen dramatically, one physical phenomenon that has attracted attention as a feasible source of energy is the tidal fluctuation along certain coasts (see Chapter 13). Where tides have a significant range from high to low, tidal power can be harnessed by allowing seawater moving in and out of coastal embayments and river mouths to flow past a constructed barrier as the tide comes in, then closing the barrier at high tide. As the tide retreats from in front of the barrier, the water trapped behind is directed out to sea through turbines that produce electricity.

Only certain locations have the necessary combination of tidal range and shoreline for tidal power generation.

Physical geography dictates that only certain locations have the necessary combination of tidal range and shoreline for tidal power generation. Coastal sites in Alaska, Argentina, western Australia, and England are some of the loca-

Fig. 15-9. ***Tidal Power Plant.*** *This small power plant on an estuary of the Bay of Fundy uses tidal fluctuations to generate electricity.*

tions regarded as favorable for this purpose. Two small experimental tidal plants are located on the arms of the Barents and White seas in the northern Soviet Union. A much larger plant capturing the energy of the tides has been operating on the Rance River estuary in Brittany (France) since 1967 and produces over 500 GWh of electricity annually.[2]

The Bay of Fundy in Canada's Maritime Provinces has a tidal variation of nearly 50 feet (15 m), the largest tidal range known. Since the 1930s people have recognized the power-producing potential of this bay, but only in recent years has development seemed economically justified. In the 1980s the Canadian government and the provincial governments of Nova Scotia and New Brunswick cooperated in a project to construct a tidal power facility in the Annapolis Basin, a small arm of the Bay of Fundy in Nova Scotia (Fig. 15-9). This test project has been in operation since 1984 and produces 50 GWh annually.

This is a minuscule amount compared to, say, the conventional hydroelectric output of the generators at Hoover Dam on the Colorado River, which each year produce 6000 GWh of energy. Nevertheless, if only the three best sites in the Bay of Fundy were developed, estimates indicate that they could produce a total of over 15,000 GWh annually. Canadians will need to consider construction costs, costs of alternative electricity production, market factors, and environmental impacts before deciding whether or not to develop these sites.

OCEAN WATER POLLUTION

The pollution of ocean water, particularly near shore and over continental shelves, poses a threat to fisheries and marine ecosystems. Little is known about the extent of ocean pollution or its long-term effects on marine life. Yet, the oceans are used as a dump for most of the world's waste, including certain chemicals and insecticides such as PCBs (polychlorinated biphenyls) and DDT. We know that these chemicals have long-term effects on organisms in the food chain. In many areas of the world, municipal and industrial wastes are dumped untreated into estuaries and salt marshes, upsetting the productivity of natural ecosystems and creating health hazards to humans.

Oil pollution from tanker spills, platform leaks or "blowouts," and a number of other sources have been prominent in the news in recent years, for example:

March 1967—The grounding of the tanker *Torrey Canyon* off Land's End in England released thousands of barrels[3] of Kuwaiti crude oil into the Atlantic.

January 1969—A blowout of Union Oil offshore drilling

[2] One gigawatt hour (GWh) is equivalent to the energy consumed in one hour by 10 million 100-watt lamps.

[3] One barrel of petroleum contains 42 gallons (159 liters).

*Fig. 15-10. **The Exxon Valdez** attended by a smaller tanker in Prince William Sound, March 1989. A massive clean-up effort followed the accident that spilled 11 million gallons of crude oil in the sound and along Alaska's shore.*

platforms near Santa Barbara, California, covered southern California beaches with oil.

December 1976—The tanker *Argo Merchant* ran aground along the New England coast, broke up, and spilled 7.5 million gallons (28 million liters) of oil that drifted into the Georges Bank fishing area.

March 1978—The *Amoco Cadiz,* disabled during a storm off the Brittany coast, struck rocks after breaking its towlines. The oil-leaking wreck was blown up, and 67 million gallons (254 million liters) spread along 200 miles of coastline.

Mid-1980s—Millions of barrels of crude oil were allowed to pour into the Persian Gulf as a result of the war between Iran and Iraq.

March 1989—The wreck of the *Exxon Valdez* spilled 11 million gallons (42 million liters) of Alaska North Slope crude oil into Prince William Sound.

The results of such spills are graphically and dramatically brought to our living rooms by television news broadcasts (Fig. 15-10). Sadly, however, most of the oil pollution in the ocean is not from accidents or wars but from the routine discharge of oil by tankers during cleaning and ballasting. Only the more sensational accidental spills make the headlines. The immediate damage to sea birds and mammals is clear; the long-term effects are yet to be determined.

CONCLUSION

Although the oceans completely dominate all other earth features and undoubtedly support vast quantities of life, we know very little about them. Deep-water sailors, shore dwellers, and fishers have amassed a considerable body of practical knowledge, which has been valuable to those who would know the sea. However, compared with our understanding of the land, or even the phenomena of the atmosphere, the total accumulation of oceanographic information from the earliest of times is relatively sparse and rudimentary.

Only in very recent years has the science of oceanography been recognized and sophisticated equipment been put to use to delve into the mysteries of the seas. Already this research has corrected many long-held misconceptions and hedged and qualified old generalizations. We may expect this process to continue for some time, as is always the case in a pioneering science.

Future use of the world's oceans must be tempered and guided by lessons learned from our past use of the land.

It behooves us to learn more about the oceans—and soon. As ocean pollution increases there is greater public outcry, particularly in the industrialized nations, about the permanent damage that may be done to our oceans. Certainly we must find ways of immediately stopping the kind of contamination that makes pollution worse. If we cannot find a remedy soon, we may very well discover that we are engaged in long-range global suicide. In a world rapidly depleting its mineral resources, increasing its population, searching for additional sources of food, and constantly requiring larger quantities of fresh water than seem to be available, the sea offers some relief. Perhaps the world's oceans will be the ultimate resource, but future use of what lies in them and below them must be tempered and guided by lessons learned from our past use of the land.

KEY TERMS

continental shelf
abyssal plain
phytoplankton
zooplankton
demersal
banks
submarine canyons
turbidity currents
upwelling
adits
reverse osmosis

REVIEW QUESTIONS

1. What are continental shelves and where are the largest ones located? Why are continental shelves important areas for ocean resources?

2. Describe two theories that may explain the formation of submarine canyons?

3. What are some of the ways that minerals are extracted from the ocean?

4. Where would you expect to find the most abundant ocean fisheries? What conditions make them productive?

5. Where is desalinization of seawater used as a way of providing fresh water? Is it a practical solution for dry, seacoast regions?

6. How is electrical energy produced from ocean tides? Where are the best sites for tidal power generation?

APPLICATION QUESTIONS

1. What large seaport is closest to your location? Describe the ocean resources that are most important to this port's economy. (If you live far inland, you might choose a port that you visited at one time.)

2. The next time you are in your local supermarket look on several packages of fish and other seafood products (or ask at the seafood counter) and see if there is information that will tell you where the product was caught and/or processed. Where do most of the products come from?

3. Check your library for recent magazine or newspaper articles about ocean or coastal environmental concerns. What issues are being debated?

4. Do some library research to see if you can find any information on the long-term environmental effects of oil spills.

PART FOUR

EARTH'S LITHOSPHERE AND LANDFORMS

CHAPTER 16

INTRODUCTION TO LANDFORMS

OBJECTIVES

After studying this chapter, you will understand

1. What is involved in the science of geomorphology, the study of landforms.
2. How tectonic and gradational processes interact to shape landforms.
3. How and why scientists believe the earth was formed.
4. Methods of determining the age of the earth and its features.
5. How concepts of the erosion cycle and dynamic equilibrium have influenced thinking among geomorphologists.

The Sierra Nevada of California. This mountain mass rises above the older, weathered and eroded hills in the foreground.

INTRODUCTION

Have you ever looked down from an airplane at the very irregular surface of the earth below? Have you wondered about the forces and processes that lifted up the mountains and formed the valleys and the ocean depths? The study of landforms can help you meet those forces and processes face to face: the truly "earth-shattering" forces that build the rugged mountain ranges and the slow-acting but inexorable processes that wear those mountains down. In combination, they shape the features on our planet's surface.

The word *landform* scarcely requires definition. It is an easily pronounced, nonscientific word that quickly brings to mind an individual's perception of hills, plains, valleys, and plateaus. When dealing with landforms, we are taking a look at the many wrinkles, bumps, and depressions on the face of the earth and examining them at various scales in order to understand their origin and character. If we need a single term to describe this pursuit, ***geomorphology***—the study of earth forms—will fit the bill.

Landforms are part of the solid earth—the rigid outer shell of the planet called the lithosphere.

Thus far in our discussion of physical geography, we have been concerned with the more fluid elements of the earth's system: the atmosphere and the hydrosphere. Now we approach the subject of landforms, which are part of the *solid* earth—the rigid outer shell of the planet called the ***lithosphere***. In Greek, *lithos* means rock or stone. Litho*sphere* refers to the zone of rock that varies in composition and density and extends from the surface to a depth of 30 to 60 miles (50 to 100 km). At and near the surface, the air of the atmosphere, the water of the hydrosphere, and the rocks of the lithosphere interact through a combination of processes to give us a variety of landscapes across the globe. From a geographer's point of view, the lithosphere is the complex and dynamic physical stage on which we live; its surface expression, the landforms we encounter, influence the pattern of human occupation of the planet. Certainly, a fundamental knowledge of landforms and the ongoing processes that shape them greatly facilitate our understanding of settlement patterns and our planning of future land use.

SHAPING PROCESSES

Two categories of processes shape the earth's surface: (1) internal processes and (2) external processes. For the most part, these categories of processes seem to work in opposition, the one seemingly canceling out the effects of the other. The first category refers to the forces within the earth's interior that produce extensive ***tectonic processes***, including volcanic activity and ***diastrophism***.[1] These forces warp, buckle, break, and distort the earth's crust, building mountains and forming depressions in the surface structure. These internal forces result in *folding, faulting,* and *vulcanism* (see Chapter 18). Beginning deep in the earth, heat-driven movement of rock material pushes up mountains and creates deep ocean trenches. All about us we see signs of tectonic activity—in earthquakes, volcanic eruptions, certain deep valleys, and young rugged mountains. In the next chapter, Chapter 17, we will present a more detailed discussion of the earth's internal structure and tectonic activity and their importance to global-scale landforms.

External processes attack earth's tectonic features and attempt to destroy them, wearing down the high places and filling in the low ones. These processes include (1) exposure to the atmosphere (*weathering*), (2) action by the agents of *erosion*—running water, moving ice, wind, and waves—and subsequent deposition of eroded material, and (3) the influence of gravity (*mass wasting*). Because these external forces help level out the earth's surface they are known as ***gradational processes***. The effect of gradational processes on landforms is the subject of Chapter 19.

Without tectonic forces, gradational processes would lower the land surfaces of the earth until the seas completely covered them.

Without the tectonic forces to build new structural features, the action of gradational processes would lower the land surfaces of the earth until, theoretically, the seas completely covered them. In certain parts of the world, such as the Amazonian lowland and western Australia, tectonic forces have not been active for a long time, and the land is low and featureless. In areas where the tectonic forces have become, at least temporarily, more active than the gradational, the landscape features include high mountains or great ocean deeps (Fig. 16-1).

Here is a summary of the main processes involved in tectonics and gradation:

Tectonic Processes	*Gradational Processes*
1. Folding, including crustal warping	1. Weathering
2. Faulting	2. Mass wasting
3. Vulcanism	3. Erosion and deposition

[1] *Tectonic* comes from the Greek word *tekton* meaning a carpenter or builder, so it emphasizes the constructive aspects of these forces. *Diastrophism* comes from the Greek word *diastrophe,* which means *twisting.* It emphasizes the forming action of these forces.

Fig. 16-1. ***The opposing influences of tectonics versus gradation.*** *(A) The busy volcano of Mount Sakurajima in Japan is still building—the tectonic forces at work. (B) The Spanish meseta stretches to the horizon, a seemingly endless, near-flat plain. This peneplain, the worn-down roots of old mountains, demonstrates that gradational forces have been dominant here for millennia.*

Because these processes produce the landforms we study, we will classify each feature according to its cause as well as its shape. All this is the science of geomorphology.

TIME AND THE EARTH

From a human perspective tectonic and gradational processes work very slowly. Geologic time, measured in millions and even billions of years, is difficult for most of us to grasp; it takes thousands, tens of thousands, or even millions of years before most landforms show significant development or change. To the average person it may seem that the processes are not at work at all.

Even though the movement of a continent takes millions of years, this is only a tiny fraction of the earth's total age.

Literature is filled with phrases like "timeless mountains" and "rocklike stability," implying that the rock is the ultimate symbol of permanence. But that is because we think in terms of the "fourscore and ten" years in a human lifetime or the few thousand years of recorded history. From this point of view, rocks and mountains may very well seem eternal and unchanging. But geologic time, the measurable age of the earth, is an immense span almost beyond comprehension. We must constantly remind ourselves that, even though the wearing away of a mountain range or the movement of a continent takes millions of years, this is only a tiny fraction of the earth's total age (see Appendix D).

Earth's Origin

As old as the landforms are, the earth itself is older still. Learning about its origin has always been tricky. Since nobody has ever been able to locate a first-hand witness, our scientific speculation about the manner in which the earth was formed is just that—speculation. This does not mean wild guessing. We are well beyond the point of calling up fire deities and water gods to account for events we don't understand. We are, however, still hypothesizing, deducing as best we can from earthly clues. The problem is not yet resolved, but our knowledge keeps increasing as our research methods and technology become more refined. We can safely assume, therefore, that our modern theories are closer to the truth than the myths of antiquity.

Most astronomers believe that the earth was slowly formed billions of years ago out of a swirling, disklike cloud of particles and gases.

At present most astronomers believe that the earth was slowly formed billions of years ago—along with the rest of the solar system—out of a swirling, disklike cloud of particles and gases. Like the other planets, the earth condensed from swarms of particles spinning about the central mass that formed the sun. The fact that the planets revolve around the sun in the same direction and roughly the same plane seems to support this explanation. It also seems to square with the Big Bang theory of the creation of the universe. This theory holds that matter and energy were originally contained in a huge nucleus, or "cosmic egg." About 20 billion years ago the "egg" exploded, sending matter out in all directions. Later, gravity drew the matter together in groups of stars and planets.

There is evidence that the matter flung outward during the Big Bang is continuing to move. In addition, some scientists have detected what they believe to be residual radiation left over from the explosion. Still, scientists do not all agree on any one explanation. We would like to think we are on the right track and that modest adjustments and additions (data supplied from analysis of moon rocks and from space probes to the outer planets, for example) will lead to general agreement, but further fundamental theorizing may still be in order.

Earth's Age

But how long ago did all this occur? What is the age of the earth? Only within the last 150 years or so have we approached the problem in a scientific manner. Before the eighteenth century, church officials made all the decisions on these matters and discouraged scientists from offering their views. Even when the religion–science deadlock eased a bit, the technology of dating remained both primitive and tentative. Almost unbelievably, at the turn of the twentieth century, most geologists agreed that the earth's true age could be no more than 100 million years. Our present information suggests that the earth is something like 4.6 *billion* years old (Fig. 16-2).

Dating Techniques

We may think of age in two different ways. First, when we think of how old something actually is in years, we are thinking of ***absolute age.*** The earth's absolute age, as we mentioned in the previous section, is approximately 4.6 billion years. Second, when we think of how old something is in relation to something else, we are thinking of ***relative age.*** We can express relative age in many ways, depending on what we are using for comparison. The earth, for example, is younger than the universe but older than the fossil of a dinosaur.

Today, we have many techniques for determining the relative and absolute age of the earth and its features. Traditionally, geologists have estimated the relative age of rocks by looking at the fossils they contain and the sequence in which they are laid down. We have incorporated this information into quite an accurate working system which we now use to determine the relative ages of landform features that developed as long ago as about 600 million years before the present, or BP (Fig. 16-3).

During this century, the development of several more precise dating techniques has allowed us to determine the absolute age of many of the earth's features. Each of these techniques is like a piece in a jigsaw puzzle: each one provides part of the answer, and all together they complete the

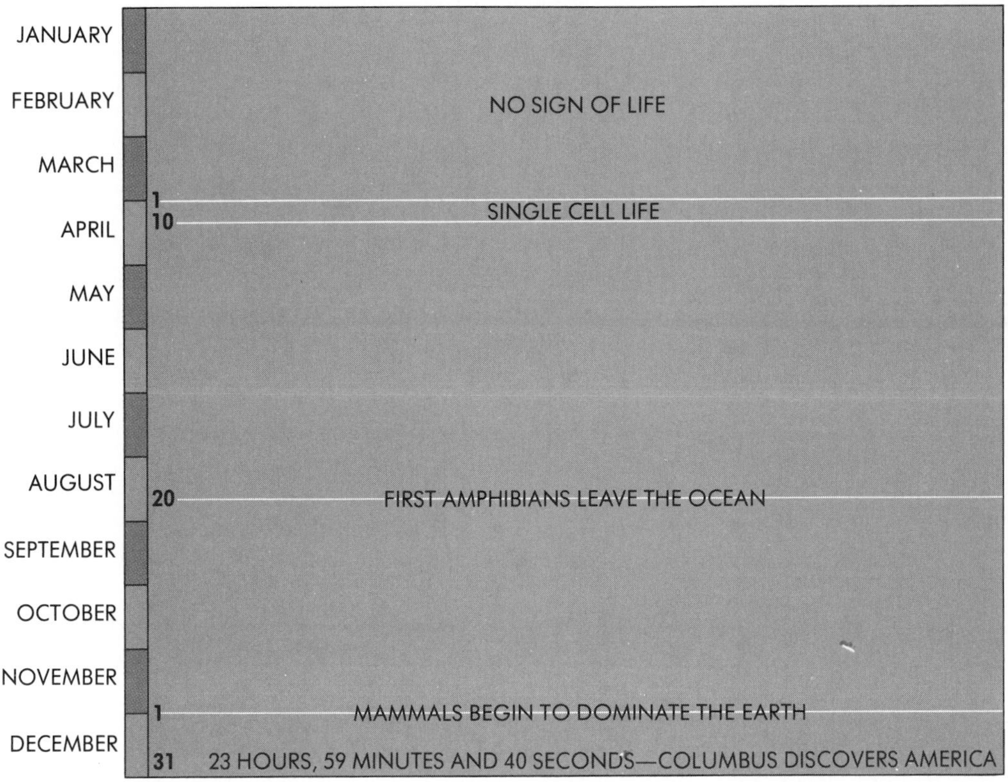

Fig. 16-2. ***Equating one calendar year with the 4.6 billion years of earth's existence.***

picture. In this section, we'll examine three of the most common:

1. Analysis of varved clay
2. Radiometric dating
3. Analysis of paleomagnetism

Fig. 16-3. ***Fossil detail.*** *The fossil fish somehow swam into a muddy spot and became part of a shale. This species no longer exists but was common in Tertiary seas. It has been identified and dated by its presence in a known rock stratum from Wyoming.*

ANALYSIS OF VARVED CLAY Every year a double layer of sediment called a ***varve*** is deposited on the bottom of certain lakes. Each varve contains a silt layer and a clay layer. The silt deposit is laid down during the summer when wave and current action keeps the finer clay particles suspended. The clay deposit is laid down when the lake is iced over in the winter and the finer particles are able to sink (Fig. 16-4).

Analysis of varve deposits from ancient and modern lake bottoms, particularly glacial lakes in the Northern Hemisphere, is a common technique for dating relatively recent geologic formations and landforms. Some careful studies of accurately dated ancient lake deposits have revealed histories of 25,000 years.

RADIOMETRIC DATING Many geological formations, however, have nothing to do with lakes and are more than 25,000 years old. To date these formations, we must rely on one of the more technologically sophisticated methods developed since the 1950s. Of these methods, ***radiometric dating*** is the most accurate. It is particularly useful for clarifying the ages of the ancient pre-Cambrian rocks (older than about 600 million years BP).

Fig. 16-4. ***Varves.*** *A sediment core revealing varves from a glacial lake bed. Contrasting layers are seasonal deposits of silt and clay.*

Radiometric dating is based on the known rate at which radioactive forms of certain elements change, or "decay," into more stable elements. When a mineral is formed, or when organic material is covered up, it often contains a radioactive *isotope,* or unstable variation, of an element. We express the rate of decay of that isotope in terms of its *half-life*—the time it takes for one-half the original isotopes to decay into a stable end-product. Because neither heat nor pressure has any effect on radioactivity, we have an essentially unbreakable clock. Through this method, rocks at several locations in South Africa and Greenland have been assigned ages of almost 4 billion years. At the opposite extreme, this method has established ages for buried wood and other organic material that have only been covered for decades.

Uranium and thorium are the chief radiometric elements for long-term dating (there are some four or five others) because they are relatively common and possess a long half-life compared with the age of the earth. Carbon-14 is the principal radiometric element for dating recent organic deposits.

ANALYSIS OF PALEOMAGNETISM Another modern technique for dating ancient landforms involves ***paleomagnetism,*** or thermoremanent magnetism—the record of the earth's magnetic field preserved in ancient rocks. Scientists analyze paleomagnetism to discover the strength and orientation of the earth's magnetic field in the geologic past and to help date rock material.

Paleomagnetism is a record of the earth's magnetic field preserved in ancient rocks.

Those of us who, as children, discovered the simple wonders of a bar magnet may recall that small pieces of iron or iron-bearing objects (especially iron filings) would become aligned, with playful manipulation, into a pattern around the magnet. This pattern at least partially revealed the otherwise invisible magnetic field around the bar, which had a "north" and "south" pole.

Surrounding the earth is a polarized magnetic field, much like that of a bar magnet, with a north–south axis that extends through the earth to protrude at points we call the *magnetic poles* (Fig. 16-5). The earth's magnetic poles are not identical to its *geographic poles* which are located on the axis of rotation. Indeed, the magnetic poles shift significantly through time, often completely reversing themselves. These reversals of polarity occur at irregular time intervals, but there is evidence that about ten have occurred in the past 5 million years alone.

Volcanic rock preserves a record of the earth's magnetic reversals because lava contains a certain amount of tell-tale

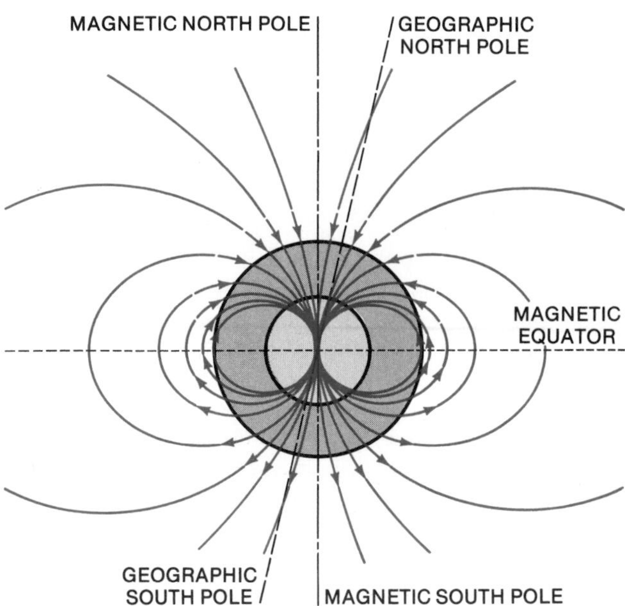

Fig. 16-5. ***Earth's magnetic field.*** *Lines of magnetic force around the earth and the position of the magnetic poles relative to the geographic poles, which are defined by the earth's axis of rotation.*

iron-bearing minerals. During the lava's molten stage, the newly formed iron-bearing particles freely orient themselves along the lines of force of the earth's magnetic field as it exists at the time. As the lava cools, a permanent record of the orientation of the earth's magnetic field is incorporated into the solid rock. This is what we mean by paleomagnetism.

By observing and analyzing selected rock samples from various locations around the world, we can determine the past positions of the magnetic poles and, hence, the positions of continents relative to one another during the geologic past. Radiometric dating techniques verify the age and interval of magnetic polar reversals. Paleomagnetism has provided important clues in reconstructing the world's past continental geography and the development of plate tectonic theory (see Chapter 17).

LANDFORM MODELS

Scientific observers do not believe that geomorphic processes occur randomly. Therefore, researchers constantly look for some kind of repetitive pattern, a way of systematizing geomorphic processes. Before launching into our discussion of specific landforms in subsequent chapters, we need to examine briefly the general models by which geomorphologists set forth organizing concepts in the discipline. Models are useful simulations, based on the most accurate information available, that describe and help to explain processes we cannot observe directly. We will look at two basic organizational concepts in this section: (1) cycle of erosion and (2) dynamic equilibrium.

Cycle of Erosion

One model for organizing and classifying the various stages of landscape development is known as the ***cycle of erosion,*** the systematic change that occurs in a landscape as it is repeatedly uplifted and then slowly worn down through time. This is sometimes called the "geomorphic cycle." William Morris Davis, a renowned Harvard geographer, developed this model as a teaching tool in the late nineteenth century. Davis's notion of landscapes changing through time was directly influenced by Charles Darwin's ideas on biological evolution that had gained attention at about the same time. It has been highly effective over the years as a way of recognizing and classifying landforms in the field. Although the cycle of erosion is theoretical and far from perfect, we introduce it here because it is a widely used fundamental concept.

One model for organizing and classifying the various stages of landscape development is known as the **cycle of erosion.**

We must first imagine a new landform, perhaps a flat-topped plateau just raised out of the sea, perhaps by tectonic uplift. No sooner is it exposed than gradational forces begin to attack and alter its outline as they attempt to lower it back to ultimate ***baselevel,*** the level of the sea. Newly established streams flow across its face, cutting as they go and carrying their debris to the ocean. But for a while at least, depending on the hardness of the rock to be worn away, the basic outline of the plateau remains intact—scarred to be sure but easily recognizable as a flat-topped block. Between the narrow stream valleys, extensive areas of the original surface still remain. This is the *youthful* stage of the cycle; erosion has made relatively little progress in altering the landform (Fig. 16-6).

But erosion never stops. Eventually, the streams cut deeper and multiply their tributaries as erosion completely alters the plateau's original appearance. It reduces the divisions between streams, or ***interfluves,*** to narrow ridges or peaks, and the only indication that the surface was once at a

WILLIAM MORRIS DAVIS

Central to a lifetime of contribution to the science of geography was William Morris Davis's (1850–1934) concept of the cycle of erosion which took the form of a theoretical model. It reduced in a single stroke the great mass of unrelated and often bewildering detail of landscape description to a systematized and predictable pattern. Accused of rigidity by his detractors, Davis vigorously defended his ideas, claiming again and again that his model was a perceptive tool and that infinite departure from it was the norm. Moreover, he readily admitted his debt to J. W. Powell, G. K. Gilbert, and others in the original formulation, although the terminology was his: *youth, maturity, old age, peneplain,* and *monadnock.* In Europe, where modern geography had its beginnings, as well as in the United States, Davis was regarded admiringly as Mr. American Geographer.

Davis graduated from Harvard (1870) with a degree in engineering, spent his early years in Argentina as a meteorologist, and returned to Harvard as a geologist/physical geographer (1876). Academic labels in those days seldom restricted a man's academic endeavor. Increasingly, however, he thought of himself as a geographer and was frustrated that so few others involved in related research did not apply that label to themselves. In 1904 he issued a call to all American scholars who were geographers at heart, no matter what their formal titles, to join him in organizing a professional society. The Association of American Geographers came into being later that year with Davis as its first president. Forty-eight geographers joined him, many newly out of the closet. Geologists were the most numerous among them, but also represented were oceanographers, botanists, agronomists, meteorologists, and economists. Davis was elected president twice more, in 1905 and 1909, a measure of his esteem among his colleagues.

---

constant level is the ***accordance*** (same height) of these ridges. Now the entire landform shows maximum relief. In other words, slopes are everywhere, and flat land is difficult to find. Such a highly dissected landform has reached a *mature* stage in the cycle; erosion has cut away and removed a significant part of it, leaving its form greatly altered (Figs. 16-6 and 16-7).

Finally, the rivers approach the completion of their work. They have cut their valleys down to a point not far above baselevel, and in so doing they have lowered their gradients, or angles of slope. As a result, the rivers have lost much of their downward-cutting ability. Now stream energy flows laterally—side to side—across broad plains, and the rolling interfluves are far below the level of the original plateau. This is an *old age* stage of the cycle; erosion has nearly planed the landform away. A nearly flat surface once again appears, but this time at a level very near that of the sea. It is called a ***peneplain*** or erosional plain—not quite at sea level and not quite flat. Although the streams can cut downward at this stage, they still expend their energy in a process of lateral erosion and deposition of stream sediment. Often, standing above this low relief, there will be a resistant bit of landscape, a hill or small mountain, that did not erode away as rapidly as the rest. This type of feature is called a ***monadnock,*** but it too will disappear in time.

Now all this is only a model—assuming that a new, uniform land surface appears abruptly; that it does not move; that the climate remains the same; and that enough time passes for the entire process to run through its sequence of stages. In nature this sequence seldom takes place. But the Davis model is still a useful tool that is widely applicable for understanding landforms and landform processes as parts of a coherent system.

Dynamic Equilibrium

Davis's geomorphic model has had a long and useful existence, but it has been criticized by geomorphologists and

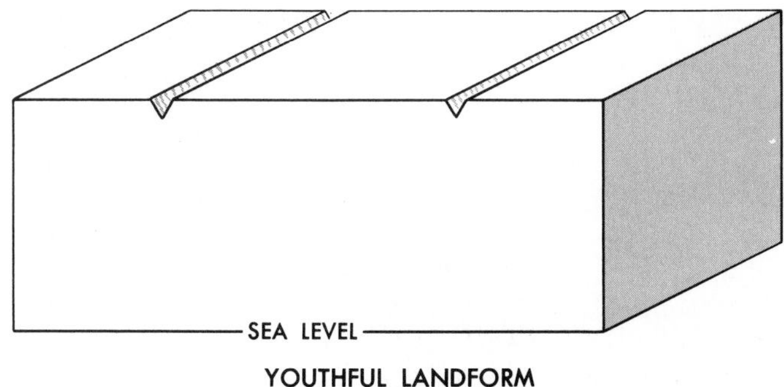

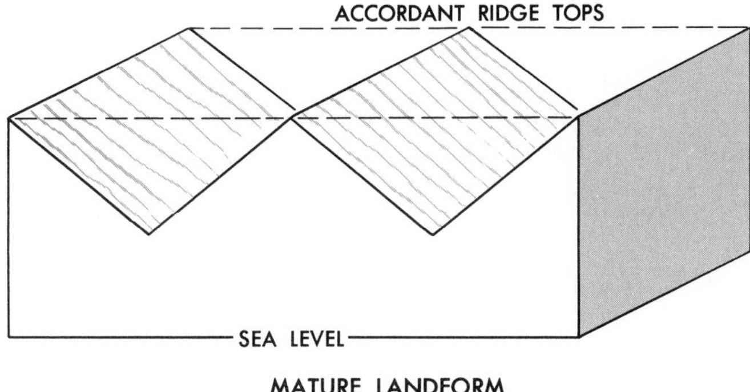

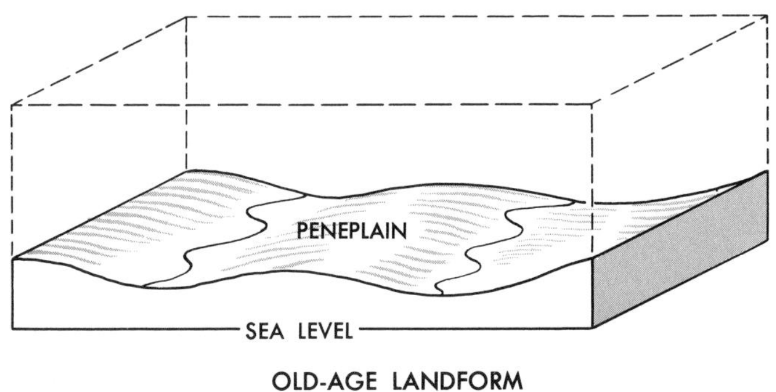

Fig. 16-6. ***The geomorphic cycle of erosion (highly diagrammatic).*** *A youthful landform is only slightly modified from the original, a mature landform is carved dramatically, and an old age landform appears as an erosional plain (peneplain) reduced to near sea level.*

challenged by alternate models and concepts regarding form and process in geomorphology. Ironically, one of the most widely accepted contemporary concepts in the study of landforms had its beginnings before Davis expounded his erosion cycle. This is the idea of adjustment and balance, or equilibrium, between landforms and the forces that work to shape them. It was suggested by G. K. Gilbert, geologist/explorer of the American West, in the latter part of the nineteenth century. Based on his field observations, Gilbert concluded that landform development responds through time to a set of given climatic and tectonic conditions and that landforms reflect a balance between local geology and dominant processes. The revival and expansion of this idea over the past few decades has led to the development of the concept of ***dynamic equilibrium***.

To illustrate dynamic equilibrium let's consider a mountain range in a semiarid environment. Tectonic forces have continued to build the range for millions of years. Two gradational agents—gravity and running water—operate on the slopes provided by the range of mountains. Over time, assuming there is no change in the intensity of the geomorphic processes shaping the range, the landform attains an average slope that reflects the balance between tectonic and gradational processes in the region. The landform

Fig. 16-7. The Caineville Badlands (photo center to lower left) at the north end of Capitol Reef National Park in Utah. There is little or no flat land here; hence it is an emphatically mature landform.

is in or near a state of equilibrium because the forces that are trying to build mountains are offset by those that are trying to wear them down.

Now, let's say that our semiarid environment begins to experience more rainfall, more runoff, and more erosion, as the result of climatic change. The rate of gradation increases, lowering the mountain slopes further than before. Until the landform adjusts to the point where the tectonic and gradational processes balance out again, a state of *disequilibrium* exists. Similarly, the landform must also adjust to a change in the rate of uplift rather than the rate of erosion. For example, if more rapid uplift of the range resulted in steeper slopes, mass wasting and the energy of runoff and erosion would increase until, once again, the slopes adjusted to the new conditions (Fig. 16-8). We rarely witness a perfect balance between the rate of uplift and the rate of erosion in nature because the forces affecting these rates vary through time.

The rate of uplift and the rate of erosion in nature are rarely in balance because the forces affecting these rates vary through time.

In explaining landscape evolution, dynamic equilibrium is not the last word, but it does account more realistically for the effect of processes on landforms than Davis's erosion cycle. At present, geomorphologists are not bound by either the erosion model or the equilibrium concept. However, most tend to view landforms as the result of the interaction of numerous climatic and tectonic variables that do not remain the same over time or space.

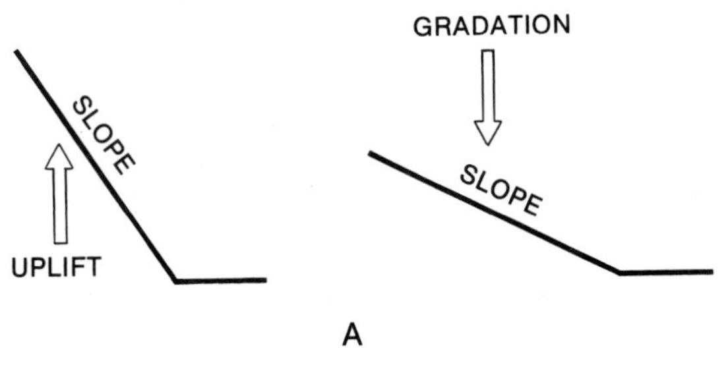

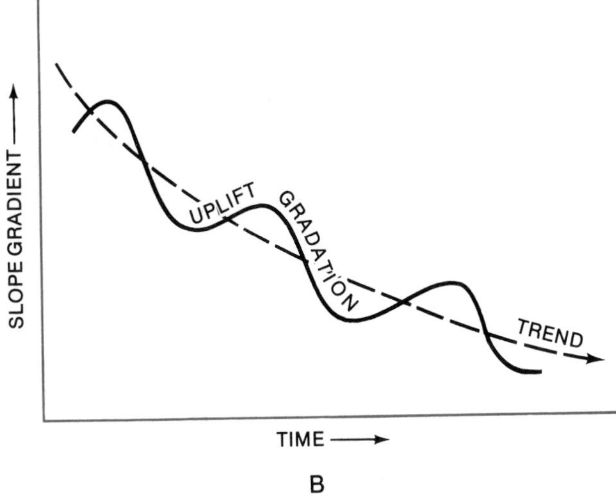

Fig. 16-8. ***A simple graphic explanation of dynamic equilibrium.*** *(A) Uplift increases slope angles, or gradients; gradational processes decrease slope gradients. (B) As illustrated here, these two opposing forces alternate periodically in influencing the character of the landform. Each time, landform adjustment to new conditions is reflected in the lowering or raising of slope gradient; the adjusting approaches a point of equilibrium but never quite reaches it. For this case, the graph shows a long-term trend where the forces of denudation are dominant.*

CONCLUSION

Landforms are dynamic features on the landscape. Most develop over seemingly incomprehensible lengths of time and as a result may appear static to casual human observers. Their study involves much more than the description of present form. Understanding landforms requires a knowledge of the unrelenting processes at work, the gradational as well as the tectonic. In the following chapter, we examine the earth's internal structure and the major tectonic processes that give shape to continents and ocean basins.

KEY TERMS

geomorphology	*absolute age*	*baselevel*
lithosphere	*relative age*	*interfluves*
tectonic processes	*varve*	*accordance*
diastrophism	*radiometric dating*	*peneplain*
gradational processes	*paleomagnetism*	*monadnock*
	cycle of erosion	*dynamic equilibrium*

REVIEW QUESTIONS

1. How are gradational processes different from tectonic processes?

2. How does the Big Bang theory help us understand how the earth was formed?

3. Name and describe three methods used to find the age of geological formations and landforms.

4. How does the known rate of radioactive decay help scientists date rock material?

5. Why is it important for earth scientists to know when the magnetic poles have reversed position?

6. Describe the appearance of a landscape in (1) the youthful stage, (2) the mature stage, and (3) the old age stage of the cycle of erosion.

7. How does the concept of dynamic equilibrium differ from the cycle of erosion model in explaining the development of landforms?

APPLICATION QUESTIONS

1. Imagine that a geologically recent mountain range is still being uplifted by tectonic forces. Would you expect the gradational processes to affect it more or less than a much older mountain range? Why or why not?

2. Take some time to observe and describe the local landscape in your area. What processes do you think are dominant at present? What is the evidence? Does your local or regional public planning agency take into account geomorphic processes when making recommendations regarding land use?

CHAPTER 17

EARTH'S STRUCTURE AND PLATE TECTONICS

OBJECTIVES

After studying this chapter, you will understand

1. How modern earth scientists picture the interior of the earth.
2. How rocks of the lithosphere are formed and classified.
3. How the theory of continental drift developed into contemporary plate tectonic theory.
4. How lithospheric plates interact to form many of earth's major surface features.
5. How moving lithospheric plates form three different types of boundaries.

Near Punta Gorda. Hikers along the northern California coast approach folded sedimentary rock of ocean-bottom origin which has been added to the continental crust by tectonic forces over the past 15 million years.

INTRODUCTION

During the past few decades earth scientists have developed new ideas about the interior structure of the earth and the processes that build the major surface features of the planet. No one has gone deep into the solid lithosphere or beyond the way the fictional Professor Von Hardwigg did in Jules Verne's classic nineteenth-century fantasy, *A Journey to the Center of the Earth.* Nevertheless, the discoveries of today's real, surface-bound investigators, using twentieth-century methods and technology to collect their data, have added significantly to our knowledge of the earth's internal character and processes.

The modern contributions in this area include:

- Accurate detection and recording of earthquakes.
- Precise measurement of gravitational and magnetic variations.
- Detailed mapping of extensive submarine ridges, fracture zones, and trenches.
- Extensive charting of differences in rock density and rigidity with depth.
- Discovery of apparent shifts in the earth's magnetic polarity through time.
- Detection and analysis of remanent magnetism in rock.

Such advances have extended our knowledge and helped us develop models of the earth's internal composition and dynamics without our traveling to its center. The more we know about the earth's interior, the more we understand the physical geography of its surface.

THE STRUCTURE OF THE EARTH

The earth has a radius of about 4000 miles (6400 km), but we have very little concrete evidence showing what the earth's deep structure is really like. Drill bores have gone no deeper than about 5 miles (8 km), and the rock that we find at the earth's surface was formed at depths of no more than about 12 to 15 miles (20 to 25 km). We have traditionally relied on studies of *seismic*[1] and gravitational data, volcanoes, and the composition of meteorites to fill out our theoretical model of the earth's structure. Data from these sources originally led us to picture the earth as a series of laminated, concentric, spherical shells surrounding a core—the hotter and denser materials (thought to be mostly nickel and iron) under great pressure at the center and the cooler and lighter materials at the periphery. Figure 17-1 illustrates the four main divisions of this traditional model:

1. The *crust*
2. The *mantle*
3. The *outer core*
4. The *inner core*

Modern geologists still accept these divisions, but recent research has suggested that the earth's interior may not be as neatly layered and static as Figure 17-1 implies. We now believe that it has somewhat overlapping zones of material that vary in composition, temperature, density, and internal motion. For example, the boundary between the outer core and the mantle has a topography all its own, and there is apparently more movement of material deep in the mantle than geologists thought just a few years ago. If the structure of the earth's interior were perfectly spherical and symmetrical, then it could not be a staging area for some of the dynamic processes that give the surface its variety of relief and form.

The earth's interior has somewhat overlapping zones of material that vary in composition, temperature, density, and internal motion.

Techniques of Discovery

Earth scientists use a variety of techniques to help them understand the overall structure of the earth. Two of the most common techniques are:

1. Analysis of seismic data
2. Analysis of gravitational data

SEISMIC DATA Earthquakes—those abrupt motions that tell us something is afoot below—cause powerful shock waves to pass through the earth. These are recorded by very sensitive seismographic instruments around the world. Think of a large bell being struck by its clapper. The shape, composition, and density of the bell help to determine the sound it makes. Similarly, the structure of the earth's interior influences the character of the seismic waves produced by an earthquake. By analyzing changes in the speed and direction of the various types of earthquake shock waves that pass through the earth, geophysicists have learned much about the earth's internal composition, density, and structure. Earth scientists have also combined the technology of ***tomography,*** the ability to use computers to generate three-dimensional interior views from multiple images, with seismic data to get an even better idea of the earth's interior.

[1] From the Greek *seismos,* meaning shock or vibration. Here it refers to earthquakes.

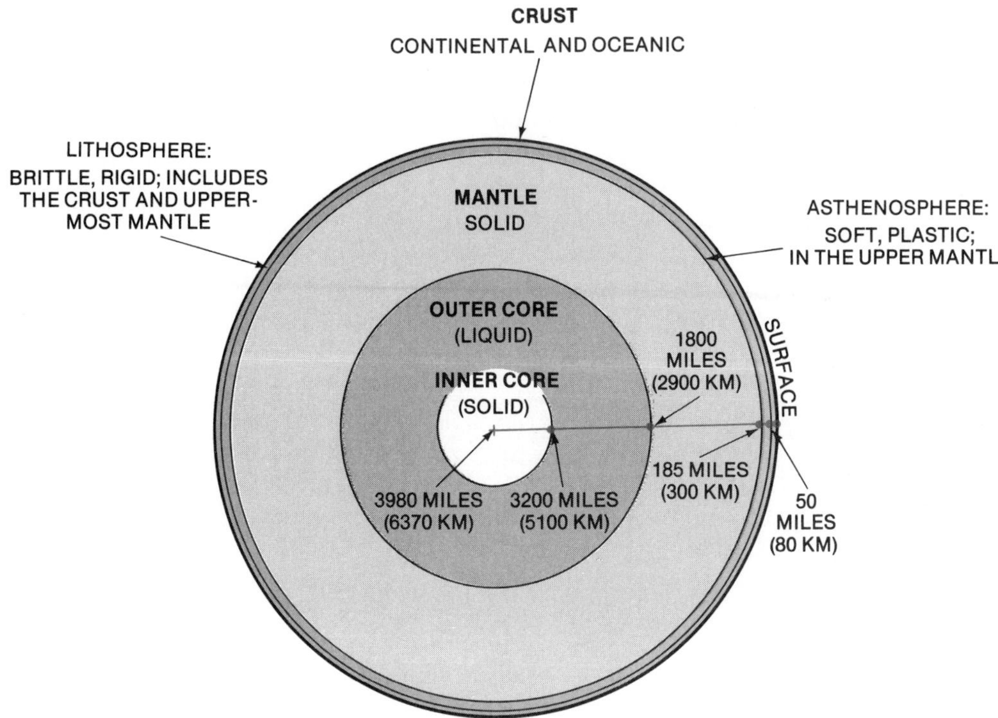

Fig. 17-1. ***The earth's interior.*** *A generalized model showing the spherical layer cake of core, mantle, asthenosphere, and lithosphere. Recent findings suggest that the boundaries between layers, particularly between the outer core and mantle, are very irregular and that with time surface material moves at great depths within the mantle.*

GRAVITY VARIATIONS Another important technique in discovering the nature of the earth's interior is the measurement of variations in gravity. The effect of gravity is related to an object's mass (volume *X* density). Therefore, after taking into account such factors as distance to the center, geophysicists can use their measurement of gravitational force at the earth's surface to infer the relative density of material at great depth.

The Core

In the center of the earth, the ***core*** is extremely hot—in the neighborhood of 5000° F (2750° C). It is also extremely dense, being under a pressure 3.5 million times greater than the pressure of the atmosphere at sea level. By comparison, the pressure under water in the deepest part of the ocean is only a few thousand times greater than that of the atmosphere at sea level.

The core has two distinct zones:

1. The *inner* core, a compressed, solid material—made mostly of nickel and iron.
2. The *outer* core—composed of essentially the same material as the inner core. However, it is in a liquid state because the pressure at this depth is slightly less (Fig. 17-1).

The Mantle

Beyond the core of the earth is the ***mantle,*** which contains the largest portion of the earth's mass. It is composed of a dark, dense rock called *peridotite,* which is made primarily of iron and magnesium. The material in the mantle slowly churns over eons of time, mainly as a result of differences in temperature, density, and pressure. Some material moves at great depths, and some circulates in the area where the upper mantle comes into contact with the lithosphere.

The upper mantle contains a zone called the* athenosphere, *where temperatures and pressure levels combine to give the rock a soft, easily deformable character.

The upper mantle contains a zone called the ***asthenosphere,*** where temperature and pressure levels combine to give the rock a soft, easily deformable character. Above the asthenosphere is the very hard and rigid lithosphere, which includes the brittle outer portion of the mantle and the crust. The highly plastic material of the asthenosphere transmits the forces from within the mantle to the rock above, breaking the lithosphere into plates and setting these plates in motion (Fig. 17-2).

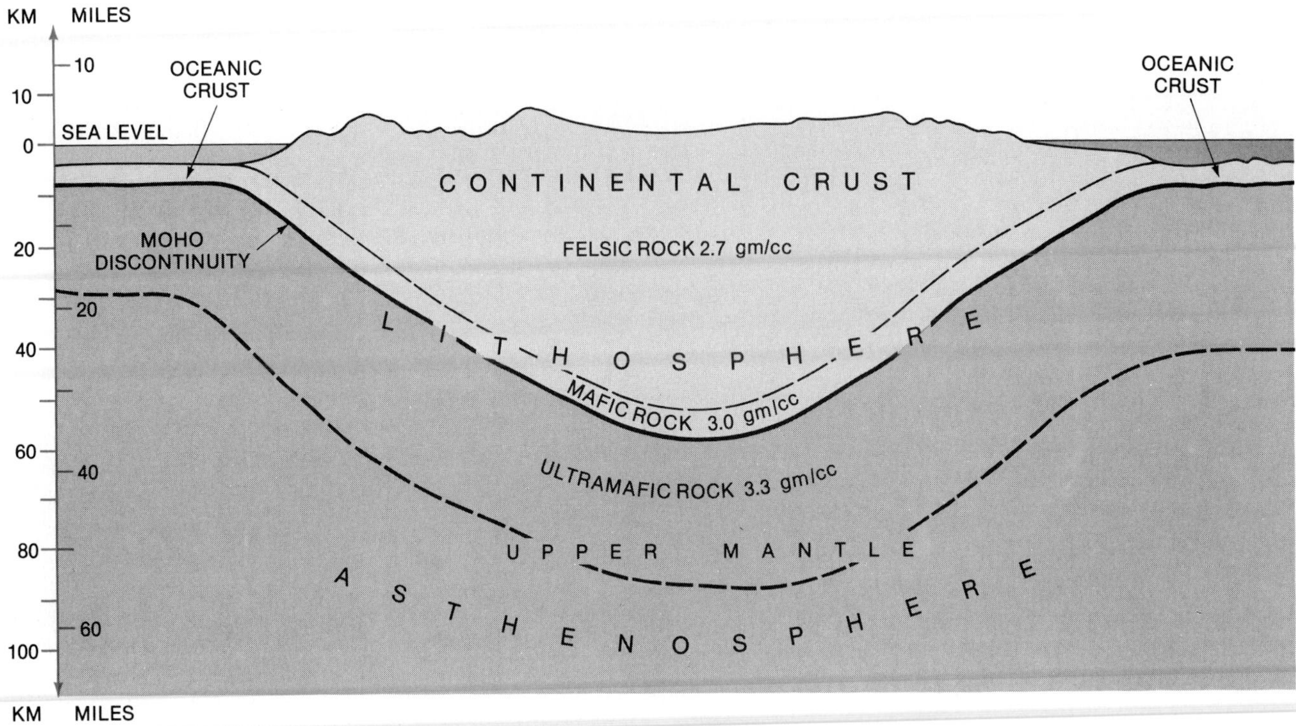

Fig. 17-2. ***The earth's crust and upper mantle.*** *The crust consists of the felsic rock of the continents and the mafic layer underlying the oceans as well as the continents. The Moho discontinuity separates the crust from the ultramafic material of the mantle. The rigid, brittle crust and a portion of the upper mantle rest on the pliable, plastic asthenosphere. Generalized densities of different layers are given in grams per cubic centimeter (gm/cc).*

THE MOHO DISCONTINUITY The mantle makes contact with the crust along a boundary known as the ***Moho discontinuity.*** *Moho* is the abbreviation for Mohorovicic, the name of the Yugoslavian geologist who first identified the mantle/crust boundary.

The crust and the mantle differ mainly in the density and mineral composition of their rock. The mantle is made of dense, dark ***ultramafic*** rock, which contains a large proportion of magnesium and iron. The portion of the crust underlying the oceans and the thin layer extending below land masses (Fig. 17-2) is less dense with somewhat less iron and magnesium than mantle rock and is referred to as ***mafic*** rock.[2] At the boundary between the two layers the distinction is strong enough that seismic waves, traveling nearly 4 miles (7 km) per second through the crust, accelerate abruptly to 5 miles (8 km) per second as they pass into the mantle. This shift in the velocity is the result of the difference in composition and density of the two layers and marks the Moho discontinuity. This also shows how the analysis of seismographic data can help determine internal structural differences.

The Crust

The earth's crust is divided into two categories: *oceanic* and *continental.* The continental crust has deep roots and is up to 40 miles (65 km) thick (Fig. 17-2). It is made up of a mixture of light-colored, relatively low-density material referred to as ***felsic*** rock.[3] As mentioned in the previous section, the oceanic crust is mafic—dark, dense, and with a high proportion of iron and magnesium. It is also relatively thin, averaging 3 to 7 miles (5 to 11 km) in thickness.

To picture the crust, imagine a two-ply veneer composed of

1. A lumpy, discontinuous outer layer that concentrates in continents the least dense and most brittle rocks in the entire earth.
2. A thin, continuous inner layer encircling the entire

[2]The term *mafic* is derived from the names of dominant elements in the rock material: **ma**gnesium and iron (**f**errous), thus *mafic* or *ultramafic,* if in relatively large proportion.

[3]The term *felsic* is derived from the names of dominant minerals found in continental rock: **fel**dspars (light-colored, relatively low-density minerals) and **si**licates (silicon-oxygen minerals), thus *felsic.*

earth, flooring the oceans and underlying the lumpy continents. This inner layer is dense, dark-colored mafic rock.

CRUSTAL ROCKS

Because we occupy the surface of the earth, the continental and oceanic crust are of great importance to us. Mineral resources exist here, and soils develop on land surfaces to support natural vegetation and agriculture. The rocks and landforms of the crust provide clues to the processes that shape the surface of the planet. But to understand the nature of the crust, we must understand the character of the rocks that compose it.

The rocks and landforms of the crust provide clues to the processes that shape the surface of the planet.

The Components of Rocks

All matter is made up of *atoms.* Substances consisting of only one kind of atom, such as hydrogen, calcium, or carbon, are called *elements.* Solid combinations of certain elements generally form *minerals,* and common rocks are mixtures of various minerals.

ELEMENTS As a percentage of mass, the most plentiful elements in the earth's crust are oxygen (49.5 percent) and silicon (25.7 percent). In fact, of the nearly one hundred naturally occurring elements, only eight make up over 97 percent of the crustal total (Table 17-1). Each of the other elements makes up less than 1 percent of our earth's crust.

Table 17-1 The Most Plentiful Crustal Elements

Element	*Symbol*	*Percentage of Crust's Mass*
Oxygen	O	49.5
Silicon	Si	25.7
Aluminum	Al	7.5
Iron	Fe	4.7
Calcium	Ca	3.4
Sodium	Na	2.6
Potassium	K	2.4
Magnesium	Mg	1.9
		Total 97.7

These elements, occurring in small proportion, include some of our highly valuable and useful nonferrous ("not iron") metals such as gold, copper, lead, and zinc.

MINERALS Minerals are ordinarily combinations of elements that occur naturally in a solid state. We can express their composition in a chemical formula. For example, NaCl is the formula for sodium chloride, common table salt, and SiO_2 is the formula for silicon dioxide, quartz. Most minerals form distinctive crystals and have their own degree of hardness. These qualities help identify them (Fig. 17-3).

Occasionally, a mineral may be a single element. For instance, the minerals graphite and diamond are two different forms of pure carbon. So-called native gold (Au) or native copper (Cu) may also occur in nature in pure form rather than in combination with other elements. But single-element minerals are relatively rare; most minerals are combinations of elements.

Fig. 17-3. ***Distinctive crystals.*** *Quartz crystals are invariably hard and hexagonal.*

Fig. 17-4. ***Surface igneous rock.*** *Here's a jumbled broken mass of recent lava that covers extensive areas of Idaho's Snake River plain.*

ROCKS The nearly one hundred naturally occurring elements can merge into a virtually endless number of combinations to produce minerals. These minerals, in turn, can combine to form rocks with an infinite number of possible chemical combinations. Of course, the earth's crust is composed mainly of eight elements, and a very large number of rock-forming minerals are compounds of oxygen, silicon, and one or more of the other six common elements. Nevertheless, an immense variety of rocks exist.

Minerals can combine to form rocks with an infinite number of possible chemical combinations.

Because a classification based on mineralogy and associated chemical composition is beyond our needs in this text, we will focus on a *genetic* approach. That is, we will base our classification on how rocks are formed. This approach yields three general classes:

1. Igneous
2. Sedimentary
3. Metamorphic

Igneous Rocks

Any rock that has cooled and solidified from a molten state is an ***igneous rock.*** Therefore, if the earth began as a superheated sphere in space, all the rocks making up its crust may well have been igneous and thus the ancestors of all other rocks. Even today, approximately 95 percent of the entire crust is igneous. Periodically, molten material wells out of the earth's interior to invade the surface layers or to flow onto the surface itself (Fig. 17-4). This material cools into a wide variety of igneous rocks. In the molten state, it is called *magma* as it pushes into the crust and *lava* when it runs out onto the surface.

All magma consists basically of a variety of silicate minerals (high in silicon–oxygen compounds), but the chemical composition of any given flow may differ radically from that of any other. The resulting igneous rocks will reflect these differences. Igneous rocks also vary in texture as well as chemistry. ***Granite,*** for instance, is a coarse-grained igneous rock whose individual mineral crystals have formed to a size easily seen by the naked eye. A slow rate of cooling has allowed the crystals to reach this size. Normally, slow cooling occurs when the crust is invaded by magma that remains buried well below the surface. We may find granite on the surface of the contemporary landscape, but from its coarse texture we know that it must have formed through slow cooling at a great depth and later been laid bare by erosion. Igneous rocks formed at depth with this coarse-grained texture are called ***plutonic.***

On the other hand, if the same magma flows onto the surface and is quickly cooled by the atmosphere, the resulting rock will be fine-grained and appear quite different from granite, although the chemical composition will be identical. This kind of rock is called *rhyolite.* Many other couplets differ only in texture; for example, *diorite* and *andesite, gabbro* and *basalt, pyroxenite* and *augite.*

The most finely grained igneous rock is volcanic glass or

Fig. 17-5. ***Cappadocia, Turkey.*** *The local residents have taken advantage of a soft sedimentary rock called* tuff *and carved out cave homes. Tuff is the product of violent volcanic explosions. The resultant ash and cinder are carried downwind and deposited selectively according to size. Buried beneath sufficient weight of overburden, the fine ash has become compacted into tuff. In some parts of the world blocks of tuff are used as building blocks because of its ease in quarrying.*

obsidian which has *no* crystals. Some researchers believe this is because of rapid cooling; others believe it is because of a lack of water vapor and other gases in the lava. The black obsidian cliffs of Yellowstone National Park are the result of a lava flow of basalt running head on into a glacier. Some of the glacier melted on contact, but suddenly there also appeared a huge black mass of glassy stone.

Regardless of color, texture, or chemical composition, rock formed from the solidification of magma or lava falls into the igneous category.

Sedimentary Rocks

Most ***sedimentary rocks*** began as an accumulation of loose, unconsolidated rock or organic materials that were compressed or cemented into solid rocks over time. Some sedimentary rocks, such as salt rock and gypsum, are precipitated from solution. It does not matter whether the individual particles were inorganic or organic, how they were accumulated, or what their chemical composition might be. All rocks that have gone through this transformation are classified as sedimentary.

The word *sedimentary* implies the settling of rock particles to the bottom of a body of water. But sedimentary rocks may form in other ways as well. Loose sand and silt may accumulate by wind, stream, or ice action on the land and eventually become sedimentary rocks; evaporating seas may form massive salt deposits; organic matter may solidify into rock (such as coal); solid particles ejected from a volcano in a violent eruption may be cast up into the wind and deposited as loose debris (Fig. 17-5). But by far the largest part of the earth's sedimentary rocks was formed (and is being formed today) by sedimentation in the sea. Therefore *sedimentary* is an apt term.

Fig. 17-6. ***Stratification.*** *Sedimentary rocks can nearly always be identified by this characteristic stratification. Deposited layer upon layer in some ancient sea bed, these imposing ramparts have been uplifted en masse and then exposed through erosion.*

This constant deposition of material on the relatively flat floors of continental shelves has resulted in the layered or *stratified* appearance of the rocks developed in this way. Stratification is a characteristic of sedimentary rock (Fig. 17-6). Incidentally, virtually all the world's fossils are found embedded in sedimentary layers. Without them the interpretation of the history of the earth would be severely hampered.

DEPOSITION Let us assume that a stream is discharging into the sea, whose bottom is of a gentle shelving character. As the stream strikes the sea, its velocity is checked rather abruptly, and since its velocity is related to its ability to carry sediment, it is forced to drop the heaviest portion of its load, pebbles and larger rocks, near the shore. Beyond this, as the stream continues to slow, it will progressively deposit coarse sand, fine sand, and, finally, minute silt and clay particles (Fig. 17-7). This sorting by size is typical of stream deposition. But beyond the influence of the stream, further sedimentation is occurring. Tiny microscopic plants and animals (called *plankton*) with a lifespan of only a few days or weeks are constantly raining down on the ocean floor to mingle with the organic remnants of bottom dwellers. Commonly, the accumulation of the shells and skeletons of this sea life, made up of calcium compounds, results in the deposition of a black, limy ooze which is high in both carbon (from the decomposing organic matter) and calcium. With compression and cementation through time, these calcium-

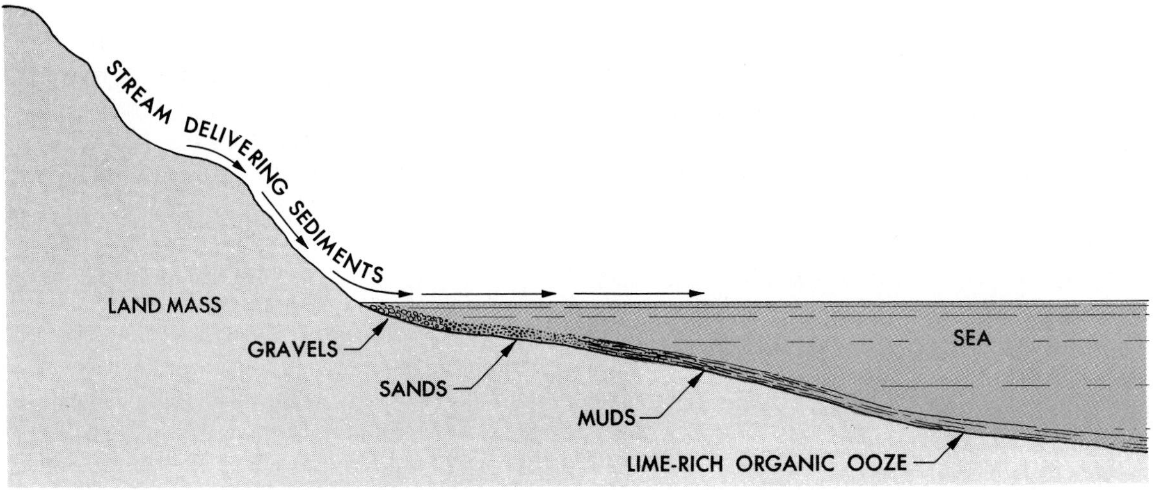

Fig. 17-7. ***Sorted, unconsolidated sediments on the sea floor—sedimentary rocks in the making.*** *If the proper sequence of events follows this simple deposition, each of these categories will develop into a distinctive rock.*

Fig. 17-8. ***Chalk cliffs on the English Channel coast.*** *Soft limestone of marine origin is exposed in the famous white cliffs on the coast of England.*

rich organic deposits become a sedimentary rock composed chiefly of calcium carbonate ($CaCo_3$) called ***limestone*** (Fig. 17-8).

COMPRESSION AND CEMENTATION Pressure alone is normally enough to transform limy ooze and muds into rock. A handful of mud, if squeezed firmly to wring out the moisture, can become a compact mud ball. If this simple pressure is multiplied many times, as layers of sediment pile on top of the original or the ocean bottom folds and flexes, the sediment will be compacted. In the case of the organic limy ooze, the liquid portion (embryonic petroleum) is squeezed out, and, in time, the solid calcium carbonate rock (limestone) is formed. Fine-grained mud sediments lose their water content and become ***shale,*** which characteristically develops *foliation* at right angles to the pressure that formed it. That is, when struck with a hammer, shale tends to break into thin planes or foliations parallel to each other.

Simple pressure, unless it is of unusual force, normally is not adequate to consolidate sands and pebbles into rock. Sand is not only larger in size than mud particles, but is also harder and more angular. The great bulk of the world's sand is silica, one of nature's hardest widely occurring minerals. When worn down through weathering and abrasion, most minerals break down into very fine particles, but silica can keep its identity at sand size and even resist smoothing and rounding. Thus, through natural selection, most particles of sand size are light-colored, angular silica. The individual grains of sand do not fit tightly together, and the many gaps make it highly porous and allow water to move readily through it. Gradually, various minerals in solution or materials in suspension in the water precipitate in the openings between the grains and set as a cement. The resulting ***sandstone*** takes a great deal of its character from the cementing agent. For instance, reddish sandstones probably have oxides of iron as a cement, and dark sandstones may have mud. These are relatively weak stones because the cement is less resistant than the grains of sand. On the other hand, if silica is deposited as the cement, the sandstone will be light colored and very hard. Seldom are all the gaps filled completely with cementing material, and sandstone is often porous and capable of holding liquid such as ancient sea brines, groundwater, or oil (Fig. 17-9).

Pebbles and larger stones are also commonly cemented together. Each pebble is a sizable piece of rock with its own characteristics of color and hardness, and it is usually worn down to a rounded form through stream abrasion before deposition. The large number of possibilities of rock character resulting from the many different cementing agents and variety of pebbles makes it difficult to generalize, but any coarse-grained rock with pebbles cemented together is called a ***conglomerate.***

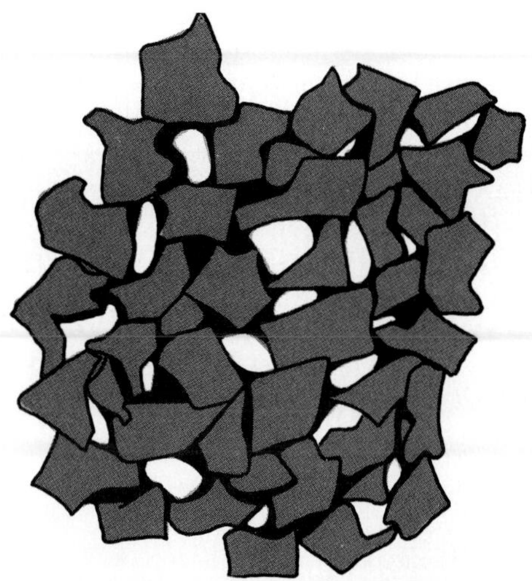

Fig. 17-9. ***Sandstone.*** *The cement (black) fails to completely occupy the large intergrain spaces formed by the angular sand particles. The hardness of the cement largely determines the resistance of the sandstone, and the degree to which the spaces are filled determines the ability of the rock to hold and transfer liquids.*

These then are the four basic types of sedimentary rocks:

1. Limestone formed by long-term pressure on limy ooze.
2. Shale, formed by pressure on muds.
3. Sandstone, formed by cementation of grains of sand.
4. Conglomerate, formed by cementation of pebbles and larger stones.

Although sedimentary rocks do not have to form from sedimentation on ocean floors or show stratification, most do. The foregoing explanation of their formation is highly typical.

Although sedimentary rocks make up less than 5 percent of the total earth's crust, they form a widespread thin covering over much of each continent.

In the surface rocks of the continents, the sedimentaries are strongly represented. Although they make up less than 5 percent of the total earth's crust, they form a widespread thin covering over much of each continent. In a few places where deposition has been continuing for long periods of time, as in the Ganges valley of India, sedimentaries are estimated to be at least 50,000 feet (15,000 m) thick.

Metamorphic Rocks

The word *metamorphism* means change of form. Thus ***metamorphic rocks*** are existing rocks that change form. Powerful forces in sufficient quantities can change sedimentary, igneous, and even metamorphic rocks in appearance, mineral distribution, size, and complete molecular structure. The forces are pressure and heat working in unison. The results are rocks of a type that did not exist before.

Nature has many ways of generating heat and pressure intense enough to cause changes in rocks. Sediments may pile up in sinking troughs to great depths, applying tremendous pressures on the bottom layers—and pressure produces heat. Deep low-angle fractures (or faults) may occur in rock, allowing skidding along the fault plane. Heat from friction generated along the fault plane combined with the pressure of overlying strata may alter the character of the rocks in the immediate vicinity. Igneous intrusions into existing rock masses will frequently furnish heat and pressure at the zone of contact (and often some chemical activity as well). The folding and deformation of flat-lying beds into involved uplands will be equally effective. Most major mountain systems have experienced all these effects in the general process of mountain building. Therefore metamorphism is not rare, and metamorphic rocks of many kinds abound in the earth's crust.

Enough heat and pressure will change fine-grained limestone into coarse-grained ***marble,*** a completely different rock. The metamorphic rock will be both harder and denser than the rock from which it was formed, as when coal changes to *anthracite.* Shale will become ***slate,*** harder and more distinctly foliated, and the metamorphism of sandstone fuses the individual sand grains into an extremely hard silica rock, ***quartzite.*** Distinctive coloring often results from the various cementing materials and impurities. For instance, carbon or iron particles in limestone cause the typical dark or reddish streaks in marble, and iron-rich cements in sandstone give quartzite a pinkish hue.

Solid igneous rocks may also be metamorphised by subsequent pressure and heat. A wide variety of rocks called *schists* and *gneisses*[4] are common results of this metamorphism. The schist is characterized by the formation of mica or similar crystals that are aligned at right angles to the pressure. This lineation is visible to the naked eye and gives the rock a rough foliation, as well as a shiny, almost metallic appearance as the light reflects off the many cleavage surfaces (Fig. 17-10). Gneiss also exhibits a lineation of miner-

[4]Schists and gneisses may also be occasionally formed by the metamorphism of sedimentary rocks.

Fig. 17-10. ***Banding in schist.*** *Visible plastic flow lines speak of its reconstructed origin.*

als, but typically it is banded. That is, it shows wide, sharply defined bands of alternating dark and light materials.

Any rock can metamorphose given enough heat, pressure, and time. Theoretically, metamorphic rocks can metamorphose, but this is often difficult to recognize without radioactive dating in the laboratory. Recently, we have had to consider a process called *granitization*—the final metamorphism of a wide range of rocks into granite. If this process occurs on any large scale at all, and many think it does, it helps to account for the great continental masses of granite that we formerly attributed to igneous intrusion alone.

The Cycle of Rock Formation

The crust of the earth as we know it today is a scrambled combination of igneous, sedimentary, and metamorphic rock. The original crust may have been igneous as the heated earth cooled at the surface (if we are willing to accept such a theory). Immediately, however, weathering and erosion began to break this crust up into small particles and individual minerals that were carried off, sorted, and deposited as sediments. At the same time, organic remains were accumulated in favored locations. As these became cemented and compressed into rock, they were frequently uplifted into mountains and continental masses. The forces of such constant crustal deformation caused many igneous and sedimentary rock to metamorphose. This cycle is repeated over and over and continues at this moment, constantly supplemented by fresh intrusions of magma. Because of this cycle, we find a great variety of rocks in the earth's crust, especially in the continental regions. But despite this great variety, all rocks can be placed in one of the three basic categories: igneous, sedimentary, or metamorphic.

The crust of the earth is a scrambled combination of igneous, sedimentary, and metamorphic rock.

PLATE TECTONICS

When we think of the sheer mass of rock that forms our earth's crust, phrases like "terra firma" and "steady as a rock" naturally come to mind. We imagine that only the most violent upheavals—volcanoes and earthquakes—could disturb such massive solidity. In fact, earth scientists have come to believe that our planet's crust is not a single immobile mass at all but a mosaic of slowly moving plates. Let's see how this concept developed and how it is presently being used to explain the awesome mechanics of world-building.

Continental Drift: Seed of Theory

By the end of the sixteenth century, cartographers of the sailing nations of Europe had mapped most of the coastlines of Africa and the Americas, as well as those of Europe. These maps revealed thought-provoking relationships between the coasts of Africa and South America, Europe and North America. If these continents could be pushed together in the middle of the Atlantic, they would fit comfortably with little alteration to their outlines. Could it be that they were once together and then split apart (Fig. 17-11)?

Sixteenth-century maps revealed thought-provoking relationships between the coasts of Africa and South America, Europe and North America.

Few Renaissance scholars took note of this curious congruency of coastlines. If they did they probably dismissed it as a coincidence, since most people then—and for the next few centuries—believed that the continents were fixed in position. They did not move about, at least not horizontally.

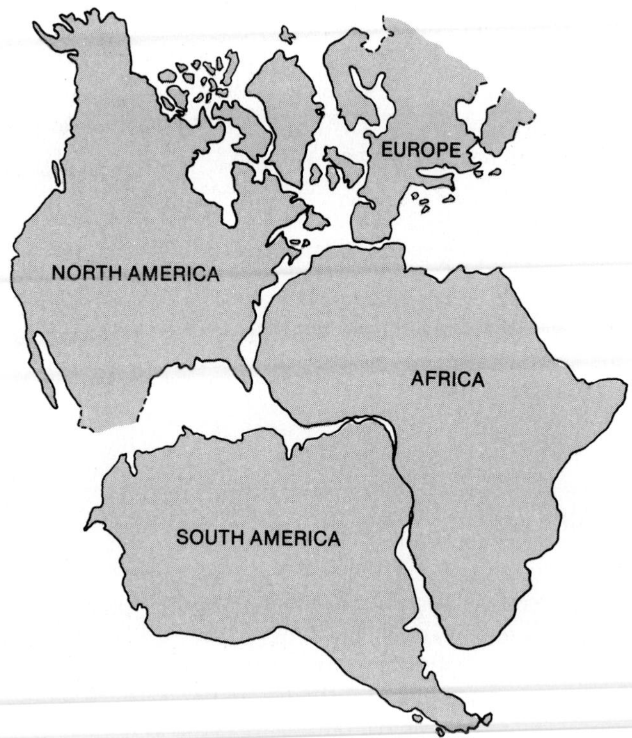

Fig. 17-11. ***Fit of the Atlantic continents.*** *Atlantic coast outlines of North America, South America, Africa, and Europe have a convincing fit when arranged next to one another.*

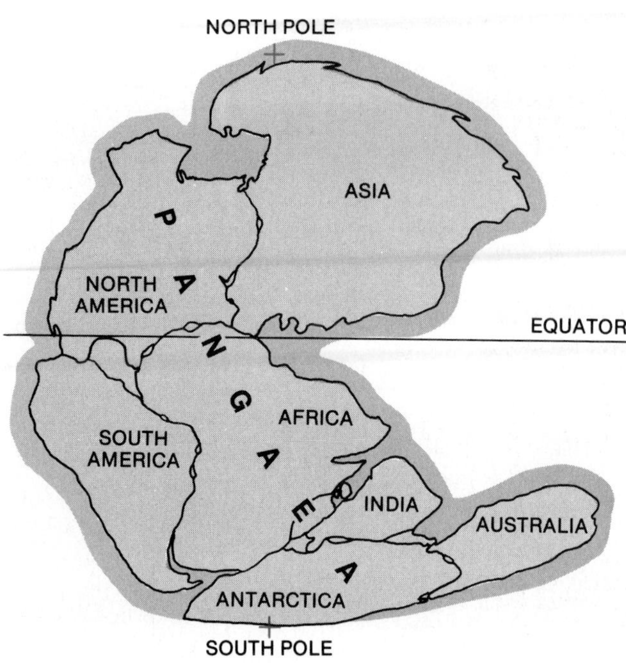

Fig. 17-12. ***Pangaea.*** *The map pictures Wegener's supercontinent 200 million years BP, before it broke up into the major land masses of today.*

Any motion that did occur was vertical, up and down, to form the ocean basins and mountain ranges.

THE PANGAEAN BREAKUP Alfred Wegener (1880–1930), an early twentieth-century German meteorologist-turned-geologist, believed that the "fit" of the continents was more than coincidence and proposed a theory of *continental drift* in 1912. He hypothesized that the present continents had separated and slowly drifted away from a single supercontinent, which he called *Pangaea* (*pan* = all, *Gaea* = Greek goddess of earth), about 200 million years ago (Fig. 17-12). Wegener based his ideas not only on the fit of continental coastlines, but also on several other pieces of information available at the time:

1. Fossils found in rocks in parts of Africa and in South America are similar in type and age.
2. Although Australia, Africa, and India are widely separated tropical and subtropical regions, they contain evidence of the same ancient glaciation (marks and deposits left by huge ice sheets 300 million years ago).
3. Large blocks of the continents of Africa and South America have similar geological patterns and would form continuous belts if the land masses were together.
4. Continental rock is less dense and therefore lighter on the whole than the rock of the ocean floor. Thus, in Wegener's theory, continents could float on, and pass through, the dense rocks of the ocean bottom.

CRUSTAL ISOSTASY During Wegener's time scientists knew that continental and oceanic crustal rock had different densities. This knowledge gave birth to the idea of crustal ***isostasy,*** the theory that a condition of buoyant equilibrium exists between the lighter continental land masses and the denser material below that allows the continents to float—like icebergs floating in the open ocean. Because land masses are high in elevation, the laws of gravity demand that they must also penetrate deep into the mantle to maintain equilibrium. The denser ocean crust, on the other hand, is a much thinner layer stretched across the mantle. Any change in the mass or weight of the crust is balanced by an uplift or depression of the affected areas. For example, when continental-scale glaciers melted thousands of years ago, they released a burden from the land, which has slowly uplifted since. Conversely, as sediments are deposited in structural valleys or ocean bottoms over a long period, the crust is depressed.

REFUTING THE LAND BRIDGE THEORY Many of Wegener's contemporaries believed that the continents, though fixed in their geographic position, were connected at one time by a land bridge that allowed the diffusion of ancient terrestrial animals and plants. Some of their remains account for the similar rocks and fossils found in Africa and South America. Eventually, the "bridge" sank, isolating the land masses.

Wegener pointed out that the difference in density between the continental and oceanic materials made this theory highly unlikely. After all, he reasoned, a land bridge would have been composed of felsic continental rock and therefore would have been too light and buoyant to sink and form an ocean basin. He suggested that separation of the continents, not a sinking land mass, explained the fossil/rock similarities.

THE SEARCH FOR A DRIVING FORCE But alas, Wegener had his own problem—he did not have a credible and proven mechanism that would prod the lighter continents to move through the sea of ocean crust. Wegener strongly believed that the mountain systems and deeper parts of the oceans were the result of horizontally moving continents colliding with one another or plowing through the ocean floor. But what could be the driving force behind the drifting? He came up with two possibilities: the centrifugal force of earth's rotation or the gravitational pull of the moon. Neither of these was satisfactory, even to Wegener himself.

Wegener's theory of drifting land masses was revolutionary but incomplete and therefore unacceptable to most earth scientists of the day. Indeed, many severely criticized his arrogance at suggesting continental movement and went so far as to question his sanity. Nearly all scientists at the time remained comfortable with the explanation that the mountain systems and ocean basins of the world were the result of the earth's cooling, contracting, and slowly wrinkling up its surface over time like a drying apple. This contracting and wrinkling process could produce the vertical motions that gave topographic relief to the crust. But Wegener's critics insisted that horizontal movement of entire continents was impossible.

When Wegener died in 1930, nearly everyone who knew of his theory had dismissed it. Although his original ideas do not exactly match today's explanation of tectonic processes, the fundamental notion of horizontally moving crustal units is the cornerstone of present global tectonic theory. It was not until the 1950s that evidence began to emerge establishing modern plate tectonics as the best model for explaining major landform features.

Plate Tectonics: The Modern View

Plate tectonics is a dynamic model of the lithosphere—a lithosphere broken into a number of large, irregularly shaped plates that move on the soft, plastic layer within the mantle called the *asthenosphere*. Through geologic time the plates have collided, pulled apart, and scraped against one another to produce the major structural features of the planet's surface.

Through geologic time the plates have collided, pulled apart, and scraped against one another to produce the major structural features of the planet's surface.

Soon after World War II, new technology such as sonar (*so*und *na*vigation *r*anging), deep-ocean drilling, and the magnetometer (an instrument used to detect the presence and intensity of a magnetic field) made it possible to map the ocean bottom and discover topographical and geological patterns and remanent magnetism in the rocks of the oceanic crust.

Three of the most important concepts related to the development of the plate tectonic model were:

1. Paleomagnetism
2. Sea floor spreading
3. Convection in the mantle

PALEOMAGNETISM Remanent magnetism, or paleomagnetism (see Chapter 16), in rocks of the ocean floor provides a major clue to the dynamics and chronology of the lithosphere. After earth scientists mapped the topography of the Atlantic Ocean floor in the early 1950s, they found that a midoceanic ridge ran nearly the full length of the Atlantic, north to south (see plate in color insert). Further investigations revealed that ocean bottom sediments near the ridge were thin and that the age of the crustal rock below these sediments was geologically recent.

SEA FLOOR SPREADING Beyond the midoceanic ridge to the margin of the continents the age of the oceanic crust became progressively older, but none of it older than about 180 million years. Geologists recorded and analyzed paleomagnetic patterns in oceanic rock from core samples taken by oceangoing scientific vessels in the 1950s and found a mirror image on either side of the midoceanic ridge

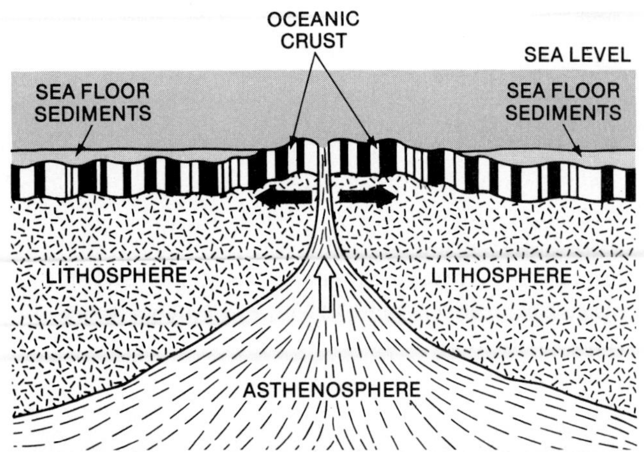

Fig. 17-13. ***Mirror image.*** *Stripes in the oceanic crust represent paleomagnetic patterns in the rocks. The pattern is sequentially the same in either direction from the midocean rise in the center, indicating the formation of new crust and the spread of the ocean floor.*

(Fig. 17-13). This discovery indicated that crust forms at the ridge and is then pushed aside as new material emerges from within the mantle at the ridge, or *rise*. As this new crust forms, the ocean floor slowly spreads away from the axis of the rise toward the continents. Evidence of *sea floor spreading* also appeared in the Pacific and Indian oceans, which, like the Atlantic, have long, linear ridges running through them.

By the 1960s earth scientists had mapped nearly all the ocean bottom relief. Deep-ocean *trenches* in the western Pacific Ocean, off the west coast of South America, and paralleling the Aleutian Island chain were part of the picture (Fig. 17-14A, B), as were numerous fracture zones running perpendicular to the ridges. The pattern supported the idea that huge lithospheric plates, some of them comprising both ocean basins and continents, were in motion—horizontal motion—a concept introduced by Wegener 50 years earlier.

CONVECTION IN THE MANTLE Where plates bound one another, earthquake activity is greatest. Along the rises (ridges) molten rock from the mantle squeezes into the ocean crust causing shallow earthquakes. Beneath the trenches the ocean crust is pushed slowly back into the mantle building stress that ultimately results in earthquakes deep below the surface.

From our experience with earthquakes, volcanoes, and the deformed rock strata in folded mountain ranges, we have long recognized that the earth has a source of internal energy. Although scientists are not certain, most believe that a major source of energy is from the heat generated by the decay of radioactive elements deep in the earth. They think that this ***radiogenic heat*** sets in motion portions of the mantle that have become hotter and less dense. The currents resulting from this heating move material upward in the mantle to push together and pull apart the lithosphere. This movement is called ***convection in the mantle.*** The force of the convection pushes and drags parts of the lithosphere with it as it spreads horizontally (Fig. 17-15).

The currents resulting from internal heating move material upward in the mantle to push together and pull apart the lithosphere.

WEGENER VINDICATED By the late 1960s the fundamentals of what we now refer to as plate tectonics were solidly in place. Evidence from paleomagnetism, sea floor spreading, and possible convection in the mantle explained the movements of lithospheric plates. Alfred Wegener was vindicated, at least in part, and most earth scientists have now accepted the concept of horizontally moving plates that grind into, slide by and plunge under one another.

The Boundaries of Plates

Visualize the brittle, rigid lithosphere broken into a series of mobile plates carrying continents and parts of ocean basins with them. The sea bottoms are either included as portions of a greater plate or, as in the case of the Pacific, may be plates in their own right. As a result of the convection forces from below, they jostle and jockey for position, interacting with each other along boundaries where tectonic activity is greatest (Fig. 17-14). Lithospheric plates have three types of boundaries:

1. Converging boundaries
2. Diverging boundaries
3. Transform boundaries

As a result of the convection forces from below, lithospheric plates interact with each other along boundaries where tectonic activity is greatest.

CONVERGING BOUNDARIES When the continental portion of a plate, such as South America, converges with an oceanic plate, such as the Nazca plate, something must give. The oceanic plate of dense mafic rock is slowly but relentlessly pushed below the lighter continent in a process called ***subduction.*** A deep, linear subduction trench just off the coast of Chile and Peru shows the steep angle at which the oceanic plate is plunging (Fig. 17-16).

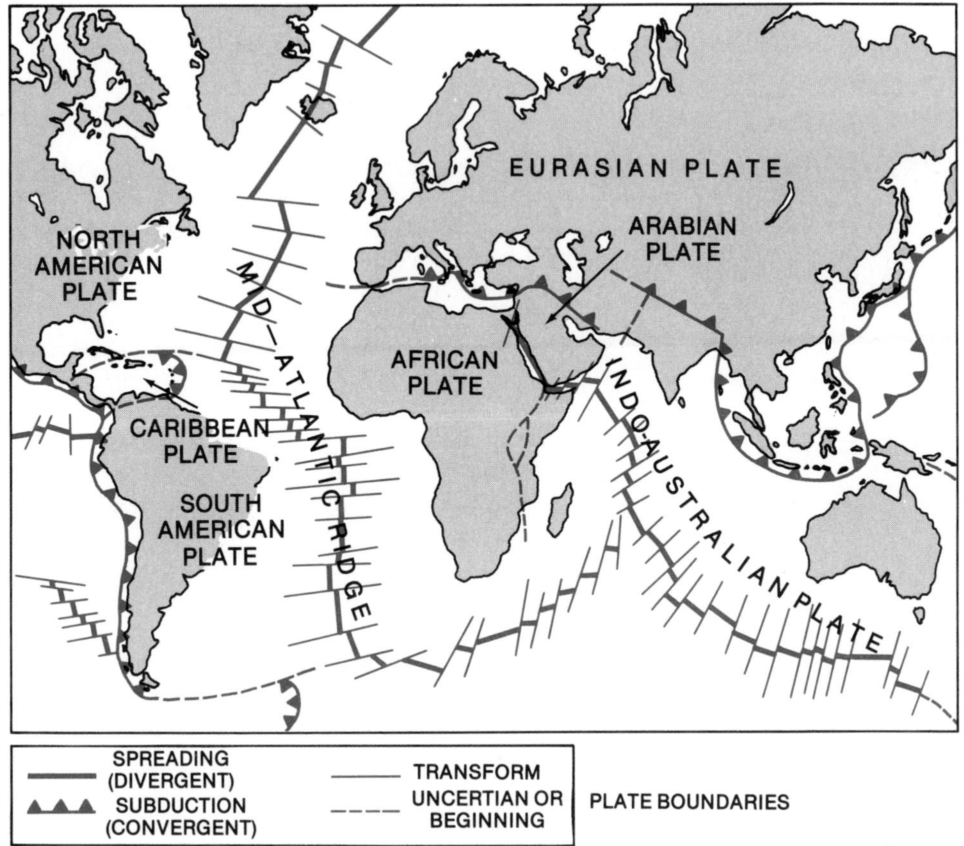

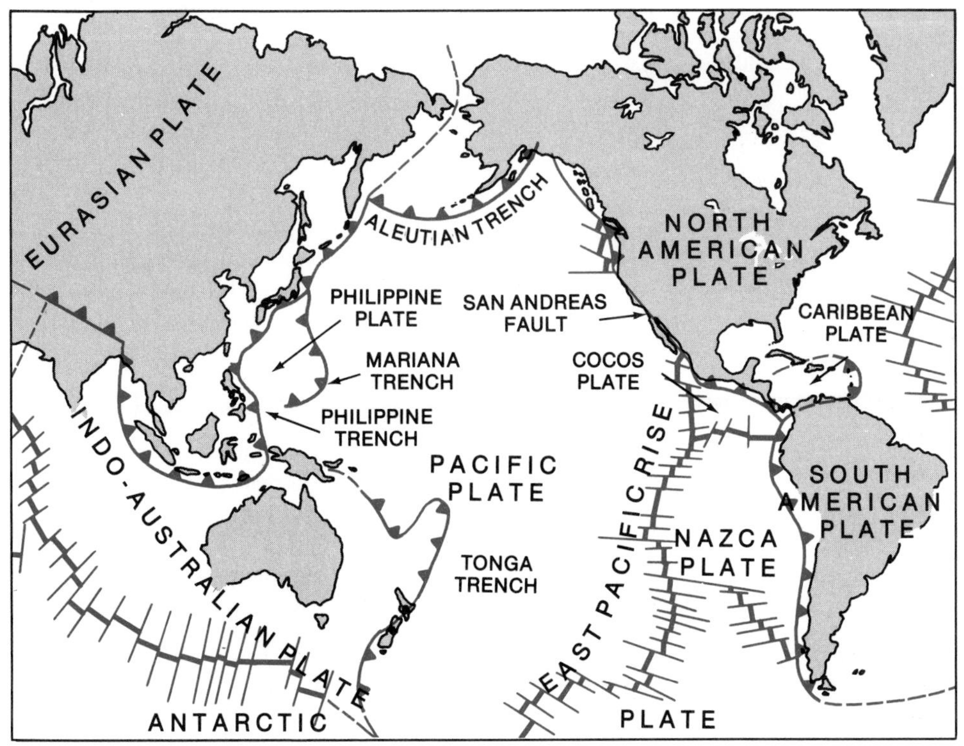

Fig. 17-14 A and B. ***Tectonic plate boundaries.*** *Major convergent, divergent, and transform boundaries are generally located in midocean or along continental margins.*

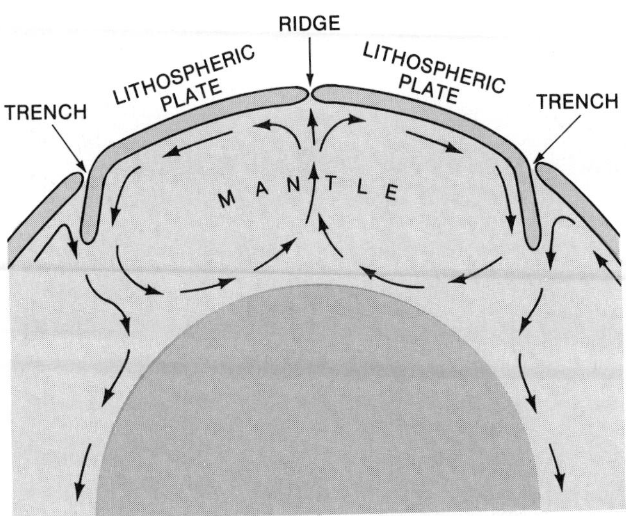

Fig. 17-15. ***Convection in the mantle.*** *This motion, shown by arrows, is thought to provide the force needed to move lithospheric plates.*

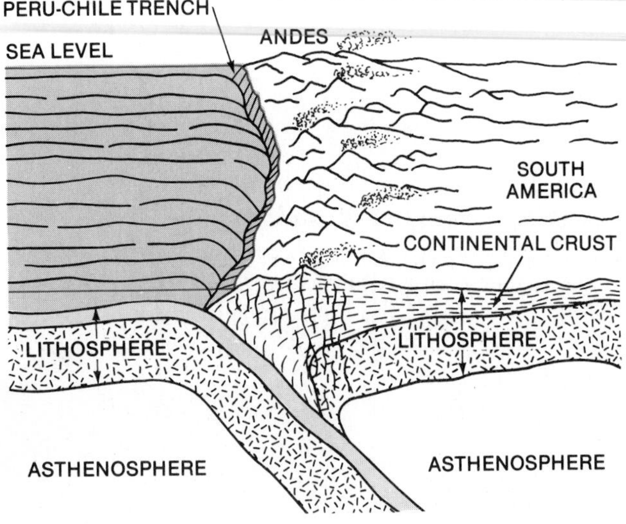

Fig. 17-16. ***Subduction.*** *A deep subduction trench has formed along the west coast of South America as the ocean plate passes at a steep angle beneath the continent. Subducted material heats and rises through the continental crust to erupt in the many volcanoes of the Andes.*

The coastline of South America and the mountain chain of the Andes run parallel to the trench along the length of the continent's western margin. The Andes are a geologically young mountain system, severely folded, with many active volcanoes. Strong earthquakes usually result in loss of

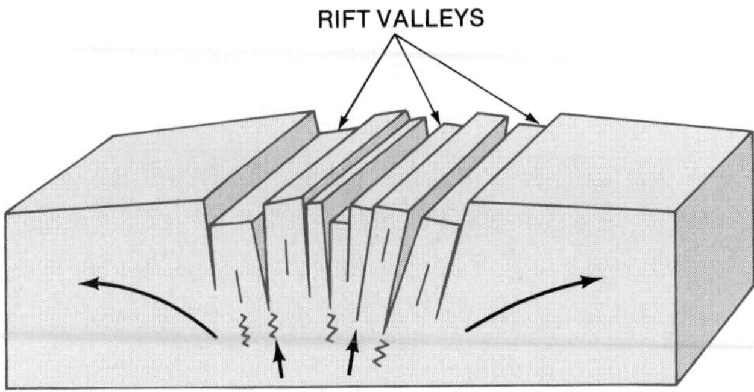

Fig. 17-17. ***Rifting.*** *As the earth's crust is pushed up and apart by convectional forces,* rift valleys *form along the axis of the divergent boundary.*

life and major damage in these relatively populated mountains.

Plates also converge beneath the western Pacific Ocean where portions of the lithosphere have broken up as the huge Pacific plate moves westward. In this region are numerous deep-ocean trenches where oceanic plates subduct against one another (Fig. 17-14). The great island arcs of the Philippines, Japan, the Marianas, and the Tongas pushed above sea level as these pieces of oceanic lithosphere were forced into the mantle, scraping off bottom deposits, bowing up the crust, and generating volcanic activity. All this activity contributed to the construction of the ***archipelagos,*** or island chains, that parallel the subduction trenches.

The subduction of oceanic plates in the western Pacific is very different from the collision of continental plates, yet both form ***converging boundaries.*** The Himalayas, the high, massive mountains that separate India from the rest of Asia, were formed as a portion of the Indo-Australian plate merged with the Asian plate. Little subduction occurred because the rock of both colliding land masses was equally dense and light. It therefore crumpled up into high, folded, and fractured mountains that contain the sedimentary rocks of the ocean basin once separating the two plates before collision. Similar collision of continental plates is occurring to the west of the Himalayas in Afghanistan, Iran, and Turkey.

Convergence is evident in the strongly compressed and fractured Zagros and Elburz Mountains in Iran and the Caucasus Mountains in the southern Soviet Union. The disastrous earthquake in Soviet Armenia in December 1988, in which approximately 25,000 people lost their lives, was the result of the sudden release of crustal stress that built up as the Arabian plate pushed against the Eurasian plate (Fig. 17-14). The Himalayan mountain system of south and western Asia, the Andes at the continental margin of South America, and the deep-ocean trenches and archipelagos of

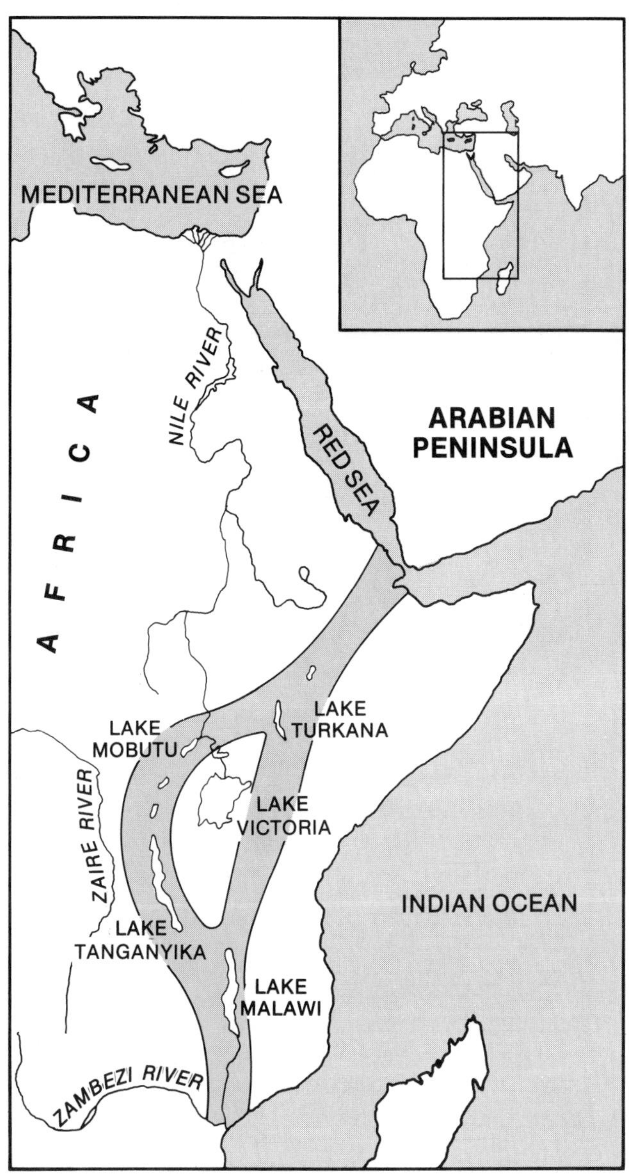

Fig. 17-18. ***East African Rift Zone.*** *Here the continent is slowly breaking apart, creating rift valleys that contain elongated lakes. If present processes continue, a new sea will open up in this area as eastern Africa is carried away from the rest of the continent.*

the Pacific Ocean are all distinctive features of converging plate boundaries.

DIVERGING BOUNDARIES In ***diverging boundaries,*** convectional motion is pulling apart the lithosphere. Such activity occurs both at midocean and within continental land masses. Tension causes the crust to fracture into huge blocks that slip vertically by one another to form structural valleys and highland masses. Typically, magma rises into the fractures between the blocks, at times flowing onto the surface as lava, solidifying and expanding the crust. This ***rift valley*** topography is typical of diverging boundaries (Fig. 17-17). For example, midoceanic ridges have rift valleys along their axes that are the result of the crust being pushed up and pulled apart at the same time.

One of the best examples of such a rift valley system is in East Africa where plates have begun to diverge. The rift zone extends 2000 miles (3200 km) from the southern end of the Red Sea to the Zambezi River in southern Africa (Fig. 17-18). A few tens of millions of years from now it is possible that a new ocean basin will have opened up separating what is now eastern Africa from the rest of the continent.

Another region of the world where the continent has been stretched and faulted into blocks is in western North America. The mountain ranges and high basins of Nevada, western Utah, eastern California, and southeastern Oregon are the result of continental divergence.

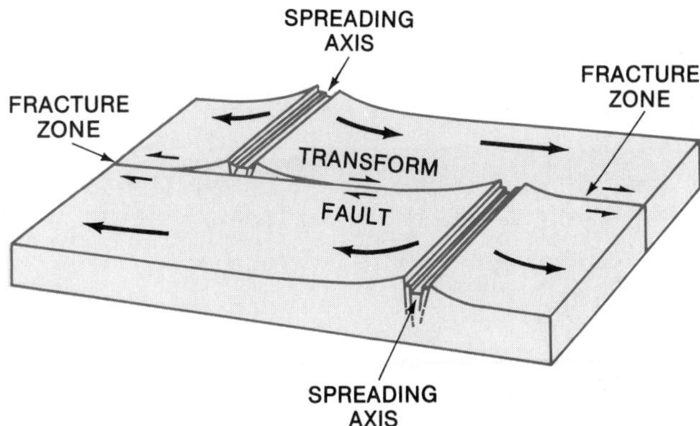

Fig. 17-19. ***Offset ridges.*** *A transform fault develops between two offset segments of a spreading axis, for example, a midocean ridge. The transform fault is where two plates are moving parallel to the fault in opposition to each other.*

TRANSFORM BOUNDARIES A third type of plate boundary is called a ***transform boundary.*** Here one plate slips horizontally past another along the boundary rather than colliding or pulling apart, as in the case of converging and diverging plates. The plates move parallel to the boundary, and therefore crustal deformation is less dramatic.

Most transform boundaries and the fracture zones associated with them are found between midoceanic ridges and subduction trenches. Here portions of oceanic plate may be moving at different velocities away from the spreading axis of the rise, or the rise itself may be offset in segments (Fig. 17-19).

MENDOCINO TRIPLE JUNCTION

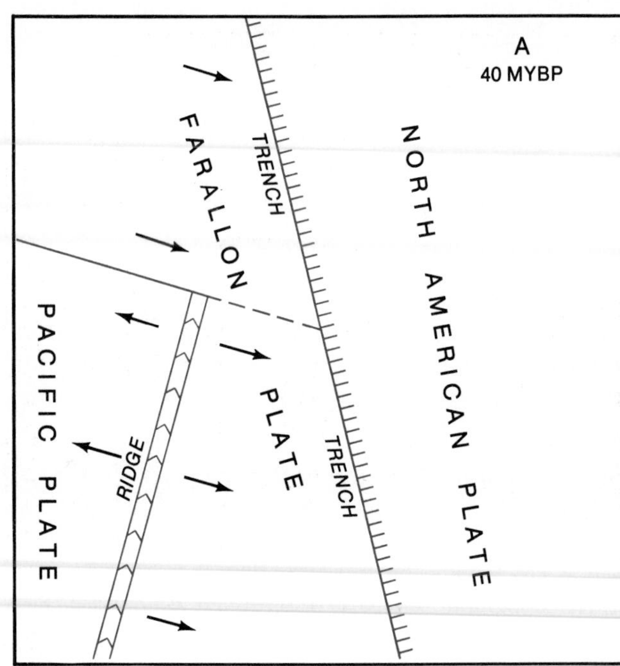

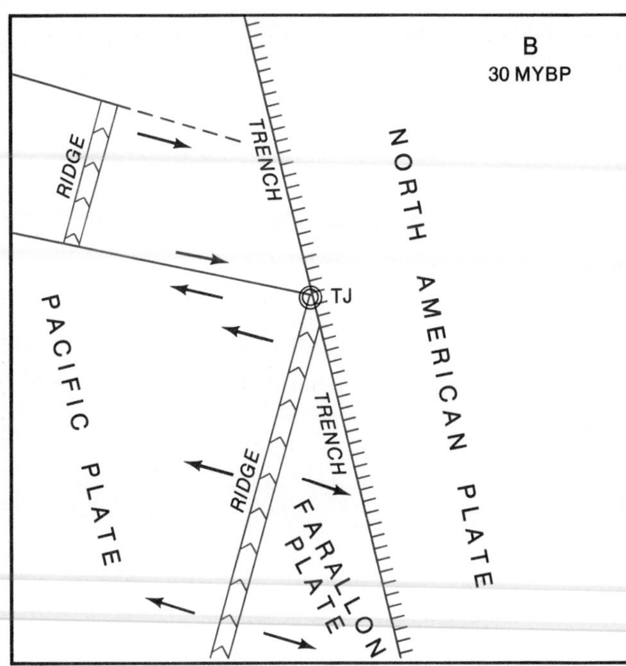

Tens of millions of years before the present, the North American plate and the Farallon plate of the eastern Pacific Ocean basin began to merge (A and B). The ocean-spreading axis was segmented and offset so that as the ridge met the continent, a transform fault, the San Andreas, appeared (Diagrams C and D). At the north end of this fault is the Mendocino Triple Junction where three different plate

Given the long, linear Peru–Chile subduction trench off the west coast of South America, it is surprising that no similar deep-ocean trench parallels the west coast of North America. This is because the North American and Pacific plate boundary is more complex than a straightforward converging boundary of oceanic and continental plates.

About 30 million years ago the ocean ridge of the eastern Pacific, the East Pacific Rise, collided with the western margin of the North American plate. The trench and much of the rise existing at the time began to pass beneath the continent. The highly generalized diagrams A, B, C, and D illustrate what researchers believe occurred during the past 30 million years as the active boundaries and moving plates interacted.

What remains of the rise is a series of offset segments, still spreading the sea floor, found off the west coasts of Mexico, Oregon, and Washington. In between lies the San Andreas Fault running through most of coastal California, one of the longest and most active faults in the world (see inset). To the north of the San Andreas Fault are parts of the ancient Farallon plate, which is still being subducted at a very low angle below the North American plate (Diagram D). This type of subduction produces a relatively shallow structural trench that fills rapidly with sediment. Hence there is little local evidence of the process.

At the north and south ends of the San Andreas Fault the boundaries of three different plates meet. Such a meeting point is called a *triple junction*. At the south end of the San Andreas a

A ***transform fault*** develops where the plates move in opposite directions between two offset ridges. For example, the mid-Atlantic ridge is not a straight line, oriented pole-to-pole. As a result, there are many offset segments and, hence, transform faults and fracture zones (see plate in color insert). The famous San Andreas Fault along the west coast of North America runs for several hundred miles along a transform plate boundary between rise segments (Fig. 17-14).

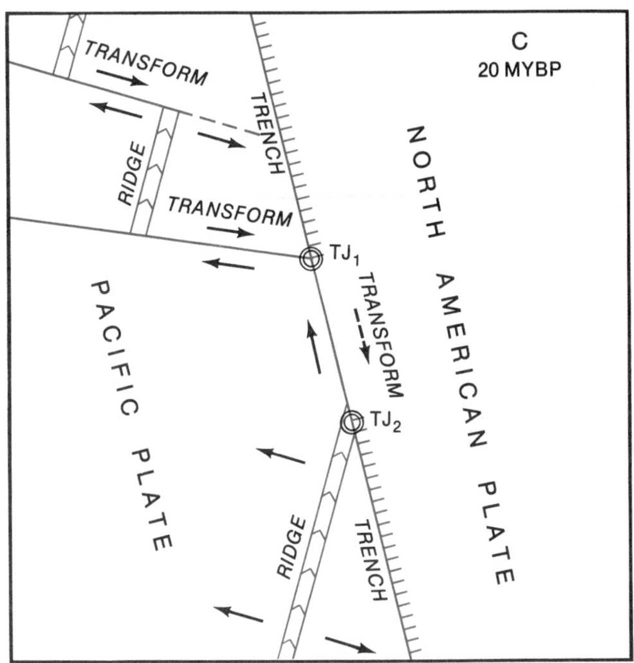

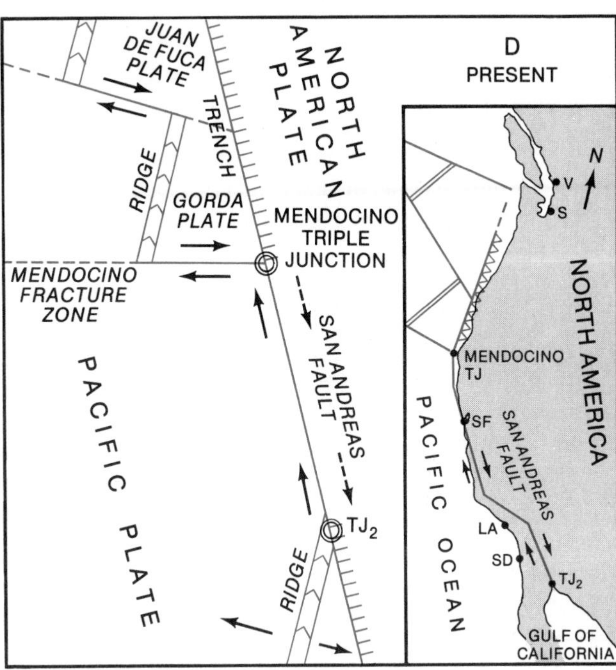

boundaries interact to produce a complex and dramatic coast (see Chapter 17 frontispiece). Source: *Allan Cox and Robert Bryan Hart, 1986.* Plate Tectonics: How It Works, *p. 79. By permission of Blackwell Scientific Publications, Inc.*

diverging boundary, a converging boundary, and a transform boundary join; at the north end a converging boundary and two transform boundaries meet. Here the subducting remnants of the Farallon plate (Gorda and San Juan de Fuca plates), the old transform fault formed between offset rises (Mendocino Fracture Zone), and the transform San Andreas Fault come together as the three plates mesh. How did this triple junction form? When the ocean rise and the continent began to merge, the trench marking the eastern margin of the Farallon plate was overridden by the continental plate (Diagrams A and B). The rise segment eventually encountered the continent as well, and as it did so, the direction of plate movement away from the rise to the west allowed the Pacific plate to slip north at the boundary with the North American plate (Diagram C). This formed an extensive transform fault, the San Andreas. The plate junction at the north end of the San Andreas Fault is located near Cape Mendocino in northern California and is known as the Mendocino Triple Junction. As might be expected, the coastal geomorphology is dramatic and the geology complex in this region.

There are other triple junctions around the world, the most notable being the Afar Triangle at the south end of the Red Sea, a junction composed of three diverging boundaries. One is responsible for the opening of the Red Sea, another separates the Arabian Peninsula from the "horn" of Africa (the Gulf of Aden), and the third is the east African rift zone. The Afar Triangle is over a hot mantle plume where rising material pushes the crust apart along three axes.

Because of the significant stress that accumulates where these plates oppose each other, earthquakes are frequent and at times severe.

At the north end of the San Andreas Fault near Cape Mendocino in northern California is a point where three different plate boundaries come together. This is known in plate tectonics as a ***triple junction.*** Here the transform fault of San Andreas meets two other plate boundaries: a converging boundary from the north and another transform boundary from the west.

CONCLUSION

To understand the global-scale geomorphology of the planet, we must have a fundamental knowledge of the materials and processes that help shape its surface. The structure and processes of the earth's interior are important in explaining the landforms we observe at the surface. Learning the ways in which rocks are formed introduces us to the character of the lithosphere and to many of the processes that give shape to the physical landscape. Explanations of landforms cannot rest on description alone; process—the action behind the form—gives us a fuller picture of the physical geography of the earth.

Understanding the global-scale geomorphology of the planet requires a fundamental knowledge of the materials and processes that help shape its surface.

Over the past few decades we have witnessed one of the most productive periods in earth science. The development and verification of plate tectonic theory has been an exciting classic case of science in action. The past 30 years has seen the consolidation of a working model that helps us to understand earth's internal energy system and the subsequent evolution of the continents, ocean basins, and major structural landforms. The tectonic processes are slow, yet they continually operate to construct a surface that is subject to the external forces of gradation which play such an important role in molding landscape.

KEY TERMS

tomography
core
mantle
asthenosphere
Moho discontinuity
ultramafic
mafic
felsic
igneous rock
granite
plutonic
sedimentary rocks
limestone
shale
sandstone
conglomerate
metamorphic rocks
marble
slate
quartzite
isostasy
plate tectonics
sea floor spreading
radiogenic heat
convection in the mantle
subduction
converging boundaries
diverging boundaries
rift valley
transform boundaries
transform fault
triple junction

REVIEW QUESTIONS

1. What are the major structural divisions of the earth's interior. Why is this traditional model changing?

2. What methods do geophysicists use to determine the interior structure of the earth?

3. Why does the earth's surface contain so many different kinds of rocks?

4. What is the difference between an igneous rock and a metamorphic rock?

5. Describe how sedimentary rocks form.

6. Earth scientists in the nineteenth century believed that the continents were once joined by land bridges, which later sank to become ocean basins. How would you refute this theory *without* mentioning plate tectonics?

7. We know that lithospheric plates make up the earth's outer layer. Why do these plates move?

8. What is likely to happen when an oceanic plate meets a continental plate? When two ocean plates meet? Two continental plates?

9. Why do earthquakes occur along the San Andreas Fault?

APPLICATION QUESTIONS

1. Imagine that you have the ability to send picture postcards back in time. They can contain any short message—and any photograph—you choose. You would really like to help Alfred Wegener support his theory of continental drift and restore his reputation among his contemporaries. What picture and message would you send?

2. You have heard the expression "tip of the iceberg" and know that most of the iceberg remains underwater. But what happens when the tip melts faster than the part underwater? What can this tell you about land masses?

3. You have just met a couple who are avid travelers but have an extreme fear of earthquakes. Make a list of the areas they should avoid. Be descriptive; they have no prior knowledge of plate tectonics.

C H A P T E R 18

TECTONIC PROCESSES

O B J E C T I V E S

After studying this chapter, you will understand

1. How the earth's crust is shaped by tectonic forces.
2. The possible causes of crustal warping and the manner in which it produces broad basins, domes, regional tilt, and uplifted plateaus.
3. How faults are formed and classified according to type of tectonic stress.
4. Why faults, earthquakes, and volcanoes occur together geographically.
5. What is being done to minimize the devastating effects of major earthquakes.
6. The two different types of vulcanism and the landforms associated with them.

Sheep Mountain in Wyoming. Tectonic forces have pushed up sedimentary rocks in an anticline. Erosion has exposed the core of the mountain and removed the less resistant rock between the narrow ridges of harder strata.

INTRODUCTION

Our attention now shifts from the global pattern of lithospheric plates to the specific regional landforms attributed to the forces of plate tectonics. Here we will examine the processes that push, bend, twist, tear apart, and add to the earth's crust to produce the landforms that give the earth's surface its fundamental features.

As discussed in Chapter 16, geomorphic processes are divided into two categories: tectonic and gradational. Internal tectonic forces raise, lower, extend, and compress the earth's crust; the external forces of gradation reduce and shape the rough tectonic landscape. Internal forces not only cause the movement of lithospheric plates and the warping, folding, and fracturing of the earth's crust, but are also responsible for the volcanic activity that transports molten material into the crust and onto the surface, sometimes in very dramatic ways.

FOLDING

The *folding* of solid rock is a slow, deliberate process. No matter how carefully you watch a rock, you cannot see it fold, and yet we know it does because we see folded rock all around us. It seems scarcely possible that strong brittle rocks may be bent into sharp folds without first being reduced to a molten mass. Certainly, if we placed a slab of rock in a powerful vise, it would become cracked and broken. But there are examples everywhere of what must have originally been flat-lying beds with, for instance, marine fossils as evidence of their formation on the sea bottom. We can now observe these beds folded into a variety of positions. Folds as we find them today may vary from mild arches and domes scarcely visible to the eye to involuted and recumbent folds where the rock strata have been squeezed into tight corrugations that have overturned and even curled about themselves. They do not have to be symmetrical; in fact, they seldom are.

It appears that if strong lateral pressures are exerted over a very long period of time, on a stratum that is buried at some depth beneath overlying layers, folding may be accomplished in all its many forms. In other words, both vertical and horizontal pressures are required—and a great deal of time is needed even to form modest folds. Folds are therefore the product of forces that laterally press against the edge of flat-lying strata.

Folding works much like pushing against the edge of a rug. If the material is stiff and heavy, a broad warp may result, but if it is light and pliable, a whole series of parallel ripples will build up, eventually collapsing over each other as the pressure continues (Fig. 18-1). In rocks, unlike rugs, however, erosion begins to bite away at the top of each fold as soon as it is exposed at the surface.

Although all types of rocks can be folded under the proper circumstances, the sedimentaries are the most striking and readily observed, for they are accumulated into horizontal beds. Once this flat-lying character is distorted by even a few degrees, it immediately becomes obvious. Furthermore, since most sedimentaries are formed in the marine environment, they almost surely contain sea-bottom fossils. Any change of sea level, any sort of crustal distortion (folding among others), will expose the fossils at high elevations. Leonardo da Vinci (1452–1519) pointed out that the fossiled shell in the Tuscan Hills was evidence of uplift, especially since the sedimentary stratum bearing it was contorted at a steep angle.

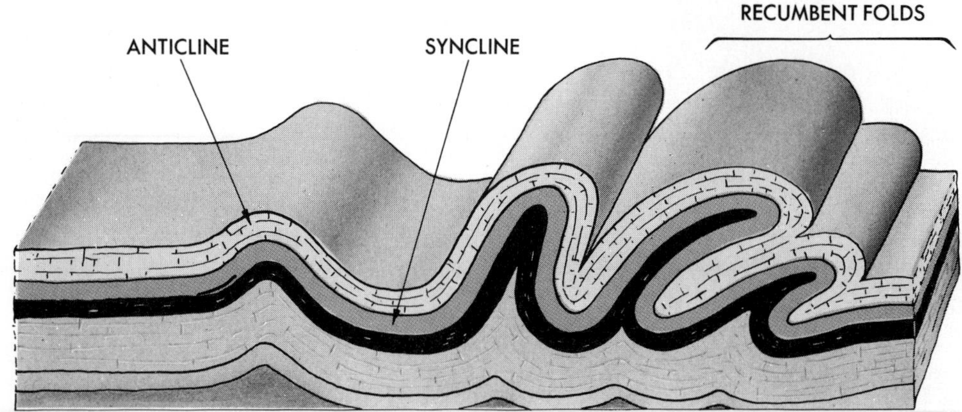

Fig. 18-1. ***Folding.***

Crustal Warping

As lithospheric plates push against each other over time, linear, tightly folded mountains rise at their margins. This is what we usually think of as the main result of plate convergence. However, a more subtle form of diastrophism (see Chapter 16), which is often a forerunner of folding, may occur well beyond the plate contact boundary, affecting large areas in midcontinent or even in midoceanic basins.

The crust in many parts of the world is gently arched upward or depressed downward over great distances.

The crust in many parts of the world is gently arched upward or depressed downward over great distances, so that elevation change is gradual and undramatic. This ***crustal warping*** does not result in severe deformation, but it does influence a region's overall topography. Some effects of crustal warping are:

1. Regional tilt that influences the general stream drainage pattern.
2. Structural sagging that forms wide depositional basins.
3. Broad domes or arches that increase potential energy for erosion.
4. Emergent or submergent coastlines.

CAUSES OF WARPING Compression resulting from the pressure of converging lithospheric plates is the obvious process behind the warping of the crust, but there may be other reasons for the flexing of the earth's surface. Not all the processes are well understood, but some other possibilities are:

1. Divergent or tensional (i.e., pulling apart) forces exerted by convection in the mantle below the region that cause a crustal thinning and sagging rather than the usual rising and fracturing.
2. Isostatic rebound or subsidence (see *isostasy,* Chapter 17) of the crust as it adjusts to removal or addition of eroded or deposited material or continental ice sheets.
3. Phase changes owing to heating and cooling in the mantle that affect the density of rock. Heating would expand the rock and raise the crust, and cooling would cause contraction and depress the crust.
4. Intrusion of large masses of molten rock into the crust that raise the elevation of the surface significantly.
5. The "bumps" encountered as lithospheric plates move over the asthenosphere. These "bumps" raise and lower the surface.

RECOGNIZING WARPS Warping is usually inconspicuous. Geomorphologists therefore depend on the sedimentary rock strata and their subsequent erosion to interpret the presence of warping, up or down (Fig. 18-2). A simple upwarped dome is easy to recognize when its top erodes away: the remains of the sloping limbs are left exposed in the form of a circle (Fig. 18-3). An excellent example is in the Black Hills of South Dakota where breaching has laid bare a hard core that stands as the highest part of the feature. The scenic sheer cliffs, always facing the center of the hills, are the eroded butts of the strata originally overlying the dome. Even the gentle warp of the Cincinnati arch, which stretches through much of Ohio, Kentucky, and Tennessee, has moderately worn away at the crest. Through the centuries "erosional windows" have widened, allowing the development of supremely fertile soils on exposed limestone in the Nashville and Bluegrass basins. Surrounding each basin is a highland rim of less productive shale-based soils (see Fig. 18-2).

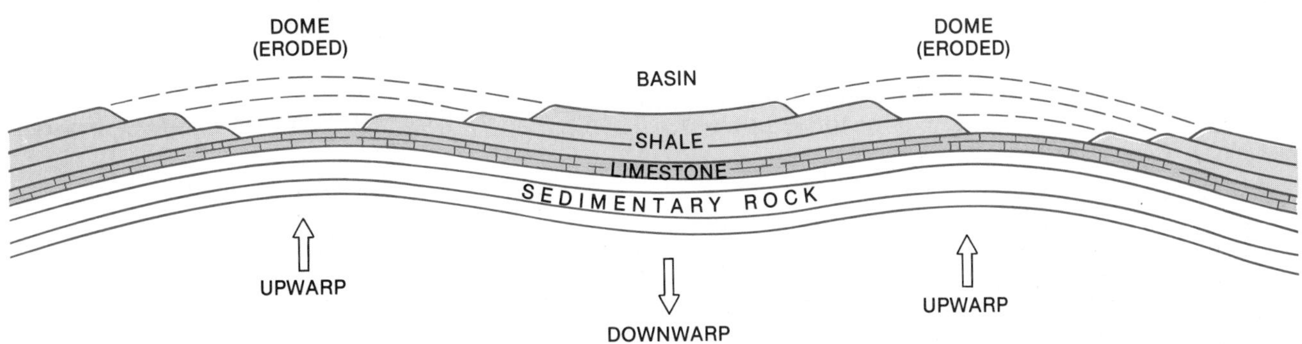

Fig. 18-2. ***Crustal warping—domes and basins.*** *Note the inward-facing, truncated rock strata in the eroded domes. Sometimes eroded upwarped areas are also referred to as basins because their base elevation is much the same as the downwarped areas.*

Fig. 18-3. ***Breached dome.*** *With the top of the dome eroded and a drainage to transport material away, the exposed butts of angled sedimentaries form a series of concentric circles—steep face toward the center and gentle slope away. See also the cross-section in Fig. 18.2.*

The extensive upland involving parts of Utah, Colorado, New Mexico, and Arizona is by any measure a plateau; we call it the Colorado Plateau. It is high, averaging roughly 7000 feet (2150 m), and its dominant rock structures display a strongly developed horizontal stratification—flat-topped buttes and mesas on every hand, canyon walls with parallel and varicolored layers. Yet, if one could stand back far enough and view the "plateau" as a single construct, it would become evident that it is in fact a modest, if extremely large-scale, upwarp. Most of those apparently horizontal rocks are angled ever so slightly.

The history of the Colorado Plateau discloses episode after episode of shallow seas accumulating layers of sediment with occasional periods of huge dryland sand dunes. As these now-sedimentary rocks were slowly uplifted, there appears to have been a minimum of contortion or sudden rending of the structure.

The Colorado River, with its sister drainage arteries, is part of the modern erosion cycle, but it is probably less than 10 million years old. Displayed along its convoluted canyon walls is a revealing segment of the earth's history. At the bottom of the mile-deep abyss, the active river is cutting into dark rocks that may be as much as 3 billion years old (Fig. 18-4). The progressive layers to the top show development of fossil forms, each more sophisticated than the ones

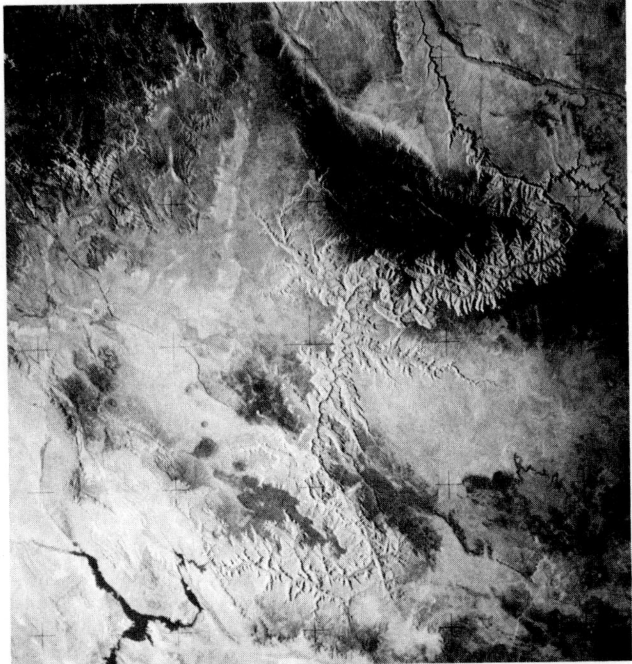

Fig. 18-4. ***The Colorado Plateau.*** *Viewed from 270 miles (435 km) high, the Grand Canyon is still impressive. Lake Mead is in the lower left, and in the upper left are elements of the Grand Stairway.*

Fig. 18-5. ***Grand erosion.*** *Bryce Canyon is not really a canyon at all; it was etched out of the edge of an escarpment by accelerated erosion. The soft clays of the Wasatch formation were deposited in an ancient lake bed, and it is their ready erosion that leads to the pinnacled panorama. The added attraction that mesmerizes the viewer is a fairyland of delicate pink, rose, and white coloration.*

before, and these allow for accurate dating. But the hard Kaibab limestone at the canyon lip was formed 300 million years ago! This is certainly much more recent than down where the river is cutting, but shouldn't there be even newer strata on top of this?

Farther north in Zion National Park, where the Virgin River etches out another deep gorge, the Kaibab limestone from the *top* of the Grand Canyon shows up here at the *bottom.* Rising above it along the canyon walls is another series of distinctive, colorful, but younger, sedimentaries that culminate in a hard capping layer. This is the Carmel formation, and it is only 170 million years old. Moreover, the Carmel formation on top of the Virgin Valley is lower by several hundred feet than the south rim of the Grand Canyon.

Another 50 miles north are the distinctively colored rock formations of Bryce Canyon National Park (Fig. 18-5). The stratum exposed at the *lowest* point in Bryce is the Carmel formation, and ranging above that are ever younger rocks. The Wasatch formation at the top of Bryce Canyon is about 55 million years old.

So there *are* more recent rocks than those exposed in the Grand Canyon, and we assume that they were in place over it. But where the Colorado River now flows was the top of an uplifted dome, and erosion was most active at the top. Everything above the Kaibab limestone has been stripped away. Along the flanks of the dome are the exposed, gently sloping residual butts of each rock stratum that was removed from above.

Northward of the Grand Canyon is a series of distinctive cliffs—those angled stubs, each younger than its predecessor. They have been called the ***Grand Stairway*** and are named according to dominant colors: from south to north, the Belted or Chocolate Cliffs, Vermillion Cliffs, White Cliffs, Grey Cliffs, and Pink Cliffs. The Pink Cliffs are the Wasatch formation, which dominates Bryce; the White Cliffs are the Navajo sandstones which form the pinkish-stained white walls of Zion Canyon (Fig. 18-6).

Anticlines and Synclines

Well-developed folds are classed as either anticlines or synclines. A series of strata that is upbowed and relatively small in scale is called an ***anticline.*** A series that is bent downward and relatively small in scale is called a ***syncline*** (Figs. 18-1 and 18-7). If this same sort of folding occurs on a larger

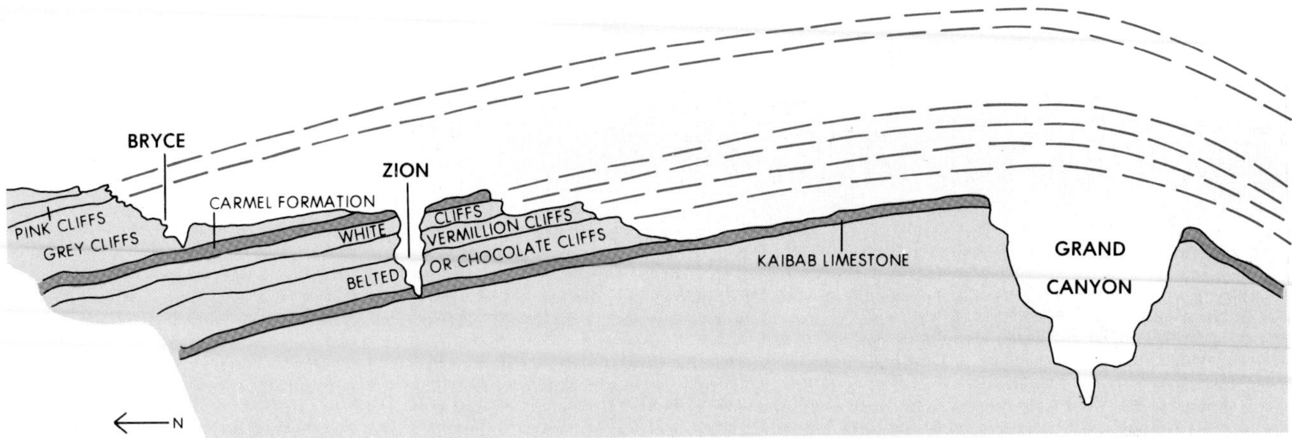

Fig. 18-6. ***The Grand Stairway.***

scale involving perhaps hundreds of miles, the forms are called *geanticlines* and *geosynclines* (the prefix from the word *geo* = earth, emphasizing the size of such structures). The word *geosyncline* also carries the implication of a sea-filled basin in which there are great thicknesses of sediment whose weight further depresses the bottom.

Well-developed folds are classed as either anticlines or synclines.

It does not necessarily follow that anticlines appear as hills on the earth's surface, for their formation is so slow that erosion may easily wear away the top of the fold as rapidly as it is uplifted. Nor is it uncommon to find a syncline at the top of a hill in present-day topography and an eroded anticline in a valley (Fig. 18-7). For instance, the Appalachian ridges and valleys of eastern Pennsylvania are some of the most striking features in the United States. Oriented northeast to southwest, these symmetrical linear ridges, all of the same height, seem to be an endless roller coaster to the traveler driving across them. Somebody shoved the end of the rug, as it were, and multiple parallel folds resulted.

Actually, the thick sediments that had piled up in the ancient sea were relentlessly squeezed by an advancing crustal plate into major convulsions. But almost as rapidly as they developed, erosion probably whittled away these immense folds, and we now see their mere roots. Thus this apparently straightforward series of dramatic but modest folds is in fact the remnant of the original—second-generation corrugations formed as the steeply tilted strata were etched by running water. Today we see the harder rock layers standing up as ridges, while the weaker soft rock strata are the valleys (Fig. 18-7).

Recent Folding

Lining up in an east–west direction across southern Europe and Asia is a series of high mountains exhibiting some of the most spectacular and severe folding in the world. The Pyrenees, Alps, Caucasus, and Himalayas are all basically folded mountains (Map III). At one time, just south of these mountains in the general area of the Indo-Gangetic Plain, was a great sea-filled geosyncline called the *Tethys basin.* For millions of years, sediments filled this basin and formed many thousands of feet of sedimentary rocks.

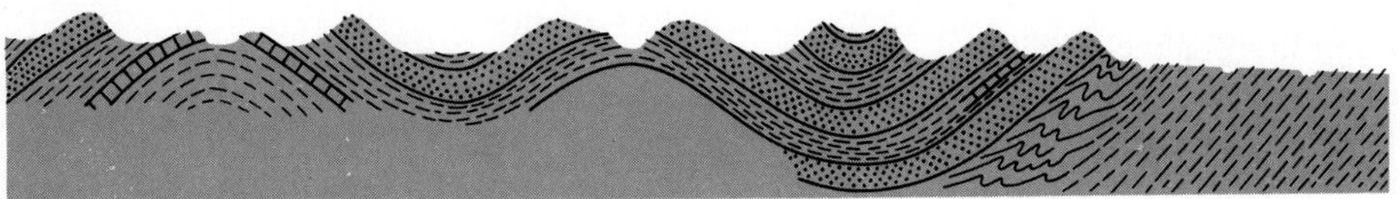

Fig. 18-7. ***A cross-section of the Pennsylvania ridges and valleys.*** ***Within the overall anticline–syncline pattern, ridges are formed where there is erosion-resistant rock strata; valleys are formed where the rock is less resistant.***

Fig. 18-8. ***The Alps.*** *It is not readily apparent from this general view, but the Alps are folded mountains. Their great elevation is the result of massive and highly involuted folding of sedimentaries. Today's rugged detail, however, derives from ice erosion—both current and of the recent past—and it has made the reconstruction of the original flexures particularly difficult.*

Beginning about 60 million years ago, these rocks were caught in a vise of the Indian and African crustal plates drifting northward against the Eurasian plate. Mighty overlapping folds, some fractured and recumbent, resulted from the great forces that compressed the original flat-lying sedimentary beds and built them up into today's great Eurasian mountain system. The extreme height of these related ranges, extending halfway around the world, shows that they are of fairly recent origin. But erosion, including intense glaciation, carves and hews at their flanks, erasing some of the more obvious signs of original folding (Fig. 18-8).

Occasionally, in a road cut or on a cliff face, a perfect fold may be easily visible. Not yet destroyed by erosion, its undisturbed outline is evidence of how folding occurs. Presumably, folding is occurring in a great many places around the world today, but its progress is so slow that it is virtually impossible to measure. From observing folds of all kinds in many places, we know, however, that folding has been a major force in the deformation of the earth's crust.

FAULTING

A fault is a break or fracture in the earth's crust along which there is or has been movement. Nearly all rocks have cracks or joints resulting from their method of formation, such as the contraction of cooling lava, or in some cases, resulting from the seasonal or daily heating and cooling of surface rocks. If a fracture occurs on a large scale, however, and readjustment takes place along the plane of breakage, it is called a ***fault***. The occasional slippage of faults is of great concern to us because this translates into an earthquake. Shocks from strong earthquakes can be extremely destructive of life and property if they occur at the right time and place.

A fault is a break or fracture along which there is or has been movement.

The compressional forces at work in the earth's crust, usually those associated with plate convergence, are responsible not only for warping and folding, but for certain types of faulting as well. If rock is too brittle or massive to relieve these pressures by folding or if the forces are exerted abruptly rather than slowly over a long period, then faulting will occur. The tension of plate divergence may pull the earth apart, opening up faults along which massive crustal blocks can drop vertically along a ***fault plane***. The fault plane is the surface along which the blocks slip (Fig. 18-9). In Chapter 17 we saw how the action of lithospheric plates moving by one another horizontally formed a boundary characterized by transform faulting. Virtually all faulting can be interpreted in terms of plate tectonics. It is a very common occurrence in the diastrophic episodes along the boundaries of moving plates.

Much crustal faulting takes place beneath the land and sea surface and is traceable only through earthquake studies. Surface faulting creates certain characteristic landforms, however. Minor vertical displacement may form escarpments or cliffs that make the fault line easily visible. Or massive blocks may be tilted up along a fault, resulting in high, uneven mountain ranges (Fig. 18-10). The Sierra Nevada in California, though slightly modified in places, is essentially a single tilted fault block. The sheer eastern escarpment marks the fault line, and when viewed from the east it is a spectacular sight rising directly up from Owens Valley. This is in sharp contrast to the long gentle slope of the inclined western side of the Sierra Nevada block that dips toward the Central Valley of California.

This displacement did not occur at any one time; it has taken millions of years. However, many of the lesser displacements must have occurred in fairly recent geologic time to exhibit such a sheer wall scarred by only moderate erosion.

It is useful to think of the landforms that go along with faulting as parts of fault zones. Faults occur in related clusters or series of parallel fractures, for rocks do not always break neatly along a single sharply defined plane. Instead, they are mangled for some distance on either side of the

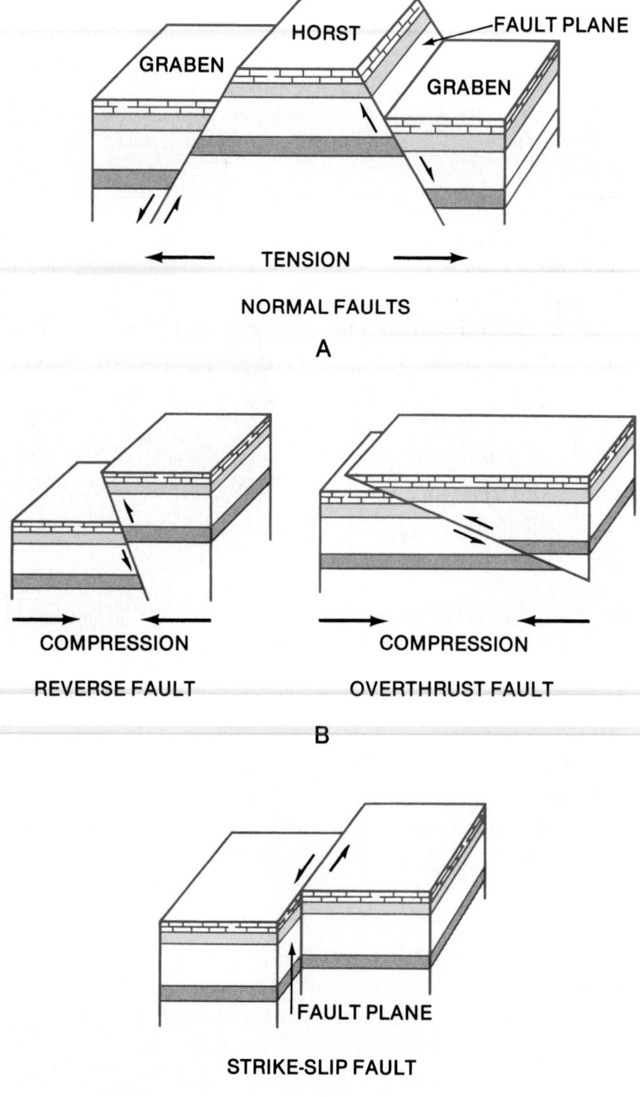

Fig. 18-9. ***Diagrammatic views of three types of faults.*** *(A) Normal faults have a high-angle fault plane and result from tensional tectonic forces. (B) Reverse faulting occurs where crustal compression pushes one block of crust up over another. Overthrust faults have a low-angle fault plane and are often found in older folded mountains where long-term compression has caused crustal overlap. (C) Strike–slip faults have a definite horizontal rather than vertical motion. They are found where plates are slipping by one another.*

Fig. 18-10. ***Sierra Nevada of California.*** *Elevated dramatically along an incisive fault line, the eastern escarpment of the Sierra Nevada (California) towers above Owens Valley.*

fault. Furthermore, once any kind of vertical motion takes place on a fault plane, softer rock previously protected below ground may be exposed to erosion, which can enlarge the fault feature far beyond the size created by any original slippage.

Drainage, too, will be attracted to the newly made valley, and rivers tend to develop fault-controlled patterns. It is more than mere coincidence that the master streams of the California coastal ranges and most of their tributaries follow southeast to northwest courses parallel to the transform faults (San Andreas being the dominant one) that mark the boundary between the North American plate and the Pacific plate. Some of these streams have opened up extensive flat-bottomed valleys, such as that of the Salinas River. These features are caused by faulting but are not exclusively fault valleys.

Because of the weakness fracturing brings to the crust, volcanic activity is often associated with faulting.

When viewed from the air or on a satellite image, a long fault zone may be an obvious landscape feature. Slim linear lakes or an invading finger of the sea may define the valley limits with great clarity. Because of the weakness which fracturing brings to the crust, volcanic activity is often associated with faulting. Volcanic landforms may appear in the fault zone. Lava flows, volcanic buttes, and hot springs may give away the presence of past or present faulting that has rather badly disrupted nature's underground plumbing system.

Fault Types and Topography

Crustal compression, stretching, and slippage caused by earth's internal forces result in a variety of fractures at the surface. However, faults may be divided into three general types based primarily on the kind of tectonic stress that produced the fracturing:

1. Normal faults
2. Reverse and overthrust faults
3. Strike–slip faults

Each type has a distinctive topography that goes along with the kind of displacement generated by the faulting.

NORMAL FAULTS As convection currents within the mantle push up and pull apart the lithosphere, the crust stretches and breaks so that deep fractures develop defining individual blocks of crust. Movement of the blocks is essentially vertical: some blocks drop relative to others as gravity and the upward pressure from below create differential stress. The slippage that eventually results along the fault plane produces a ***normal fault*** (Fig. 18-9A). This is a rather common type of fault, and the related landforms can be very dramatic.

As convection currents within the mantle push up and pull apart the lithosphere, the crust stretches and breaks so that deep fractures develop.

Fig. 18-11. Downfaulted graben. Rising abruptly from the Dead Sea shore is the eroded, almost sheer west wall of the Jordan graben. The Dead Sea occupies the graben at a record 1300 feet (396 m) below sea level at its surface. With no outlet, the Dead Sea is strongly saline, virtually bereft of life. Along the sea's margin are extractive operations that mine potash and other useful minerals from the brine.

The flat-bottomed valley formed by a down-dropped block is called a ***graben***, after the German word for grave or ditch. An up-thrust block is a ***horst*** (no English equivalent). The Jordan River and the Dead Sea (Fig. 18-11) occupy the northern end of an extensive series of grabens that includes the Red Sea and the rift valleys of East Africa (Fig. 18-12; see also Figs. 17-17 and 17-18). In the upper Rhine Valley, where the river forms the boundary between Germany and France, the Rhine flows northward through a graben, flanked by the horsts of the Vosges on the west and the Black Forest on the east.

The Sierra Nevada is not a horst because it is merely tilted along a single fault plane, but the Great basin, especially in Nevada and southeastern Oregon, displays many horsts. Here is a remarkable collection of faulted features extending hundreds of miles over a wide area and all oriented

Fig. 18-12. The distinctive Y-shaped head of the Red Sea graben with the Sinai Peninsula in the center. To the east, the Gulf of Akaba is a mere extension of the greater Jordan graben, with the Dead Sea and Lake Tiberius/Sea of Galilee (almost obscured) both below sea level.

LAKE BAIKAL

One of the world's great grabens is occupied by Lake Baikal in southern Siberia. The lake itself is 400 miles (744 km) long and 30 miles (48 km) wide, its surface is 1495 feet (456 m) above sea level, and its bottom has been measured at 4250 feet (1295 m) below sea level. The towering sheer walls of the graben extend to 6600 feet (2012 m), displaying a total crustal relief of well over 2 vertical miles (3 km). So steep are its sides that for many years the Russian engineers constructing the trans Siberian Railway were thwarted in their every attempt at track construction along the western lake margin. Rather, they chose to approach the northern shore via the Angara River where it exited the lake and to lay temporary tracks across the winter ice to the valley of the entering Selenga River. In the summer freight and passenger traffic reverted to barge and ferry. The western shore track was eventually completed but only with great effort and expense.

Baikal's claim as the world's deepest lake at 5745 feet (1751 m) stacks up impressively alongside those in the United States: Crater Lake (volcanic origin) 1932 feet (589 m), and Lake Chelan (glacially scoured valley) 1419 feet (433 m).

roughly north–south. Apparently, tensional forces pulled the crust apart in an east–west direction, and the resulting clusters of faults were formed at right angles to the direction of tensional pull. Slippage along these fault planes over millions of years has produced many horsts, grabens, tilted blocks, and enclosed basins at various levels. Continuing erosion has subdued the stark outlines of some of the uplands and mountain ranges and partially filled the basins. However, in this arid region where erosion is somewhat retarded, the more recently formed features retain much of their original character and remain easily recognizable.

Often cut off one from the other and at different levels, each down-faulted basin in this alternating basin–range topography separately collects the runoff from infrequent rains and the meltwater from winter snow in the mountains. Periodically, broad sheets of water cover the basin floor, but they may attain only a foot or two of maximum depth. They will evaporate in a very short time, leaving behind a dry lake bed or *playa,* a word that literally means beach. Indeed, these lake basins are more often beach rather than lake since the normal landscape of the basin bottom, except immediately following a rain, is a blinding white, alkaline flat (Fig. 18-13).

REVERSE AND OVERTHRUST FAULTS Reverse and overthrust faults are usually found where lithospheric plates have been

Fig. 18-13. Sevier Lake, Utah—a playa. In the spring, melting snows from the surrounding mountains will sometimes supply sufficient water to support a thin sheet in the basin bottom. But it will usually evaporate in a few days or weeks, and the lake will revert to its normal character as an alkaline flat.

Fig. 18-14. ***Tracing the San Andreas Fault.*** *From the air it is easy to trace the San Andreas Fault across the landscape, in this case through the Carrizo Plain in central California.*

converging. As a region undergoes compression, the earth's crust sometimes slowly warps or folds as we described earlier in this chapter. But often the crust will not deform easily under pressure because of its rock type or the intensity of the compression. Instead, it will break to form faults. The simplest of these faults is the ***reverse fault,*** which has a high-angle fault plane like the normal fault, but one block is pushed up the fault plane by compressional pressure (Fig. 18-9B).

Reverse and overthrust faults are usually found where lithospheric plates have been converging.

The ***overthrust fault*** is a much lower-angle fault in which one portion of crust has passed over another in a nearly horizontal movement (Fig. 18-9B). Overthrust faulting sometimes occurs after folded mountains have formed, and continued compression forces the folds to fracture and overthrust. Some examples of overthrust faulting are found in the European Alps and in a portion of the Rocky Mountains in Wyoming.

STRIKE–SLIP FAULTS Along the transform boundaries where two lithospheric plates move past one another in opposite directions (see Chapter 17), the surface displays long, linear fault lines with deep, vertical fault planes. When slippage occurs after stress buildup, it is horizontal along the strike, or surface direction, of the fault line—hence the name ***strike–slip fault*** (Fig. 18-9C). The terms *transform* and *transcurrent* are sometimes used to describe this type of fault.

The San Andreas Fault running several hundred miles from the Sea of Cortez in western Mexico to Tomales Bay north of San Francisco is the longest, most active, and most famous of the strike–slip faults. Major earthquakes along the fault in 1906 and 1989 were responsible for loss of life and destruction in the San Francisco Bay area. The losses in the 1906 quake were not so much from the shock but from subsequent fires, which were difficult to extinguish. Evidence of the San Andreas Fault on the landscape consists of linear valleys, hummocky hills, slump zones, traces of small paralleling faults, and disrupted drainage patterns (Fig. 18-14).

Earthquakes

The margin of the Pacific Ocean in its present outline is a very unstable region marked by widespread faulting and continuing slippage and readjustment. From the Andes and the Sierra Nevada to Alaska, and from Kamchatka, Japan, and Indonesia to New Zealand and Antarctica, continuing seismic shocks are an unalterable part of the natural environment—much as tornadoes are in other parts of the world. Another major region that experiences frequent earthquakes follows the Alpine-Himalayan mountain system that stretches from southern Europe to southeast Asia. Wherever lithospheric plates interact—either through subduction, rifting, collision, or transform faults—earthquakes are common (Figs. 17-14 and 18-25).

The margin of the Pacific Ocean is a very unstable region marked by widespread faulting and readjustment.

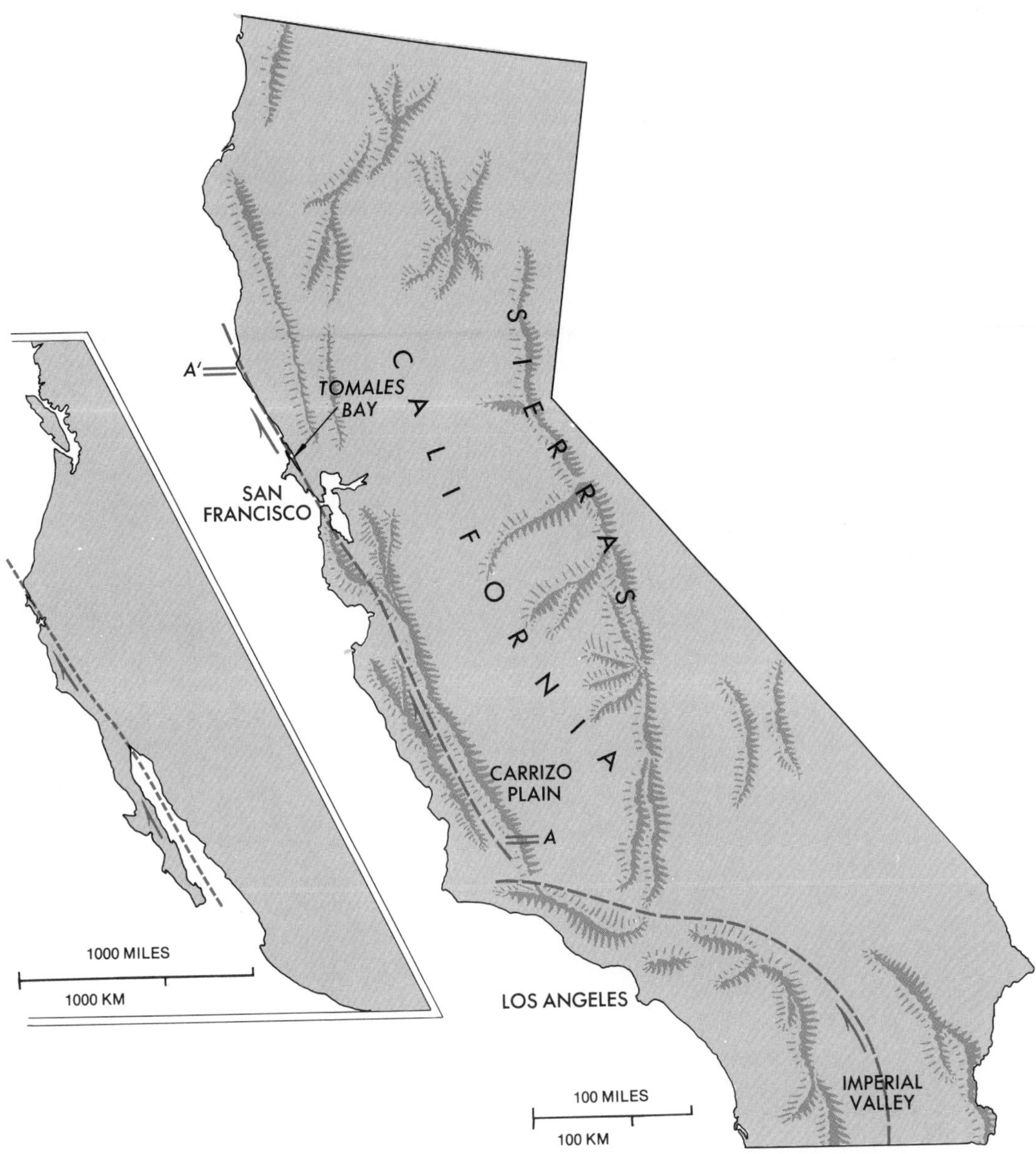

*Fig. 18-15. **The San Andreas Fault.** That the west side of the fault is progressing north is clear. Reasonable correlation of the same rock structure has been made at A and A′.*

Crustal movement along existing faults or the formation of new faults is not gradual. Pressures may build up slowly over many years, but when the rocks can no longer sustain that pressure, the break or the slippage comes with explosive suddenness. Part of the shock comes from ***elastic rebound*** in which rocks that have been somewhat deformed by the increasing pressure snap back to their original shape after the sudden slippage. The result is an earthquake with the shock waves radiating out in all directions from its *focus*—the point of maximum energy release. The ***epicenter*** of an earthquake is the point at the surface directly above the earthquake's focus.

Displacement of mass may occur vertically, horizontally, or both. If the fault is at the surface, vertical displacement is sometimes visible in the sudden formation of cliffs; if displacement is horizontal, fences, streams, and roads crossing the fault will be forced out of line. Horizontal displacement is typical along the San Andreas Fault in California (Figs. 18-14 and 18-15).

The west side of this deep-seated transform fault is the

Pacific plate that moves northward as it apparently has been moving for an extended period of time. A glance at Figure 18-15 will show that Baja California, separated from the coast of western Mexico proper by the narrow Gulf of California, would fit rather well into the configuration of that coast if it were shifted 200 to 300 miles (325 to 475 km) southeastward. The Gulf and its northward extension, the Imperial Valley, are both part of the greater San Andreas Fault system. It would appear that Baja and the entire western margin of California have been drifting northward in abrupt spurts along the line of the fault zone. Several attempts have been made to establish the rate of movement by matching rock types on either side of the fault. Studies have indicated a rate of about 3 miles (5 km) per million years.

Several other strike–slip faults have been recognized and measured. One in the Philippines and another in New Zealand are currently active, but perhaps the best studied of them all, the great Glenmore of Scotland, seems to be quiescent—although statements of this kind should be made with caution. A quick look at the map of Scotland will show that a deep trench virtually cuts off the northern highlands. Occupying the distinctive northeast–southwest linear valley are Loch Ness and Loch Lochy, which are connected with each other and the sea at either end by the Caledonian Canal

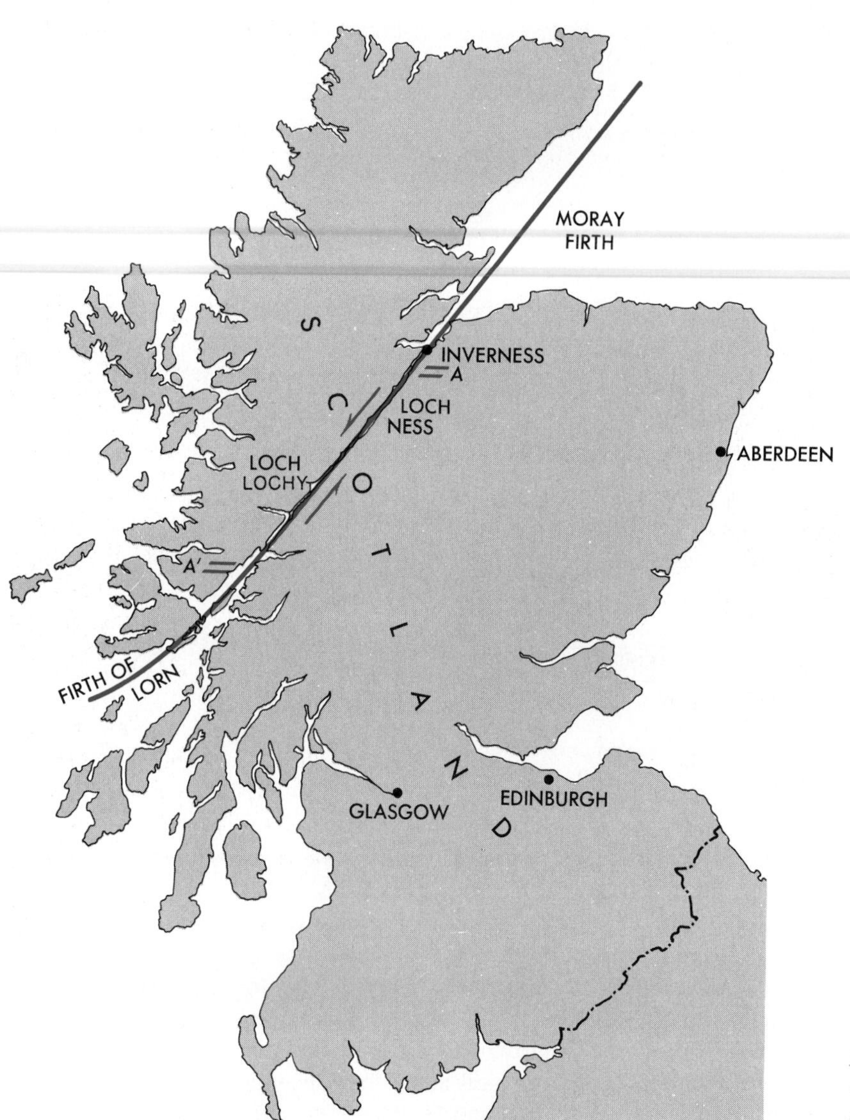

Fig. 18-16. ***The Glenmore Fault.*** *At one time rock strata and A and at A′ were contiguous across the fault.*

to form a navigable route across the heart of the highlands. The trench, though rasped out deeply by moving tongues of ice, is a fault zone. Proof of extreme horizontal movement along its plane has been demonstrated in many locations (Fig. 18-16).

SEISMIC SEA WAVES Coastal areas, even those well outside of seismic zones, are subject to the effects of earthquakes at the sea bottom or, in some cases, land-based quakes that trigger massive undersea landslides. The shock waves from these events are transmitted through the water as well as the earth. The energy coming from great depths is carried in the form of waves that travel at speeds of hundreds of miles per hour and may devastate coastlines thousands of miles away. This kind of wave is called a seismic sea wave or *tsunami*.

The shock waves from coastal and undersea earthquakes and related massive slide events are transmitted through the ocean in the form of rapidly traveling waves.

A major problem associated with tsunamis is detection and warning. Ships at sea, for example, are completely unaware of a passing tsunami, for the wave is only a few feet high and may be indistinguishable from other waves at sea. It may also be traveling at a great speed (the deeper the water, the greater the velocity), yet the ship has no sense of abnormal sea behavior. As it approaches the shore, the leading edge of the tsunami slows, while the mass behind runs up its back to build a high offshore swell. Moments later it breaks on the beach and surrounding ocean frontage to heights of 50 to 100 feet (15 to 30 m).

The Pacific with its great depth and many earthquakes is the most active seismic wave ocean; at least one per year has been detected since 1800. On the average, a major tsunami can be expected once every ten years, but Hawaii, in its vulnerable midocean position, has experienced 37 in the past 130 years. In April 1946 an Aleutian trench earthquake triggered a wave that five hours later broke over Hawaii centering on the city of Hilo. The damage and loss of life was so great that Hilo set up a warning system essentially monitoring all seismic activity around the Pacific basin. As a result, within two hours of the great Chilean earthquake of 1960 sirens warned city residents to evacuate. Some left but others didn't. The tsunami arrived in three echelons, each about three hours apart. The first two were 4 and 9 feet (1 to 3 m) high, respectively, and the last was a 20-foot (7-m) killer. Most of the 61 dead were those who had rushed downtown to view the excitement following the first two waves.

Twenty-four to thirty-six hours after this same Chilean quake, coastal areas as far distant as New Zealand and Japan experienced destructive high seas. Similarly, the Anchorage quake in March 1964 caused damage along the Oregon and California coasts.

LANDSLIDES Landslides are another destructive result of earthquakes. In the Hebgen Lake, Montana, quake of August 1959, landslides blocked a sizable river to form a lake, and in the process, a number of people were killed (Fig. 18-17). Or, in another variation, landslides may thunder down steep mountain slopes and into narrow arms of the sea or lakes behind confining dams. The damage is not the immediate result of the moving earth but the secondary effect of the resulting wave generated in the body of water. Many Alaskan fjords exhibit a sharp line well above the water's edge below which no trees are visible. The tree line marks the highest point reached by a recent devastating wave. In the case of a dam-formed lake, the waves may either rupture the dam or overrun it. In either situation settlements in the valley below are swept away in minutes.

Fig. 18-17. ***Montana earthquake of 1959.*** *Just to the west of Yellowstone National Park, a 1959 earthquake induced a mammoth slide into the Madison River canyon. A new lake (left) was impounded behind the sudden new dam, imaginatively named Earthquake Lake. Over in the park, the jolt cleared up some underground plumbing clogs, and new and dormant geysers gushed to life.*

By far the greatest disaster of this sort occurred in the Peruvian earthquake of May 31, 1970. Of the 50,000 estimated deaths from all quake causes, 20,000 were in the two mountain resort villages of Yungay and Ranrahirca. A U.S. Geological Survey report tells the story:

> In the beginning the avalanche, triggered by the quake, started with a sliding mass of glacial ice and rock about 3000 feet (915 m) wide and a mile long (1.6 km) on the nearby sheer slopes of Nevadas Huascarán (at 21,800 feet (6650 m), Peru's highest mountain).
>
> It swept downward, dropping 12,000 feet (3660 m) vertically in a distance of 9 miles (14 km) and hit the town of Yungay at speeds of up to 248 miles (400 km) per hour. As the ice and rock fragments rushed down the mountain, frictional heat changed the ice to water and the debris to a muddy mixture. By the time the mass reached Yungay it is estimated to have consisted of about 80 million cubic feet (2,265,340 m^3) of water, mud, and rock. A mud flow of such proportion originating from an ice mass, indicates a geologic process never before recorded.

EARTHQUAKE FORECASTING Wouldn't it be helpful if we could monitor physical events or measure something concrete that would help us predict earthquakes in specific places? (Fig. 18-18). In the past some California old-timers talked knowingly about "earthquake weather"; or a more bona fide expert would break into print with a learned generalization that "pressures have been building up ever since the big Frisco shake in '06 so another is likely anytime now." This historical approach, which bases earthquake occurrence on patterns in past records, seems to have some validity in light of the severe earthquake in the San Francisco area in October 1989 (Fig. 18-19). But in the past several years advances have been made in the recognition of measurable phenomena. With the Russians, Japanese, and Chinese leading the way, researchers are now examining a number of promising lines of investigation.

In the past several years advances have been made in recognizing measurable phenomena.

One approach is in the area of ***dilatancy***—tiny cracks that open up in rocks under extreme pressure along a fault plane. The effects of dilatancy on groundwater level, rock volume, speed of human-induced seismic waves, and magnetic field are quantifiable in an earthquake-prone area, but the usefulness of these measurements in forecasting the exact time and location of a quake remains illusive. Delicate tilt meters are now used to detect even minor changes in the level of the land or well water. Sudden increases in trace amounts of radon (a gas resulting from radioactive decay) in groundwater also appear to indicate impending earth movement. The Chinese have spoken very seriously about abnormal animal behavior prior to a seismic episode. The world took a second look at Chinese methods on February 4, 1975, when they predicted a major earthquake at Haicheng (northeast China), evacuated the city, and saved an esti-

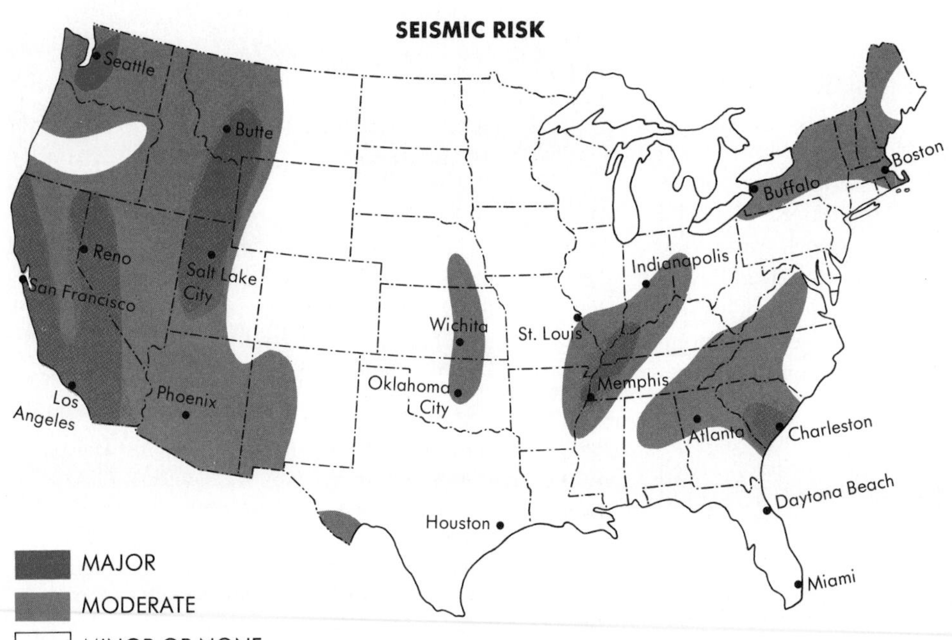

Fig. 18-18. ***Seismic risk.***

Fig. 18-19. ***San Francisco Marina district, October 17, 1989.*** *A severe earthquake (Richter Scale 7.1) caused structural failure and fires in this fashionable neighborhood in San Francisco. Throughout the San Francisco—Santa Cruz region of California—several dozen people lost their lives as walls crumbled and freeways collapsed. San Francisco has suffered even more destructive quakes in the past and will most likely do so in the future.*

mated 100,000 lives as a severe shock destroyed it. But forecasting is not yet a precise science. Only about a year later, a quake of strong magnitude shook Tangshan (150 miles or 240 km west of Beijing) to the ground, killing 240,000 people with no warning at all.

Nonetheless, we appear to be on the right track. We now seem to possess some valid indicators, and others will undoubtedly surface. The U.S. Geological Survey along with several universities and research institutes have set up seismic monitoring grids using networks of automated sensors.

THE RICHTER SCALE A numerical scale devised by C. F. Richter in 1935 has become the definitive measure of earthquake magnitude. It expresses as closely as possible the energy released as deformed rocks, no longer able to withstand the strain, rebound. Seismologists express earthquakes on the Richter scale from a magnitude of 1.0, near the lower limits of detection, to about 9.0, about the level of the most severe earthquake recorded to date. Some of the world's greatest earthquakes, those with a magnitude greater than 8.0 on the Richter scale, are listed in Table 18-1. Each level on the scale, 1.0, 2.0, 3.0 and so on, represents an increase by a factor of 10 in the amount of energy released by an earthquake. Thus a 4.0-magnitude earthquake (a mild shock detected by most people near the epicenter) is ten times stronger than a quake of 3.0 magnitude, and a 5.0 earthquake is one hundred times stronger than a 3.0-magnitude quake. Many earthquakes that occurred prior to 1935 were measured by a primitive means and have been assigned an estimated Richter magnitude. The scale is open-ended, but no quake has yet reached 9.0.

Table 18-1 Earthquakes with a Magnitude Greater than 8.0 on the Richter Scale

Year	*Site of Quake*	*Richter Energy Quotient*
1556	Shensi, China	—
1755	Lisbon, Portugal	—
1811	New Madrid, Missouri	—
1857	Southern California	—
1872	Owens Valley, California	—
1886	Charleston, South Carolina	—
1899	Yakutat, Alaska	—
1906	Andes of Colombia and Ecuador	8.6
1906	Valparaiso, Chile	8.4
1906	San Francisco, California	8.3
1911	Tien Shan, Sinkiang, China	8.4
1920	Kansu, China	8.5
1923	Sagami Bay, Japan	8.2
1927	Nan-Shan, China	8.3
1933	Japanese Trench	8.5
1934	India, Bihar—Nepal	8.4
1939	Chillan, Chile	8.3
1946	Honshu, Japan	8.4
1950	North Assam, India	8.6
1960	Central Chile—3 major shocks	8.3, 8.4, 8.9
1964	Anchorage, Alaska	8.6
1970	Yungay, Peru	8.4
1976	Tangshan, China	8.0
1977	Northwest Argentina	8.2
1977	Sumbawa, Indonesia	8.3
1979	Pacific, N.W. of New Guinea	8.0
1985	Mexico City, Mexico	8.1

Fig. 18-20. ***Leninankan, Soviet Armenia, December 7, 1988.*** *Thousands of people were killed and injured as a strong earthquake reduced many unreinforced buildings to rubble. Many thousands more were left homeless in the winter chill.*

The energy released by an earthquake may be great, say, 8.0 or greater on the Richter scale. Fortunately, most earthquakes of this severity occur in unpopulated areas. On the other hand, a lesser quake of moderate magnitude may cause great loss of life and substantial damage if its epicenter is in a populated area. The tragic Soviet Armenian earthquake of December 1988 was 6.9 on the Richter scale, but it occurred in a mountainous region with midsized cities and numerous small towns and villages. Estimates of deaths range up to 25,000, most of them being the result of collapsing buildings (Fig. 18-20). An earthquake of magnitude 7.7 shook northern Iran shortly after midnight on June 21, 1990. Some 40,000 people are thought to have died—a death toll that would have been much lower if the quake had not struck when most people were indoors asleep. Such loss of life strengthens the resolve to find ways of minimizing the effects of dangerous earth tremors. We might, for example, design buildings that are more earthquake resistant, attempt to forecast quakes, or, as some have advocated, control the intensity of shocks by setting off explosions or injecting water into fault planes to trigger minor quakes. It is thought that this method would relieve stress buildup, thereby avoiding a major shock.

VULCANISM

The term *vulcanism* implies volcanoes—and they are, to be sure, a spectacular product of it. But vulcanism involves much more than mere volcanoes. Any invasion of the earth's crustal zone by magma from below is properly called vulcanism. If the magma pushes its way far up into the crustal strata but does not reach the surface, it is termed ***intrusive vulcanism.*** If the molten material flows out onto the surface via volcanic or other vents, it is called ***extrusive vulcanism.***

Why does molten material from the earth's interior force its way into and through the hard rock of the crust? This question has vexed geologists for many years. Not only have they asked "Why does magma invade the crust?" but also "Where precisely does the magma originate?" It seems reasonably certain that the release of pressure on the interior of the earth by deformation and fracturing of the crust triggers the outward movement of magma, but whether it originates at the earth's core, the mantle, or even the crust itself is difficult to determine. The core is the least likely origin. Radiogenic heat (Chapter 16) in the upper mantle or the lower crust is probably the energy source for vulcanism.

The release of pressure on the interior of the earth by deformation and fracturing of the crust triggers the outward movement of magma.

Thus intrusive masses underlie nearly all major mountain masses, and frequently some sort of extrusive vulcanism exists as well. Mountain-building mechanisms such as folding and faulting cause vulcanism, but once set in motion, vulcanism can be a potent mountain-building mechanism in its own right. Massive intrusions may lift and warp the surface layers, and extrusive magma, finding its way to the surface through fault zones, can build huge volcanic forms. In addition, intrusions may be exposed as erosion wears away the softer overlying strata, and the hard igneous structure will remain as peaks or highlands.

Intrusive Vulcanism

BATHOLITHS AND STOCKS The largest intrusive mass is the ***batholith,*** which underlies nearly every large mountain system and is frequently exposed at the surface as the roots or cores of ancient mountains that have eroded away. The batholith is not merely the upper margin of the monolithic lower crust. It is also a lobe that has forced its way into the uppermost sedimentary (or altered rock) layers that blanket the crust, probably because of the lessened pressure as a result of diastrophism (Fig. 18-21). Arbitrarily, anything covering less than 40 square miles (100 km^2) is called a ***stock,*** although the feature may resemble a batholith in every other way. At least this gives us some idea of minimal size, for batholiths are considerably larger.

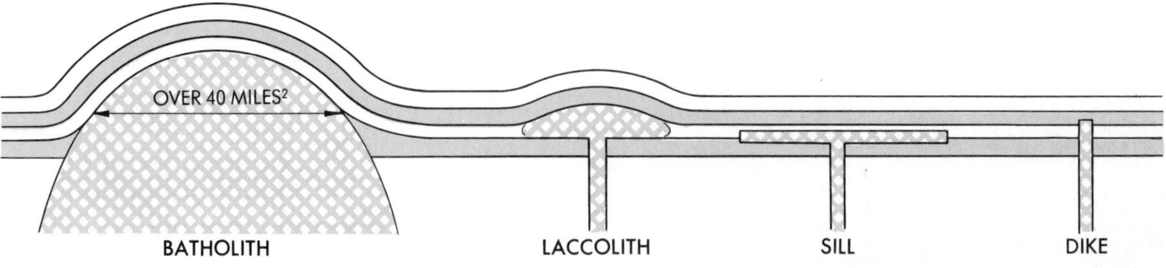

Fig. 18-21. Types of intrusive vulcanism.

The batholith and stock are coarse-grained plutonic rock that has cooled slowly, deep under the surface. At their point of contact with the rock above is a metamorphosed *aureole* or *baked zone* where the combination of heat, pressure, magmatic fluids, and gases has commonly formed a concentration of useful minerals. Much of our knowledge of the boundaries of the batholith has been gained by mining these zones.

But nobody has seen the underside of what must be regarded as an essentially bottomless structure, nor has anyone seen intrusive rock crystallized as extrusive lavas have. All we can do is conjecture about the reason for batholith intrusion as well as the whereabouts of the rock replaced by the intrusion.

LACCOLITHS On a much smaller scale than the batholith are several other kinds of magmatic intrusions. One of these, the *laccolith,* is a small batholith or stock except that it is fed from below by a tube cutting across the overlying strata like the conduit of a volcano. However, when the volcanic bore gets into the surface rocks, magma rising up the laccolith feedpipe loses some of its impetus and spreads out between two flat-lying rock layers, eventually forming a lens-shaped mass that lifts and bows the surface into a dome (Fig. 18-21).

SILLS AND DIKES Similar to the laccolith but much more common is the *sill,* in which the invading magma spreads out as a flat sheet, often for many miles, but does not throw up overlying layers (Fig. 18-21). Sills come to our attention when erosion exposes an edge as a sheer wall, often darker and harder than the surrounding rock. The downcutting Hudson River reveals a sill in the Palisades that is particularly striking because of the columnar jointing of the cooling magma. Another major feature that stands out in a moderately subdued landscape is the Great Whin Sill of Northumberland, England. The black cliff describes a 100-mile (160-km) crescent from the Pennines to the sea. Since its steep side faces northward, the Romans used it as the foundation for Hadrian's wall (Fig. 18-22).

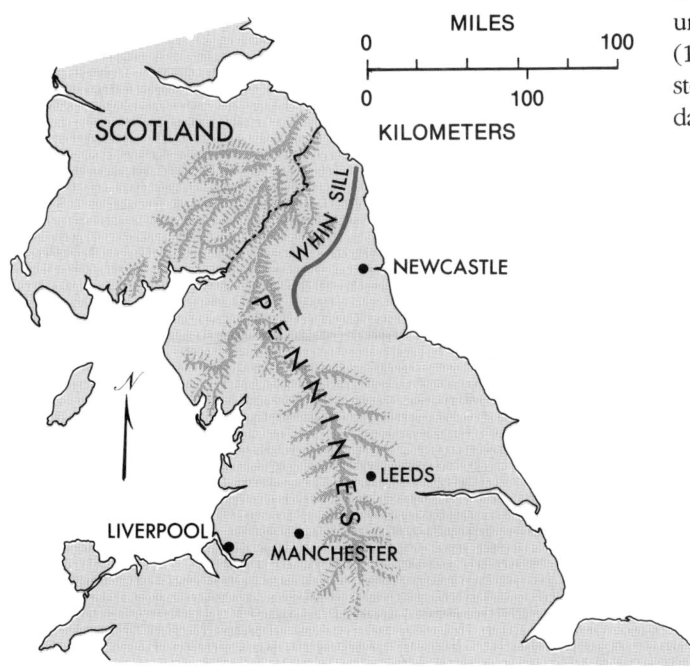

Fig. 18-22. The Great Whin Sill.

Fig. 18-23. ***In the Massif Central of France.*** *Ancient volcanoes have eroded away completely, leaving behind only their former vents filled with solidified lava. These volcanic necks (sometimes called "plugs") now stand dramatically above the current landscape. At LePuy, the Saint Michel d'Aigulhe monastery has taken advantage of this prominent locale. There are 265 stairs in the ascent.*

If the invasion of the surface layers occurs in the form of a thin sheet cutting sharply across all horizontal strata but not quite intersecting the surface, then we have a ***dike.*** Visualize a sill turned on edge, except that the dike draws magma from its entire lower edge rather than from a bore or conduit. Dikes often occur in swarms, frequently radiating from a volcanic vent. If the resulting igneous rock is harder than that which surrounds it, erosion may lay bare the solidified magma in the volcano conduit as a ***volcanic neck*** (Fig. 18-23). Ship Rock in New Mexico is an outstanding example of a volcanic neck with radiating dikes forming long, narrow ridges (Fig. 18-24).

Extrusive Vulcanism

VOLCANOES Volcanoes are of particular interest not only because of their unpredictable, explosive habits but also because of their impressive structure. They tend to follow fault lines, and thus the world's great volcanic zones coincide with the earth's great seismic zones. Faulting, earthquakes, and volcanoes go together, and faulting is the cause of the other two. The Pacific's unstable margin is often called the "Pacific Ring of Fire" because of its extensive volcanic activity. Everywhere cones of all sizes, both active and dormant, are evidence of continuing vulcanism (Fig. 18-25).

Faulting, earthquakes, and volcanoes coincide geographically.

Fig. 18.24 ***Ship Rock, New Mexico.***

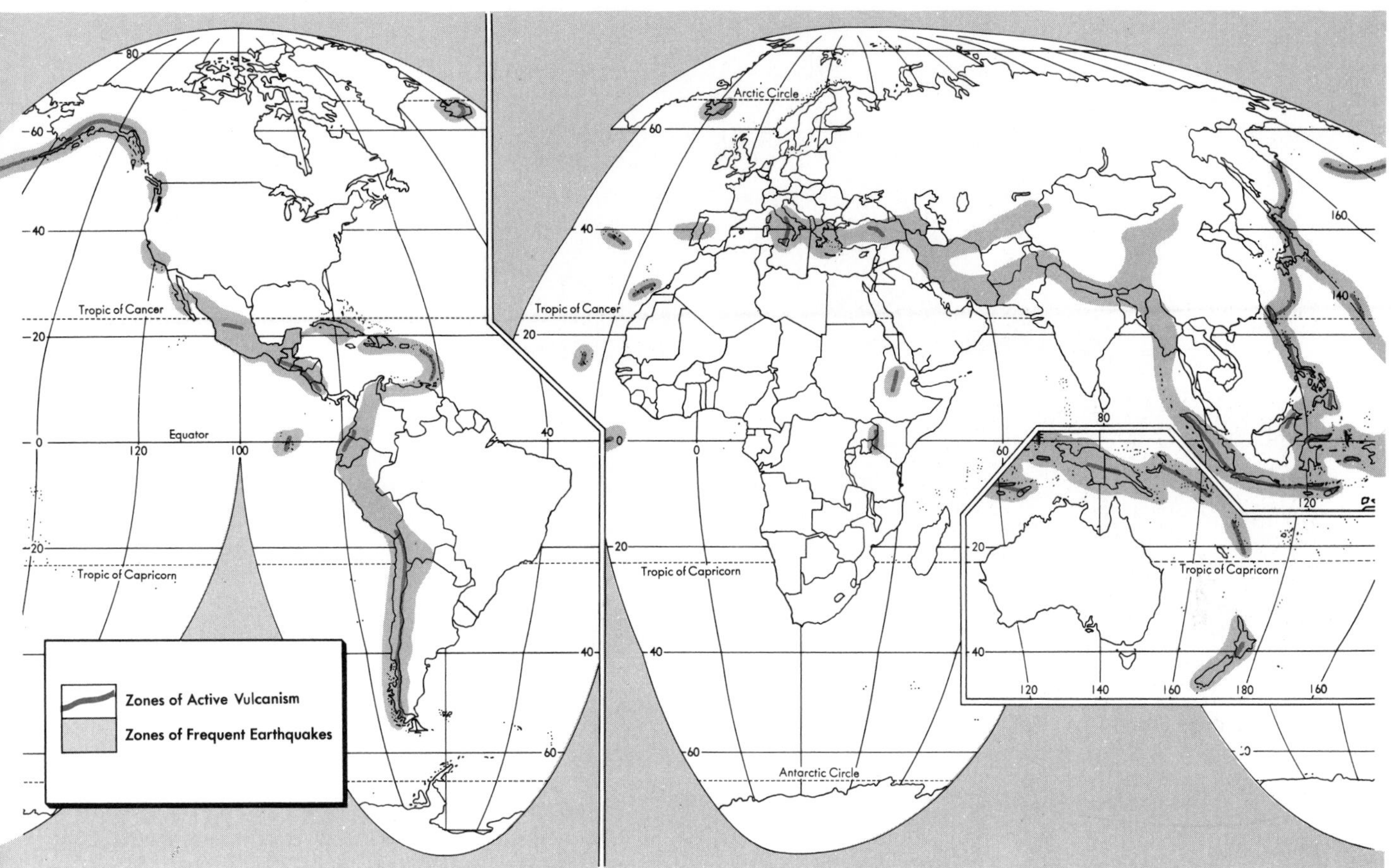

Fig. 18-25. The coincidence of volcanic and earthquake zones.

It is never accurate to say a volcano is "dead." *Dormant* is the better term and should be used if there has been no eruptive activity in recorded history. But even historic time is not always sufficient. Above the bay at Naples, Italy, in 79 A.D., Vesuvius, a not very imposing landform half hidden in the broken remnant of the older Mount Somma, had never erupted in human memory. Yet in that fateful year Pompeii was buried in a blast of fiery ash, and nearby Herculaneum was overrun by repeated flows of hot mud. Then for sixteen centuries Vesuvius experienced only ten additional major eruptions. In 1631, after 130 years of inactivity, it began its more frequent modern eruptive cycle marked by increased lava flows that had been of little consequence earlier. So today Vesuvius is by any measurement active, its personality having changed considerably over the centuries. If Somma is any evidence, many times in the distant past Vesuvius and its precursors have surprised the folks along the Bay of Naples.

The Cascade Range, paralleling the U.S. West Coast from northern California to British Columbia, is pimpled with a host of alabaster volcanic cones brooding in silent majesty. Mount Rainier, at 14,410 feet (4392 m), is the mightiest of the lot, but there are other sizable peaks as well, visible in good weather for hundreds of miles. Mount Saint Helens in the southwest corner of Washington was the smallest—only 9671 feet (2947 m). Then in May 1980 it abruptly became an 8400-foot (2560-m) peak with the north side blown completely away and a yawning new crater 3000 feet (915 m) deep (Fig. 18-26). Saint Helens is a hot-blooded youngster that has slumbered beneath a mantle of snow with uncharacteristic calm for the past 120 years. It is so new by geologic measures that it displayed an almost perfectly symmetrical form, not even sculpted by glaciers that disfigure its Cascade neighbors. But this perfect shape tells the vulcanologist that there has been relatively recent activity. The pieces of evidence are many—ancient cinder layers 37,000 years old and

Fig. 18-26. ***Mount Saint Helens.*** *On May 18, 1980, the eruption of Mount Saint Helens blew ash into the atmosphere.*

mudflows and lava eruptions repeated over the past 18,000 years. The modern cone is probably about 2500 years old, built of alternating depositions of cinder, ash, and lava. John C. Fremont witnessed an eruption in 1843, and that was followed by over a decade of annual eruptive activity. Its last recorded outburst was in 1857.

The Saint Helens eruption in 1980 was a ***pyroclastic*** (*pyro* = fire, *clastic* = broken) event. That is, rather than a liquid lava flow, there was a release of immense gaseous pressures as the north face of the mountain collapsed (Fig. 18-26). The result was solid ejecta of all sizes, from dust to jagged boulders, blown both laterally and vertically out of the mountain. Some have estimated that the roughly 1.5 cubic miles (4.2 km^3) of solid debris was blasted to the north at nearly 250 miles per hour (400 km per hour). Huge clouds of dust moved downwind to the east. In an area extending northward 5 miles (8 km) from the site of the explosion, nothing of the forest remained; for 7 more miles (11 km) trees were down, all pointing north. The eruptive mass had pushed out as far as 17 miles (27 km), and ash and other fine debris flowed several miles down the Toutle River valley. In the decades and centuries to come Mount Saint Helens will most likely construct its peak again, punctuated by periods of destructive activity. It is the nature of volcanic tectonism to build landforms sporadically.

Variations in Eruption. No two volcanoes are exactly alike in their eruptive habits, and, as we have seen from the Vesuvian example, an individual volcano is often unpredictable. Some are spewers of ash, like Irazú in Costa Rica, which erupted continuously for almost two years (1966–1968), subjecting the residents of the capital city, San Jose, to a seemingly endless rain of thick dust (Fig. 18-27). Others eject lava, but with variations. For example, Vesuvius goes off violently, with loud explosions and gushes of lava streaming down its flanks; Kilauea, the Hawaiian crater, quietly spills lava over its lip with filmy curtains of fire along its radiating fissures; Strombolian eruptions, named after the volcanic island of Stromboli off southern Italy, exude lava that cools and crusts over lightly in the crater to trap gas beneath it. The ultimate explosion throws glowing clots high into the air.

No two volcanoes are alike in their eruptive habits.

Then there are the spasmodic eruptors whose conduits become clogged with hardened igneous rock between eruptions. With such a cork in the bottle, it is not uncommon for the increasing pressures to blow out suddenly through a weak spot on the side of the mountain. This is

LAVA

Newly formed Hawaiian basalt.

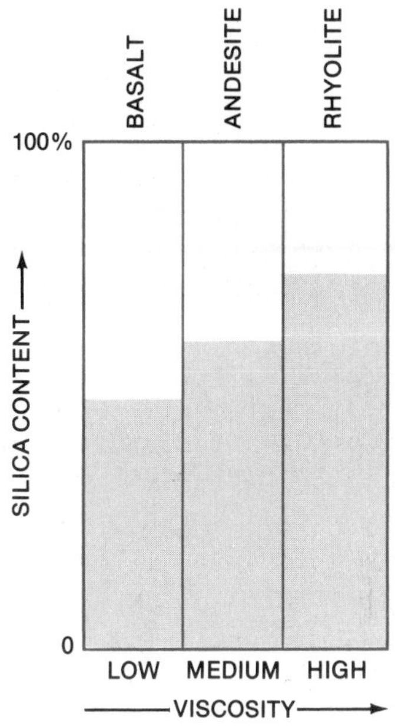

The viscosity of lava is related to its silica content.

The most common component of all lavas is silica, but the amount differs—from as little as 45 percent to as much as 75 percent. Silica-poor lavas (although they may average 50 percent) are *mafic*. They and their igneous end-products display a relatively high content of magnesium, iron, and calcium oxides. These are fairly alkaline lavas/igneous rocks that break down into a particularly fertile soil. The silica-rich acid lavas (65 percent or more silica) are *felsic*. They cool into rocks that are rich in quartz as well as aluminum, sodium, and potassium oxides.

Mafic rocks and lavas are dark because of their manganese and iron content, and felsics are lighter in color. ***basalt*** is black to dark gray; felsic ***rhyolite*** is buff to yellow to even pink. In between the two extremes are the intermediaries, grayish ***andesite*** being the typical example.

The geographic significance of all these data, putting aside for the moment the parent rock's contribution to soil fertility, is that the amount of silica in lava essentially controls its viscosity. The less silica, the more fluid the lava. So basalt is extremely fluid, andesite only moderately so, and rhyolite barely. Volcanoes featuring mafic lavas are not nearly as prone to destructive explosion as those with felsic lava because magmatic gases attempting to dissipate into the atmosphere encounter no great difficulty as fluid lava wells out onto the surface, but increasing viscosity magnifies the problem of gaseous diffusion.

When Kilauea on the island of Hawaii erupts, people rush to the crater to view the fireworks. "Curtains of fire" along radiating fissures are not really dangerous if one stands back at a safe distance. And it is always prudent not to stand in the path of rapidly flowing rivers of molten basalt. There is even a tourist hotel, Volcano House, on the rim of the crater where one can sit in the lounge and safely enjoy Pele's fury through the picture window.

Pre-1980 Mount Saint Helens was an explosive composite cone made up of alternating cinder/ash and andesite layers. The fact that the 1980 explosion blew away the whole side of the mountain points to an increase in silica content in its magma supply. Support for this notion is the rapid growth of a pasty volcanic plug in the maw of the new crater—a volcano within a volcano. Plugs are produced by viscous lava which almost wholly contains the trapped gases within, allowing the pressure to build to the danger point.

Fig. 18-27. ***Eruption of Irazú volcano near San José, Costa Rica.*** *Volcanic ash covers the area.*

what happened to Montagne Pelée on Martinique in 1902. Internal pressure started moving the cork, pushing it out of the vent in the form of a high spire, but before it could wholly clear the passage the side blew out. A dense cloud of intensely hot gases and self-exploding lava mist swept down the slope to the sea, completely wiping out the town of Saint Pierre in an instant and killing more than 30,000 people. This type of explosive debris, both superheated gas and solid particles, is airborne but heavy enough to respond to gravity, following the surface terrain. It is called a ***nuée ardente*** (glowing cloud).

Calderas. Yet another kind of event involves the reverse of eruption: a sudden emptying of the magma chamber. When this event occurs, the entire top of the mountain, lacking support, collapses and is engulfed into itself. The result is a huge, craterlike pit called a ***caldera.*** Crater Lake, Oregon, is an example (Fig. 18-28). It is not a true crater because glacial evidence on the contemporary slopes indicates that a peak over 12,000 feet (3700 m) must have existed to support full-blown glaciers. This reconstructed volcano has been posthumously named Mount Mazama.

But volcanoes can blow their tops; how, then, do we know that it collapsed rather than blew up? We know because the entire surrounding countryside would have to be covered with debris from the old peak. There is some debris but not enough to support a theory of major explosion. What causes the abrupt emptying of a magma chamber? We have no sure answer.

Of more recent origin than Crater Lake is the caldera of Krakatau (Rakata Island, Indonesia). On August 27, 1883,

Fig. 18-28. ***Crater Lake.*** *Despite its name, Crater Lake is cradled in a caldera. Wizard Island, a recent cinder cone, came to life sometime after the great summit collapse some 6000 years ago.*

there were four tremendous explosions, the last of which was heard 3000 miles (4800 km) away in Australia. A towering dust cloud was thrown up, noticeable around the world for two years, and a great sea wave drowned 36,000 people. Subsequent investigation revealed that a deep pit had replaced two-thirds of the island. Despite the ash and dust ejected, the bulk of the island had collapsed and the sea had rushed into the resultant caldera. Even the original Krakatau was merely a remanent of an earlier, much larger island—like Vesuvius standing in the breached caldera of Somma.

When we think of the great loss of human life from volcanic disaster, we might wonder why anyone would choose to live with volcanoes as neighbors. But if a given volcano can be conveniently classified as dormant, local residents come to love it, are inspired by its majesty, and even regard it as divine. Mount Shasta in the Cascades of northern California is such a mountain; it has attracted many groups and individuals who attach special significance to its imposing grandeur. That "the mountain" could be venomous and turn on its friends is unthinkable—until it does.

Farmers, drawn to frequently fertile lava soils, willingly take an unknown risk to derive a livelihood. If the adjacent giant should cough up showers of ash periodically, so much the better. Anyway, as the thinking goes, absolute security is not a guarantee of living anywhere. Why would anyone choose to live in earthquake country, tornado alley, or along hurricane coasts?

Classification by Shape. The simplest classification of volcanoes is on the basis of the shape of the volcanic mountain. Most fall into one of three or four easily recognizable categories. Shape also gives some indication of the type of ejecta and the eruptive habits of the volcano.

First there is the ***cinder cone.*** This is the product of a violently explosive volcano where the lava has solidified in the vent, forming a plug. The accumulation of steam and magmatic gases gradually develops sufficient pressure to blow the plug with such force as to shatter it into tiny fragments. These, along with cinders and ash, are deposited in a symmetrical pile around the vent, the larger particles nearest the vent and the finer ones farther away. The resulting cinder cone is steep-sided (up to 37°, the maximum angle of repose of unconsolidated material) and usually symmetrical (Fig. 18-29). Although evident in many parts of the world, the cinder cone seldom achieves any great size, as erosion rapidly wears away what is essentially a pile of loose material. The disappearing islands of the Pacific are often mere cinder cones, where an eruption may build a pile of solid ejecta above the ocean level to be visible for a few days or months until wave action removes it.

The second type of volcanic mountain shape is called a ***shield volcano.*** Here a quiet flow of fairly fluid lava issues from the vent, forming a vast low-angle cone as it cools. Mauna Loa/Mauna Kea, whose 13,000-foot (4000-m) summits form the island of Hawaii, is an excellent example of this type of volcano. If the low angle of the island's slope is

*Fig. 18-29. **The cinder cone.** There was a lava spill in the foreground not too long ago, but the most recent event has been the formation of an almost perfectly shaped cinder cone.*

Fig. 18-30. ***The Basin of Mexico.*** *This area is nearly surrounded by impressive snow-capped volcanic peaks, the highest of which is Popocatépetl at 17,887 feet (5451 m), here viewed from Puebla. The concave flanks and pointed crest are classic signs that the great mountain is a composite cone built from alternating explosive and quiet eruptions.*

traced to the sea bottom some 5 miles (8 km) deep, the full extent of the multilayered volcano becomes apparent.

Most volcanoes pass through several changes in their history, alternating between explosive eruptions and lava flows. The resulting cone develops a combination of the low-angle shield and high-angle cinder cone. A ***composite cone*** displays concave slopes with a sharp peak (Fig. 18-30). Imagine first a cinder cone, but before it can erode away, it is overlain by lava flows. Built on top of this form is another layer of cinder and ash, followed again by lava. The end result looks like Fujiyama, Shasta, Rainier, or Egmont, the world-famous volcanoes that have inspired poems and legends. These four are now dormant, but El Misti in Peru is equally impressive and still active, as is Mayon in the Philippines. Mayon, despite its generally low elevation and lack of picturesque snow cap, is reputed to be one of the most perfectly shaped of the world's composite volcanoes.

Most volcanoes undergo several stages in their history, alternating between explosive eruptions and lava flows.

Composite cones can build to great height, but only after millions of years of activity. A ***plug volcano,*** on the other hand, may develop very quickly, although it is so self-destructive that it seldom survives intact for long. Plug volcanoes are constructed of lava so thick that it is barely plastic at all; it pushes upward into a bluntly rounded form. As long as the plug slowly extrudes, its surface cools, becomes grayish in color, and igneous boulders cascade down its steep slopes. The only visible indications that it is flowing lava are glimpses of a glowing red interior through occasional cracks, and its constant growth. But such lava does not effectively dissipate trapped gases, and its usual fate is a massive explosion, often out the side of the mountain. A plug volcano is a very dangerous neighbor.

Lassen Peak in California is a plug, a second-generation volcano in the corner of a huge caldera. Its ancestor, ancient Mount Tehama, was an 11,000-foot (3500-m) composite mountain supporting large glaciers as recently as 20,000 years ago, but like Mount Mazama, it collapsed into a caldera. Lassen is its successor. After a relatively short and violent life, Lassen is far from finished, though badly deformed at the moment.

FISSURE FLOWS Lava does not always issue from volcanic vents; it may well out of faults or fissures many miles in length. If it is highly liquid and the terrain fairly subdued, such ***fissure flows*** have been known to cover thousands of square miles and to build up extensive plateaus as in eastern Washington, eastern Oregon, southern Idaho, and northern California. The bulk of peninsular India is also of this origin.

A

B

Fig. 18-31. *(A) A view across the Grand Coulee in eastern Washington reveals the multiple fissure flows that built up the Columbia Plateau. (B) As new lava cools and begins to weather, pioneer plants invade the little crannies that hold moisture.*

In Washington the Columbia River Gorge and the Grand Coulee reveal along their sides a banded layering of differing colors and textures, each of which represents a separate flow (Fig. 18-31).

On occasion, the hot lava from fissures will be heavily charged with gases, and as it cools, the escaping gases leave holes in the rock, making it extremely porous and permeable. Groundwater will occasionally flow through the rock as though it were a pipe. In a moderately viscous fissure flow, with the surface cooling and congealing through contact with the atmosphere, it is also possible for the hot, still-liquid lava underneath to run on and out an igneous cave or tube. These features are common in Hawaii and in northern California's Modoc Plateau.

CONCLUSION

Warping, folding, faulting, and vulcanism are all processes that derive from earth's internal tectonic engine. The landforms that result from these processes are varied. Some are dramatic and, in terms of geologic time, rapidly formed, whereas others are slow to develop and may undergo severe gradation even as they are being constructed. Faulting can be relatively rapid, and impressive escarpments can be formed. The folding of crust into high mountains may take tens of millions of years to accomplish, and the end result may be subdued by ongoing erosion. Volcanic eruptions are dramatic whether they are low-viscosity lava flows or highly explosive blasts from volcanic peaks. They may be temporarily destructive, but are nevertheless building and shaping some of the more distinctive landforms we see on the surface of our planet. Earthquakes are also dramatic—and dangerous as well—but these flashy events should be considered only small reminders of the much less visible yet pervasive processes at work in the depths of the earth. In the next chapter we will investigate the external processes that wear down and shape the surface products of tectonism.

KEY TERMS

folding
crustal warping
Grand Stairway
anticline
syncline
fault
fault plane
normal fault
graben
horst
playa
reverse fault
overthrust fault
strike–slip fault
elastic rebound
epicenter
tsunami
dilatancy
intrusive vulcanism
extrusive vulcanism
batholith
stock
aureole
laccolith
sill
dike
volcanic neck
pyroclastic
basalt
rhyolite
andesite
nuée ardente
caldera
cinder cone
shield volcano
composite cone
plug volcano
fissure flows

REVIEW QUESTIONS

1. What are some of the effects of crustal warping?

2. How has the Cincinnati arch eroded through time to produce the Nashville and Bluegrass "basins?"

3. Describe how earth scientists explain the formation of the Colorado Plateau region with its many canyons and exposed sedimentary formations.

4. Under what tectonic conditions would you expect the development of anticlines and synclines? Where in the world are there good examples of these landforms?

5. Explain the tectonic process that results in normal faulting. What sort of landscape would you find in a region where this type of faulting has been prevalent?

6. In terms of crustal movement, how does a strike–slip fault differ from a normal fault?

7. Name and describe the types of faults and resulting topography you would expect in parts of the world where compressional forces have been deforming the crust for millions of years.

8. Explain why faulting, earthquakes, and volcanoes usually occur in the same geographic region.

9. Seismic sea waves are almost imperceptible at sea. How and why do they become so destructive along exposed coastlines?

10. Name and describe two features caused by intrusive vulcanism and two features caused by extrusive vulcanism.

11. Identify at least three different types of volcanic landforms and explain how each was formed.

APPLICATION QUESTIONS

1. The next time you travel in a mountainous region take a close look at the road cuts, those engineered slots that keep the road gradient reasonable, to see if you can detect any kind of deformation in the exposed rock. It will be most obvious in rock made up of sedimentary strata.

2. Sketch a series of four or five cross-sections showing sedimentary rocks progressively being deformed as compressional forces squeeze the crust.

3. If you were assigned the job of designing a superhighway across the Sierra Nevada in California, how would the basic form of this mountain range affect your design? Describe your proposed route for this highway from east to west across the Sierra Nevada.

4. On a good-sized map of California (a road map would do) follow the trace of the San Andreas Fault. The small-scale map in Figure 18-15 may help locate the fault line. Identify the larger cities and towns that lie on or near the fault.

5. List the countries that are on the "Pacific Ring of Fire."

6. From reports in newspapers or magazines collect information on the location and magnitude of earthquakes occurring throughout the world over a period of several weeks or months. Make a chart with a list of the locations affected and the magnitude of the quake, and give a short account of any loss of life or damage. Use a world map to pinpoint locations.

CHAPTER 19

GRADATIONAL PROCESSES

OBJECTIVES

After studying this chapter, you will understand

1. How chemical weathering alters rock.
2. How physical processes break down rock.
3. How different forms of mass wasting help to lower land surfaces.
4. Why erosion and deposition are important parts of the dynamic equilibrium concept.

Moraine Lake, Banff National Park, Alberta, Canada. Weathering in a mountain environment. Water, ice, temperature variation: either alone or together, they effectively weather hard rock into a variety of forms that are susceptible to removal. Above the tree-line, mountain slopes are deeply littered with coarse, broken rock fragments. In colder times a glacier once occupied this site to push debris downslope. Even now it is still an active arena for ice-wedging in rock cracks and fractures, and judging from the piles of debris it is a highly effective weathering mechanism.

INTRODUCTION

As tectonic processes continually build landforms, gradational processes are at work to reduce and reshape them. Unlike tectonics, gradational processes are external: they are part of the interaction of the atmosphere, hydrosphere, and biosphere with the lithosphere; they are powered by the energy of the sun and by gravity rather than by the heat energy in the earth's interior. The earth's surface is a vast interface where deliberate forces work to reduce the relief of the continents through weathering and transport of rock material.

With gradation comes deposition. Structural depressions on continents and ocean basins are filled in as the weathered rock material is carried to lower and lower elevations. Along the way, depositional landforms may develop, adding variety to the landscape. But in the long run, the rate of tectonic construction is great enough to ensure that the forces of gradation do not reduce the land masses of the earth to sea level. Indeed, tectonic and gradational processes operate in different, and at times opposite, ways to give distinctive form to surface features.

In the long run, the rate of tectonic construction is great enough that the forces of gradation do not reduce the land masses of the earth to sea level.

Fig. 19-1. ***Columnar jointing.*** *Formed as igneous rock cooled, joints in the rock allow access to water and ice. As the tall columns are pried loose, their size is further diminished by free-fall impact at the cliff base.*

To preview what we are going to discuss in this chapter, imagine yourself as a highway engineer building a road through rugged terrain. Your job is to modify the landscape so that vehicles may pass through the region easily and efficiently. So you cut through ridges, fill in valleys, and sometimes follow a route perpendicular to the slope. To do this you must break up rock with heavy equipment or explosives, pick it up and transport it from one place to another, and smooth out the roadbed so that it is as level as possible. Natural gradational processes work in much the same way: weathered rock is broken up slowly over time, carried by running water or some other agent or pulled downslope by gravity, and then deposited somewhere or transported again. Keep in mind that weathering, mass wasting, erosion, and deposition act in concert and continuously. Without tectonic deformation or uplift, the result would be a graded landscape with little relief.

WEATHERING

Weathering breaks down solid rock into a form that can be more easily transported. In other words, it *prepares* the rock for gradation, but it doesn't do the grading itself. Therefore virtually any rock can begin to disintegrate from the moment it is formed (or at least from the moment it appears on or near the surface). Although massive stone seems indestructible, every rock has flaws.

For instance, most rocks are cracked. The contraction of cooling in nearly all igneous rocks results in some kind of jointing, which is often very pronounced (Fig. 19-1). Other varieties of rocks break to some extent in response to even minor forces of uplift and pressure. These breaks can range from out-and-out faults to minuscule stress and fatigue fractures. Breaks in basic structure lessen the rock's ability to sustain its own weight and that of the strata above it. They also allow outside forces to intrude, widening the crack and increasingly attacking the interior. Tree roots, water, and acids—each is effective in its own way, and they often work together.

Certain rocks disintegrate in characteristic ways. Sedimentaries tend to separate back into their original units. If sandstone or conglomerate is nothing more than chunks of rock stuck together, the urge to come unstuck is built in. Because sedimentaries are almost surely stacked up in flat layers, one atop the other, every bedding plane becomes a

Fig. 19-2. ***Bedding plane weakness in weathered sedimentary strata.***

zone of weakness (Fig. 19-2). In igneous formations, the size of the crystal is frequently the critical factor. Obsidian is reasonably resistant to weathering; granite is not. The larger crystals of granite make it vulnerable to selective attack, some crystals proving much less resistant than others. But what about the metamorphics, which are denser and harder than most rocks? As hard as it is, slate foliates into thin sheets cleanly and easily, and the banded schists are inclined to follow suit.

Because many types of rocks make up the surface of the earth's crust, weathering must function in a variety of ways. Most introductory treatments simplify this variety by separating all weathering into two very different kinds—chemical and mechanical. In fact, chemical and mechanical weathering usually operate in conjunction, but there are extensive regions where one strongly dominates the other. Chemical weathering requires water and is most effective under conditions of high temperature, so it usually dominates in warm, moist climates. Mechanical weathering, on the other hand, tends to dominate in deserts, high altitudes, and high latitudes. The general configuration of the landscape is usually a good clue as to what kind of weathering is present. Rounded, muted, and subdued outlines result from chemical weathering; sharp-edged, jagged landforms are of mechanical origin.

Chemical and mechanical weathering usually operate in conjunction, but there are extensive regions where one strongly dominates the other.

Chemical Weathering

Chemical weathering is, in a sense, simply the rotting, mouldering, and decay of rock materials. Chemicals react with a wide variety of minerals to alter their form, and in the process they weaken or destroy the rock. The most familiar example of the chemical decomposition of a seemingly indestructible material is the rusting of steel. In a short time, handsome, shiny steel, resistant to abrasion and other mechanical wear, can be changed to a reddish brown powder and blown away. All that is required is oxygen and water. The iron in the steel chemically interacts with atmospheric oxygen and water molecules in processes called *oxidation* and *hydration*. Oxygen and water are present in almost all

Fig. 19-3. ***Chemical effects.*** *Selective solution weakens the entire structure.*

natural environments, and these chemical processes, as well as others, work effectively on most exposed rocks.

Rocks are simply combinations of minerals, and most minerals have a chemical Achilles heel. Each can be attacked and broken down by a different chemical process. If clear water is not effective, then perhaps the acids contributed by vegetative decomposition will be. If these fail, then minerals that have already broken down may add still other corrosive properties to the groundwater. Rare is the mineral that can stand up to an all-out chemical attack (Fig. 19-3). One that does much of the time is grasslike silica, a widely occurring rock mineral. The chemistry professor always makes his or her speech at the beginning of each term beseeching students to keep their acids in glass containers (silica is a major component of glass) because they will eat their way out of anything else. So it is with natural acids—silica resists. Silica's hard, unaltered crystals appear in great profusion in beach sands as they are released from their rock bond by the decomposing neighboring minerals and carried off to sea.

The feldspar family of minerals is an ingredient in many rocks (making up 60 percent of all igneous rocks), and feldspars break down relatively easily. The result is soft, flaky clay. The widespread evaporites (table salt is an example), which form as evaporation leaves behind minerals, are particularly prone to solution when in contact with plain water. These kinds of reactions weaken a rock by creating soft spots and cavities. The surface crumbles or sometimes peels away in rounded layers, and the underlying fresh surface is exposed to the weather for its round of treatments.

Another common rock is limestone, made up almost entirely of calcium carbonate. In arid country this is considered a highly resistant rock and often stands above the surrounding countryside as mesas or ridges. Limestone is only moderately soluble in clear water, so it is not just the lack of moisture in the desert that makes limestone difficult to break down. Rather, it is the lack of vegetation. Once groundwater filters through a carbon dioxide-rich mat of decaying organic matter, it becomes carbonic acid, and carbonic acid destroys limestone in short order by taking it into

Fig. 19-4. ***Karst topography in southern China.***

solution. In humid areas limestone exposures tend to be the low places.

Where groundwater is mineralized, it is taking into solution and removing mineral matter from the rocks through which it passes.

Where groundwater is mineralized or even brackish, it is taking into solution and removing mineral matter from the rocks through which it passes. Such solution weathering can produce very specialized landforms. Percolating through the cracks in massive limestone formations, the carbonic acid of groundwater can etch out sizable cavities in a relatively short time. Given a longer time, the result is such huge caverns as Carlsbad in New Mexico and Mammoth Cave in Kentucky.

Eventually, the terrain above these caves can be affected. The roof supports are removed, and slumping and cave-ins occur, forming deep holes called ***sinks*** or *dolines* at the surface. Normal surface streams may disappear into such sinks only to reappear many miles away, having flowed as an underground river through the cave. In central Florida, where the groundwater level is high, these sinks become lakes. Further solution within the cave will cause increased slumping until the original surface all but disappears and the landscape becomes a solution valley with a scattering of striking, blunted spires (Fig. 19-4).

At last, even these spires become subdued through the action of wind and rain, weathering, and erosion. Such a landscape, whose character derives from the continued expansion of solution caverns beneath it, is called ***karst,*** after the region in Yugoslavia where it was first described. Karst landscapes, like all other landscapes, display the typical sequence of youth, maturity, and old age. Scattered sink holes only slightly deform the original surface in youth; major collapse results in maximum local relief in maturity;

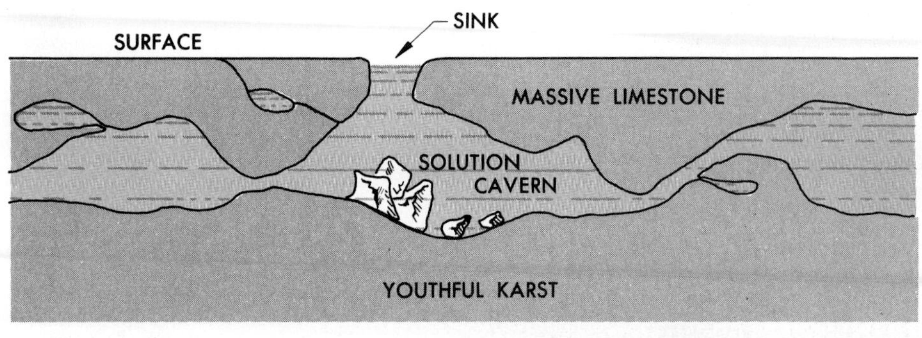

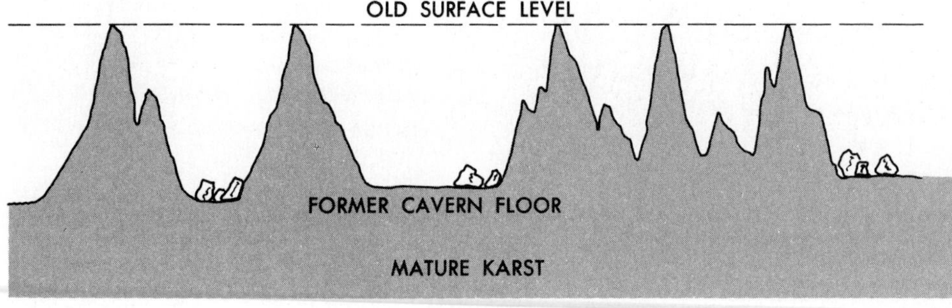

Fig. 19-5. ***The developmental stages of karst landscape.***

and virtual flattening occurs at a lower level in old age (Fig. 19-5).

Mechanical Weathering

Mechanical weathering is the reduction of rock to smaller and smaller fragments by physical factors, and it involves no chemical change. However, it is not always easy to sort out the purely mechanical influences that pry away chunks of rock and shatter them, for, as often as not, chemical processes are involved at the same time. Surely the wedging effect of ice in a rock crack is mechanical; ice has a greater volume than water and exerts a tremendous force on the rock. It is akin to the force exerted on the solid metal of an engine block, which ruptures after the antifreeze has been forgotten. But ice is water, and, where there is water, chemical change is also present. To what extent did chemical decomposition make the rock more vulnerable to the force of the ice? A steel water pipe suffers from rusting as well as freezing.

It is not always easy to sort out the purely mechanical influences that pry away chunks of rock and scatter them, for chemical processes are frequently involved at the same time.

Fig. 19-6. ***Desert weathering.*** *Boulders appear stacked on one another. Many processes are at work here: wind scouring, spalling caused by salt crystals forming in the spaces between grains, and strongly contrasting day and night temperatures. This was once a single rock mass before weathering began eating away along the joints of the rock.*

Nonetheless, part of total weathering (and often a very large part) is unquestionably mechanical. In a closed system at 32° F (0° C), it is estimated that ice will exert at least 2000 pounds per square inch (141 kg per cm^2), and this pressure can build to 3000 pounds per square inch (211 kg per cm^2) at a few degrees below 0° F (−18° C). Of course, this closed system does not occur in nature, but water in a crack freezes first at the top, which does form a quasi-seal. As the water freezes downward in the crack, hydraulic pressure of the remaining liquid water works to open the crack more. Even if we deal with only a very conservative 1000 pounds per square inch (71 kg per cm^2), few rocks can withstand this kind of pressure in a water-filled crack—an established point of weakness. Any accompanying chemical reaction is essentially unimportant. The evidence will be found scattered at the base of the weathered rock: an accumulation of coarse, sharp-edged debris, demonstrating that the rock's disintegration is being accomplished largely by mechanical means.

Frost wedging works best at high altitudes. The daily rather than the seasonal sequence of freezing and thawing tremendously speeds up the process of rock destruction. It is not at all uncommon for the ground at tree-line to be absolutely covered with angular rock fragments of many sizes. The Swiss have given this the highly descriptive name ***felsenmeer*** (rock sea). Ice can be an effective weathering mechanism in high latitudes and deserts too. Near the poles, however, alternate freezing and thawing is limited to a short fall and spring, and in middle-latitude deserts where winters are severe enough for ice, water is not always present even in the form of dew.

The desert displays another type of crystalline wedging that sometimes goes unrecognized: the formation of salt crystals. The action is much like that of ice and also requires water. As the water evaporates at the surface of a porous rock, moisture from deeper down is drawn outward by capillary action until finally a great concentration of ever larger saline crystals builds up in the surface openings. Intergrain and shallow crack rupturing results.

In the desert, where the daily temperature contrast for most of the year is quite severe, rock surfaces can be made to crumble away by rapid alternation of intense heat and cold. Each mineral expands in its own way, and, since every rock is a mixture of minerals, radical differences in rates of expansion and contraction can tear the surface of a coarse-grained rock apart (Fig. 19-6). Some people play down the influences of temperature change on rock disintegration. But anyone who has witnessed aluminum barn siding pulling away from the walls because it was affixed with steel nails expanding and contracting at a different rate cannot be a total unbeliever.

The gradual removal of overburden from deeply buried rocks through erosion or the melting of glacial ice causes the exposed bedrock to adjust to the conditions of lessened pressure. This process is called ***unloading***. Rocks are massively compressed in their original state, and, as the overlying pressure is gradually eliminated, the surface rock tends to pull away from that below. This scaling off of the exterior in broad sheets is called ***exfoliation*** and is beautifully displayed in the granitic domes in Yosemite National Park (Fig. 19-7).

Tree roots can also exert great pressure when they insert

Fig. 19-7. ***Exfoliation.*** *Massive granite released from a great compressing overburden of rocks by erosion expands most rapidly at the surface and peels away in layers. Here on Liberty Dome in Yosemite National Park, ice has been the latest eroding agent and has a considerable weight in its own right.*

themselves in any sort of available rock crack. We are tempted to feel sorry for the giant gnarled pine perched atop a rocky battlement, its roots forced to search for soil and water in the seemingly sterile rock apertures. But individual Bristlecone pines in just such a difficult mountain environment in eastern California and Nevada have survived for perhaps 5000 years. In that time, we might transfer our sorrow to the rock, for measurable wedging and breakage have occurred along the course of every probing root. Evidence of the root's lifting power is commonplace: in every city we find street trees dislodging ponderous slabs of sidewalk concrete (Fig. 19-8).

Even burrowing animals contribute. This is not to say that hordes of hungry chipmunks, gnashing their tiny teeth, chew away stone. But their tunnels or warrens do remove support for rocks on slopes, causing them to tumble. These same burrows open up channels of attack for roots and groundwater.

Fig. 19-8. ***Weathered coast.*** *This sea cliff was already jointed both vertically and horizontally and doomed to eventual breakdown, but probing tree roots and groundwater acids from mats of vegetation are greatly aiding its fragmentation.*

MASS WASTING

All loose particles on the earth's surface are attracted by gravity, and each makes every attempt to respond. If we look around us as we roam the earth, we will find much evidence of movement. Everything is headed in the same general direction—downhill. Some things move slowly, some pell-mell with a maximum of sound and fury, some at a steady pace, some spasmodically and unpredictably—but all materials progress downslope. Chunks of bedrock break off from undercut cliffs or are pried away by ice; water-lubricated volcanic ash and fine dust swirl by as a muddy matrix carrying a cargo of large boulders; soil creeps constantly downward, causing terracing and wrinkling of sod that tries to hold it in place. A single bounding pebble may upset the delicate equilibrium of a piled apron of debris, and the

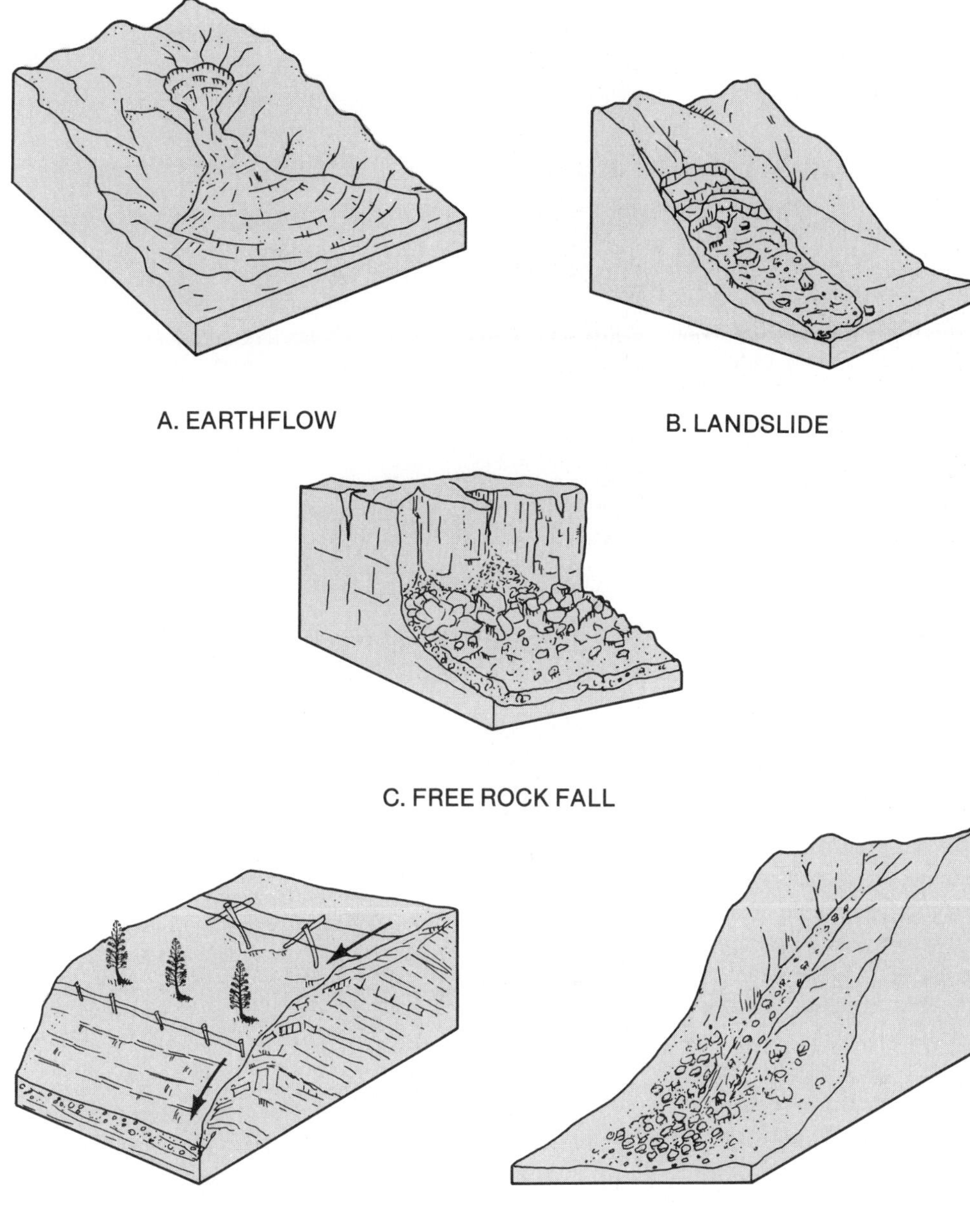

Fig. 19-9. ***Major types of mass wasting.*** *(A) Earthflow—a rapid flow of water, soil, mud, and rock fragments after heavy rain. (B) Landslide—a moderately rapid movement on an unstable slope of relatively dry, weathered rock material. (C) Free rock fall—rapidly falling, mechanically weathered rock fragments accumulate at base of cliff, often as* talus *cones. (D) Creep—soil and finely weathered rock slowly move downslope causing deformed trees and leaning posts. (E) Rock avalanche—sudden movement of accumulated rock fragments in a high mountain valley.*

entire accumulation will rumble off to find a more stable angle of repose. Similarly, snow may react violently as an unconsolidated mass and carry with it tons of rock debris in an avalanche. Humans play a part in these processes by increasing slope instability through removal of vegetation, by introducing water on slopes where infiltration adds weight and lubrication to loose material, and by carelessly engineering roadways in mountains.

Everything is headed in the same general direction—downhill.

All this downslope motion influenced by gravity is part of the universal cycle of gradation in which the high places are whittled away and the accumulated residue is transported to the low spots. This transfer of loose materials downslope—this creep, slide, flow, glide, skid, slip, and fall of all kinds and sizes of rock material—is only the first and somewhat limited part of the transportation system that ultimately moves each particle to the sea (Fig. 19-9). ***Mass wasting,*** therefore, moves the end-products of weathering to the bottom of the slope, placing them in an advantageous position to be carried off by the final agents of gradation.

Slow Movement

The slow, steady movement of materials downslope (as opposed to their sudden runaway slide or fall) is everywhere, but the actual motion is usually difficult to detect. On a clean-tilled field in rolling terrain, it is clear that the soils upslope are increasingly thinner: the hilltop may even lack soil altogether. When a slope is wooded or supports brush or grass cover, however, we must infer soil creep from the cracking of retaining walls and the downhill lean of fenceposts or trees. If winter freezing takes place, then ice crystals replace the soil water and an upward swelling action called ***frost heave*** takes place. Melting completely collapses the soil into a mass of extremely fluid mud, which sags a step downward. Every freeze and melt (and there can be several each year) alternately lifts and drops the soil with a net movement downslope (Fig. 19-9D).

The Role of Water

Water is a major ally in the efforts of masses of earth to relocate at a lower level. This is not to say that water is indispensable, for landslides may occur without water, as when earthquakes trigger the action. Nonetheless, water, whether released from ice or snow, precipitated as rain, or introduced as groundwater, is an excellent lubricant and, in saturating soil, adds tremendously to its weight. After every rainstorm, the highway department in mountainous country is a busy agency, clearing away slides of all sizes. The simple arrival of spring in many parts of the world, with its sustained melting of ice and snow, will bring the same result (Fig. 19-10).

Fig. 19-10. ***Mountain slide.*** *In the building and maintenance of mountain roads, there is always a certain risk. Forests on the steep slopes are slide stabilizers, but they are not infallible if the earth movement is deep-seated. Repairs here may consume six months, and the work crews are fairly safe in assuming more trouble before long.*

Landslides and Earthflows

Both landslides and earthflows, though technically not identical, are relatively abrupt movements of large masses of soil and rock downslope, usually with the help of water (Fig. 19-9A and B).

Landslides occur in a variety of circumstances—for instance, when soil and debris have piled up on an inclined hard-rock or impervious clay stratum. Because groundwater penetrates only to the impervious level, it becomes a line of moisture accumulation and an obvious glide or slip plane for the sudden movement of surface material.

When soils are deep and no slip plane exists, (1) degree of slope, (2) soil character and depth, (3) vegetative cover, and (4) water are all part of the answer to "why did the hill come down into my backyard?" The initial motion in deep soils is frequently a sudden slump in a concave or cup-shaped pattern (Fig. 19-11). The "head" pulls away and drops, but the mass movement is a rotation that causes the soil to lift farther down the slope. Both the deep-seated surface cracks and the holding-basin formation allow the retention and penetration of surface water. This effect adds greatly to total weight, making further slippage even more likely.

A rotating slide, however, tends to move sporadically. It gives us time to institute some remedial action, the chief one being to drain away as much groundwater as possible. Retaining walls, plastic covers to repel the rain, and pilings

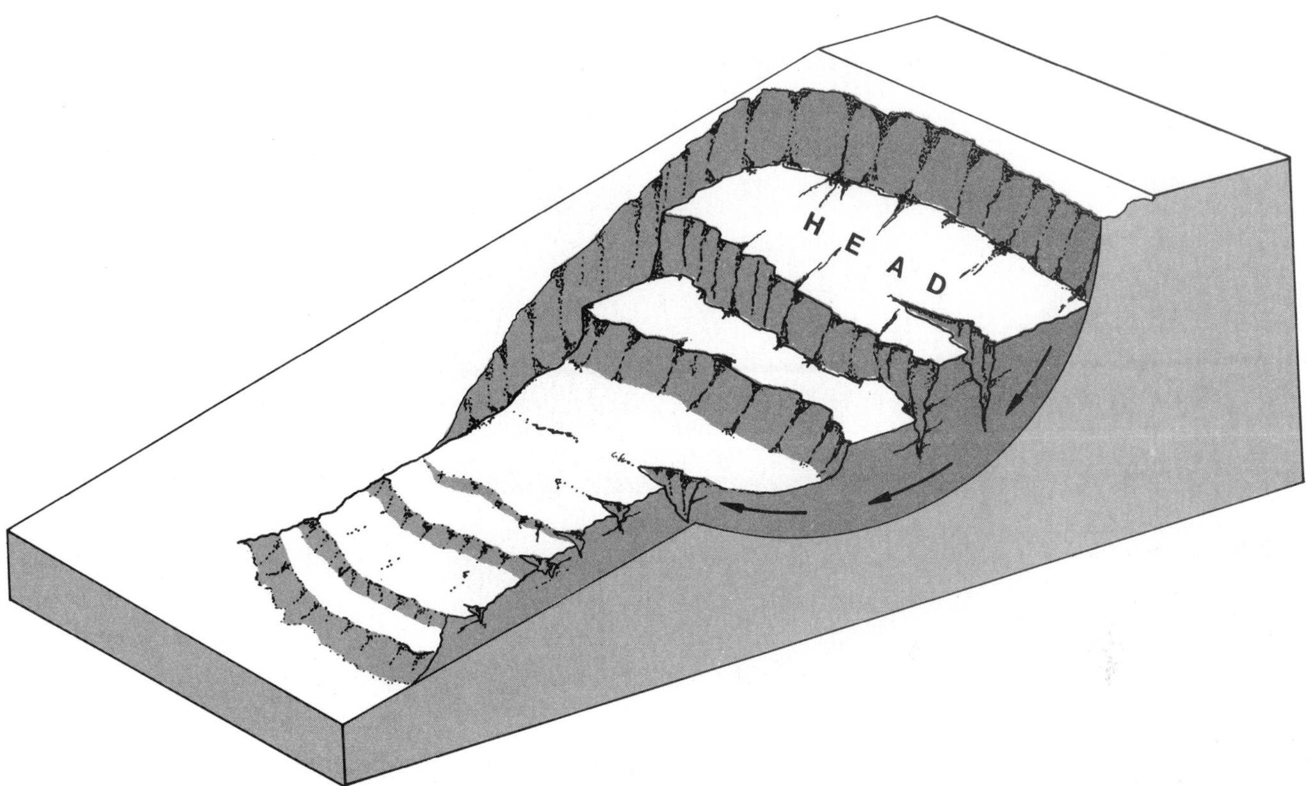

Fig. 19-11. ***A slumping, rotating slide.***

driven into the slide proper may all help a bit, but the slumping mass has now developed as a major zone of weakness and can seldom be controlled permanently. Next year's rainy season or next year's snowmelt may set it off again.

With increasing urbanization and the urge to seek the view site (or to get above the smog), contractors, architects, and unhappy homeowners have all become intimately involved with landslides. In the residentially exclusive canyons surrounding the Los Angeles basin, the story goes, "If the summer brush fires don't get you, the winter mudslides will." Periodic fires clear away the vegetation on the surrounding slopes, exposing the loose soil to occasional torrential winter rains. Runoff, concentrated in the canyons, can bring with it such a heavy sedimentary load that the consistency is that of thick gruel. It is called a mudslide rather than a flood (Fig. 19-12).

Rock Avalanches

In a narrow mountain valley, if we ignore the snow avalanche (see focus box), the chief villain of mass wasting is the rock slide or rock avalanche (Fig. 19-9E). For the most part, we are dealing with mechanically weathered angular rock debris that often builds to a considerable depth in steep draws for lack of any flat surface. Here is a potential disaster for the region below, for broken rock fragments have little cohesion with the ground (often bare rock) or each other, and the slope is likely to be excessively steep. There develops, then, a loosely organized and extremely heavy mass hanging poised above the mountain basin. It takes very little to set an avalanche in motion; one more bounding rock from above, or one whistling marmot scrambling across its face, and off it goes thundering into the abyss.

Free Rock Fall

Free rock fall is common from high-standing crags and cliff faces. Big chunks pried free by mechanical weathering either smash themselves into tiny pieces on impact (actually a process of mechanical weathering itself) or go bouncing, rolling, and skidding well beyond the cliff base (Fig. 19-9C). Lesser pieces, all the way down to fine powder, arrange themselves at the foot of the cliff in cone-shaped aprons according to size: the finest at the top and the coarsest at the lower margin. This ***talus*** slope contains no binding material

Fig. 19-12. **Fire and mud.** *In southern California, fire on vegetated slopes is often followed by winter mudslides down canyons during torrential rains.*

Fig. 19-13. **Talus slopes.** *Free-fall debris from the high cliff face piles up in coalescing talus slopes.*

AVALANCHE

Summer visitors to scenic Swiss mountain villages are enthralled by the beauty and tranquility of the Alpine countryside. They are not there when the fearsome avalanches roar. In every Swiss valley these winter scourges are a fact of life, and simple survival requires a critical knowledge of their behavior. What might appear to be a random pattern of town and farm building location is in fact a very carefully arranged strategic plan based on centuries of experience with the character of local avalanches. Certain ridges act as avalanche "splitters," and carefully protected forest plots on the hillsides, known as ***bannwalds*** (forbidden forests), impede, if they do not completely halt, runaway snow slides. The villages are placed in lee positions, protected from the well-defined avalanche tracks, and much of the valley floor and lower slopes is allocated to permanent fields and pasture.

Given a mountain grade well above the snow line with a degree of slope that is something less than sheer so that it can accumulate a depth of snow, a potential avalanche is in the making. Snow piled up over a winter season is layered. At the bottom is old snow, near-solid ice from percolating meltwater and great pressure from above. Interspersed throughout the depths are hard, slick sun-crusts that were subjected to freezing and thawing on the surface over periodic weeks of good weather. Light, fluffy layers from recent storms form an unstable base for heavy, wet, new snow on top. And overbalanced above all of this are massive projecting snow cornices, fashioned by strong winds whipping across the ridge top.

At a certain point the equilibrium becomes so delicate that one more snowy day or a single skier can trigger an avalanche. And once the snow mass breaks loose, it picks up speed, quickly skidding along an icy interior plane. Rising above and ahead is a great cloud of billowing ice particles that spills down the mountainside. Essentially frictionless, some dry powdery snow clouds have been clocked at 200 miles per hour (322 km per hour), followed by the avalanche proper at over 100 miles per hour (161 km per hour). And propelled out ahead is an invisible blast wave that can be very destructive in itself.

Deep-seated avalanches, scouring at their base to bare rock, move more slowly with friction supplying a great deal of meltwater at the point of contact. During the summer in high mountain basins, avalanche tracks are visible on every slope where they have completely stripped away the trees and moved great quantities of rock and debris, sometimes clear across the valley and up the opposite side.

In 218 B.C. the story goes that half of Hannibal's 38,000 troops and elephants were killed by avalanches as they attempted a winter crossing of the Alps. The worst U.S. disaster was the complete burial of two trains in the Cascades in 1910 with the loss of 96 lives. Today the danger is greater than ever before as snowmobiles and skiers invade the mountains in ever increasing numbers. Prevention teams are out in populated mountain regions, using carefully placed explosives to set off little avalanches before big ones can develop. But the system is far from foolproof. Winter vacations in the mountains carry with them a certain calculated risk, while a lifetime in a Swiss village or a place like Juneau, Alaska, at the base of sheer mountain walls, is a continuing adventure.

of any kind and is totally unconsolidated, a situation that dictates a maximum slope of about 37°. Such a delicate balance is easily disturbed, certainly by additional rockfall. A talus environment is a restless one, constantly shifting and readjusting and endlessly rejuvenated from above (Fig. 19-13).

EROSION AND DEPOSITION

Erosion and deposition are perhaps the keys to understanding the modification and shaping of landforms by gradation. Up to this point we have been discussing the preparation of crustal material for transport: weathering and mass wasting are preliminary to ***erosion***, a general term signifying all aspects of removing and transporting earth materials. Weathering disintegrates the rocks, and mass wasting moves the resulting debris into place for easy pickup and removal; they undoubtedly aid and speed the entire erosion operation.

Erosion and deposition are the keys to understanding the modification and shaping of landforms by gradation.

The great rock masses pushed up by warping, folding, faulting, and vulcanism present a formidable challenge to the agents of erosion. Yet, given the proper implements and, especially, unlimited time, the world's mountains can and do wear away. The four agents of erosion are:

1. Running water
2. Moving ice
3. Wind
4. Waves

The action of each of these agents involves some form of ***deposition***. Rock debris, stream sediment, soil, and anything else carried by water or wind come to rest at some point, perhaps only to be picked up and relocated once again. Deposition is therefore an inseparable part of the erosion process. Each of the four erosional agents has a set of landforms attributed not only to its erosional phase, but to its depositional phase as well.

In the introduction to landforms, Chapter 16, we discussed the concept of dynamic equilibrium in landform development. Dynamic equilibrium implies that forces building landforms and forces tearing them down are engaged in a battle to shape the earth's surface. Given no change in gradational and tectonic processes, a state of equilibrium is approached. When there is a change in intensity of one or more of the variables involved, an adjustment will bring the land *form* into balance with the change in process.

Climate and geology change in both time and space; therefore the level of modification and the fundamental shape of landforms differ from place to place.

Erosion and deposition are affected by many variables, but the most important are climate and geology. For example, climate affects the nature of stream erosion and deposition; landforms in deserts will be very different from landforms in the wet tropics, given the respective climate present in these environments. The geology of the underlying rock can have a significant influence on geomorphology; mountains composed of granite erode differently from mountains of limestone, all other factors being equal. So, if conditions for erosion change, the landform adjusts to the new conditions in an attempt to balance form with processes. Climate and geology change in both time and space; therefore the level of modification and the fundamental shape of landforms differ from place to place.

CONCLUSION

Gradational processes begin with the physical and chemical weathering of rock at the surface of the lithosphere where it is exposed to the elements of the atmosphere and hydrosphere. With the aid of gravity, mass wasting and stream erosion continually remove material from land masses by transporting decomposed and disintegrated rock material to lower elevations. Deposition of material in flood plains, sand dunes, glacial troughs, or ocean basins are examples of the end-products of gradation. Running water, moving ice, wind, and waves act as agents of erosion and are intermediary between weathering on one hand and deposition on the other.

Human beings have had an important impact on the gradation processes. Agriculture, especially over the past two or three centuries, has added significantly to the erosion of soil and other unconsolidated material. The cultivation of land has caused the removal of soil by both wind and water, but mostly by streams that carry the fine sediments to sea or deposit them behind dams, which cut off the supply to deltas and beaches. Acid rain caused by air pollution in, and downwind from, industrialized and populated regions accelerates the chemical weathering of exposed rock—in addition to its adverse effects on lakes, soils, and vegetation. Slopes are sometimes made less stable by construction of buildings and roads or the addition of groundwater, and mass wasting becomes a problem. So, as humans we all affect land-shaping processes.

Remember that all the gradational processes—weathering, mass wasting, erosion, and deposition—can and do operate together and are continually working to shape landforms. While material is being eroded, it is also weathering; when wind or water carries abrasive particles, mechanical disintegration occurs at the same time; as streams cut into their banks, mass wasting provides more material for transport. Also remember that most gradational work is slow; some exceptions are huge, sudden mass wasting events or devastating floods. Changes in landforms may be imperceptible in a person's lifetime, but after thousands, perhaps millions, of years, the effect is enormous.

We are now ready to bring together our knowledge of earth-shaping processes to focus on specific categories of landforms. The following four chapters examine landforms that develop as the result of running water (Chapter 20), glaciation (Chapter 21), wind (Chapter 22), and waves (Chapter 23).

KEY TERMS

chemical weathering
sinks
karst
mechanical weathering
felsenmeer
unloading
exfoliation
mass wasting
frost heave
bannwalds
talus
erosion
deposition

REVIEW QUESTIONS

1. Explain why gradational processes are considered "external."

2. Why do we say that weathering and mass wasting are preliminaries to gradation?

3. Where is mechanical weathering more likely to dominate?

4. Describe one way in which each of the following substances contributes to chemical weathering: plain water, oxygen, carbon dioxide, and vegetation.

5. Describe one way in which each of the following factors contributes to mechanical weathering: ice, salt, temperature contrast, unloading, tree roots, and burrowing animals.

6. List and describe four types of mass wasting.

7. Explain two important ways in which water contributes to mass wasting. Give an example of each.

8. What are the agents of erosion?

9. What two variables most affect erosion and deposition? How?

10. What role do erosion and deposition play in the concept of dynamic equilibrium?

APPLICATION QUESTIONS

1. Do you think the landforms in your region were influenced more by chemical or mechanical weathering? Why?

2. Visit a local cemetery, preferably an older one, and observe the degree of weathering on the various types of headstones and grave markers. Which types of materials hold up the best? What does this tell you about the weathering of different rocks in nature?

3. Take two boards that can be handled easily and place equal piles of sand on each one. However, make one pile wet sand and the other dry. One at a time, slowly incline the boards until the sand falls off, making note of the angle at which the sand slides off the board and the manner in which the sand moves. What does this demonstrate about mass wasting?

4. Describe two ways in which the climate and geology of your area have affected erosion and deposition.

CHAPTER 20

STREAMS AND LANDFORMS

OBJECTIVES

After studying this chapter, you will understand

1. How running water acts as an agent of erosion and deposition.
2. How streams are classified and how each type can be recognized.
3. How stream-related landforms are developed.

Marble Gorge, Colorado River. Over time the river has cut a deep canyon into the Colorado Plateau and transported huge amounts of sediment downstream.

INTRODUCTION

By all odds, the most important gradational agent is running water. Wave erosion of the shorelines is very effective and widespread, but waves merely nibble at the edges of the continents, whereas running water operates over their entire surface. Even in the desert where surface stream courses carry water only now and again after infrequent storms, they may move an enormous amount of material in a very short time. The hard-baked earth and the violence of the typical desert downpour mean maximum surface runoff, leaving the land scarred and eroded into numerous deeply cut channels and dry washes (Fig. 20-1). Elsewhere, every little rivulet and creek is a gradational agent, delivering its load of material to the master stream to carry away and gullying as it goes.

The most important gradational agent is running water.

You may want to review the material presented in Chapter 13 on how water moves across the surface and in channels before proceeding with this chapter on streams and landforms. As streams of all sizes do their work on the landscape, they leave their mark in a variety of landforms: steep-sided valleys; deep gorges; broad, sediment-filled valleys; depositional fans; and deltas. We will now explore the processes by which streams create these landforms.

GRADATION BY RUNNING WATER

Slope is a major factor in stream erosion. Without slope, the stream ceases to flow. And unless the stream is flowing, it cannot (1) pick up, (2) transport, or (3) deposit materials—the three basic functions of all gradational agents.

Stream gradation has two separate components. It begins with ***degradation,*** cutting away and eroding the materials to be transported. It ends with ***aggradation,*** depositing or building up materials in their new location. Transportation comes between these two components: it takes the materials from the area of degradation to the area of aggradation. As a result, high spots are carved away and the low ones are filled in.

Sediment Transport

All streams carry sediments. Nowhere on the surface of the earth can we find absolutely clear running water. Obviously,

Fig. 20-1. ***Riverine erosion in Imperial Valley, California.*** *There is little water in this entire desert landscape, where less than 3 inches (8 cm) of rain falls annually, yet the marks of riverine erosion are cut into the land. Obviously, stream runoff is very effective following the occasional torrential downpour. This type of symmetrical drainage pattern is called* dendritic *because it resembles the branching pattern of a tree.*

a big, muddy-looking river such as the Hwang Ho or the Missouri River is moving a great deal of sediment in suspension, but even sparkling mountain cascades are transporting surprisingly large amounts of material. They only *appear* clear because they are capable of carrying much more sediment than they do.

Every stream, large or small, transports three categories of load:

1. The ***solution load***—those invisible components that water dissolves from the rocks it encounters.
2. The ***suspension load***—all the fine material easily carried by running water, usually well up in the main current.
3. The ***bed load***—the segment of the solid sediment that is heaviest and impelled along in contact with the stream bed.

Swirling sands and skipping pebbles are in constant motion, but even sizable boulders advance spasmodically during the highest water period when increased velocity and buoyancy are temporary aids.

Stream Types

The ability of a stream to transport material depends on many factors, but stream velocity and volume are of overwhelming importance. The *discharge* (Chapter 13) of a stream—its rate of flow measured in cubic feet per second (m^3 per second)—increases downstream as more and more water enters the trunk stream from tributaries. The discharge depends on the average velocity of the stream and the cross-sectional area of its channel. Low-discharge, high-gradient mountain streams transport large amounts of bed load and suspended load for their size, but they also expend much of their energy actively eroding their channels vertically. Although they appear to be sluggish, gentle gradient streams with greater discharge can move across relatively flat land carrying huge amounts of sediment, particularly suspended sediment. These streams typically work to transport and deposit sediment and have little active downward erosion—although during high water and flood events, significant erosion of streambanks and channels will occur. Streams vary in their capacity to erode, transport, and deposit material. Therefore we attempt to classify them according to these characteristics.

The ability of a stream to transport material depends on many factors, especially stream velocity and volume.

Try to visualize a stream as a dynamic system trying to reach a balance between its sediment load and its carrying capacity. A balanced stream in which the load supplied from the drainage area is exactly equal to the capacity of the stream to carry that load is called a ***graded stream***. Such perfect equilibrium is virtually impossible to achieve in nature, but many streams come fairly close and all work to achieve this goal. A high-velocity stream with tremendous energy but too little to carry will expend energy to erode more materials. Such a watercourse is called *youthful*. On the other hand, a placid, low-gradient stream may be heavily overloaded with sediment, especially if it is fed by vigorous tributaries. As the increased volume of water and sediment load of such a stream enters lower gradient flatlands, it attempts to rid itself of its surplus load. This type of stream is classed as *old age*. Between these two extremes are many watercourses with moderate discharge that approach the graded ideal. From season to season they may occasionally exhibit youthful or old-age tendencies, but because they normally have a reasonable balance between sediment load and discharge, they are called *mature*.

Caution: you may remember that we used this same classification terminology of youth, maturity, and old age to characterize the various stages of the geomorphic cycle (Chapter 16). Here we use these terms to describe streams as they evolve toward a graded ideal. Although there are exceptions, youthful streams do tend to flow in youthfully eroded landscapes, mature streams in maturely eroded landscapes, and so on.

YOUTHFUL STREAMS The high-gradient watercourse with its tremendous energy is attempting to carry as much sediment load as possible. It has only one arena of operation—its channel—so it works with great vigor on that channel. The friction of running water alone is scarcely capable of eroding the rocks of the channel bottom. All the same, water can take some of the minerals into solution, thereby weakening the rock and allowing pieces to tear away and be carried off. These rock fragments and others delivered from the valley slopes vary in size from boulders to sands and silts. They become the "teeth" of the stream and give it great cutting ability. Especially effective are the hard, sharp-edged quartz fragments in sands that allow a youthful stream to corrade (scape away by abrasion) its channel downward at a very rapid rate. In doing so, the stream begins to supply sediments from its bed while at the same time gradually lowering its gradient. Eventually, as the gradient lowers, the amount of material supplied to the stream network will approach the stream's capacity to carry that material.

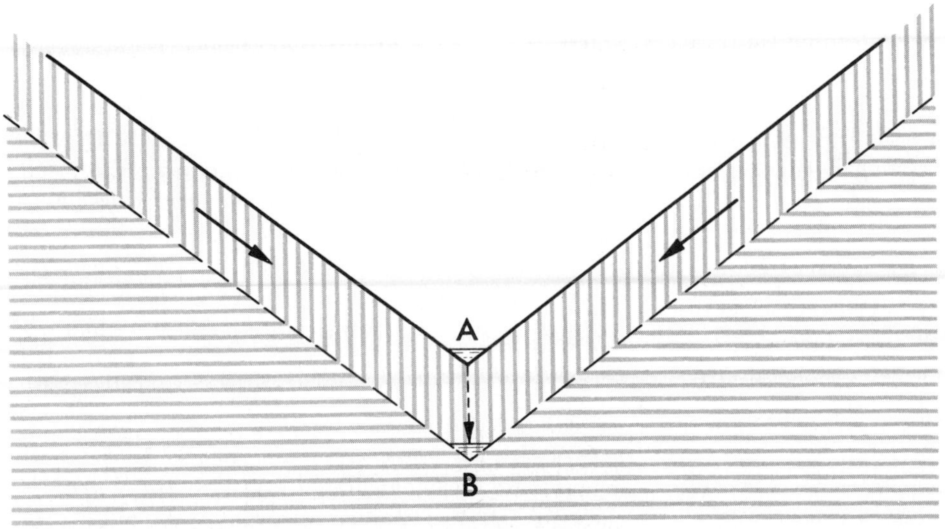

Fig. 20-2. ***V-shaped valley and youthful stream erosion.*** *As the stream works its valley down from A to B, the material on the valley sides, with their now-steepened slopes, is delivered to the stream via gravity.*

Valley Shape. If it is cutting through soft rock or soil, the youthful stream typically forms a V-shaped valley and occupies the entire narrow valley bottom (Fig. 20-2). The slope of the V is determined by the steepest angle at which loose soil will remain in place (the angle of repose). As the river works its way rapidly downward, that angle increases and great volumes of soil and weathered rock slide into the valley and supply the stream with added material. In addition, any tributary creeks and rivulets find their gradients becoming steeper as they attempt to maintain their mouths at the level of the constantly lower master stream. Thus they become more effective eroders and deliver increased quantities of sediment.

The youthful stream typically forms a V-shaped valley and occupies the entire narrow valley bottom.

Where the youthful stream flows across hard-rock strata, it cuts a vertical-sided gorge or slot rather than a V, since, unlike soil, rock can maintain sheer cliffs without sliding. However, weathering begins to work immediately on newly exposed rock along the sides of the gorge. Eventually, it is weakened and undercut by the stream, and the debris is delivered to the water via gravity.

Many of the spectacular gorges of the Colorado Plateau have developed in this manner. They are commonly a half-mile or more in width across the top and narrowing via terraces toward the bottom. Each terrace surface represents a hard-rock layer overlying softer strata, while far below in the bottom of the canyon, the river occupies a narrow slot as it busily cuts its way downward (Fig. 20-3). This type of canyon development, of course, depends on the exposure of beds with differing resistances to weathering as in the Grand Canyon and many others. The Snake and Columbia

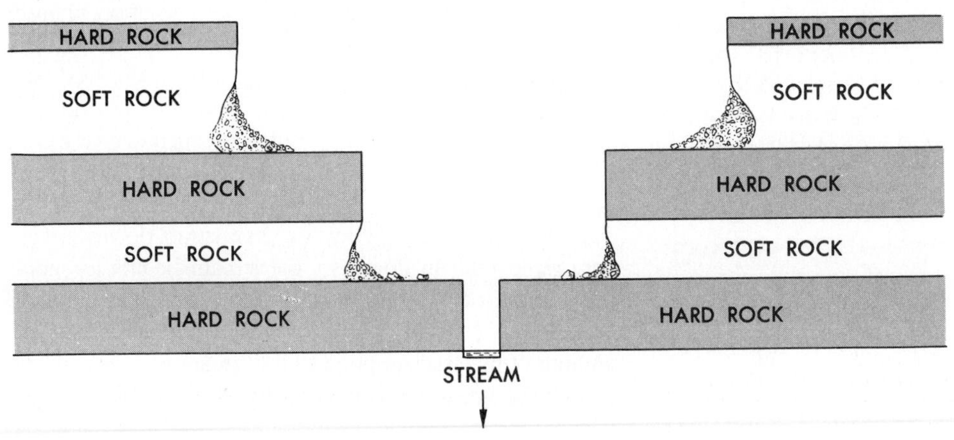

Fig. 20-3. ***Youthful stream erosion—canyon.*** *By cutting its way across alternating hard and soft strata, the stream exposes soft rock to weathering and erosion.*

rivers, on the other hand, have worked their way down through hard rocks of similar resistance, and their gorges tend to be narrow and sheer sided.

Youthful Stream Course. As youthful streams charge across the countryside, they tend to maintain relatively straight courses, because they typically overrun and cut away obstacles rather than detour around them. Slides may choke the channel for short periods, but the energetic stream will rapidly chew its way through.

Lakes, falls, and rapids are common features in early youth as a drainage pattern establishes itself over newly uplifted terrain, but these are short-lived features. The upper end of a lake will silt up rapidly, whereas the downward erosion at its outlet cuts away the lip that maintains the water level. As a result, the lake will gradually drain away. Similarly, falls and rapids will be erased within a short time (Fig. 20-4). Close to home, we have the example of the Saint Lawrence with many rapids, Niagara Falls, and the Great Lakes in its drainage system. The rapids are being smoothed out, Niagara Falls is retreating rapidly, and left to itself (which is unlikely in light of its economic benefits), the river would drain the lakes within a very short geologic time span.

Do not confuse the length of time that a stream has existed with its category of youth, maturity, or old age. Its behavior is what determines its classification. Like people, streams are only as old as they act. Obviously, a stream attempting to erode a hard-rock channel will maintain its youthful characteristics much longer than one cutting through soft strata. But given sufficient time with no tectonic interruptions, in the normal course of events all youthful streams will lower their gradients to the point where they can achieve maturity.

Fig. 20-4. ***Youthful stream.*** *As it hurries by, often over falls and rapids desperately searching for added materials to transport, the river scours, abrades, and consumes its channel bottom.*

MATURE STREAMS Mature streams come closest to the graded ideal. Remember that a perfectly graded stream does not exist. Even when carrying capacity matches sediment load most of the year, increased volume during springtime's melting snows or a sustained rainy season will cause a stream to sweep its channel clean and cut into its bed in an attempt to find a larger load to match its greater capacity. Increased discharge will allow it to carry somewhat more. Conversely, during a dry season volume will be reduced so that the stream must drop part of its load, thereby choking the channel with gravel and sand bars. These bars make it difficult for the main current to maintain itself in a single channel, and the flow may break up into many lesser courses, following twisted routes around the barriers. Such a multichanneled stream is called a ***braided stream.*** Braided streams often occur in semiarid regions or in mountain valleys that have melting glaciers. In these cases, runoff is seasonal, and there is an oversupply of sediments for the rate of discharge (Fig. 20-5). In other words, there is not enough water to maintain an open channel. Nonetheless, any stream that normally has a reasonable balance between load and carrying capacity is considered mature.

Work of Mature Streams. Because the mature stream has typically lost some of its youthful ability to overrun obstacles in its path, it will develop moderate bends or meanders in its course. For instance, if a slide were to block its

Fig. 20-5. ***A braided stream.***

course in early maturity, the stream would detour around it. In doing so, the current, which attempts to run in straight lines, would be diverted against the bank. Here it would cut horizontally on the outside of the bend, expanding it farther to the side. Then, bouncing off this side, it would be diverted again to the other side and cut laterally to form another bend, and so on downstream.

Once these bends or meanders are established, the main current, moving alternately back and forth across the river, cuts at the outside bends and enlarges the meanders. Here we have something quite different from the youthful stream, which focuses its erosional energy downward. During high water a mature stream cuts moderately downward, but it also erodes laterally, forming meanders and widening its valley.

A mature stream cuts moderately downward, but it also erodes laterally, forming meanders and widening its valley.

The flowing water of the mature stream consists of two segments:

1. The main current, which has the greatest velocity as it cuts away at the bank of the outside bends and keeps its channel clean and deep.
2. The more slowly moving bulk of the stream, which flows along the inside bends. Here all the sediments cannot be carried and it drops a part of the load along the bend, thus leaving deposits.

Inside bends display a gently shelving ***slip-off slope*** built up by deposition, whereas outside bends are deep with a cliffed ***cutbank*** (Fig. 20-6).

This lateral cutting combined with matching deposition across the river means that a mature stream occupies a broad valley floor, as opposed to the channel of the youthful stream, which is the valley floor itself. The outside bends touch and cut away the sides of the valleys on both sides, continually expanding the width of the valley bottom and developing an extensive ***flood plain***—the flat, frequently flooded area adjacent to the stream channel (Fig. 20-6). Thus the meander belt (the zone occupied by fully developed meanders, outside bend to outside bend) is the exact width of the valley floor (Fig. 20-7). Any road attempting to take advantage of such a valley through rough country must bridge the same stream repeatedly. In addition, flooding may well be a hazard during high-water periods.

OLD-AGE STREAMS Any watercourse flowing across a low-gradient plain and attempting to transport a load of sediment beyond its carrying capacity is called an old-age stream. No matter how long it has existed, it is categorized

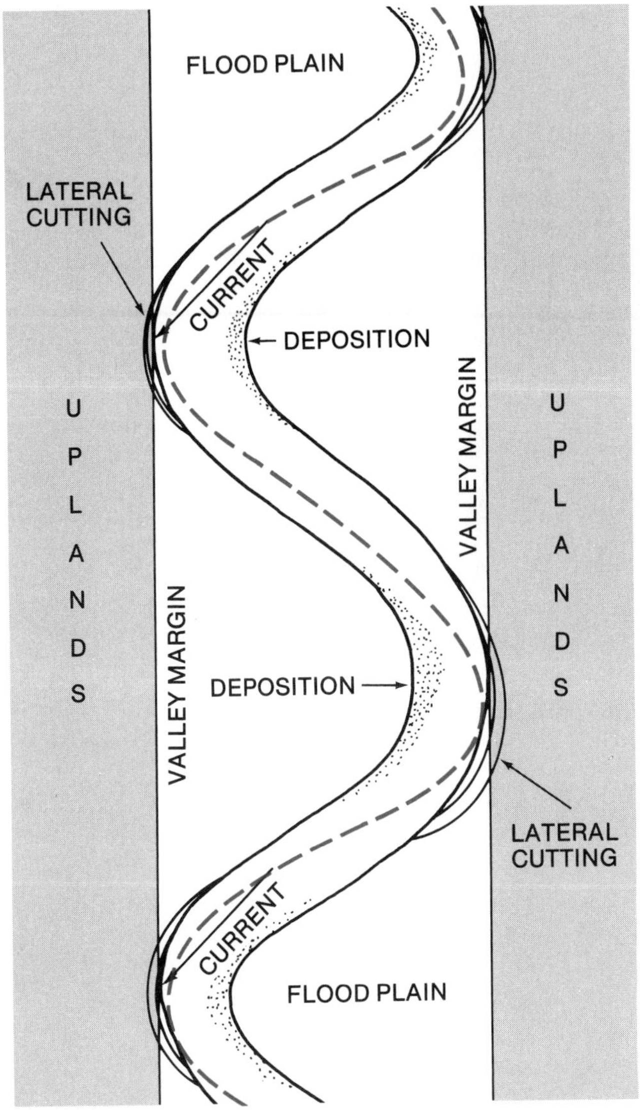

Fig. 20-6. ***Mature stream process.*** *Valley sides are eroded as the flood plain receives deposits.*

as old age for it rapidly acquires a set of old-age characteristics. Basically, it is an overloaded stream flowing through its flood plain and reworking its own deposits. An old-age stream is actively eroding and depositing in its flood plain as it attempts, like all other streams, to balance sediment load with the capacity to carry that load.

An old-age stream is actively eroding and depositing in its flood plain as it attempts to balance sediment load with the capacity to carry that load.

The place of obvious deposition is the stream channel itself, and typically it has a number of sand and gravel bars and islands. Beyond the stream channel are extensive, deep layers of sediments deposited by the stream over time as it has worked its way back and forth across the flood plain. Such deposits of rock material of all sizes that are transported, laid down, and reworked by streams are referred to as ***alluvium.*** Old-age streams expend much of their energy shifting alluvium around in their flood plains.

Because most old-age streams have developed from the mature stage, they already have moderate meanders. These are enlarged during the old-age phase by the action of two stream components: (1) the swifter flowing main current attacking the outer bends, and (2) the slow-moving portion of the inside bends.

Although an old-age stream is overloaded, it continues to deposit a great deal of sediment on the inner bends and at the same time to erode the outer bends. The meander pattern thus becomes extremely sinuous and involved.

When this looping course has reached its ultimate development, increasingly two bends work their way toward each other until they join and the stream flow cuts across. In this way they establish a new channel and leave an abandoned meander called an ***oxbow lake*** (Fig. 20-8).

Unlike the meander belt of the mature stream, the meander belt of the old-age stream occupies only a small part of its wide valley. Scars on the bluffs on both sides of the valley indicate that the stream has been at work widening its valley. These scars give evidence that the meander belt shifts back and forth across the flood plain through time.

Flooding. The old-age stream carries all the sediment that it can to the sea and deposits it in the form of a delta. It also loses part of its excess load through channel deposition. Still this does not provide complete relief. The stream will normally overflow its banks during even relatively minor high-water periods. Several days of heavy rain in its drainage area are usually enough to cause some overflow, and, during the spring, snowmelt in the uplands drained by its tributaries often results in extensive flooding.

Natural levees exist because, as the river overflows its channel, its velocity is suddenly checked and heavy deposition occurs here along the channel margin.

The low banks that keep the stream in its channel during normal flow are easily overrun during high water. These banks, called ***natural levees,*** exist because, as the river overflows its channel, its velocity is suddenly checked and heavy deposition occurs here along the channel margin. Then the

Fig. 20-7. ***Evidence of stream maturity.*** *The modest meanders signal maturity in this stream. So does the steep bank on the outside bend balanced by shelving deposition on the inside. The valley width has expanded a great deal from the earlier youthful V-shape.*

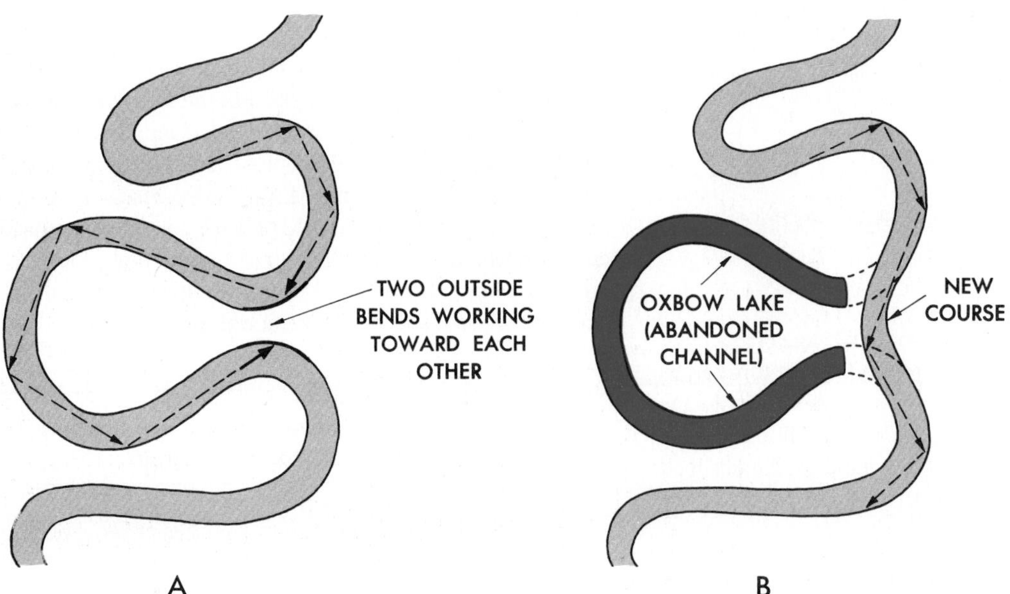

Fig. 20-8. ***Formation of an oxbow lake.*** *The lateral cutting on the outside bends is carried to its extreme when two involved meanders join to form a new channel, leaving the old abandoned loop as a lake.*

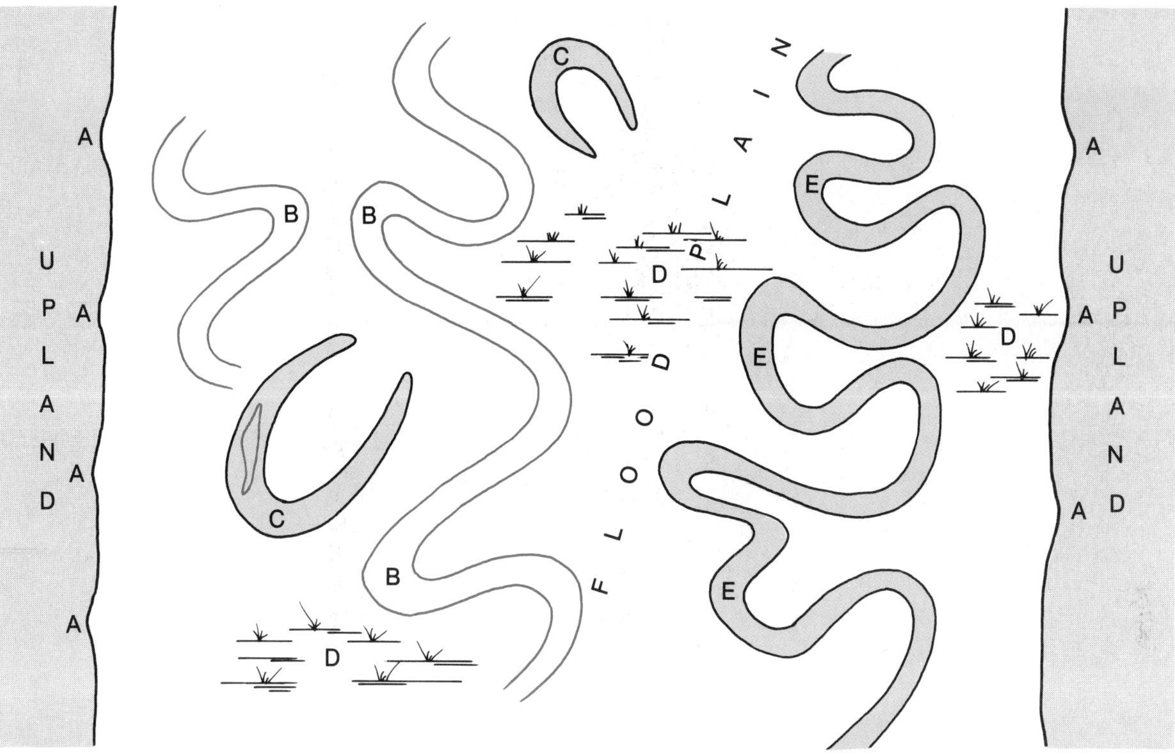

Fig. 20-9. **Old-age stream valley.** *(A) Crescent-shaped erosion scars on the bounding bluffs. (B) Abandoned channels. (C) Oxbow lakes. (D) Swamps. (E) A serpentine river course occupying only part of a broad flood plain.*

water spreads out for miles across the entire flood plain as a still lake and drops its load of sediment. In this way the old-age stream gets rid of its overload and at the same time builds up the valley floor to make the gradient to the sea temporarily slightly steeper. This increase in gradient helps the stream to carry its load farther downstream.[1]

Once the old-age stream has flooded, it may abandon its established channel, for usually it is slightly higher than the general level of the valley because of constant channel deposition. So an entirely new channel may develop, or more than one, only to repeat the overflow and abandonment over and over again. Thus the broad flood plain of an old-age stream features permanent swamps, oxbow lakes, and constantly changing channels (Figs. 20-9 and 20-10).

The valley of the lower Mississippi was like this before many people settled there. De Soto found it to be a 300-mile (483-km) malarial swamp. But such valleys have a certain attractiveness to an agricultural civilization, provided the

[1] In lower courses of rivers emptying into the sea, this is seldom successful, as delta building at the stream mouth offsets the added height of the valley floor.

Fig. 20-10. **Old-age stream.** *Meanders don't come any more twisted than these. A careful examination of the wide valley bottom will reveal oxbow lakes and abandoned channel loops.*

THE YELLOW RIVER

The Yellow River (Hwang Ho) in flood.

Cold winter monsoon winds sweeping out of Mongolia and the Gobi carry with them a cargo of fine, amber-hued desert dust. Particle size is on the order of flour, small enough to be easily transported by the wind but heavy enough to drift like dry powder snow against the slopes of broken terrain. In the provinces of Shaanxi and Shanxi in eastern China, the depositional process has gone on so long that great piles have accumulated, some to as much as 450 feet (137 m) deep. This is loess, wind-transported soil.

The Yellow River, on its long southerly reach through the loess plateau, experiences no difficulty at all in cutting a deep gorge in the soft, compacted soil, acquiring as it goes a huge sedimentary load—several large tributaries add their contributions along the route. Here is the fine, yellow desert dust on the move again, this time as water-transported soil. The running water, even though the gradient is fairly steep at about 4 feet (1.31 m) to the mile, is muddy in appearance and silted to near capacity. At this point the river course shifts sharply to the east and heads out on a 500-mile (805-km) run across the North China Plain to the sea, but simultaneously its gradient drops abruptly to no more than 10 inches (0.3 m) to the mile. The river finds itself attempting to transport a substantially larger sedimentary load than the laws of physics will allow. Three-quarters of the solids in suspension must either settle out in the channel itself or be distributed over thousands of square miles by repeated large-scale flooding.

Coveting the inherent fertility of the twice-transported alluvium, Chinese agriculturalists moved out onto the swampy North China Plain with their first order of business to make the river "behave." They acquired a great deal of experience over the long centuries in building the dikes ever higher. But in the process they have forced the Yellow River to flow along a sinuous ridge of its own making, inevitably magnifying the flooding in the event of a single break anywhere along the hundreds of miles of levees. They call the river "China's Sorrow" for there have been innumerable floods with huge loss of life and crops. And still they struggle to force an old-age stream to behave abnormally so that they might cultivate the soil born of the flood.

A major problem for the levee menders after each serious break has always been to get the escaped river back up onto its channeled ridge when water's normal inclination is to flow downhill. Sometimes they succeed and sometimes they don't. In the past 3500 years there have been 26 major channel changes, a discomforting situation for those whose fortunes are directly tied to stream behavior.

In 1960 the situation was significantly altered for the first time in history. A massive dam completed at San-men Gorge near the Yellow River exit onto the Plain blocked the sediments, and they piled up behind it. All was well for a few years. The official estimate had been that the San-men dam was good for 50 years before it completely silted up. However, the rate of accumulation behind the dam increased because the Chinese government encouraged many farmers to settle in the uplands to the west of the plain. Erosion from the agricultural activity has added significantly to the river's sediment load. The level of the reservoir behind the dam has been lowered to half-capacity to reduce siltation, but this has also reduced the dam's flood-control function. Once again, the Yellow River downstream is filled with more solids than it can handle.

The government plan is for a system of as many as 30 dams on the master stream and its tributaries. During the short time gained before they all silt up, the effort will be made to terrace, irrigate, and reforest most of the semiarid loess watershed—not much time to accomplish a huge project. It will be interesting to watch.

river can be tamed. The deep and constantly rejuvenated soils and nearly flat terrain make exceptionally productive farmland. So, the idea is to clear the tangled vegetation, drain the swamps, and then above all make the river stay in its channel where all self-respecting rivers are supposed to flow.

The most obvious and simplest way to keep the water in its channel is to build up the natural levees so that at next year's high water they will be just high enough to contain it. This approach works nicely except that much of the sediment that used to be deposited on the flood plain now are deposited on the channel bottom. A few years of this activ-

REVIEW QUESTIONS

1. Why must the action of a stream include transportation as well as degradation and aggradation? How do the three processes work together to change the landscape?

2. Name and describe three types of sediment load in streams.

3. What are the qualities of an "ideal" graded stream? Why and how do streams try to become graded?

4. What are the three types of streams? How would you recognize each type?

5. Why do some youthful streams form a V-shaped channel while others form a slot?

6. How do youthful streams become mature streams? Give an example.

7. Why do some mature streams have braided channels? What kinds of mature streams would be unlikely to form braided channels?

8. How are meanders formed? How are the meanders of a mature stream different from those of an old-age stream? What different landforms are likely to result from old-age meanders?

9. How are natural levees formed?

10. Why is it dangerous and difficult to farm on the flood plain of an old-age river? How have people attempted to control such rivers?

11. What is the difference between a delta and an alluvial fan?

APPLICATION QUESTIONS

1. Identify a stream close to your present location. What type of stream is it? Why do you think it fits into that category?

2. If a dam is placed across a mature stream and the water released from the dam is regulated so that stream flow does not change throughout the year, what might happen to sand bars downstream from the dam?

3. If artificial levees are constructed for several miles along the banks of an old-age stream, what effects would you expect downstream?

4. If climate became significantly wetter over a long period of time in a region where alluvial fans had been well developed, what might be some of the changes in the alluvial fan landforms?

5. Research the role of the U.S. Army Corps of Engineers in controlling and changing the flow of several rivers, such as the Mississippi, Rio Grande; or Tennessee rivers. What economic, political, and environmental effects has this work had?

Fig. 21-8. ***Glacial moraines.*** *The receding glacier leaves ridges of debris. Recessional moraines are breached and material is reworked by meltwater. Lateral moraines line the valley sides.*

seeks a new one, eventually forming convex aprons of sorted alluvium stretching far down the valley beyond the terminal moraine. This feature is called a ***valley train.*** It can be distinguished from moraines by the absence of sorting in the ice-deposited moraine material as opposed to the careful sorting according to size in the meltwater alluvium.

The other area of deposition at the lower end of the glacier is under the ice itself. Near the nose of the glacier, the underside of the ice is heavily charged with the debris it has picked up on its journey, and the holding capacity of the ice decreases as the rate of melt increases. Thus, like an overloaded old-age stream, the glacier must drop some of its load. This deposition of all sizes of materials occurs under the ice, the weight of which compacts it into what is described as boulder-studded clay. It is called ***till*** or *ground moraine* (Fig. 21-10).

Fig. 21-9. ***A moraine lake, Lake Wanaka, New Zealand.*** *The glacier that once moved downvalley from the snowy peaks in the distance deposited its morainal debris in the direct foreground of the photo, forming a dam. Houses and roads now take advantage of the modest elevation.*

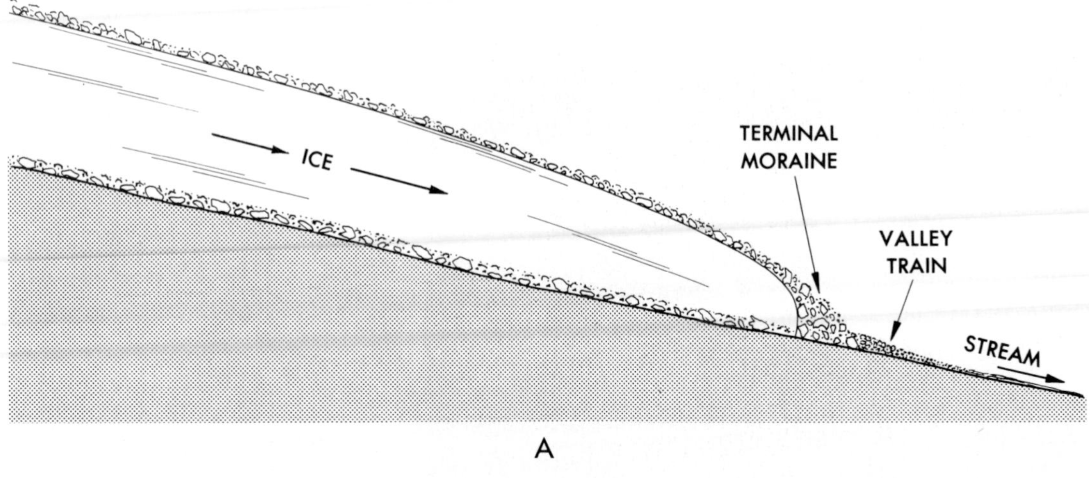

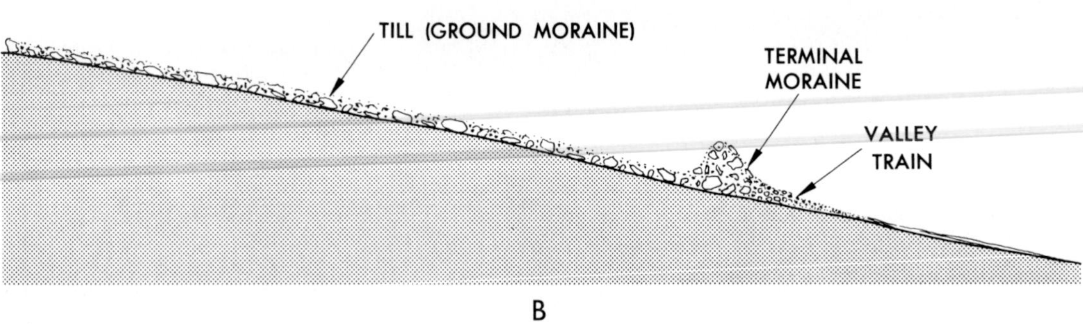

Fig. 21-10. ***Glacial deposition.*** *(A) A valley glacier brings material downslope, building up a complex of depositional forms at its terminus. (B) After the ice has melted back.*

So three types of deposits are related to valley glaciation:

1. The till deposited as a flat mass under the ice.
2. The terminal and lateral moraines formed at the end and the sides of the glacial valley.
3. The valley train as sorted alluvial deposits usually downslope from the terminal moraine.

A word of caution is in order here. The entire concept of moving ice as an agent of gradation depends on understanding the time involved. Even running water, which requires considerable time to accomplish significant gradation, is more obvious than ice, for water moves more rapidly, can be observed carrying and depositing materials, and, on a small scale as in the formation of gullies, can demonstrate its power before our very eyes. But ice does not appear to flow and must be measured carefully over a period of years to ascertain even slight movement. The best way to appreciate what it is capable of doing is through a before-and-after comparison of glacial mountain valleys. Here we can find ice in all its phases and fully appreciate its capabilities.

CONTINENTAL GLACIERS

It is one thing to point to the continental glaciers of Greenland and Antarctica (Fig. 21-11) and classify them as the only ones in the world, but it is quite another to be persuaded that much more extensive ice sheets covered large parts of North America and Europe in the none too distant past. It is difficult to accept this proposition, and a good deal of evidence must be presented to convince us. Today we not only have proof that such continental glaciers did indeed exist, but we know all the details of their many advances and retreats. The history of the ice is written on the land for

Fig. 21-11. ***Antarctic icebergs.*** *The margin of the Antarctic ice sheet advances into a chilly sea. Icebergs here are relatively thin and flat-topped in contrast to the craggy Greenland variety.*

all to see, but observers were blind until they could imagine that such a thing was possible. It was Louis Agassiz, a Swiss naturalist, who in 1834 first suggested that the ice-abandoned valleys in the Alps must have at one time supported glaciers. He came to this conclusion by comparing the deposits found in an empty valley and those found in a valley in which an active glacier was present. In other words, ice left evidence of its former existence. Some of this evidence had been puzzling geologists in both Europe and the United States for many years. Why did the large boulders present in Iowa have no relationship to the local bedrock but resembled the bedrock in Canada? What about those deep grooves in solid rock as on Kelley's Island, Ohio? Agassiz's theories answered these questions and many more very logically, and through further investigation he gradually built up a body of proof that continental glaciers must have existed. It is entirely fitting that Louis Agassiz, the champion of continental glaciers that no one ever witnessed, should have his memory perpetuated in the name of a glacial meltwater lake that existed once but vanished thousands of years ago. ***Lake Agassiz*** at one time covered a large part of North Dakota, Minnesota, and Manitoba (Fig. 21-12).

The history of the ice is written on the land for all to see, but observers were blind until they could imagine that such a thing was possible.

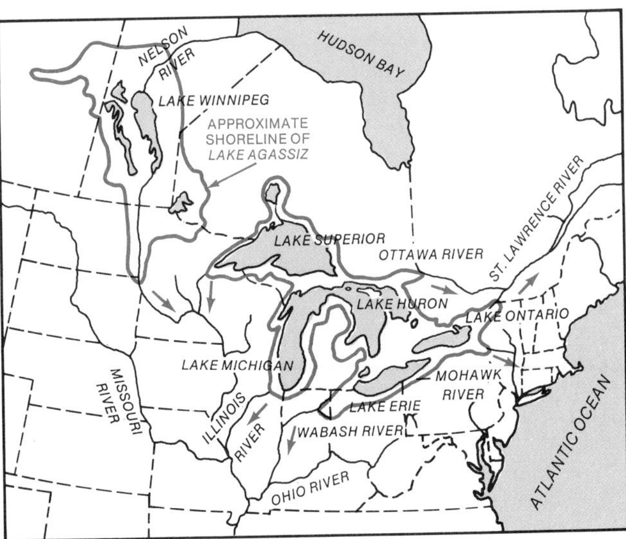

Fig. 21-12. ***The Great Lakes region and ancient Lake Agassiz during the late Pleistocene epoch.*** *Huge lakes formed as continental ice sheets began to melt and recede about 20,000 years ago. Today the major outlet of the Great Lakes is the Saint Lawrence River. Remnant Lake Winnipeg drains north.*

About a million years ago ice began to form in Canada and Northern Europe in much the same manner as has been described in mountain basins—that is, a continuing accumulation of snow gradually changing its lower elements into ice over a very broad area. As the depth and pressure increased, the ice was forced outward in all directions. In some cases this was downslope or, as in much of North America, across generally subdued terrain, but in Europe the pressure was sufficient to cause ice, centered in the Baltic Sea region, to overrun such major obstructions as the Scandinavian mountains and Scottish highlands. An extensive sheet of continental ice overwhelmed even North

LOUIS AGASSIZ

The problem was ***erratics,*** great boulders out of place in England which could be proved to be of Scandinavian origin. Early nineteenth-century experts on such matters were in fairly unanimous agreement that Noah's flood had washed them over there. The pioneer British geologist Charles Lyell even suggested, with unknowing foresight, that icebergs afloat on the great floodwater could have carried a cargo of rocks frozen into them to be released upon melting. But the fanciful notion that immense sheets of ice now long gone had the capacity to transport loads of debris and that Noah was not in the act simply didn't enter anyone's mind.

Swiss farmers live and function daily cheek-by-jowl with the glaciers. To them it came as no surprise that ice moves forward and then retreats, leaving erratics, piles of unsorted rocks and soil, and scratched and polished valley walls, as evidence of its passing. But simple, everyday peasant wisdom had somehow eluded the famous scientists who were only now beginning to free themselves from long-ingrained, traditional explanations of perplexing problems in nature.

In 1829 Ignatz Venetz, a Swiss bridge engineer, speaking to the Swiss Society of Natural Science, offered the first truly scientific postulation that a continental ice sheet had overrun much of Europe. Venetz had been in the mountains and talked with the inhabitants of glacial valleys. His own followup research convinced him that they knew whereof they spoke, but the Swiss Society was not impressed. Neither was it five years later when amateur naturalist Jean de Charpentier presented a somewhat refined version of the Venetz hypothesis. Throughout history it would seem that nonprofessionals have always found the "experts" a difficult audience to convince.

Nonetheless, in that audience in 1834 was 27-year-old Louis Agassiz (1807–1873), professor of natural history at the University of Neuchatel and a well-regarded expert on fossil fish and marine biology. Intrigued, he accepted Charpentier's invitation to spend the summer at Bex in the Rhone Valley. The first-hand evidence that Agassiz observed of every type of glacial activity thoroughly convinced him of the rightness of the cause, and he became a vigorous researcher, writer, and apostle.

Agassiz proved to be supremely adept in pursuing research grants, so that his year-round projects were well-funded and staffed. His two books, *Studies in Glaciers* and *Glacial Systems,* were handsomely printed and widely distributed, sparking great interest and considerable controversy. But his theories were supported by a huge volume of impressive research too—for several years his home base was a permanent research station on the edge of Unteraar glacier. The doubters were at last coming around, eventually including even Lyell. It should be remembered that talk of continental glaciers had to be a radical theory in this era since Antarctica had not been discovered; no one even knew that Greenland was capped by a single huge ice sheet.

In 1846, Agassiz arrived in America as a lecturer in the new science of glaciology. Everywhere he looked he found evidence of continental glaciation. At Harvard, where he filled a specially created chair, he was much respected, and his courses and lectures were in great demand. Ultimately, far from his beloved Unteraar, Agassiz's interest in ice behavior waned. Now a permanent American resident, he continued his scientific research but reverted to his old biological interests. During the last decade of his life, still exhibiting great enthusiasm for a cause, he gleefully attacked the new theories of evolution being espoused by Charles Darwin. It is ironic that the early scoffer at the constraints of traditional dogma on true science should end his career as the defender of creationism against new ideas of evolution.

After Agassiz's death in Boston at the age of 66, a 2500-ton (2270-t) erratic, striated and polished by glacial ice, was brought from the Swiss Alps to mark his grave.

America, notably the Adirondacks, the New England uplands, and the northern Appalachians.

As the depth and pressure of the continental glacier increased, the ice was forced outward in all directions.

Numerous major advances and retreats have been traced, extending roughly as far south as the Ohio and Missouri rivers in the United States (Fig. 21-13) and southern England to central Germany and the Ukraine in Europe, a result of fluctuating temperatures over the centuries. On the whole, however, the behavior of these continental glaciers and the

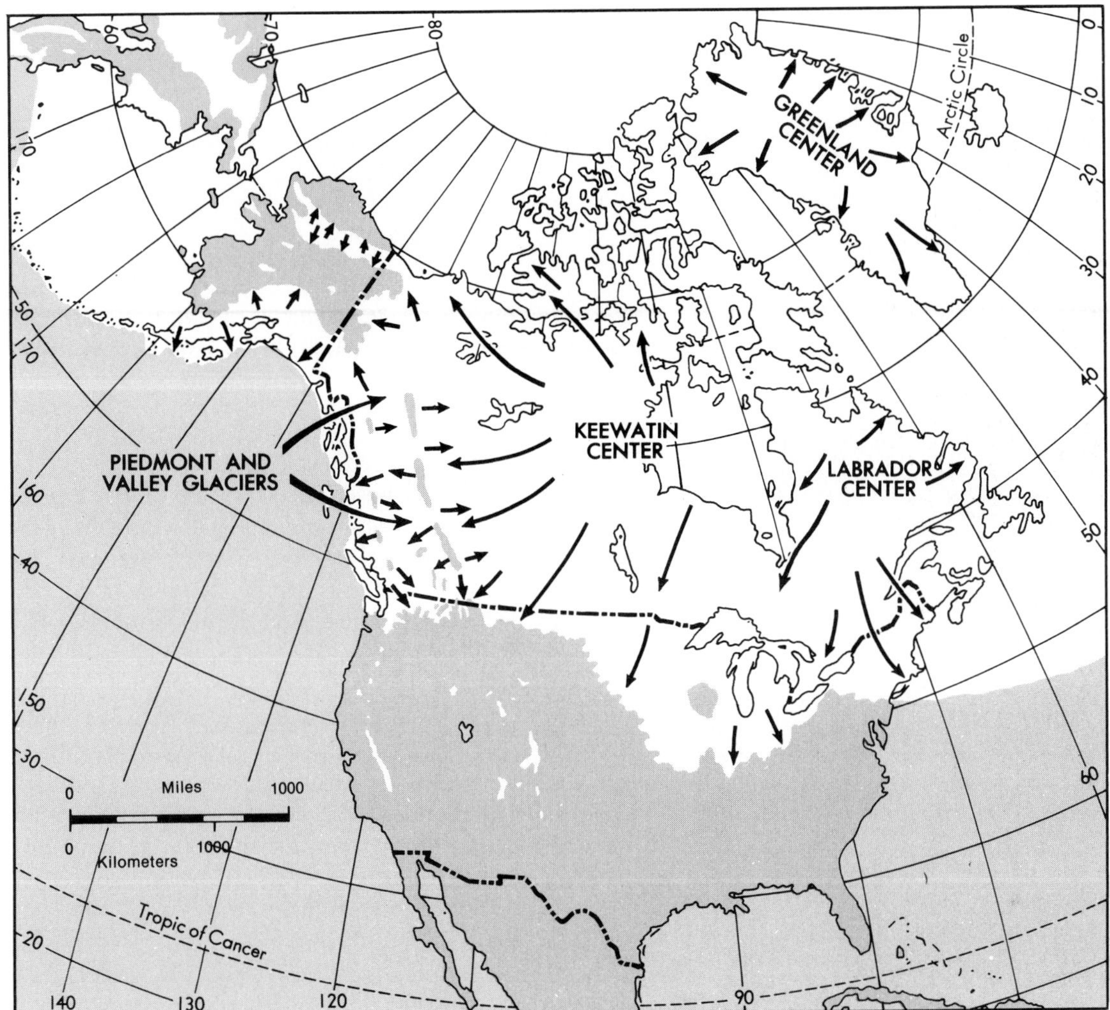

Fig. 21-13. ***Extent of the continental glaciation in North America.***

gradational work they accomplished were similar to those of ice in alpine glaciers. The one difference was that the scope of the continental glaciers' influence was much larger because they were not confined to limiting valleys.

By carefully examining the countryside over which the continental glaciers must have swept again and again, we may establish two somewhat differing zones: the zone of active erosion and the zone of active deposition.

Erosion

The *zone of active erosion* occupied an inner area extending a few hundred miles out in all directions from the center or centers of ice origin. Here the ice was thickest as it passed over the land and therefore exerted a great deal of vertical pressure. The underside was fresh, clean, and relatively unclogged with soils and rocks, so that, as a youthful stream, it had a great capacity for freezing materials into it.

As we observe the surface today, it is obvious that the glacier not only removed soil right down to bedrock, but also gouged and abraded the rock itself, cutting out many deep cavities that have since become lakes (the Great Lakes, for example). There is a great deal of similarity in the general aspect of landscape in northern Minnesota, Michigan, Wisconsin, and eastern Canada and that of Finland and northern Russia. The thousands of lakes at varying levels,

Fig. 21-14. ***Glaciated landscape.*** *Parts of coastal Greenland are no longer under the ice sheet, but the glacial grinding, cutting, and polishing of the recent past is evident in today's landscape.*

scrambled drainage, and great exposure of bedrock are typical everywhere. The uplands overrun by such ice sheets are rounded, striated, and polished (Fig. 21-14).

Deposition

The second zone is beyond the first, far out from the center of activation, and especially near the southern edge of the glacier. Here the ice was thinner, melting much more rapidly, and heavily choked with the waste it had accumulated in passing over the erosional zone. This is the *zone of active deposition.* Like the old-age stream, the ice was not only overloaded but had increasingly less capacity for picking up and carrying material. Here, too, the leading edge of the glacier had reached a point of equilibrium between melt and advance, and even minor temperature changes could cause violent fluctuations in the position of the ice front. Extensive moraine belts, great loops extending for hundreds of miles, occur throughout the zone of deposition and indicate temporary pauses and readjustments of the glacial margin. In front of each moraine is the alluvial apron laid down by meltwater, here called the *outwash plain* (similar to the valley train except that it is much more extensive). Throughout the region there are great depths of ice-compacted till effectively masking the original terrain details. This is the history of the North German Plain and much of the American Middle West.

Extensive moraine belts occur throughout the zone of deposition and indicate temporary pauses and readjustments of the glacial margin.

People often ask if continental glaciation has helped humans use the land. To those who occupy the zone of deposition, the answer is probably "yes." Deep new soils and generally subdued landscape have helped make this region one of productive farms. The soils may be too sandy or gravelly in places and clayey in others, and disrupted drainage has caused considerable swamp development. On the whole, however, such soils can be made productive with proper management. An excellent comparison with "what might have been" is provided by the *driftless area* of southwestern Wisconsin where, for unknown reasons, the glacial ice completely surrounded but did not move over several thousand square miles.[1] Here is a sample of what the Middle West might have been like without glaciation—rough hill country with thin soil, except in the river bottoms, and generally unproductive.

In the zone of erosion, glacial action has ruled out agriculture by removing the soil, but at these northerly latitudes, agriculture could only be marginal at best. In laying bare the bedrock, the glacier exposed a good many mineral deposits at the surface, the disorganized drainage has produced hydroelectric sites, and the thousands of lakes make a fisher's paradise. Even here glaciation might be interpreted as a blessing.

SECONDARY EFFECTS OF GLACIATION

Continental glaciation also has some secondary effects beyond the direct effects of the passage of ice over land. The general lowering of the world's sea level has already been mentioned, and it is obvious that the coastlines everywhere

[1] Drift is a general term referring to all types of glacial deposition; so "driftless area" denotes no drift and thus no glaciation.

Fig. 21.15. ***Bed of former glacial lake.*** *Ancient Lake Agassiz once covered all this land near Portage la Prairie, Manitoba, and much more. We need considerable imagination to visualize this flat and fertile plain at the bottom of a lake of glacial meltwater.*

have felt the repercussions of such an effect. For instance, the movements of animals and humans across land bridges that must have existed in Southeast Asia and the Bering Strait are related to the glacial lowering of the sea.

The movements of animals and humans across land bridges that must have existed in Southeast Asia and the Bering Strait are related to the glacial lowering of the sea.

Another consideration is the great volume of meltwater released on the land. Existing rivers must have carried more water than they do today and were more effective eroders. In many places, enclosed basins such as the one occupied by the Great Salt Lake were filled to the brim with huge lakes. Old beach lines of ancient ***Lake Bonneville*** are easily visible on the slopes of the Wasatch Mountains near Salt Lake City, and the salt deposits of the evaporated lake are evident in the well-known Bonneville salt flats. Lake Agassiz was formed when normal discharge to the north was blocked by retreating ice, and, until the glacier disappeared, the lake maintained itself (Fig. 21-12). Today only Lake Winnipeg remains as a tiny remnant of the original (Fig. 21-15). Ice also blocked the northeast drainage of the Great Lakes via the Saint Lawrence River and caused them first to drain southward via the Illinois River to the Mississippi and later through the Mohawk Gap and the Hudson River (Fig. 21-12). The only good break through the Appalachians occurs here, formed by a tremendous stream of meltwater. The Erie Canal, so influential in our history, was virtually foreordained by this great channel. The present courses of the Ohio and Missouri rivers, which drain much of the central United States from the Appalachians to the Rocky Mountains, generally coincide with the southernmost extent of the continental ice sheets. Because of the disruption and diversion of drainage and the deposition by the glacial ice, these large rivers developed at the margin of greatest glacial advance.

Equally important in the Pacific Northwest is the Grand Coulee, a drainage channel dug out by the ice-diverted Columbia River, carrying many times the volume of water it does today (Fig. 21-16). After the ice melted back, the Columbia abandoned the coulee and resumed its normal course, but the Grand Coulee Dam at the head of the coulee can now direct Columbia River water once again down this ready-made channel to irrigate the desert (Fig. 21-17).

*Fig. 21-16. **Location of the Grand Coulee and Grand Coulee Dam.** During the Ice Age, ice diverted the Columbia River and its tributaries at many points. The Grand Coulee was one of the major disruptions of flow. The Columbia River returned to its former channel after ice receded.*

THEORIES OF THE ICE AGE

What caused the cooling that allowed massive accumulations of ice to advance over the earth's surface at various times throughout prehistory? Perhaps in the distant geologic past shifting continents and the opening of oceans affected the climate enough to bring on glaciation over parts of the continents. What about the most recent "Ice Age"—the Pleistocene—which is believed to have begun about 2 or 3 million years ago? During this relatively brief epoch, there is evidence that dozens of cold periods alternated with warm periods: glaciers formed, advanced, and then receded, leaving a gouged, scraped, and debris-littered landscape behind, primarily in middle and high latitudes and higher mountains of the lower latitudes. Several theories have been advanced to explain this phenomenon:

1. Meteors or asteroids may have collided with the earth, sending up into the atmosphere large amounts of suspended particles that blocked the sun's energy long enough to cool the climate and for ice sheets and mountain glaciers to form and do their work. Extensive and long-term vulcanism may have produced atmospheric ash that reflected enough solar radiation to have a similar effect.
2. The sun's emitted energy may have varied enough to change the earth's climate.
3. The eccentricity of the earth's orbit around the sun may have caused variations in climate that allowed ice to accumulate.
4. Continental plate movement and related changes in interoceanic circulation may have influenced climate and the amount of moisture available for the accumulation of glacial ice.

None of these explanations has emerged as the best. There is little correlation of meteor impact or volcanic events with Pleistocene glaciation, and data on solar energy output are inadequate to explain glacial advance and recession.

There is some acceptance of the third theory, earth's orbital and rotational eccentricities, as a cause. As pointed out in Chapter 2, the earth's orbit around the sun is not perfectly circular but elliptical. The earth is closer to the sun at certain times of the year than it is at other times. On a cyclical basis, the ellipse shifts its position around the sun. Some scientists maintain that this noncircular, shifting orbit and the irregularity in the earth's spinning on its axis (it wobbles) combine to cause variation in earth–sun relationships, which in turn influences global climate.

Some researchers believe that changing patterns in oceanic circulation triggered by continental movement induced a cycle of ice accumulation and melt.

Advocates of the fourth theory believe that changing patterns in oceanic circulation triggered by continental movement induced a cycle of ice accumulation and melt. In the tens of millions of years before the beginning of the Pleistocene, the continents approached their present positions. The Arctic Ocean was nearly cut off from the Pacific, and openings to the Atlantic remained modest, restricting the flow between the large warmer oceans and the polar ocean. Initially, the Arctic Ocean provided the moisture for continental glaciers on the northern land masses. As ice sheets grew on land, however, sea level lowered, further restricting interoceanic circulation. Arctic water temperatures cooled, providing less water vapor for precipitation on land. Eventually, the Arctic Ocean froze over. As glacial ice that had moved equatorward over the continents receded from lack of snow supply, sea level rose and Arctic–Atlantic circula-

Fig. 21-17. ***Looking south down the Grand Coulee.*** *The contemporary Columbia River flows away to the right, but an ice-blocked, meltwater-swollen Columbia once surged southward, cutting a massive channel. Water in the Coulee bottom today is stored for irrigation.*

tion was restored, reducing the Arctic ice cover. Then, according to this theory, a new cycle of glaciation would begin as snow and ice accumulated again on the continents.

No explanation for the Ice Age is certain then. But regardless of why widespread glacial ice developed, we do have abundant evidence that it did a tremendous amount of work in forging many of the dramatic landforms we see today. We should note that many geologists and climatologists believe we are experiencing only one in a series of warmer periods that have occurred over the past 3 million years and that episodes of glaciation will probably continue to occur over the next few hundred thousand years.

CONCLUSION

Through its gradational effects, moving ice is a powerful earth-shaping agent. Over a long period of time, meteoric impact, changes in oceanic circulation triggered by shifts in continental position, the shifting, eccentric pattern of earth's motions about the sun, or variations in the sun's energy caused periodic cooling of the global climate, and snow accumulated, and subsequently ice, as summers failed to melt the seasonal cover. We have insufficient information to be conclusive about which of these theories best accounts for the Ice Age. Today, many scientists believe that we are headed for a short-term period of global warming caused principally by our input of carbon dioxide into the atmosphere. Others believe we are in an interglacial period of relative warmth that is part of the ongoing Ice Age cycle and that we will soon, perhaps in a few thousand years, be back in a cold period of advancing glaciers. From a geographical viewpoint, we recognize that much of the fertility of the plains of the midlatitudes, the shape of lakes and courses of many streams, and the scenery of the mountains and subarctic regions are a consequence of glaciation in the not so distant past.

KEY TERMS

Pleistocene
valley glaciers
continental glaciers
plucking
cirque
crevasses
striations
hanging valleys
tarn
fjords
terminal moraine
recessional moraines
lateral moraines
moraine lake
valley train
till
Lake Agassiz
erratics
zone of active erosion
zone of active deposition
outwash plain
driftless area
Lake Bonneville

REVIEW QUESTIONS

1. Under what conditions does glacial ice form? Why is glacial ice able to flow?

2. Make a list of landforms resulting from each of the following: (1) valley glacier erosion, (2) valley glacier deposition, (3) continental glacier erosion, and (4) continental glacier deposition.

3. How does a cirque develop? How can a cirque be hundreds of feet deep if the ice did not fill it completely?

4. How are fjords similar to ice-carved valleys such as Yosemite? Why? How and why are they different?

5. Why is it important to understand that a glacier always moves forward in order to understand how it deposits its sediments? Give examples of landforms that point up the relationship between forward motion and depositional pattern.

6. How can you tell the difference between a valley train and moraines in a glacial valley?

7. What is the significance of the so-called driftless area in Wisconsin?

8. Where was Lake Bonneville located, and why was it formed?

9. Name and describe at least three theories that attempt to explain climate change that brought on the glacial advances of the Pleistocene epoch.

APPLICATION QUESTIONS

1. If you lived in the middle of the North American continent 25,000 years ago in an area just south of the continental ice sheet, what would the climate, vegetation, and soils have been like?

2. Do you live in an area that was once covered by a glacier? If so, how can you tell? If not, what would your area be like if glacial ice had scoured it?

3. Find the name of large towns located in the Wisconsin "driftless area." Is any one of these towns noted for anything that might be related to the fact that it is in the unglaciated area?

4. Assume that global temperatures will cool over the next few thousand years to levels that would trigger the onset of another glacial advance. Describe what you think might be the impact on human occupation of the midlatitudes if this were to happen.

C H A P T E R 2 2

WIND AND AEOLIAN LANDFORMS

OBJECTIVES

After studying this chapter, you will understand

1. How wind can create landforms by moving large amounts of dust, silt, and sand.
2. Why many areas of the world are covered with wind-deposited material.
3. How sand dunes are formed and classified.
4. How deserts are classified.

Desert dunes. The desert and the beach are the great arenas for wind-motivated sand, which is an effective eroding agent, although each grain is too heavy to stay airborne for very long. The relentless desert dunes in this photograph are constantly in motion.

INTRODUCTION

Aeolus is the Latin name of the ancient Greek god of the winds, and so we often use the term ***aeolian*** (sometimes spelled eolian) to describe landforms that develop as the result of wind erosion or deposition. The dusty streamer trailing out behind a car driven along an unpaved road, the rising cloud above a tractor plowing a field, or the swirl of dust across a baseball infield are everyday evidence that wind transports minuscule bits of rock and soil. But can it move significant quantities? After all, the huge boulder, which the glacier, the raging torrent, or the storm-driven wave can handle with impressive ease, is obviously beyond the capacity of even tornadic winds to so much as jostle. Small pebbles edged along the ground are ordinarily the largest objects that strong winds can move.

One generally accepted estimate by A. K. Lobeck gives us a statistical base with which to work. He calculated that the average dust storm suspended one ounce (28.35 g) of solids in each 10 cubic feet (0.28 m^3) of air—not a very imposing figure. But take some of these 10-foot (0.28-m) cubes and stack them up for several hundred feet, then extend them along a front of 300 miles (483 km), and finally pile them in depth for another couple of hundred miles (322 km) to cover the area involved in a typical dust storm. In other words, form a three-dimensional air mass 300 feet (91 m) high, 300 miles (483 km) long, and 200 miles (322 km) wide. Add a shallow bottom stratum of much heavier sand so that now we are talking in terms of tens or even hundreds of millions of tons (or metric tons) of suspended solids moving along at a rapid rate. This is one average dust storm. With the element of time included, multiply this many times and we have an impressive operation (Fig. 22-1). Any medium that can transfer such massive quantities of earth materials simply has to be classified as a gradational agent.

Although not proscribed by gravity in the same way as are rivers and glaciers, there are limitations beyond mere size of particle to what moving air can accomplish in the realm of gradation. To be an effective agent, wind must work on loose, dry surface sands protected by a minimal vegetation cover. This means that, to a large extent, wind erosion is limited to the arid regions of the world where vegetation is scarce to nonexistent and the lack of moisture in the soil keeps the particles from adhering to one another. Yet, even in humid areas beach sands can dry out rapidly above the high-tide line and thus be subject to wind attack. In addition, it has been demonstrated in several parts of the world that the fine material in glacial debris or alluvial deposits may be removed by wind if its velocity and the vegetation and moisture conditions allow. Nevertheless, all things considered,

Fig. 22-1. ***Dust storm.*** *This photograph was taken in semiarid western Queensland, Australia, in 1965 during one of its periodic droughts. Government soil conservationists, running tests during this dust storm, estimated that 50,000 square miles (80,465 km^2) of countryside were involved and that over a billion tons (910 million metric tons) of soil were taken aloft.*

wind may be the most widespread gradational force with which we are dealing, but one of the least effective in producing landforms of any extent.

Wind erosion is largely limited to the arid regions of the world where vegetation is scarce to nonexistent and the lack of moisture in the soil keeps the particles from adhering to one another.

WIND EROSION

The carrying capacity of wind, much like that of running water, is determined by its velocity. Almost any slight breeze can sweep up dust and fine particles, but it takes a steady wind of at least moderate velocity to move sand. Once these sharp-edged particles are in suspension, the moving air has armed itself with effective cutting tools to wear away anything in its path. This supplies it with even more materials. The original load may have come from mechanically weathered rock or the loosely deposited alluvium of an intermittent desert stream. Their removal lowers the surface, often exposing the bedrock, which is polished and scoured by the wind-blown materials.

If bedrock is hard and massive, it resists abrasion and will acquire a high polish, but if it is softer, it will be cut away bit by bit. Sandstone is often exposed on the desert floor and in cliffs, its hard sand grains cemented with softer material. As the cementation is eroded, the sharp-edged grains are freed to aid in cutting. On the other hand, when sizable stones and pebbles are mixed with finer material on the desert floor, the wind erodes selectively and removes only particles up to sand size. Under these circumstances, the desert surface eventually becomes paved with a mosaic of cobbles left behind by the wind.

When sizable stones and pebbles are mixed with finer material on the desert floor, the wind erodes selectively and removes only particles up to sand size.

Any surface exposed to the constant sand-blast effect of wind-carried debris, especially sand, will inevitably be cut, chewed, and polished (Fig. 22-2). Even automobiles caught in sandstorms have their paint removed, metal surfaces pitted, and glass frosted in a remarkably short time. Rocky cliffs facing prevailing winds in the desert display tremendous honeycombing as the softer elements are etched away, leaving the hard materials in high relief. Among the peculiar features of some desert landscapes often attributed to wind erosion are top-heavy balanced pinnacles. Whether wind is responsible for the original pinnacle is questionable, but the more rapid cutting at the bottom is typical of wind action, for sand, the most effective cutting tool, is relatively heavy and seldom suspended more than a foot off the ground. You

Fig. 22-2. ***Angel Arch, Utah.*** *In addition to major mechanical and chemical weathering, the processes involved in the formation of Angel Arch included pitting, polishing, and abrading of rock strata of varying resistance by wind-transported material. The landform began as the result of vertical faulting of sedimentary rock, but the arch's current smoothed outline has almost certainly been influenced by natural sandblasting.*

may have noticed that telephone poles in dry country will have either metal shields at ground level or a pile of rocks to protect them from being cut down by sand-laden wind over time.

Although hard and generally resistant to wear, the individual grains involved in this continuous sand blasting eventually lose their sharp corners and become rounded. Beach and riverine sands, though also used as cutting tools, do not display quite this degree of wear. In fact, a trained eye can often distinguish between ancient desert sand deposits and those of the beach on this basis.

As a transporting agent, wind may carry tiny dust particles high into the atmosphere and transport them great distances. Colored sunsets and muddy rainfalls often occur thousands of miles from major dust storms or explosive volcanic eruptions. The plowing of the prairie grass east of the Rocky Mountains followed by an extended drought gave rise to the extensive "dust bowl" of the 1930s, when prevailing westerly winds carried the fine-grained topsoil eastward in great clouds. Much the same experience has been endured along the southeastern semiarid margins of the Australian desert and in the so-called virgin lands of southcentral Asia.

Deflation

Wind does not always move over the surface of these arid regions from a single direction or at a constant rate. Often it is gusty and capricious, and the shape of the surface terrain sets in motion channeled air streams and eddy currents. For instance, a swirling eddy in the lee of a minor ridge may cut downward and dig out a sizable basin. This general digging ability of swirling wind currents is called ***deflation***. Once the basin is begun, its conformation encourages stronger eddy currents, and, if the subsurface material being worked is loose sand, soil, or easily eroded rock, huge basins of great depth may be excavated.

A swirling eddy in the lee of a minor ridge may cut downward and dig out a sizable basin.

Channeled Air Currents

In addition to directing its eddy currents downward on the lee side of ridges, wind moving across broken terrain has other effects as well. Canyons, slots, and narrow gorges capture the moving air and funnel it down the chute. Such channeled winds blow constantly from the same direction and move at high speeds because they are squeezed through restricted openings. The scouring and polishing of both the floor and the walls of such landforms by debris-carrying high-velocity winds are strikingly evident.

WIND DEPOSITION

Dry, fine material carried from deserts or dried-up glacial outwash plains or stream flood plains by prevailing air currents is inclined to drift like snow against obstacles in its path and to fill in any low spots. The extensive outwash plains left after the last glacial recession are major source regions for transportable material. Silt-sized particles (between clay and sand size, see Fig. 24-4) are most readily picked up and transported by winds and are collectively known as ***loess***. Certain regions around the world have had large loess depositions:

1. Northern China, where prevailing winds over the past few hundred thousand years have deposited material picked up from the deserts and drying outwash plains of eastcentral Asia.
2. The coastal areas of southern France and northern Italy, where silt dust has been brought in by seasonal winds from the Sahara over the past few hundred years.
3. Central North America, where wind erosion and deposition from dried up lake and stream beds were associated with receding glaciers.
4. Central and Eastern Europe, where materials were transported and deposited as the Scandinavian ice sheet disappeared.

Today these areas are noted for the richness of their soils, for loess provides nutrients and develops a soil that is easily cultivated. The one drawback is that this soil is easily eroded by running water, or again by wind. The Yellow River (Hwang Ho) and its tributaries, as noted in Chapter 20, carry great amounts of sediment eroded from the loess deposits of northern China.

Sand Dunes

Wind requires higher velocity to take into suspension particles, especially sand, that are larger than those that make up simple loess deposits. In open desert terrains and on beaches, strong winds move dry sand short distances along the surface. Over time, sand masses pile up, are moved slowly along the ground, and drift up and around obstacles. The result is the formation of dunes on beaches and in

DEFLATION IN THE SAHARA

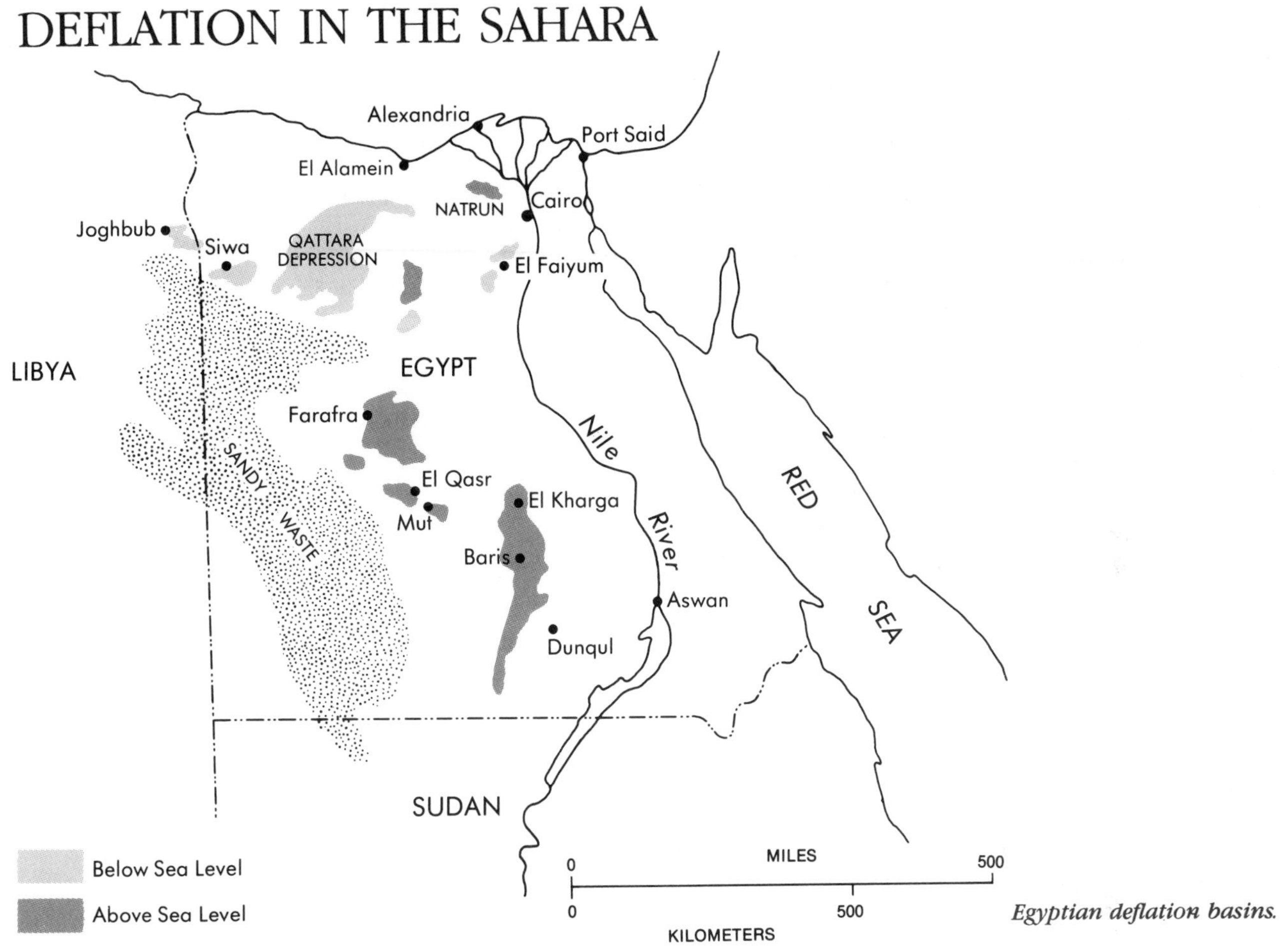

Egyptian deflation basins.

Perhaps the largest deflation basin in the world is the Qattara depression in the Egyptian Sahara west of the Nile—some 185 miles (298 km) long, 75 to 100 miles (121 to 161 km) wide, and 420 feet (128 m) below sea level. Nobody suggests that the wind alone cut out this huge cavity. Surface runoff during desert storms probably had a hand in originally breaching the rock, but the fact that the bottom is below sea level is significant. Intermittent streams flowing to the sea cannot erode below that base level. And it is definitely not a structural feature. Clearly, then, the wind is the basic factor. It worked the surface downward until it reached the groundwater table. The entire bottom of the basin is an extensive salt marsh and has stabilized at this level since the wind cannot remove the damp soil. But it is a repellent region to humans; the marsh is so treacherous that in World War II the British employed the Qattara as an effective defensive barrier against Rommel's Afrika Corps.

Today Egyptian authorities have a tentative plan to channel Mediterranean water from El Alamein, less than 50 miles (80 km) away, into the depression. In theory the plunging stream would activate turbines for power generation, and the huge new inland sea would raise water tables, change the local climate, and in general rejuvenate its Saharan neighborhood.

The Qattara is not the only wind-excavated depression in this part of the Sahara; a whole family of lesser depressions may be found in the immediate vicinity. Most of them are oases, some below sea level and some above, but nearly all are at a groundwater base level and therefore allow flowing springs, seepages, and wells. They nourish life and limit further deepening by wind action. To the south and west of this extensive region of deflation is a large area of sand dunes and ridges that are at least partly the product of the material blown out of the basins upwind.

Fig. 22-3. ***Sand ripples.*** *On the surface of massive wind-driven dunes to myriad ripples, each a tiny dune unto itself.*

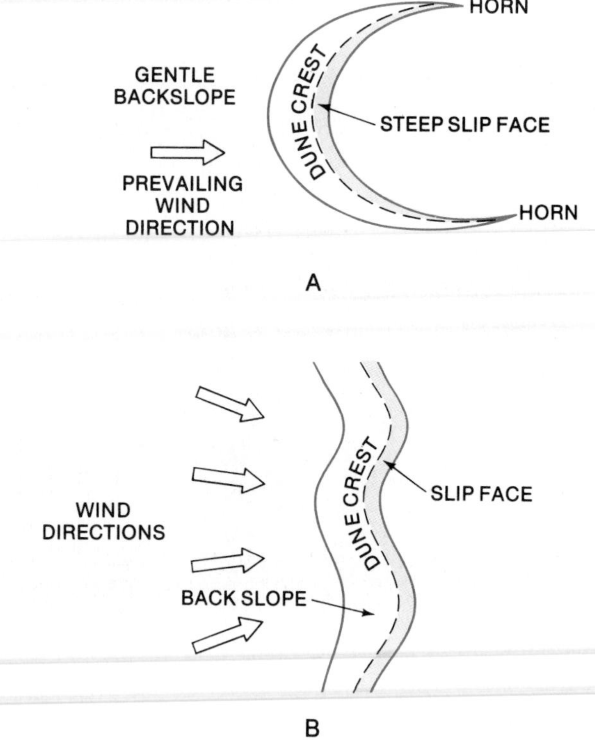

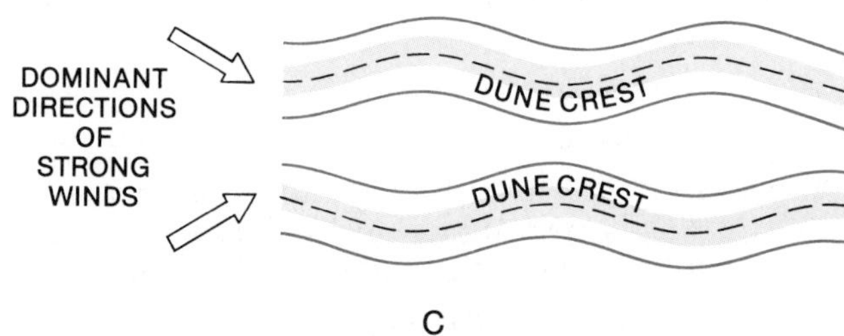

Fig. 22-4. ***A map view of three general types of sand dunes.*** *The form of sand dunes is affected by the velocity of the prevailing wind, the amount of sand available, and the constancy of wind direction.*

deserts where there is available sand. Sand dunes are moderate-sized landforms that may be hundreds of yards (meters) wide and up to several miles (kilometers) long. Yet over the surface of most dunes tiny rows of ripples form at right angles to the wind, separated by only a few inches (centimeters) (Fig. 22-3). Ripples migrate as the wind-driven sand particles at the dune surface sail and skip from one ripple to the next.

In open desert terrains and on beaches, strong winds move dry sand short distances along the surface.

If the wind prevails year-round from one direction and the sand supply is only moderate, the dunes are likely to take on a crescent shape, with the convex exposure upwind and rather gently sloped, and the concave exposure downwind and very steeply sloped (Fig. 22-4A). These individual crescent-shaped dunes have their horns pointed downwind and are called ***barchan dunes.*** As long as the sand on the windward side of the dune is not stabilized by vegetation, it is free to be transported up the gentle ***back slope,*** and the entire dune migrates slowly forward as the wind dumps sand, grain by grain, over the dune's crest. As each grain falls into the lee of the dune, it drops down the abrupt ***slip face,*** which is a much steeper slope.

If the wind displays less constancy of direction and intensity and there is a large supply of sand, dunes will typically take on a more ridgelike form. These dunes are classified as ***transverse dunes*** because their linear crests form at more or less right angles to the general direction of the wind (Fig. 22-4B). However, these long-ridged dunes normally retain the back slope and slip face that are characteristic of the

THE WAIKATO DUNES

Monterey pine on the dunes.

At the mouth of the Waikato River, 25 miles (40 km) south of Auckland, New Zealand, are the billowing Waiuku sands. Energized by boisterous winds off the stormy Tasman Sea, these actively migrating beach dunes have threatened to engulf the nearby countryside as they march steadily inland. But since 1940 experiments with planting the California Monterey pine (*Pinus radiata*) have proved successful in controlling dune migration. Through the years a mature pine forest has developed on the beach dunes. This forest (1) stabilizes sand movement, (2) produces a continuing supply of merchantable saw timber, and (3) provides the harried urban vacationer a pleasant quasi-wilderness recreation site.

Currently, the Waiuku sands support a sizable forest of mature trees that are being cut and replanted in a controlled sustained-yield system. But today the planners are accommodating their forestry and dune-stabilizing efforts to yet another complication—the sands of Waiuku are titanium-rich ironsands suitable for use as commercial ore. A few years ago an integrated steel mill was established just 8 miles (13 km) distant. So the forest is in the midst of a mineral as well as a wood-products development.

In order to use this dual resource efficiently, planners have divided the rich region into 250-acre (101 ha) blocks for clearing and mining one at a time. Mature timber goes to the mill, and the ironsand goes to a separator for preparation of blast furnace charges. The process is a simple scrubbing with water, followed by magnetic separation of the useful titano-magnetite. All residue, which makes up a very high proportion of the initial product, is returned to the quarry site and a new forest is begun. So a sustained-yield, block-cutting, lumbering regime has been initiated, complicated only slightly by the short-term borrowing of the soil so that some of its useful elements may be removed. Curiously, preliminary research tends to indicate that the trees grow slightly better on the processed sands than on the original.

---

barchan dune: the back slope on the windward side and the slip face on the leeward side.

In sandy regions where winds are strong and somewhat variable, yet another general type of dune will develop. These are the ***longitudinal dunes,*** and, as the name suggests, they form nearly parallel to the prevailing winds (Fig. 22-4C). In this case, sand is pushed up into a sharp crest from one side of the dune and then the other as the wind shifts through a narrow directional angle. There is no distinctive back slope or slip face, although the slope of the longitudinal dune tends to be steeper near the crest than at the base.

These are generalized categories; in reality, there are many variations in dune form and process. Because not all dune-forming processes are well understood, research continues on the effects of wind erosion and deposition in regions where sand and other wind-blown materials are a major part of the landscape. Along many beaches and desert margins moving sand dunes have been known to march relentlessly across virtually anything in their path. Roads, oases, even woods and buildings have been overrun (Fig. 22-5). The dunes can be stopped if some kind of vegetation is encouraged to grow on their surfaces, but sterile sand is not an ideal bedding ground for plants, and dunes are con-

Moving sand dunes have been known to march relentlessly across virtually anything in their path.

Fig. 22-5. ***Wind-billowed sand at the Souf Oasis in Algeria.*** *Groundwater is at the base of the sand so that each date grove must be excavated to this point. Sand is not necessarily beneficial to palms as the usual desert photo might imply, but given no frost, great heat at the crown, and plenty of water at the root, date palms will prosper. But danger lurks as the dunes relentlessly advance, and cultivators labor constantly to keep the groves clear.*

stantly being formed (Fig. 22-6). In the Landes district of southern France, dunes from the Bay of Biscay shore have eventually been stabilized after overrunning hundreds of square miles of territory and even a sizable village. Today, planted pine trees in the old dune region produce valuable timber products.

Types of Deserts

It is a common misconception that most of the world's deserts are made up of endless miles of billowing sand. Such deserts do exist and are admittedly spectacular, but they are not common enough to be typical. For sand dunes to form, the supply of sand must develop more rapidly than it can be blown away—for example, along a seashore or in an area where soft sandstone erodes readily (Fig. 22-7). The conditions for sand dune development occur only here and there, so that sparsely vegetated rocky deserts are more common than sandy deserts. Not more than about 20 percent of the world's deserts are sandy.

Not more than about 20 percent of the world's deserts are sandy.

Three types of desert landscapes have been recognized in the Sahara—the largest desert in the world. These landscapes also occur in many other places around the world.

1. ***Hamada***—an upland with a wind-scoured, polished bedrock surface, sometimes thinly layered with coarse material.
2. ***Reg***—a low desert plain of alluvial deposits overlain by a mantle swept clean of fine particles to leave a cover of coarse pebbles and stones.
3. ***Erg***—a wide expanse of deep, shifting sand, usually exhibiting a complex variety of sand dunes; a "sand sea."

The combination of wind and sheetwash erosion of desert surfaces, particularly in reg deserts, sometimes produces a nearly flat surface layer of tightly knit, smoothed, and often cemented coarse material called *desert pavement* (see Chapter 25). Here the eroded surface has lost much of its fine material, and the remaining coarse particles form a residual armorlike cover that protects the underlying mixture of silt, sand, gravel, and cobbles.

CONCLUSION

Wind is an underrated agent of erosion and deposition. Although its influence on landscapes is limited, many distinc-

Fig. 22-6. ***Stabilizing sand.*** *Planting perennial grasses imported from the North Sea coast of Europe on these shifting beach dunes in New Zealand is an attempt to stabilize the dunes or at least slow the migration of sand.*

Fig. 22-7. ***Ancient dunes.*** *Typical horizontal beds of sandstone in the distance are contrasted with the rock formations in the foreground, which are derived from sand dunes of the geologic past, now hardened into sandstone over time. These ancient dunes, now eroding rocks, are the source of sand particles picked up once again by the wind, perhaps eventually to be deposited in another dune.*

tive and somewhat temporary landforms, particularly sand dunes, result from aeolian action. We sometimes take the geomorphological dynamics of wind for granted, but where people occupy areas significantly affected by wind it is an environmental element we should take seriously.

For example, at the margins of deserts periodic drought,

the removal of groundwater and vegetation by humans, and overgrazing by domestic animals have acted together to extend the desert (see Chapter 27, on *desertification*). Wind quickly deflates dry, unvegetated surfaces and transports sand and silt along an advancing front into former grassland and savanna environments at the desert's edge, making it even more difficult to recover these areas.

Coastal areas are also affected by migrating sand. This situation is often aggravated by careless removal of vegetation and construction of roads and structures in or near sandy areas. The burial of roads and houses by sand is common along developed coastlines.

Regions of stabilized loess deposits with good soil development are important for their agricultural productivity and must be protected to prevent loss of surface material through wind and water erosion.

Atmospheric motion, wind, is initially powered by the solar energy received by the earth, which establishes thermal and pressure differentials over the globe to cause wind (see Chapter 5). Wind transfers this energy of motion to water surfaces to form waves. In the next chapter we will explore the processes and landforms that occur along a portion of the surface interface between the lithosphere and hydrosphere—the continental shores.

KEY TERMS

aeolian
deflation
loess
barchan dunes
back slope
slip face
transverse dunes
longitudinal dunes
hamada
reg
erg

REVIEW QUESTIONS

1. How is wind different from and similar to water and ice as an agent of erosion?

2. Under what conditions is wind erosion most likely to occur? Why?

3. How is a deflation basin formed?

4. What are the sources of loess? Why are loess deposits significant?

5. How does the leeward slope of a migrating sand dune differ from the windward slope?

6. How is a barchan dune different from a longitudinal dune?

7. Describe the three major types of desert landscapes you might encounter in the Sahara.

APPLICATION QUESTIONS

1. Get a small amount of clean, dry sand from a local plant nursery, building supply store, or other source. Put the sand on a large table in a suitable location and hold a blowing hair dryer so that the sand begins to be transported across the table. Do you notice any of the processes and forms discussed in this chapter? Suggestion: put a few obstacles in the path of the moving sand.

2. What sort of recreational activities are usually associated with sand dunes, either in the desert or at the seashore? Do you think these activities affect dune landscapes? Why or why not?

C H A P T E R 2 3

WAVES AND COASTAL LANDFORMS

O B J E C T I V E S

After studying this chapter, you will understand

1. How waves develop in the open ocean and how wave energy is expended along shorelines.
2. How shoreline processes produce erosional and depositional landforms on coasts.
3. How coasts are classified by changing sea level, position relative to lithospheric plates, and geomorphic processes.

The active margin of the restless sea. Waves are most effective during storms when they assail the base of sea cliffs with a maximum of energy. Harder rock is left as small islands to be subdued in time; rock debris and sand are deposited to form beaches which are continually rearranged by tide and surf.

INTRODUCTION

Waves can move against an unprotected shore in such massive assaults that the sheer weight of the water is a powerful agent of gradation. Only someone who has seen storm waves battering a cliff front or beach face can fully appreciate their violence. Tons of water repeatedly dashed with great force against even solid rock can accomplish a tremendous amount of erosive work in only a few days or hours—and loose sand offers virtually no resistance. Water also compresses air in cracks, vents, or caves in a cliff. This action, along with the weight of the water, helps to tear loose great chunks of rock. Like running water, seawater can take certain minerals into solution, thereby removing them from rocks and weakening the entire structure. The water is also buoyant: waves can lift massive rocks up to several tons in weight once they have been loosened and either carry them away immediately or jostle them until they gradually work free.

All this can happen without the aid of sands, gravels, and boulders as grinding agents. But often waves do have agents of this kind, some of them very large, derived from the wearing away of shorelines. These greatly multiply the erosive effect of the waves.

Along bedrock coasts that have some relief, the weaker or softer rock is cut away first. Ocean waves, armed with an array of tools from huge boulders to tiny, sharp-edged fragments of silica, can destroy less resistant rock at a rapid rate. Caves, arches, and deeply undercut bluffs result, while harder material is bypassed and left behind. These hard-rock residuals, though a common feature of eroding coasts, are themselves transitional, for they, too, are eventually cut away by vigorous wave action in a short time. Waves perform the bulk of their work during storms, but they are continuously active at other times as well, attempting to lower any land that stands above sea level. Even on a calm day this work is visible, though at a subdued rate.

Waves perform the bulk of their work during storms, but they are continuously active at other times as well.

Low-lying islands and coastlines are subject to continual wave action that often has far-reaching effects. A strange example is Falcon Island in the Southwest Pacific Tonga archipelago. First discovered as a tiny islet in 1885, it disappeared in 1894 only to reappear successively in 1896, 1909, and 1927. Each time it reared its head above the sea, waves reduced it to a mere reef, the last time in 1949. Born of vulcanism and destroyed by the surf, this island provides a good, if not typical, instance of a tectonic/gradational standoff. Other classic cases are found in northwest Europe, where the retreat of inhabited island coastlines has been carefully recorded from Roman times. The north York coast of England has worn back as much as 3 miles (5 km), flooding several towns and villages and destroying many square miles of productive farmland. Out in the stormy North Sea is the island of Heligoland. A tiny fraction of its former self today, it would have long since disappeared except for its military value which led to the reinforcement of its margins by cement breakwaters and seawalls (Fig. 23-1).

Fig. 23-1. ***Heligoland Island.*** *Known since early Roman times, Heligoland was once a sizable island. We have a long record of its accelerated decline to the current area of less than 42 acres (17 hectares). Today, its tortured red sandstone cliffs continue to be attacked by North Sea waves in the few places where protective devices have yet to be installed.*

WAVE ACTION AT SEA

The waters of the sea respond to the gravitational pull of the sun and moon (tides), salinity and temperature differences (upwelling and convectional currents), Coriolis effect (surface currents), sea bottom seismic and volcanic activity (*tsunamis*), and wind (waves). Friction and drag from the last of these, wind, transfer kinetic energy to the water. This energy manifests itself in waves.

Friction and drag from the wind bring kinetic energy to the water, an energy that manifests itself in waves.

Because the ocean contains a huge liquid mass, energy applied on one side is readily transmitted via waves to the other. The high-wind energy from a Caribbean hurricane, for example, is translated into a frenzied sea. A few days later that same energy has traveled thousands of miles to be expended in the crashing surf against the beaches of Scotland, Brazil, and Morocco.

The waves that move out of a storm region at sea are no longer the short-lived, wind-whipped white caps and irregular, violent tongues of water seen at the storm's center. They have become smoother, evenly patterned waves that, because of an increased distance between their crests, are nearly invisible to the casual observer at sea (Fig. 23-2A). These ***swells*** move rapidly across the ocean, carrying energy destined for the margins of land. Swells are large, long open-ocean waves, with a regular spacing pattern and usually a wave height of a few feet (1 or 2 meters), as Figure 23-2B shows. Smaller waves, the ones we usually see jostling against one another across the water's surface, are more or less superimposed on the swells. Generated by localized winds, their energy is quickly used up or added to the larger waves.

There are, then, two general types of waves: (1) the large and predictable great ocean swells and (2) the many disorganized, choppy, smaller waves atop the swells. These wave forms may not exist at all places at all times, but the pattern is common.

The water is not going far as a result of wave action; rather, it is the energy that is traveling.

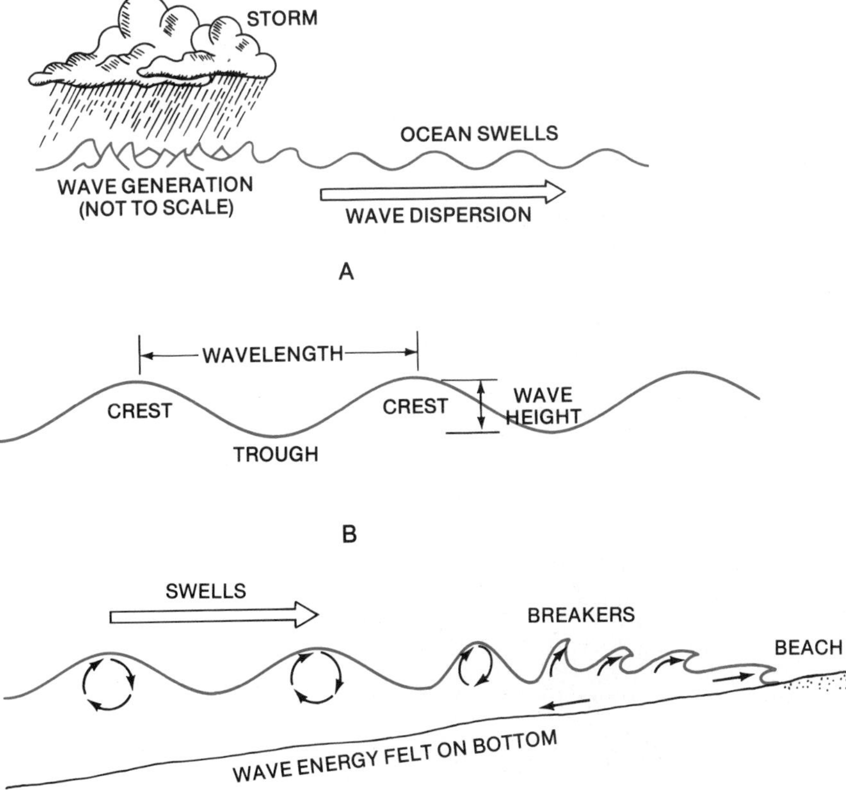

Fig. 23-2. ***Ocean swells.*** *(A) The generation of ocean swells that disperse across the ocean from a turbulent storm center. (B) Measurement of waves is based on wave length and wave height. Typically, swells are much greater in length than in height; therefore they are hardly noticeable to a ship at sea that is slowly raised and lowered by the passing swell. (C) In the open ocean, water moves in a circular fashion as swells pass by. As swells approach shore, the wave energy is felt along the bottom, and the circular motion is disrupted. Wave length becomes shorter, the waves steepen, and breakers form, sending water crashing onto the beach.*

Whatever type of wave we may see, however, it is important to know that the water is not going far as a result of wave action; rather, it is the energy that is traveling. Waves certainly appear to make the water move horizontally, but, for the most part, the water is merely sloshing up and down as waves pass. Waves are often compared to tall grass rippling in the wind; there is motion but no forward progress. The wave form (energy) moves swiftly but not the wave medium (water).

SHORELINE PROCESSES

In the open ocean, any given particle of water will move in a circle, traveling roughly as deeply below the median sea level as the waves are above it and eventually returning to its original point of departure. But when wave forms encounter a shoaling bottom, the water cannot circulate easily and a frictional drag sets in. The waves show they are "feeling bottom" by shortening up or crowding as the forward ones are slowed and those coming up behind step on their heels. This action causes the waves to become oversteepened, and they form breakers (Figs. 23-2C and 23-3).

There is also a strong tendency for bottom drag to pull the wave fronts around to something like a parallel approach to the shoreline, no matter how broken that shore may be. This is called ***wave refraction***. (As with light through a prism, refraction simply means a deviation from a straight line.) On headlands and irregular coastlines, wave refraction directs the energy of the waves against the points of land; that makes erosion more intense than on straight stretches of shoreline. In embayments wave action is quieter, and deposition is the dominant process, creating some fine isolated beaches (Fig. 23-4). In the long run, this process tends to straighten a shoreline.

Beyond the breakers, where frictional drag and wave refraction begin, the initial erosion of the shore occurs. The

Bottom drag tends to pull the wave fronts around to something like a parallel approach to the shoreline, no matter how broken that shore may be.

Fig. 23-3. ***Breakers.*** *As waves approach the shore and begin to "feel bottom," their crests close in on each other and their height increases correspondingly. This eventually results in instability, as shown by the distant plunging breaker. At this point bottom erosion begins.*

wave energy bites into the bottom, which is eventually worn away. The breaker line then moves shoreward as the schoaling bottom is slowly eroded. Breaking waves carry some of the eroded particles up onto the beach face and move others out to sea. The rocks and sand that repeatedly advance and retreat with each wave aid in smoothing and abrading the beach, while at the same time they are reduced in size so that they, too, may be carried into deeper water.

Water retreating down the beach face creates an ***undertow*** that, particularly on steeper beaches, can at times present a danger to people enjoying the surf. Very often ***rip currents*** of short duration will develop on certain beaches when successive waves keep pushing water higher up on the beach face in defiance of gravity. This buildup of excessive water is relieved by currents perpendicular to shore that systematically and rapidly channel the mass of water through the breaker zone back to the sea.

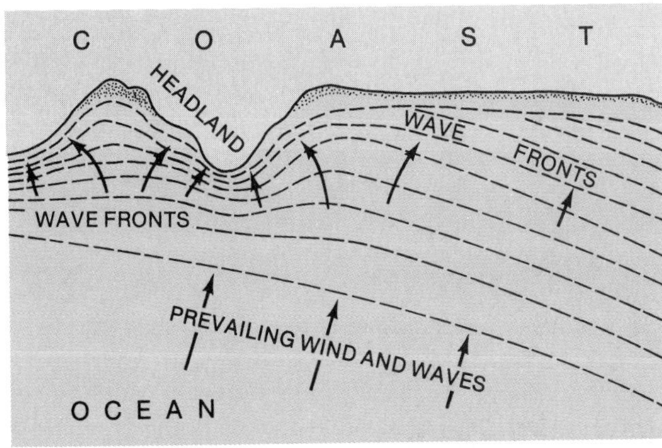

Fig. 23-4. ***Wave refraction.*** *As waves enter shallow water, they begin to break and their crestlines begin to bend, or refract, to approach nearly parallel to the shore. Because of wave refraction, erosive energy is directed at headlands rather than at coves and embayments.*

The "in-and-out" action of a typical surf zone is not the only process that affects material transport along the shore. ***Longshore drift*** is a process by which sand and other sediment are moved parallel to the shore. Despite the general effects of refraction, waves rarely strike the shore from a parallel orientation. Thus, if the winds that form the waves come from a particular direction for any considerable length of time, sediments and debris will move along the shore as well as onshore and offshore (Fig. 23-5). The undertow moves its load away from the shore downslope in answer to gravity, but the succeeding wave will carry the same material back up the beach at an angle, so that beach-forming sands will be gradually transported great distances along the shore. Often they will come to rest in quiet coves, while cliff headlands, which are producing much of the material as the waves beat against them, will have their bases swept clear (Fig. 23-6).

Longshore drift is a process by which sand and other sediment are moved parallel to the shore.

Beaches as well as other sandy landforms parallel to the shore are products of longshore drift. Of course, the winds may shift, often seasonally or even daily, and reverse the direction of drift. Locally, tidal currents set in motion by some peculiarity in shoreline may complicate the pattern. Periodically, storms will with great violence disrupt the formation of drift features.

During the construction of the breakwater enclosing the Los Angeles/Long Beach harbor, new currents set in motion by the partially completed breakwater almost entirely re-

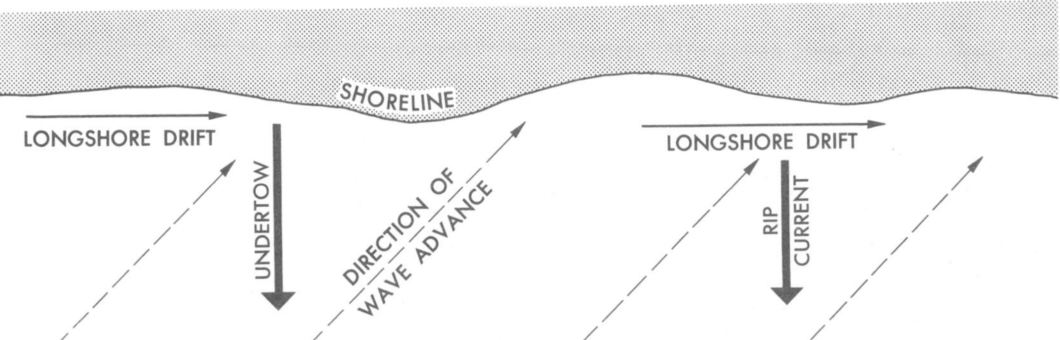

Fig. 23-5. ***Longshore drift.*** *Longshore drift moves beach and nearshore sand and other sediments parallel to the coast.*

Fig. 23-6. ***Effects of shoreline processes.*** *Headlands are eroded, and the products of erosion, primarily sand, are carried by longshore drift into coves and stretched out to form beaches.*

moved the famous beach for which Long Beach was named. For a number of years, the residents and city officials were greatly chagrined, living in a city called Long Beach with no beach at all. With the completion of the breakwater, however, the beach returned and all was well. This is not a rare occurrence, for longshore currents are fickle and subject to change as normal erosion alters the coastal outline. Beachfront residents commonly construct ***groins*** and ***jetties*** at right angles to the shore in an attempt to catch and hold the sand moving longshore. They are sometimes the subject of lawsuits by their neighbors downshore whose beaches would receive those sands if they were allowed to move normally.

EROSIONAL COASTAL LANDFORMS

Erosion by ocean waves is a serious problem for seaside dwellers, engineers, planners, and scientific investigators. Both coastal erosion and deposition constantly work to ruin the plans of those who choose to live and build by the sea. To those of us who take the time to observe and interpret coastal landforms, the erosion of sandy beaches, the undercutting of cliffs on which people have built their homes, and the storm overwash of settled coastal islands should come as no surprise.

Both coastal erosion and deposition constantly work to ruin the plans of those who choose to live and build by the sea.

Along rugged coasts where relief begins immediately at the shore and headlands jut seaward, the ***sea cliff*** is a common feature (Figs. 23-6 and 23-9). Sea cliffs are undercut by crashing waves and usually recede inland at a rate that is rapid for a geomorphic process (Fig. 23-7). As waves eat away at the base of the cliffs, the steep slope becomes unstable and gravity takes over, sending chunks of the cliff into the sea. The fallen rock is broken up by the pounding waves, and the debris is transported along the shore. Eventually, much of it is carried off, allowing the waves to apply the full force of their work on the cliff once again. Large particles of debris linger in the turbulent water at the cliff base and add an abrasive element to the erosion process.

Wave erosion along cliff coasts often produces a ***wave-cut bench***. As headlands and cliffs retreat, the bedrock in the breaker zone is abraded to a specific depth related to the energy level and circulation pattern of the continuously attacking waves (Figs. 23-7 and 23-8). The flattened surface is relatively level and kept clear of excess debris by undertow and longshore drift. Wave-cut benches are found along the cliff coasts of northern California and Oregon. Along central and southern California, tectonic activity has raised wave-cut benches well above present mean sea level. A depositional ***wave-built terrace*** is often found seaward of the bench (Fig. 23-7). The width of the bench and terrace is limited because, as they expand, incoming waves must travel a longer and longer distance over them. Their energy is therefore soon dissipated, and cliff retreat is slowed.

Another common landform associated with erosional coasts is the ***stack***. Remnants of headlands are left standing offshore after erosive, refracting waves have whittled away at either side of a coastal promontory. Many of these rocky columns are much higher than they are wide and resemble a stack of rocks (Fig. 23-9). As the headland narrows in the early stages of stack formation, sea caves may be cut into its sides by the force of waves. If the conditions are right, the wave action may cut through the narrow headland without collapsing the overlying rock. Thus, a ***sea arch*** is created.

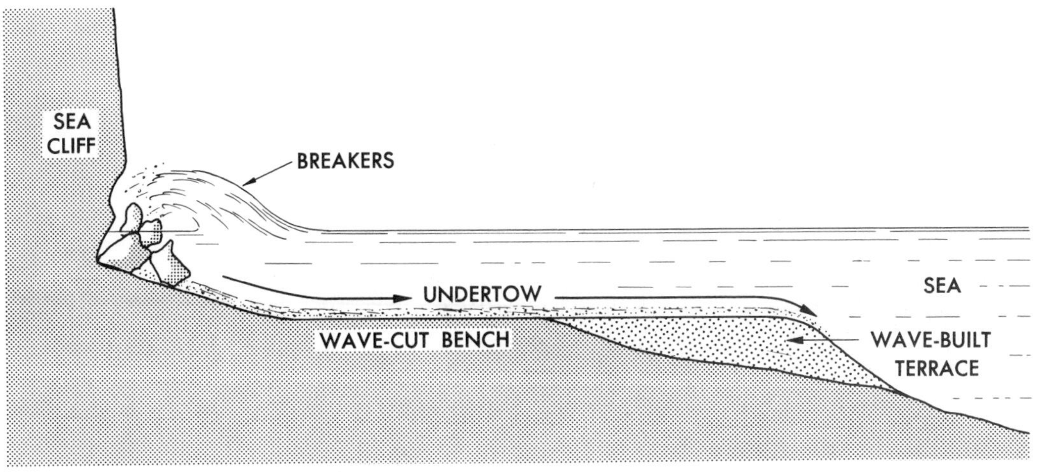

Fig. 23-7. ***The undercutting of a sea cliff.*** *Commonly found at the base of sea cliffs are wave-cut benches (see Fig. 23-8) and wave-built terraces.*

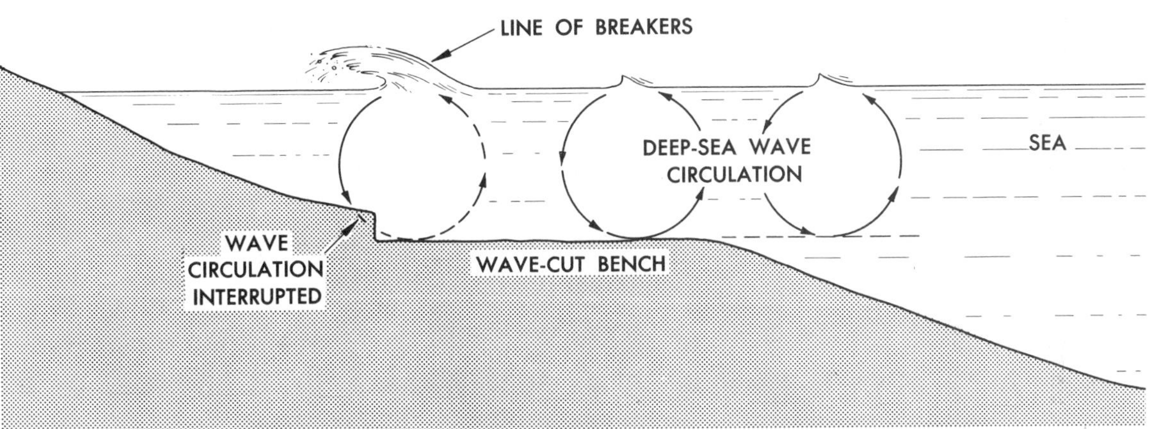

Fig. 23-8. ***Wave-cut bench.*** *Here we see the process by which a wave-cut bench is excavated in bedrock to a depth that depends on the size and strength of advancing waves.*

Fig. 23-9. ***Stacks, a result of selective wave erosion.***

We must not forget that, even as these landforms of erosion are being carved out of coastlines, their debris—gravel, sand, and silt particles—is being transported to be deposited elsewhere, perhaps at sea, on a beach, in the inlet to a bay, or on a developing island.

DEPOSITIONAL COASTAL LANDFORMS

Depositional processes can cause as many problems as erosional processes, if not more. Navigators and engineers, for example, are kept busy figuring ways to avoid the shoaling of seaways, inlets, and harbors as longshore drift rearranges sand. The material that is eroded on one part of a shoreline is likely to be deposited along another. Wherever there are a large enough supply of sand and powerful enough shoreline processes, depositional landforms will occur.

Wherever there are a large enough supply of sand and powerful enough shoreline processes, depositional landforms will occur.

One of the most common depositional coastal landforms is the ***sand spit.*** Sand carried by longshore drift, as we have seen, moves parallel to the shore and may eventually extend across coastal embayments or extend between headlands, particularly where average conditions are relatively calm. Spits are typically narrow and elongated (Fig. 23-10), with elevations that wind, and waves may build up to several dozen feet (a few meters) above mean sea level. As material continues to be deposited at the end of a spit, it may reach entirely across a bay entrance. But tidal currents or wave action are usually too strong, and sand either is swept to sea or settles in the embayment. In some locations the moving sand can completely close off the bay entrance. In this case the spit becomes a ***baymouth bar*** and blocks the bay entrance.

Fig. 23-10. ***Two examples of spit formation.*** *(Top) The mouth of the Russian River north of San Francisco, California. Longshore drift from the right, reinforced by stream sediments, threatens to seal off the river from the sea. (Bottom) Port Angeles, Washington, on the north coast of the Olympic Peninsula. The open Pacific is 50 miles (80 km) to the right; if it were not for the spit (Ediz Hook) built by Pacific-activated longshore drift, the protected port could not exist.*

A special kind of spit, a ***tombolo,*** reaches out to connect small islands to the land. In this situation, the island may produce enough shelter from sea waves that sand is deposited between the shore and the island, eventually bridging the open water (Fig. 23-11).

Barrier Islands

Long, narrow islands parallel many coasts that have a wide, gently sloping coastal plain and plenty of shore sand available for transport. The southeastern coast of the United States is a good example. Here dozens of sandy islands stretch from near Long Island in New York State and continue, with only a few major breaks, around Florida and along the coast of the Gulf of Mexico all the way to south Texas. These are ***barrier islands*** (Fig. 23-12).

Geomorphologists do not agree on how these islands

Long, narrow islands parallel many coasts that have a wide, gently sloping coastal plain and plenty of shore sand available for transport.

Fig. 23-11. ***Tombolo.*** *The small volcanic cone in the foreground, Mount Maunganui, was captured by a spit working its way out from the mainland to form a tombolo. Inside the entrance are the spacious joint harbors of Tauranga and Mount Maunganui, New Zealand.*

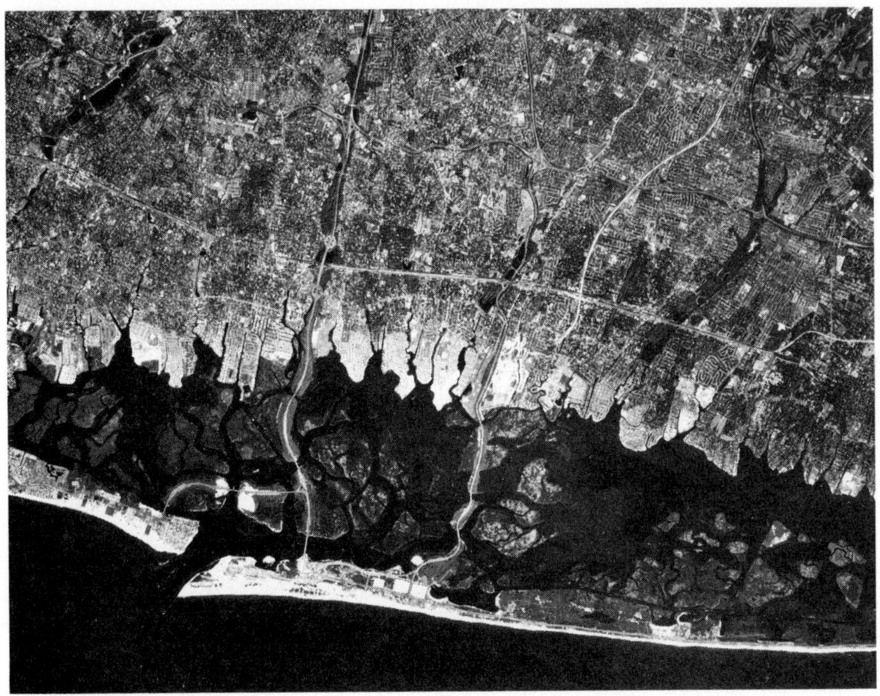

Fig. 23-12. ***An example of a barrier island.*** *A long stretch of sandy island parallels the mainland. A marshy lagoon formed behind the barrier is filling in over time. Note the tidal inlet between the two island segments. Scouring tides keep the inlet open and the islands from joining.*

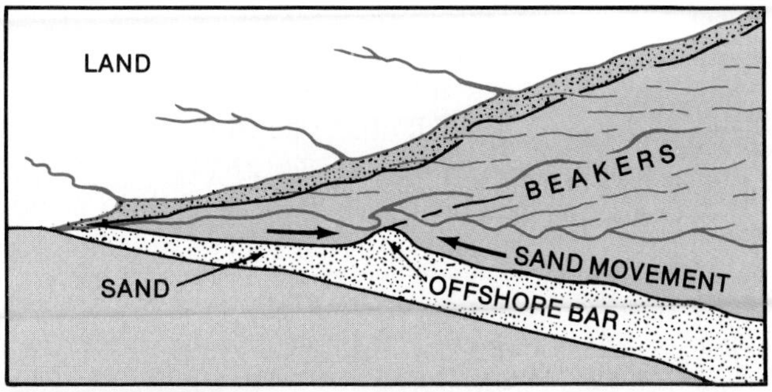

LAND
BEAKERS
SAND
SAND MOVEMENT
OFFSHORE BAR

Fig. 23-13. Formation of barrier islands. Offshore bars commonly develop parallel to the shoreline as a result of waves pushing sand toward the beach in the breaker zone and then pulling sand off the beach in the backwash to accumulate in a bar deposit.

originated, but many believe they began as offshore bars built by waves breaking onto a shallow shore. At the point where advancing waves begin to be felt on the bottom, they push sand toward the shore to develop a submerged ridge, or bar. After the waves break, the return undertow carries sand from the beach to settle on the bar (Fig. 23-13). During the last Ice Age, the level of the world's oceans was significantly lower than it is today, perhaps as much as 300 to 400 feet (90 to 120 m), because a great deal of water was tied up on land in the form of ice. Offshore bars that developed at the time were pushed toward land as sea level initially dropped and then rose again, fed by melting ice and runoff into the ocean. The waves of the slowly rising sea kept the sand of the former offshore bar in front of the breaker zone and above mean sea level. The former offshore bars, now barrier islands, gained height as wind and storm waves built up the sand even higher.

The processes of longshore drift have strung out barrier islands for long distances, some of them up to 100 miles (162 km). Their width varies, but typically they are from a few hundred feet (a few dozen meters) to 3 miles (5 km) wide. A ***tidal inlet*** occasionally breaks the island strand, and quiet, shallow bodies of water called ***lagoons*** nestle between the barrier island and the mainland (Fig. 23-12).

CLASSIFICATION OF COASTS

The geomorphology of coasts is complex, and coastal processes and resulting landforms are numerous. To discuss them in any meaningful way some system of classification is necessary. A traditional system divides shorelines into (1) drowned or ***submerged shorelines*** and (2) uplifted or ***emerged shorelines***. No matter what the local detail, we can assign most coasts one of these labels and make reasonable generalizations regarding the action of ocean waves on it.

Some system of classification is necessary to discuss the complex geomorphology of coasts and the numerous coastal processes and resulting landforms.

Submerged Shorelines

Imagine a dam built across the course of a stream and backing up the water to form a lake. In finding its own level in an eroded terrain, the lake develops a shore featuring deep embayments and prominent points of land. This is a drowned or submerged shoreline; the same description applies for either a fresh or salt water body (Fig. 23-14). On the coast of Norway, the sea is advancing up the valleys while the ridges stand out as bold peninsular headlands. The same processes are occurring on the coasts of southeastern Alaska, Newfoundland, the Mediterranean, and the mid-Atlantic United States. Perhaps the land is sinking rather than the water rising; it does not matter. Sea level is rising, and the shoreline of even a moderately eroded land surface will show the same broken character.

Wave activity along the typically ragged and broken submerged shore is consistent in its general pattern everywhere. First, it removes the indentations by both cutting and filling, and then it drives the entire coast back relentlessly. Waves attack the exposed headlands with great vigor, especially if the sea is deep immediately offshore and their full force can be concentrated directly on the land without friction with the bottom. Under the force of this attack, the point of land is cliffed and pushed back rapidly.

Emerged Shorelines

The emergent shore, on the other hand, is in theory smoothly shelved and featureless. It is the kind of shore we find along the margins of evaporating lakes such as the Great Salt Lake or the Caspian Sea. Here the former bottom is exposed, smoothed, and planed by wave action and often floored with water-laid sediments. This is in direct contrast to the steep and jagged character of the drowned coast.

Classification should therefore be easy: any variation on a smooth and gentle shore derives from emergence, and the rough and broken coast must result from submergence. However, from what we are learning of the recent history of

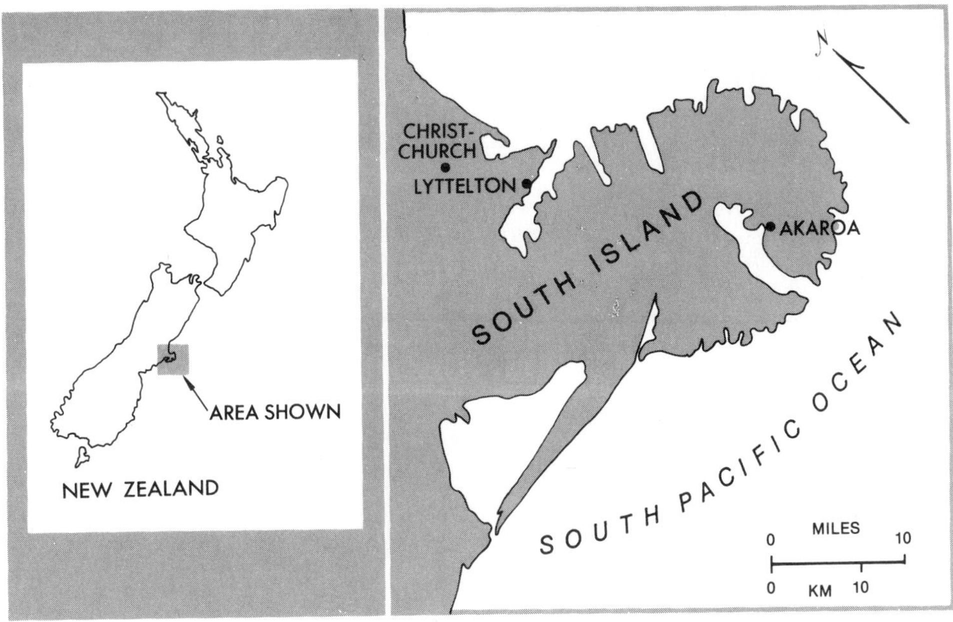

Fig. 23-14. ***Submerged shoreline.*** *The sketch map indicates the location of Akaroa on Banks Peninsula, a partially foundered volcano. Originally, it was a perfect land-based cone complete with crater and symmetrical drainage pattern. As it slowly sank (or sea level rose), fingers of the sea pushed their way up each stream valley, even to the point of breaching and flooding the crater of Akaroa. The photo is taken from the hills behind the village with the Pacific to the left.*

HOLLAND'S DELTA PROJECT

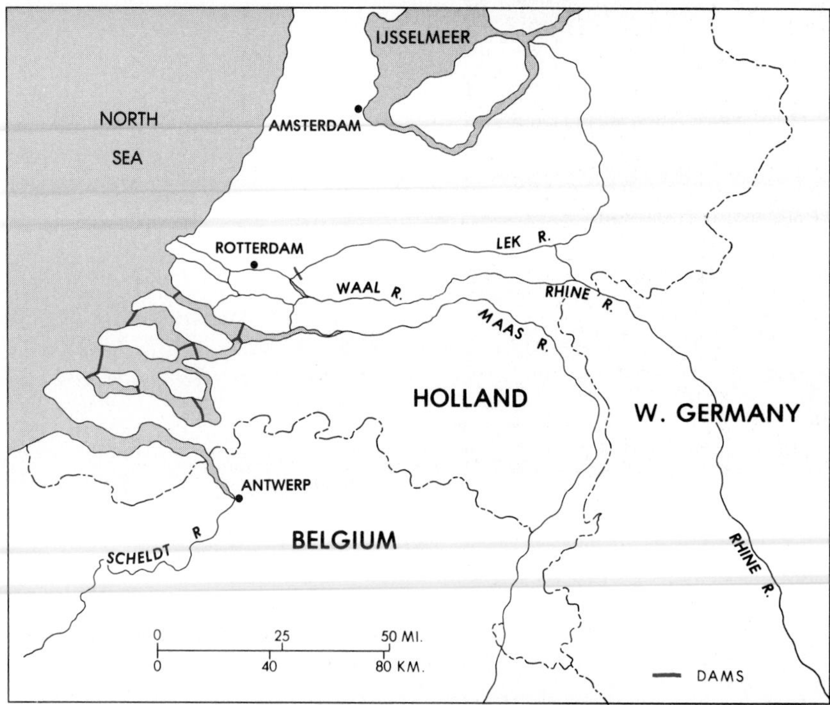

The great, involved marshy delta of the Rhine and a bevy of lesser streams are of major significance to the Netherlands for two reasons. First, at its northern periphery the modern port of Rotterdam controls the chief navigable access to the upper Rhine (West Germany, France, and Switzerland). Second, the drained and reclaimed agricultural land of Zeeland makes up more than one-third of the nation's productive polders (land reclaimed from flooding). Yet vicious North Sea gales and flooding rivers are a constant threat. At best, with 434 miles (698 km) of vulnerable dikes, a delicate equilibrium has been achieved.

A disastrous storm in February 1953 severely damaged nearly all the diked land in Holland and disrupted the delta's complex balance almost as thoroughly as the German bombs had during World War II. This was not the first sea-borne disaster, but in the wake of the 1953 storm it became obvious that a major effort would have to be made if this strategic and heavily populated region was to escape recurring disaster. Merely shoring up the dikes once again was not the long-term answer.

The plan that emerged encompassed four major dams across the delta mouths, which in essence cut off 440 miles (708 km) of estuary coastline. These dams would be backed up by a series of seven secondary dams upstream to regulate river outflow. This enormous project costing $5 billion was finally completed in October 1986. Converted now to fresh water lakes, the old estuary channels function as a buffer against tidal salt water that threatens to ruin arable land. At the north the canalized channel to the sea from Rotterdam is open to serve Rhine traffic. Similarly, to the south the Scheldt River mouth, Antwerp's deep water access, remains undammed.

The Delta Project has been a mammoth effort, requiring huge expenditures of capital and labor and the development of whole new processes in design and engineering. Its scale is so large that the machines to make it work had to be built on the spot and ingenious new methods invented along the way. It is in place now—the world's largest project to stabilize a coastline whose natural tendency is to remain in constant flux.

the earth, especially events that would affect variation of sea level, it begins to appear that pure emergent shorelines are moderately rare—much more so than at one time was supposed to be the case. Given the melting of continental ice sheets and the general rise of sea level from 300 to 400 feet (90 to 120 m) in the past 20,000 or so years, it is unlikely that many coasts have escaped some sort of modern drowning.

Far up on a mountain slope we may encounter a series of

Fig. 23-15. ***Marine terrace.*** *Facing the sea with a steeply cliffed frontage is a well-defined terrace. Its flattish surface was once the smooth offshore sea bottom, and its inner margin was a wave-cut bluff.*

marine terraces that appear as giant steps. Each platform, backed by a steep wall, is the product of past planation by the surf and sea cliff erosion (Fig. 23-15). This is good evidence of uplift, as are elevated coral remnants that were nourished during their formation at or below sea level. Even these observations, however, do not absolutely prove that the most recent movement has been a relative lowering of the sea, although they certainly make it appear likely.

The emerged shoreline does, however, exist. The Baltic Sea and Hudson Bay are becoming more shallow as the land responds to the unloading of its massive weight of ice by rebounding upward at a measurable rate; lava cascading down the slopes of Hawaii forms a new shoreline; live coral and mangrove extend the coast out to sea in the tropics, and so does the active river delta.

Other Classifications of Coasts

Geomorphologists and oceanographers have developed other classification schemes for coastal regions as well. Two noteworthy systems point out the differences in the scope of processes that affect the geographic character of a coast.[1] The first links coastal configuration with global plate tectonics (see Chapter 17) and assumes that there is a strong connection between the geomorphic character of a coastline and the movement of lithospheric plates. The second classification focuses on processes that directly affect characteristic landforms along a coast.

The first classification identifies three major types of coast:

1. Collision coasts, which are on the collision edges of continental plates.
2. Trailing-edge coasts, which are on noncolliding sides of continents.
3. Marginal coasts, which are protected by island arcs.

Examples of collision coasts are the west coast of South America and northwest North America. This type of coastline is relatively straight and is marked by volcanic and seismic activity. Mountains of high relief rise directly from the shore; there are numerous sea cliffs and raised wave-cut benches. Island arcs such as the Aleutian Islands and Japan are considered collision coasts, because the arcs were formed by the collision of parts of oceanic plates.

Trailing-edge coasts are located on the part of a continent that faces a spreading center, such as the east coast of North America. This particular coast is now at some distance from the mid-Atlantic ridge, where continental separation began. It is a mature trailing-edge coast modified by long-term erosion and deposition from the highlands of the continent. The coasts on either side of the Red Sea, where there is a relatively new spreading center, would also be classified as

[1] These classifications are from (1) D. L. Inman and C. E. Nordstrom, 1971, "On the Tectonic and Morphological Classification of Coasts," *The Journal of Geology* 79:1–21 and (2) F. P. Shepard, 1976, "Coastal Classification and Changing Coastlines," *Geoscience and Man* 14:53–64.

trailing-edge coasts, but landforms related to erosion and deposition are limited to the immediate coastline.

Coasts sheltered by island arcs, such as in China and Southeast Asia, face marginal seas (South China Sea, Sea of Japan). These marginal coasts have a great variety of landforms, are curved in general outline, and are modified by large rivers and their deltas.

The second classification system separates coastal types and related landforms on the basis of their origin in terrestrial (land-based) processes such as land erosion and deposition or marine processes such as wave erosion–deposition. Coasts affected by land-based processes include those with glacial erosion features (fjords), drowned river mouths (estuaries), and coasts with deltas and wide coastal plains of river-deposited material. Marine processes create wave-cut and eroded landforms as well as depositional landforms. Examples are sea stacks and cliffs, sand spits, and barrier islands.

CONCLUSION

A tremendous amount of energy is expended at the interface of land and oceans. Waves, wind, tides, and nearshore currents work vigorously to shape the land into landforms that are subject to unrelenting changes. On the whole, geomorphic processes are slow and subtle, but at the coast the effects of battering waves, longshore currents, and strong winds work quickly to change the landscape. Most people think of the seashore as a place of unchanging beauty, destined to remain constant through time. Only those who are especially observant or who must combat the ocean's power respect and appreciate the constant shifts that occur along the shore.

The many attempts needed to check longshore drift, undercutting waves, beach erosion, or a rising or lowering sea level usually lead to frustration. Artificial structures such as seawalls, jetties, and groins are destroyed by waves or seem only to add to the problems they were meant to solve. Jetties may stop sand deposition in a navigation channel, but they may also cause loss of sand and beach recession farther down the shore. Groins may protect the beach in one location but make neighboring beach segments virtually disappear by depriving them of sand.

Coastal processes are confined to a relatively small portion of the earth's surface, but much of the world's population resides at or near coasts in numerous large cities and densely populated areas. Take the case of the Netherlands where millions of people live on land diked off from the sea. The tens of millions of inhabitants in and around the cities of Shanghai and Tokyo live very near sea level on coastal plains. Bangladesh is a low-lying country at the head of the Bay of Bengal with a large rural population that grows rice on the delta of the Ganges and Brahmaputra rivers. Storm surges and tides in the bay occasionally combine with heavy river discharge during the wet monsoon to cause frequent devastating floods. Stockholm and Venice are European cities, both laced with canals, that have an intimate relationship with the sea. People in all these coastal locations have an interest in the processes that affect their destiny. Although we can begin to understand the physical processes that give us coastal landforms, we need to pay more attention and give more respect to these same processes when we practice our engineering arts at the shore.

KEY TERMS

swells
wave refraction
undertow
rip currents
longshore drift
groins
jetties
sea cliff
wave-cut bench
wave-built terrace
stack
sea arch
sand spit
baymouth bar
tombolo
barrier islands
tidal inlet
lagoons
submerged shorelines
emerged shorelines

REVIEW QUESTIONS

1. How are open-ocean waves, or swells, generated?

2. What do we mean when we say that, in waves far out at sea, the water is not going far, but the energy is?

3. Describe the process of wave refraction. Why is this significant to coastline development?

4. How do sea arches and sea stacks form?

5. Along what type of coast would you expect to find wave-cut benches? Why?

6. Explain the longshore movement (drift) of sand. What major landforms result from this process?

7. Describe the theory that explains how barrier islands develop.

8. Are classifications of coasts and coastal landforms useful? In your opinion, why or why not?

APPLICATION QUESTIONS

1. Find two or more recent newspaper or magazine articles on the subject of beach or shoreline erosion. Do the articles suggest any solutions to this erosion? Summarize them. Are these solutions reasonable in light of what you know about shoreline processes? If no solutions are given, can you think of some?

2. What would happen if a jetty or seawall were built out into the ocean perpendicular to a sandy beach? Draw a map or diagram showing how the jetty construction might affect the beach and the near shore.

3. If you owned a home on a high headland overlooking the sea and you discovered that wave erosion was advancing at a rate that would soon undercut and destroy the part of the headland on which your home was located, what action would you take, short of moving the house, to protect against the erosion process?

4. Contact a major insurance company and find out what their policies are regarding beachfront property.

5. If you live near the shore, find out what the local ordinances or laws are concerning building on the shore. What concerns do you think are behind these laws?

PART FIVE

SOILS AND VEGETATION

CHAPTER 24

UNDERSTANDING SOILS

OBJECTIVES

After studying this chapter, you will understand

1. How minerals contribute to the formation of soil.
2. What gives soil its texture.
3. How organic matter enters the soil and helps to form different soil types under different conditions.
4. How gases enter and exit the soil.
5. How the alkalinity, acidity, and salinity of soil water are affected by climate and vegetation.
6. Why soil color can mean different things in different locations.
7. Why soil profiles are important and how they are formed.
8. How human culture can both damage and improve the soil.

This was once a productive agricultural field. Today we understate the case when we simply classify it as severely eroded. Soil erosion is insidious at first, nobody really notices that the topsoil is being carried away bit by bit, so that by the time moderate gullying begins much of the damage has been done. Erosion is a cancer on the land—if caught in the very early stages it can sometimes be controlled; but more often than not it is much too far along to heal when the danger is recognized. Obviously this farm in Bolivia has reached the terminal stage. If the inhabitants of the little farmhouse step out the front door they are in danger of falling into a 50 foot (15m) deep badland chasm.

INTRODUCTION

Have you ever wondered why some soils are red whereas others are nearly black? Or why some support thick forests and others will grow nothing at all? Or why some soils are soggy long after rain, but others are dry and firm after a short time? We answer these and many other questions about soils here in Part Five. In this chapter we discuss the general nature of soils and how they mature over time. In the next chapter we introduce the major soil types and world patterns of distribution.

Never shrug off soil as "just dirt"; it is an important element in the human physical environment. Certainly, its ability (or inability) to produce living growth affects our ability to use earth effectively as a habitat. So we should not only be interested, but at least rudimentarily informed as to its character. What is soil? Is it merely rock broken into tiny fragments? If we were to take a rock and crush it mechanically, we would not be able to make soil, no matter how fine the particles. We would have only the inorganic raw material from which soil is formed.

The breakdown of rock by natural means is the first step in the development of soil, but other steps must follow. Other elements, besides simple rock fragments, must be present before true soil evolves. So it is best to think of soil as developmental—gradually changing and maturing over time into a series of horizontal layers in tune with prevailing climate, weathered bedrock, and plants and animals living in and above it.

Soil is developmental, maturing over time into a series of horizontal layers.

In soil science (***pedology***), we traditionally regard soil as the result of the interaction of four elements (Fig. 24-1):

1. Mineral matter (rock fragments or ***regolith***)
2. Organic matter (both plant and animal)
3. Gases
4. Moisture

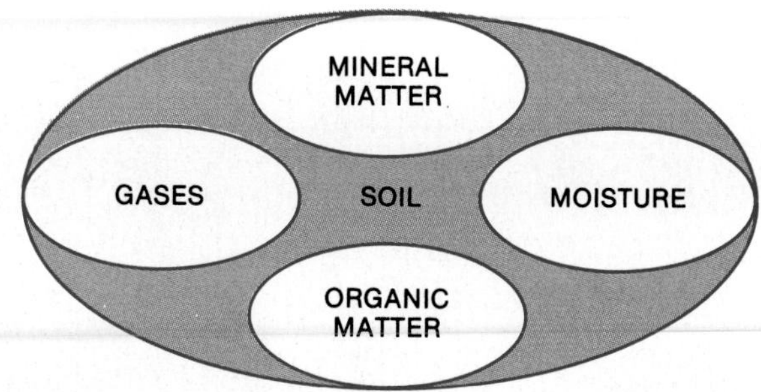

Fig. 24-1. ***Soils are the product of four interacting components.***

MINERAL MATTER

Mineral matter, including many of the nutrients needed for plant growth, comes from the disintegration of rock through weathering. It provides the ***parent material*** for soil development. Parent material may be either ***residual*** or ***transported.***

Residual Parent Material

Residual parent material comes from the disintegration of rock in place. It accumulates gradually and overlies decomposing bedrock (Fig. 24-2). Nutrient content—levels of iron, calcium, sodium, phosphorus, and so on—reflects the richness of bedrock composition. Thus if the bedrock is not well balanced in mineral content, which is sometimes the case, the residual soils that develop may be nutritionally imbalanced.

Bedrock that is not well balanced in mineral content may result in nutritionally unbalanced residual soils.

Transported Parent Material

A great deal of weathered material lying above bedrock is removed through erosion and is deposited elsewhere. It is therefore not surprising that much parent material does not come from underlying rock. Instead, as we have seen in Part Four, it is transported by gravity, moving water, wind, and ice.

Transported materials make up many kinds of deposits (Fig. 24-3).

- Colluvium at the base of slopes
- Alluvium in fans, flood plains, and deltas
- Marine terraces
- Deposits in lakes
- Wind-blown sand and loess deposits
- A whole array of glaciofluvial deposits left by glaciers

Several mineral sources may contribute to these deposits. The result is often a better mix of plant nutrients and thus

Fig. 24-2. ***Residual parent material.*** *Rock masses slowly disintegrate to provide parent material for soil development. Here on the Island of Hawaii, an older lava flow (lighter surface) is partially covered by a younger flow (darker surface). Note the fractures in the rock and scattered grasses that denote the early stages of rock disintegration and soil development on the older surface. In time, the residual volcanic soils that develop may be several feet deep and support a dense tropical forest.*

Fig. 24-3. ***Transported parent material.*** *Soils in this valley along the Colorado River have developed in deep alluvium.*

higher plant productivity than certain imbalanced residual soils.

Climate and vegetation have far greater control over the nature of soil than parent material alone, but these factors certainly work together.

Soil Texture

Soil texture, or the mix of particle sizes, is closely related to the original parent material and its mineral composition. As bedrock disintegrates, smaller and smaller particles are left. Particle sizes range from boulders to clay (Fig. 24-4). Crystalline rocks with high quartz content such as granite break down into sand and silt-sized grains, whereas basalts often weather to clay-sized particles.

Most soils contain some sand, silt, and clay. The particular mix of these sizes is the basis for soil texture classification (Fig. 24-5). For example, soil with 40 percent clay, 30 percent silt, and 30 percent sand is called a *clay loam.* (*Loam* is a mixture of sand, silt, and clay, and in this case there is more clay than silt or sand.) Soil with 10 percent clay, 70 percent silt, and 20 percent sand is a *silt loam;* and so on.

	Particle Size	
Class	*(Millimeters)*	*(Inches)*
Boulders	256–4096	10–160
Cobbles	64–256	2.5–10
Pebbles	2–64	0.08–2.5
Sand	0.05–2	0.08–0.002
Silt	0.002–0.05	0.00008–0.002
Clay	below 0.002	below 0.00008

Fig. 24-4. ***Soil particle sizes.*** *Particle sizes range from huge bedrock boulders to minute clays. Only the sand, silt, and clay sizes (collectively called "fines") are considered in descriptions of soil texture.*

The soil's particular mix of sand, silt, and clay is the basis for soil texture classification.

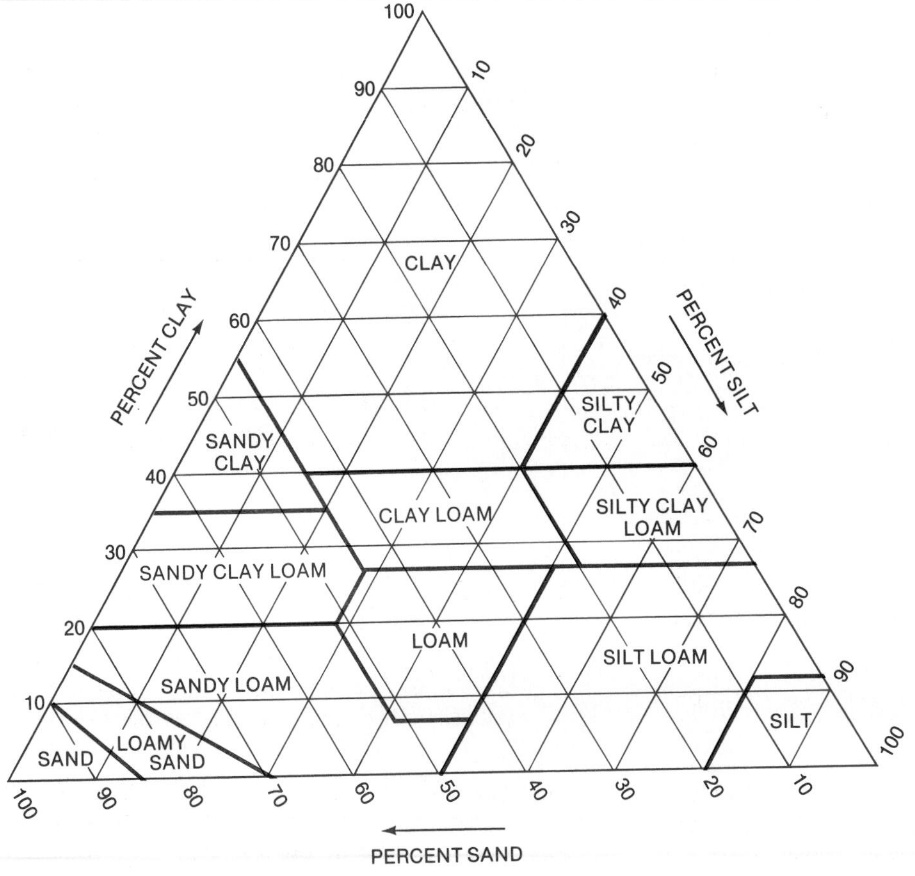

Fig. 24-5. ***Soil texture.*** *Soil texture categories are based on the relative proportions of sand, silt, and clay.*

Fig. 24-6. ***Poor drainage.*** *The clay soils of this wetland slow drainage so much that water ponds on the surface throughout the wet season.*

Although we can find the exact proportions of sand, silt, and clay in the laboratory, we can perform fairly accurate estimates of soil texture in the field:

- Moist soils with high clay content can be rolled with the hands into long rods that do not break when bent.
- Moist silty soils break when rolled into rods and "squeak" when rubbed between the thumb and forefinger.
- Sandy soils cannot be rolled into a rod, and they feel gritty to the touch.

Knowledge of soil texture is important for two reasons:

1. Texture is one of the characteristics used in soil classification, and it is an important distinguishing characteristic of layers in mature soils.
2. The mix of particle sizes is directly related to the soil's water-holding capacity and has a strong effect on vegetation and potential for agricultural use.

In general, the coarser the texture (the greater the percentage of sand), the faster water will drain through; the finer the texture (the greater the percentage of clay), the slower water will move through. Clay has much more surface area than sand. Once the surfaces of minute clay particles are wet, water is strongly held as a surface film. In wet areas, sandy soils may be droughty with sparse vegetation, but clay soils in the same climate may make drainage so slow that marsh and swamp plants develop (Fig. 24-6).

The rate at which water drains depends on the texture of the soil.

Agriculturists prefer "lighter" (silty, loamy) to "heavy" (clayey) soils. Lighter soils are easier to work, that is, to plow, disk, and make into a seed bed. Even though they may have to be irrigated more often, water does not pond on the surface for long periods and smother emerging seedlings.

ORGANIC MATTER

Organic matter gives the soil its character, its link with life. Plants and animals are the source of organic matter, but how do they become part of the soil? Roots penetrate the soil, and soil animals live within it. When they die, they are already below the surface and mixed with mineral matter. In contrast, most animals and living plant structures are above the soil surface. When they die, they must be partially decomposed before entering the soil matrix. Bacteria and fungi are responsible for decomposition, and a whole host of organisms traveling between the surface and the soil mixes in the organic matter.

Organic matter is the soil's link with life.

Stages of Decomposition

Organic matter passes through four stages, beginning with the deposition of recently dead material on the soil surface and proceeding through time and depth to its total decomposition (Fig. 24-7).

1. ***Litter*** consists of recently deposited leaves, branches, and dead animals. It is the uppermost organic layer, derived from the current season. Decomposition has begun, but parts are still recognizable.
2. Below litter, but still above the soil surface, is ***duff***—the accumulation from previous growing seasons. Because of advanced decay, the duff layer grades from distinguishable plant and animal parts to shapeless particles lower down. The soil proper (solum) begins at the base of the duff.
3. The organic matter within the solum takes a finely divided, but not totally decomposed, form: ***humus***. Humus is washed into the soil by rain, brought in by burrowing animals, or left by the death of roots and soil animals. It gives a dark, sometimes nearly black, color to the upper part of the soil where it is most concentrated, but the darkening decreases with depth as humus decays further to its elemental parts.
4. In the end, the only remains of organic material are ***nutrient ions***—chemical elements and compounds in ionic form. They can either be moved to greater depth by the downward movement of water (percolation) or be taken up by plant roots to make new organic matter.

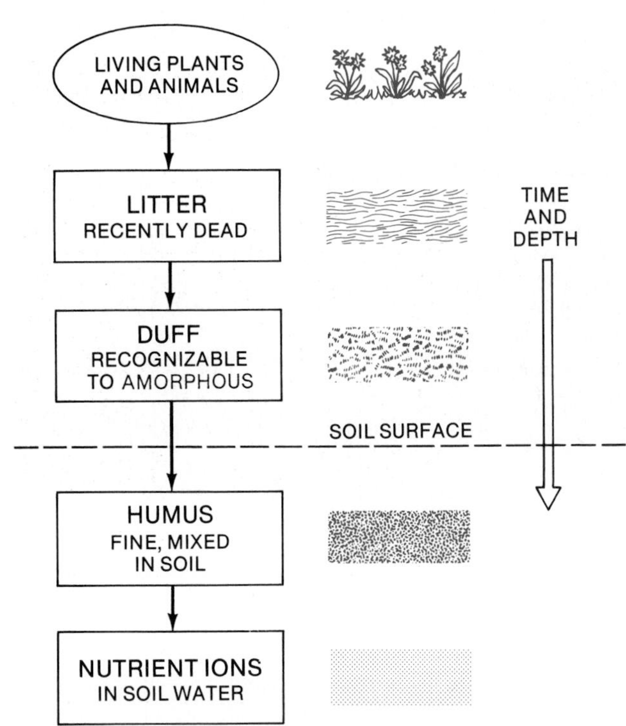

*Fig. 24-7. **Space–time continuum of organic matter.** As dead plant and animal material deposited on the soil surface decomposes, it goes through several stages that occur at increasing depth until only nutrient ions remain.*

Factors Influencing the Rate of Decomposition

Not all organic materials decompose at the same rate. The rate is influenced primarily by (1) chemical composition of vegetation, (2) temperature, and (3) moisture.

The rate of organic decomposition is determined by chemical composition of vegetation, temperature, and moisture.

The chemical composition of vegetation is especially important. For example, needles, branches, and bark from softwood trees such as pine and spruce contain acid-forming compounds and resins that resist decay. Once decomposed, these materials yield only small quantities of key nutrients for plant growth like potassium, calcium, and magnesium. On the other hand, hardwood trees such as oak and maple, and many other plants including grasses, contain little resin, are generally nonacid, and consequently decay more rapidly. These also yield larger amounts of potassium, calcium, and magnesium.

Decomposition rates are also influenced by temperature and moisture. Cold and dry conditions slow bacterial and fungal activity, whereas warmth and moisture encourage it.

Mor, Mull, Moder, and Mineral Soils

As you might expect from the differences in type of vegetation and rates of decomposition, the thickness of litter, duff, and humus layers—and the concentration of nutrient ions—is not the same in all locations. Soils with large amounts of litter and duff are called ***mor soils***. (*Mor,* from

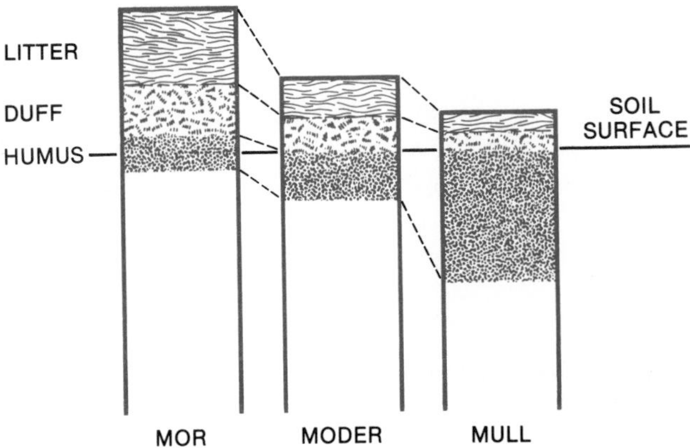

Fig. 24-8. ***Soil classification based on organic matter.*** *Relative amounts of litter, duff, and humus serve as the basis for distinguishing mor, moder, and mull soils.*

Danish, means an accumulation of forest organic matter abruptly distinct from the mineral soil beneath) (Fig. 24-8). Coniferous forests of cool, moist regions like those in much of Canada, Alaska, and Northern Europe, for example, have thick litter and thick duff but very little humus. This is because bacterial decomposition is greatly slowed by cool temperatures and acid conditions. Slower acting fungi do much of the breakdown work (Fig. 24-9), and so there is a backlog of superficial organic matter waiting to be decomposed. This condition makes walking cross country in these cluttered forests difficult (Fig. 24-10). Earth worms and other soil organisms are not abundant under high acid conditions, so that little organic matter as humus is incorporated into the upper soil.

Mull soils (*mull,* also from Danish, meaning a mixture of organic and mineral materials) have little litter and duff but much humus. Such conditions are most prevalent in grasslands of the midlatitudes where temperatures are warmer and there is enough moisture to convert the annual dieback of above-ground growth rapidly to humus. Numerous organisms, including earth worms and gophers, then incorporate this material into the soil.

Moder soils (from *moderate*) are midway between the mor soils of coniferous forests and the mull soils of grasslands. They contain moderate amounts of litter, duff, and humus and are common in moist, temperate, and tropical areas occupied by deciduous and evergreen broadleaf forests.

Mineral soils have very little litter, duff, or humus. They occur under arid conditions, where plant growth is at a minimum and organic matter of any form is not abundant, and in recently deposited parent material, where little organic material has had a chance to accumulate.

Fig. 24-9. ***Forest decomposition.*** *This bracket fungus on a dead tree in Washington's Olympic National Park is an important decomposer in needleleaf evergreen forests.*

GASES

Soils breathe. Air diffuses into the soil when atmospheric pressure is high and out of the soil when pressure is low. The air passageways are ***pores*** that exist between ***peds*** (Fig.

Fig. 24-10. ***Thick forest litter.*** *Slow decomposition in this needleleaf evergreen forest on the west slope of the Washington Cascade Range results in a buildup of litter and difficult cross-country travel.*

24-11). If you take a clump of soil and gently pull it apart, the pieces that resist further breakage are the peds. These clumps of soil have similar size and shape, are made up of many soil particles (including sand, silt, clay, and organic matter), and are held or cemented together by mucus from soil organisms, iron oxides, carbonates, or clays. Ped shape, generally referred to as ***soil structure,*** may be granular, platy, blocky, or prismatic. Shape varies from one soil type to the next and even from one layer to another in the same soil.

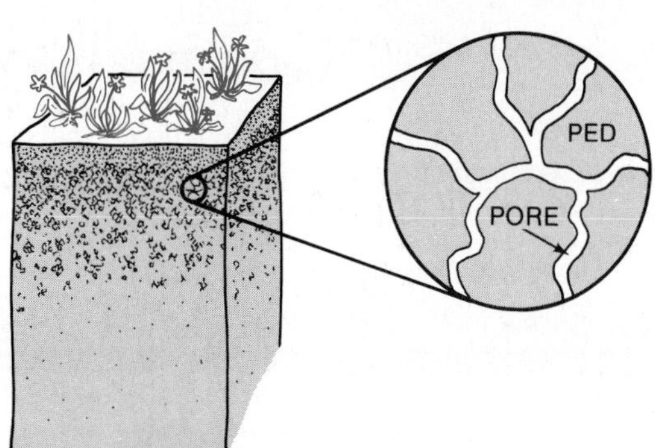

Fig. 24-11. ***Gases.*** *Soil structure consists of aggregates (peds) between which are passageways for air and water (pores).*

Air diffuses into the soil when atmospheric pressure is high and out of the soil when pressure is low.

The pores between peds are well established in undisturbed soils. In such soils, oxygen can easily reach plant roots for use in respiration, and the resulting carbon dioxide travels through pores on its way to the atmosphere. Other gases, such as methane and nitrogen from decomposition, may also escape through pores.

Pores are important passageways not only for gases, but also for water from rain or irrigation as it moves through the soil. (See Chapter 13 for a discussion of soil-water movement.) In a saturated soil, all the pores are filled with water. If this condition persists, the lack of oxygen will slow decomposition and smother plant roots. In moist soils, water and air coexist in pore spaces without adversely affecting plant life.

Water and air can coexist in pore spaces without adversely affecting plant life.

Cultivation often disturbs soil structure. One reason why fields are plowed and disked prior to planting seeds is to disorganize the pores. This procedure slows percolation, lengthens water retention, and thus makes more water available to young plants. Prolonged use of heavy tractors and implements, however, may compact soils so much that it destroys the preexisting soil structure and seriously reduces pore spaces.

MOISTURE

Water is an important constituent of the soil. It is vital in forming soil and maintaining its fertility. Water, itself an essential plant nutrient, dissolves other nutrients in the soil, creating a rich solution drawn in by plant roots. Let us consider how two of the main characteristics of soil water—acidity/alkalinity and salinity—are affected by climate and vegetation.

Acidity Versus Alkalinity

The acidity or alkalinity of the soil water, expressed as a pH value, is a valuable indicator of soil fertility. The *pH scale* of soil normally falls in the range 4 to 10 with a value of 7 neutral. Smaller values indicate increasing acidity, and greater values indicate increasing alkalinity (Fig. 24-12).

The acidity or alkalinity of the soil water is a valuable indicator of soil fertility.

The pH value is actually a measure of hydrogen ion concentration. Ions are atoms and compounds that take on an electric charge because they have lost or added electrons. Hydrogen in the soil water has lost an electron and is thus positively charged. Why should we be concerned with hydrogen ions? To answer this question we must consider soil chemistry and the effect of water moving through the soil.

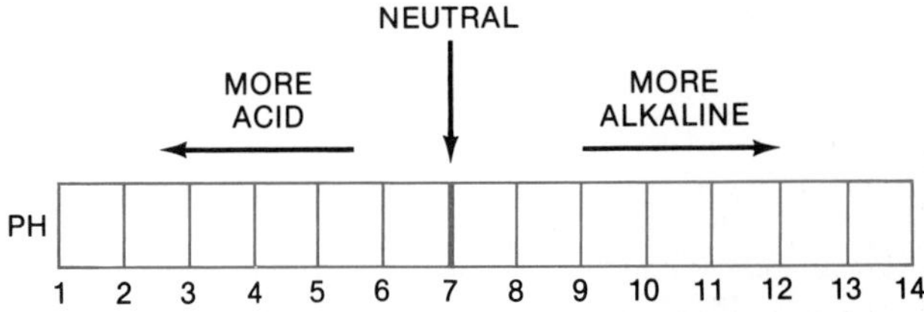

Fig. 24-12. The pH scale. The pH scale measures hydrogen ion concentration. Low values indicate higher concentration and acidity, high values indicate low concentration and alkalinity, and 7 is neutral. Most soils fall in the pH range from 4 to 10.

THE CLAY–HUMUS COMPLEX Most soil contains some clay, which is composed of silicon and aluminum. Because of the way these two elements are held together, minute clay particles carry a rather strong negative charge. Tiny particles of humus are also negatively charged. Together the clay and humus are referred to as the *clay–humus complex.*

Nutrients like calcium, phosphorus, ammonium, and magnesium, which occur as ions in the soil water, are positively charged and thus attracted to the clay–humus complex (Fig. 24-13). Hydrogen ions are similarly attracted. The resulting multitude of clay and humus particles with their attached positive ions reflects the nutrient composition of the soil water and functions as nutrient storage.

LEACHING *Leaching* is the process by which nutrient ions are washed downward and out of the soil. As rain falls through the atmosphere, it absorbs some carbon dioxide to form a weak carbonic acid. In areas of air pollution, acidity may increase as the rain absorbs oxides of sulfur and nitrogen. This "acid rain" may pick up additional acids from litter and duff before it enters the soil (Fig. 24-14). In rainwater, acids break up into positive hydrogen ions and negatively charged ions like carbonate, sulfate, and nitrate. As the water percolates through the soil, hydrogen ions exchange places with nutrient bases on the clay–humus complex. This happens because great numbers of hydrogen ions swarm around the clay and humus particles, and, under acid conditions, hydrogen is more attractive to the particles than are the bases. Continued percolation carries away the displaced nutrient bases resulting in stronger acidity. It follows, then, that soils in areas of high rainfall tend to have lower pH values (in the acid range) and have lower nutrient levels because of leaching, whereas arid and semiarid soils with little percolation are alkaline and tend to retain nutrients.

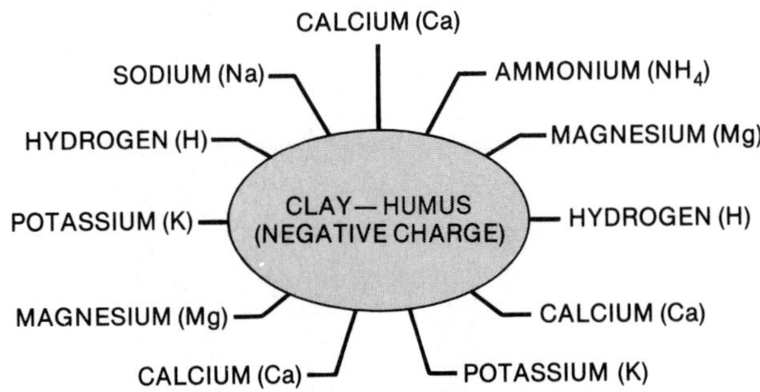

Fig. 24-13. Generalized diagram of a clay–humus complex. Clay and humus soil particles are negatively charged. Each attracts and holds positive ions. These tiny units in the soil store nutrients but can lose them by leaching.

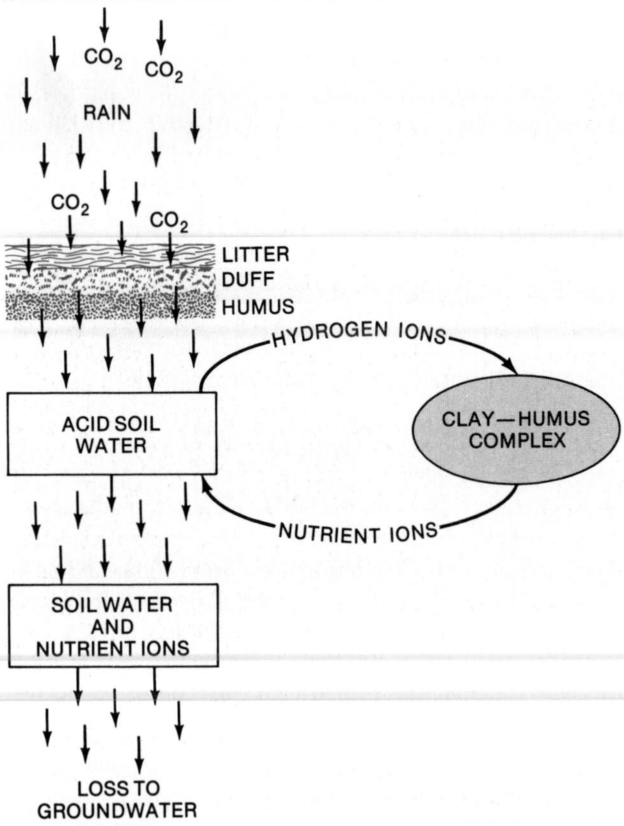

CO_2 CO_2
RAIN
CO_2 CO_2
LITTER
DUFF
HUMUS
HYDROGEN IONS
ACID SOIL WATER
CLAY—HUMUS COMPLEX
NUTRIENT IONS
SOIL WATER AND NUTRIENT IONS
LOSS TO GROUNDWATER

Fig. 24-14. ***Leaching.*** *Leaching removes valuable nutrient ions from the soil and increases soil acidity by leaving hydrogen ions to take up vacant positions on the clay–humus complex.*

Indeed, in some desert locations downward soil-water movement is temporary because surface evaporation draws it upward. At the surface, water is lost as vapor to the air and salts are left behind (Fig. 24-15).

Through leaching, nutrient ions are washed downward and out of the soil.

Salinity

Plants in general and nearly all agricultural crops are intolerant of sodium salts in the soil. Fortunately, with any leaching at all, sodium ions are among the first to be removed from the clay–humus complex. Why, then, do we worry about salinity? The problem is that areas of little leaching—the world's deserts—not only have many needed plant nutrients but also clear skies and, often, warm temperatures. In short, they make good agricultural areas *if* water is available for irrigation. Many countries (for example, the United States, Egypt, the Soviet Union, and Israel) have constructed major water-diversion projects to irrigate dry lands for agricultural use.

With few exceptions, irrigation in areas of high evaporation makes salts move upward to the root zone and soil surface, reducing productivity and killing plants. Agriculturists respond by applying more water in an attempt to leach away the salts. As a result, many arid regions develop new problems such as waterlogging and salt pollution of rivers. If

Fig. 24-15. ***During infrequent showers this desert basin fills with a shallow sheet of water.*** *As it evaporates, its salts are deposited on the surface. In addition, for several weeks thereafter further soluble minerals are pulled from the damp soil by capillary action, and the result is a soil that is toxic to plants.*

salt tolerance cannot be increased in agricultural plants and water pollution reduced, many of these croplands will probably have to be abandoned. This would have serious effects on world food supplies.

Irrigation in high-evaporation areas makes salts move upward to the root zone and soil surface.

SOIL COLOR

The color of soil is one of its most obvious characteristics, and it is tempting to try to relate certain vivid or striking colors to fertility. Locally, it may be possible to do so, but few generalizations work on a larger scale. For example, farmers of the Deccan district in India know that their black soils are much more fertile than those of the surrounding areas because they have been derived from dark alkaline volcanic parent material. Japanese farmers, on the other hand, have learned to shun a dark volcanic ash soil that is highly acid despite its color and the popular conception that volcanic soils must be fertile. Wheat farmers of the Ukraine find their black prairie soil to be somewhat alkaline, whereas bog farmers of the Sacramento Delta find theirs to be black but acid. Desert alkali wastes may be either black or white. In Brazil and Vietnam certain bright red soils (*terra roxa* and *terres rouges,* respectively) have a reputation for great productivity, whereas elsewhere in the tropics a reddish hue often means excessive leaching, and the darker alluvials of the river bottoms are preferred. And so it goes. Color can be a clue to fertility, but only for those who know the peculiarities of the local situation.

Color is a clue to fertility only for those knowledgeable of local peculiarities.

One very common source of dark color in soil is fixed carbon, an end-product of organic decomposition (Fig. 24-16). Indeed, carbon often dominates soil color so much that it masks the coloring of parent materials. In the prairie, for instance, the grassy turf with its myriad of hair roots dies back each year. This dead material (as seen in our discussion of mull soils) is rapidly converted to soil humus, and as humus breaks down carbon is released to color the soil. The Russian term for such soils, ***chernozem,*** meaning black earth, is especially fitting.

Iron, a common element in soils, is the most important inorganic coloring agent, but the color it gives the soil depends to a great extent on drainage and the presence or absence of oxygen. With good drainage and plenty of oxygen, iron appears in its oxidized or "rusty" form and produces red colors. In the absence of large humus content, as under conditions of high rainfall and high temperature (for example, in the Humid Subtropics), soils are especially red. Where drainage is poor and oxygen is in short supply, as it is in marshes and swamps, reduced iron gives a bluish-gray coloring. Where the soil is moist and aerated, as in the tropical rain forest, iron may be both oxidized and hydrated, giving a yellow color. Additional coloring agents, often associated with whitish colors, are quartz, white clays, carbonates, gypsum, and salts.

Fig. 24-16. ***Black and friable in texture, this field is rich in humus from repeatedly plowed inorganic debris.*** *Bits of last year's waste stems and straw are visible in the foreground, not yet decomposed into humus.*

THE SOIL PROFILE

It has been said that a soil does not deserve the name "soil" until it has developed some sort of vertical zonation or ***soil profile*** (Fig. 24-17). Only then has it evolved into a proper

IRRIGATION

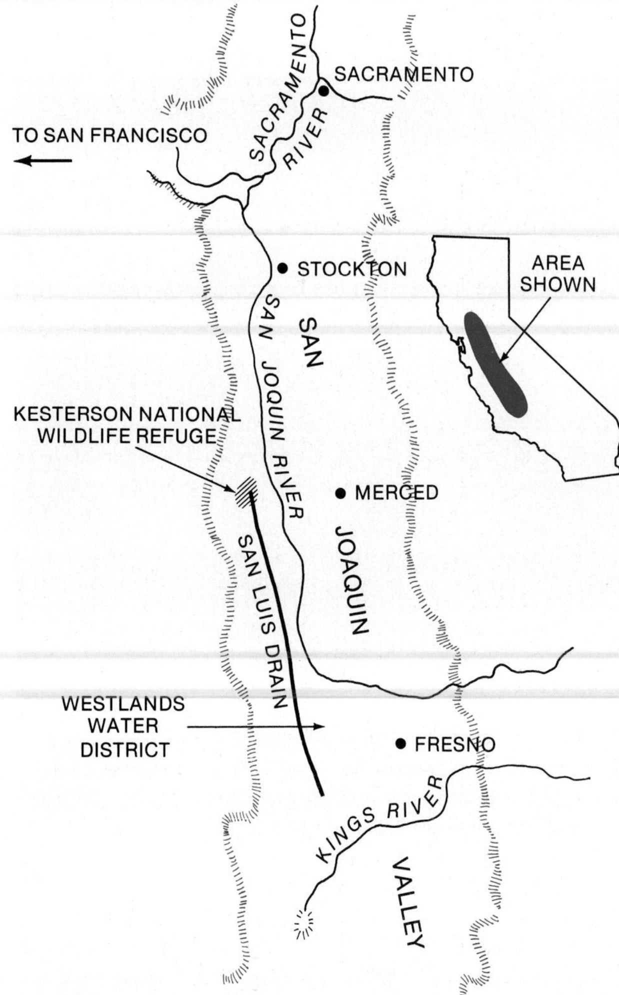

San Joaquin Valley, California.

Irrigation of arid lands offers the prospect of abundant harvests. Crops ranging from grains to vegetables and citrus do well under the desert sun. It has been said, however, that irrigation is at best a temporary expedient—that no matter how carefully the operation is controlled, it is doomed to fail. The problems are waterlogging and salinity buildup, and no project old or new has escaped damage from both to some degree. Irrigation in California's San Joaquin Valley is an example.

Large-scale irrigation in the San Joaquin Valley began with statehood in 1848. By 1860 water diverted from the San Joaquin River irrigated over 400,000 acres, and soon new turbine pumps brought groundwater to the surface to spread on thousands of additional acres. In 1937 Congress authorized a grand scheme to further expand irrigation—the Central Valley Project—which called for a massive dam on the Sacramento River, a large reservoir behind the dam, and a system of canals to distribute water to farm fields.

The eastern half of the San Joaquin Valley drains naturally to the San Joaquin River and on to San Francisco Bay. Waterlogging occurs in some fields, but most of the salt is washed away before it can accumulate. Salt pollution of the San Joaquin River, however, is a continuing problem.

The west side of the valley—much of which is in the Westlands Water District—is more arid and does not have natural drainage to the main river. Dense clay below the surface inhibits percolation, and soils are naturally salty and alkaline. Irrigation on the west side has always caused severe waterlogging and salt accumulation.

When additional water for Westlands farms became available after 1960 as part of the California State Water Plan, state officials and the United States Bureau of Reclamation planned a drainage ditch to carry salty wastewater all the way to San Francisco Bay. However, they constructed only 85 miles of the drain (San Luis Drain), with discharge into Kesterson National Wildlife Refuge instead of San Francisco Bay.

Kesterson National Wildlife Refuge, containing over 6000 acres of wetlands, is an important resting area for migratory birds on the Pacific Flyway and is home to nine endangered species, including the blunt-nosed lizard and the Southern Bald Eagle.

Irrigators began heavy dumping of wastewater into the refuge in 1981. By 1982 biologists discovered a sharp decrease in the refuge's populations of catfish, carp, and bass; only mosquito fish survived in measurable numbers. The following year researchers found a high rate of birth failures and deformities among birds. Some of the chick embryos were missing eyes and beaks; others had only stubs instead of legs and wings.

Investigators from the Fish and Wildlife Service determined that drainage water from Westlands farms was to blame. Their tests showed that the water contained high levels of dissolved salts and metals—boron, selenium, arsenic, lead, copper, and chromium—that can cause sickness and birth defects in wildlife. Most troubling, however, were the high concentrations of selenium, a trace element that is known to cause birth defects at relatively low levels and death at high levels. Between January and April 1984 alone more than 15,000 birds died at Kesterson. In 1985 the federal government closed the refuge to all drainage water dumping.

Westlands farmers must now retain all their drainage water. Soil salinity and waterlogging are increasing with no real solution to the problem in sight. Environmentalists warn that extending the drainage ditch to San Francisco Bay will harm wildlife there, and neither farmers nor government agencies want to pay for treatment plants to remove salts and other pollutants from wastewater. To clean up Kesterson, researchers are now exploring the possibility of using soil fungi to convert selenium salts into harmless volatile compounds.

The situation at Kesterson may be only the "tip of the iceberg," according to wildlife experts. A recent report by the Fish and Wildlife Service listed 84 national wildlife refuges as having problems with contaminants from runoff. In at least 32 of these, selenium, arsenic, boron, or farm pesticides leached out by agricultural drain water were found to be the major factors in the contamination. Meanwhile, the promise of sustained high yields from irrigated farms in arid lands is fading.

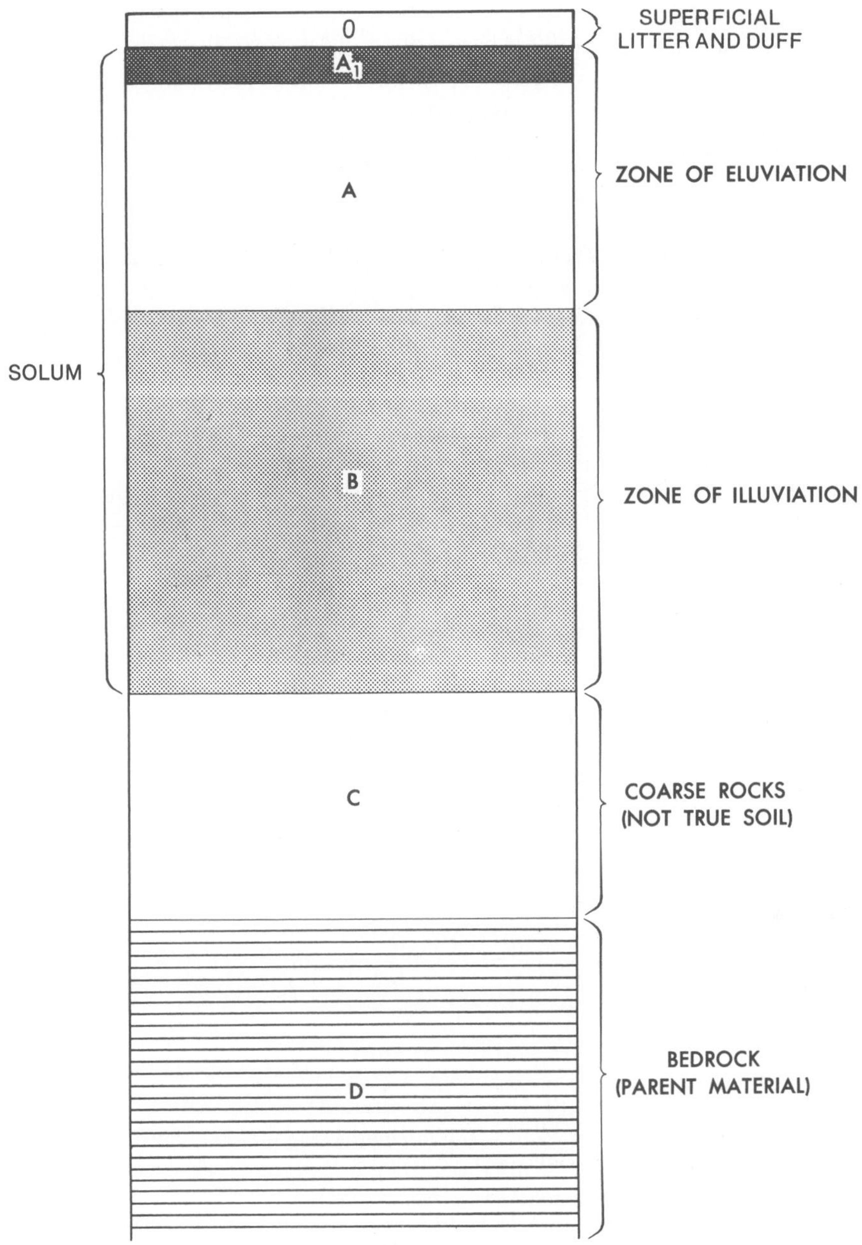

Fig. 24-17. ***Idealized and simplified soil profile.*** *These horizons are encountered, at least to a degree, in nearly every mature soil.*

life-supporting material, and the soil profile is the measure of its maturity.

A vertical cut through most soils, from the surface to unaltered parent material, will display a cross-section with prominent horizontal bands called ***horizons.*** At the top, but not a true part of the soil, is the O horizon, which contains litter and duff. The soil proper begins with the A horizon, the layer often called the ***zone of eluviation*** (from *e* meaning out and *luv* meaning washed). This layer shows greater leaching and loss of finer soil particles through percolation

A vertical cut through most soils will display prominent horizontal bands called horizons.

than any other layer. It thus tends to have a somewhat coarser texture and lighter color than the horizons below it. However, the upper part of the A horizon (A_1) may be black to brown owing to the presence of organic matter, especially in grassland soils.

The B horizon, below the A, receives the minerals and

clay particles washed out of the upper layers and is the ***zone of illuviation*** (from *il* meaning in and *luv* meaning washed). It is often of higher density than the A layer, and it may contain iron-rich deposits. In some soils, it develops into a *hardpan,* a layer of closely packed soil cemented together by silica or clay that slows the downward movement of water.

Still farther down is the C horizon, which is only one step removed from solid rock and thus not yet true soil. It is made up of weathered and broken rock particles and, as time goes on, will become part of the B horizon at its upper margin. Below this layer is the parent material itself, sometimes called the D horizon.

Obviously, no two soils will have identical profiles. By recognizing their differences, we can sort out soil types and draw boundaries between them. The A_1 horizon, for instance, is often missing where vegetation is minimal, as in the desert, or where decomposition is too rapid for it to develop, as in the Humid Tropics and Subtropics. Mull soils of the prairie, however, are so dominated by the A_1 horizon that the remainder of the A horizon is seldom recognizable.

No two soils have identical profiles.

In some soils the margin of each horizon is sharply defined, whereas in others it is blurred and shapeless. Yet most soils have some kind of profile, and it is a recognizable feature of soil development. Only where parent materials have been recently deposited or where some influence such as a steep slope interrupts their proper formation do soils entirely lack profiles.

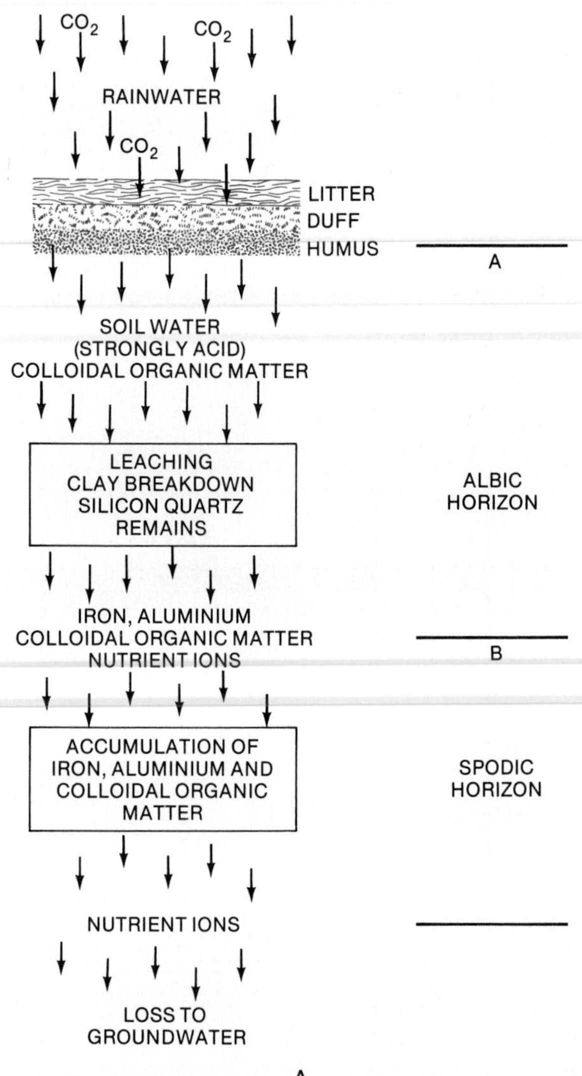

Fig. 24-18. ***The three main soil-forming processes and resulting profiles.*** *Podzolization* (A) *occurs in cool, wet regions with coniferous forest; laterization* (B) *characterizes warm, wet regions with broadleaf forests; and calcification* (C) *is found in semiarid grasslands. Calcification is also the dominant process in arid regions, but grass does not do well under extreme dryness and the humus-rich mollic epipedon does not form. Salinization (not shown here) is an additional soil-forming process of local importance in dry areas (see text).*

PROCESSES OF SOIL FORMATION

Under the influence of climate and vegetation, the four elements of the soil—mineral matter, organic matter, gases, and moisture—interact to produce typical profiles. Let us look at four processes of soil formation in detail.

1. Podzolization
2. Laterization
3. Calcification
4. Salinization

Podzolization

In the forest soils of the northeastern United States, heavy precipitation creates the conditions needed for forest vegetation and leached soils. The presence of coniferous trees and a cool climate result in thick, acidic litter and duff.

Precipitation percolating through the soil not only removes nutrient ions, but also mobilizes iron, breaks down clay, and removes aluminum leaving a concentration of silicate (silicon dioxide; quartz) in the A horizon (Fig. 24-18*A*). The A horizon may be dark at the very top because of humus (A_1), but it is otherwise very light, often being described as ash colored. The Russian word ***podzol,*** meaning ash-colored soil, has long been used to describe this general soil type. The whitish A layer is the ***albic horizon.***

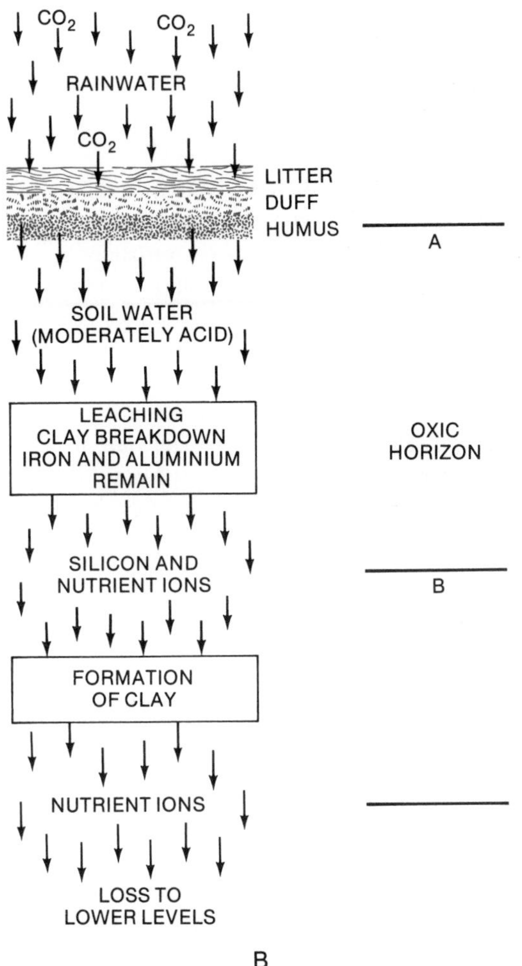

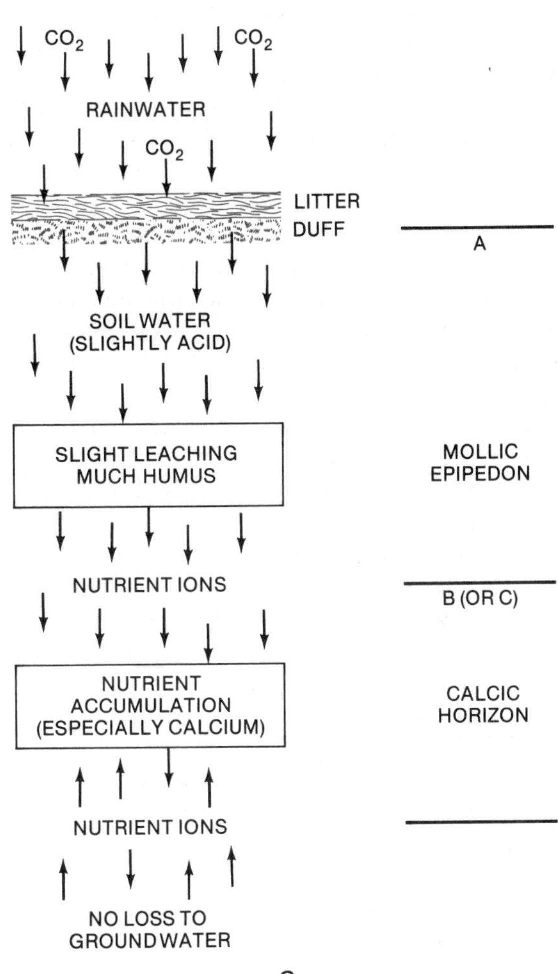

Fig. 24-18. (continued)

Rather than being washed from the soil completely, iron, aluminum, and some suspended organic matter as well accumulate below the albic horizon to form a reddish-stained or ***spodic horizon*** in the B layer. This process is called ***podzolization.*** The resulting soils, called ***spodosols,*** are typical of the middle- and high-latitude coniferous forests of North America, Europe, and Asia.

In spodosols, iron, aluminum, and some suspended organic matter form a reddish-stained or spodic horizon in the B layer.

Laterization

The humid Southeast of the United States also has heavy precipitation resulting in forest vegetation and leached soils. In this area temperatures are higher and winters are shorter, and decomposition of forest litter and duff (primarily from broadleaf trees) is more rapid and continuous. The warmer conditions and lower acidity of percolating water from precipitation result in a process called ***laterization,*** yielding soils traditionally called ***latosols*** but now more frequently called ***ultisols*** or ***oxisols.*** (Soil types are described in the next chapter.)

Here again, the clay in the upper soil is broken down, but in this instance silicon rather than aluminum is washed downward (Fig. 24-18*B*). Aluminum and iron remain, forming a reddish or, with extreme leaching, a yellowish, ***oxic horizon*** (oxidized, A horizon). Some of the washed out silicon unites with aluminum and oxygen present at depth to form clay, giving a light-colored B horizon. An oxic horizon that is very high in iron and aluminum oxides is called ***plinthite.*** With repeated wetting and drying, as occurs after the natural forest cover is removed, plinthite may become a nearly irreversible hardpan or "ironstone" that is of little or no agricultural value. In some tropical locations, the oxic horizon is so high in aluminum that it constitutes aluminum ore (bauxite) and is actually mined.

Calcification

The processes of podzolization and laterization enable us to see how temperature affects soil formation. Let us now look at a third process, *calcification*, which is primarily a response to moisture control (Fig. 24-18*C*). This process is typical of midlatitude grasslands and dry areas in general.

Calcification is primarily a response to moisture control.

Mull soil, with abundant humus from plant leaves and roots and numerous soil animals, characterizes midlatitude grasslands. Annual precipitation of around 20 inches (50 cm) supplies enough moisture for decomposition but not enough for heavy leaching. In fact, the depth of percolation is usually only about a meter or so below the surface. Here nutrient ions, especially calcium, can accumulate as a whitish deposit to form a ***calcic horizon***.

The A layer, with its high organic content and its great supply of nutrients, is very *friable*; that is, it crumbles easily and is soft to the touch. These grassland soils have a "soft upper-layer" or ***mollic epipedon*** and are therefore called ***mollisols***. (The words "mollic epipedon" and mollisols come from the Latin word for soft, *mollis*.) Because of their nutrient content and friability, mollisols are often regarded as the best agricultural soils, especially for grains such as wheat and barley.

Salinization

Salinization is a soil-forming process that occurs locally in arid and semiarid regions. It is most typical of valley bottoms in deserts where short-lived streams bring dissolved salts washed from upland soils. The salty water forms a shallow lake that soon evaporates leaving a salty crust behind. Soil water, including some groundwater if the water table is near enough, is drawn to the surface by evaporation, and additional salts accumulate. Without leaching, the sodium, calcium, potassium, and other salts simply stay in the upper part of the soil profile. Because concentrated salts are toxic to most plants, the resulting mineral soils are usually unvegetated.

Salinization is a soil-forming process that occurs locally in arid and semiarid regions.

Other Profiles

Although calcification, laterization, and podzolization characterize certain climatic extremes and salinization occurs only in arid and semiarid regions, great variation in soil profiles exists. For example, climatic conditions midway between that of the cool spodosols and warm oxisols, as in the Middle Atlantic United States, result in mature soils that share characteristics of both types. In addition, with greater dryness than that associated with mollisols, as in desert locations, mineral soils lack humus but may still have a calcic horizon. Areas of poor drainage in general may develop characteristic profiles that are nearly independent of prevailing climatic conditions.

THE HUMAN ELEMENT

Once human culture evolves beyond simple hunting and gathering, the delicately balanced soil ecology is disrupted. Herders place large numbers of domesticated animals on the natural grasslands. If wealth is measured in cows, people often allow their stock to overgraze the range. They also burn the coarse growth, encouraging new green shoots and temporary better feed. This practice alters the natural vegetation, burning away humus and encouraging erosion. It also modifies the soil-forming processes, closely linked to their covering vegetation. Loggers who remove forest trees are not the only ones responsible for soil changes. Even the smallest farmer must clear the land as the first step. Immediate soil changes are inevitable here too.

The evolution of human culture beyond simple hunting disrupts the delicately balanced soil ecology.

Let us examine specifically what causes these changes. The plow turns up the A horizon (and part of the B horizon on shallow soils), forming ready-made channels for surface runoff to carry away loose soil. Whereas the original vegetation returned nutrients to the soil when it died and decomposed, humans replace this vegetation with their own in the form of crops, and then, each year, take the crops off the land to be consumed elsewhere. Even if the product is run through a cow and returned as farm manure to the soil, much of the original fertility is lost, for there stands a healthy animal—evidence that some of the nutrition has been used up.

It is possible, of course, to improve soil fertility. Crushed lime to counteract acidity, cover crops plowed back regularly to offset lack of humus, and tilled fields to facilitate drainage—all are methods that will make the soil more productive (Fig. 24-19). In addition, specially blended chemical fertilizers, designed to meet the needs of particular crops, are being applied. In every case, however, humans become an element in soil formation and change the original soil.

Soil erosion, the physical removal of soil by wind and runoff, is probably the most striking and best publicized of the damaging effects of humans on soil. Fundamentally, the removal of natural vegetation is responsible for soil erosion. Logging, mining, or agriculture may require this removal, but once the bare soil is no longer anchored by plant roots, it can be very effectively attacked by the elements. Great gullies form rapidly on steep slopes—miniature canyons biting down to bedrock. More insidious, and therefore in some ways more dangerous, is sheet erosion that occurs on clean-tilled gentle slopes following a heavy shower. The surface runoff carries soil with it in a sheet rather than in a well-defined channel. Wind can also carry large quantities of fine soil following overgrazing or the unwise plowing of grasslands in semiarid regions.

Removal of natural vegetation is responsible for soil erosion.

Erosion can be countered, at least to a degree, if it is treated as an urgent priority and if the resources and techniques that will combat it are known and available. In more enlightened societies, strip miners are now required to fill their cuts and to plant trees, and hydraulic mining has been declared illegal (Fig. 24-20). Lumber operators replant as they cut, and reforestation of barren slopes, whose timber was removed centuries ago, is more widespread than ever. On the farm, contour plowing, terracing, strip cropping, and cover crops are all helping to reduce erosion (Fig. 24-21). Yet even in the United States, where we regard ourselves as technologically advanced, we can still stand beside the tawny Mississippi and watch somebody's farm go by every few minutes.

CONCLUSION

Soil is much more than just loose material above bedrock; parent material, organic matter, gases, and water are its main components. The production of mature soil depends on a

Fig. 24-19. ***Aerial topdressing (fertilizing) of pastureland in New Zealand.***

Fig. 24-20. ***Strip mining simply destroys the land.*** *Even leveling and reforesting do not return it to its original state (although they help), for the soil horizons have been scrambled and the drainage rearranged.*

Fig. 24-21. ***Strip cropping and contour plowing can both defeat the efforts of surface drainage rivulets to run at high velocity directly downslope (across contours).*** *These are the little streams that carry off the topsoil and eventually create gullies in the field.*

complex interaction of climate, topography, geology, hydrology, vegetation, and animal life over a long time period. Gradually, as a soil comes into equilibrium with these many variables through one of the soil-forming processes, it takes on certain characteristics such as distinguishable horizons, a particular profile, and a certain fertility level. The study of soils, then, offers us a view of how the various parts of the physical system interact. It is hoped that our knowledge of these interactions and how they affect the soil will help us avoid misusing this valuable resource.

KEY TERMS

pedology
regolith
parent material
residual
transported
soil texture
litter
duff
humus
nutrient ions
mor soils
mull soils
moder soils
mineral soils
pores
peds
soil structure
pH scale
clay–humus complex
leaching
chernozem
soil profile
horizons
zone of eluviation
zone of illuviation
podzol
albic horizon
spodic horizon
podzolization
spodosols
laterization
ultisols
oxisols
oxic horizon
plinthite
calcification
calcic horizon
friable
mollic epipedon
mollisols
salinization
soil erosion

REVIEW QUESTIONS

1. Why is it wrong to think of soil as "dirt"? What is a better way to think of it?

2. How can parent material influence the soil nutrient balance?

3. What gives soil its characteristic texture? Why is texture important?

4. How do plants and organisms that live above ground make their way into the soil when they die? Why is this process important in the formation of soil?

5. What conditions are necessary to produce mor, mull, moder, and mineral soils?

6. Why do we say that soil "breathes"? Why is "breathing" important? Why do farmers disturb the structures that allow soil to breathe?

7. What is acid rain and how can it be harmful to the soil?

8. How can irrigation be both helpful and harmful to the soil? Under what conditions is it most harmful?

9. Why does podzolization occur in the forested areas of the American Northeast, whereas laterization occurs in the forested areas of the American Southeast?

10. What is the *main* thing that humans have done to cause soil erosion? How can this problem be remedied?

APPLICATION QUESTIONS

1. Using a sample of the soil in your area, write a description of its texture, structure, and profile. Then see if you can explain why it exhibits these characteristics.

2. Describe three different areas—real or imaginary—where climatic conditions and vegetation type would lead you to expect different soil profiles. How and why would these profiles differ?

3. What is the color of your local soil? How is its color related to its fertility? Can you account for both color and level of fertility?

4. Why do soils in areas with high rainfall tend to have lower pH values than soils in arid or semiarid areas?

5. In a good world atlas, look up the areas in the world where bauxite is mined. What do they have in common? Can you describe what their soil might be like? Why?

6. Find out if your area has a soil erosion problem. Does your state have a soil conservancy program to deal with it? What is being done—and what still needs to be done—to prevent further erosion?

CHAPTER 25

CLASSIFICATION AND SOIL REGIONS

OBJECTIVES

After studying this chapter, you will understand

1. How our modern concept of soil classification evolved.
2. How we now classify soils.
3. The characteristics of the various soil orders.
4. How the soil orders have evolved.
5. How the soil orders are distributed worldwide.

Plowing the field and garnering the crop is accomplished with thousands of variations around the world; the irrigated oasis in Algeria is profoundly different from an Iowa mechanized farm. With the onset of the summer monsoon, an Indian farmer turns up the soil to begin yet another speculative cropping cycle. That the methods are ancient and the scale of the enterprise small does not mean that this farmer's aims are so very different from every other—a primary concern that the soil maintain the ability not only to produce a current crop, but endless ones into the future.

INTRODUCTION

From our discussion of world climates in Part Two, we know that numerous regional types together make up global climate patterns. There are numerous soil types as well, many of which relate closely not only to climatic regions, but also to hydrology, landforms, and vegetation. In this chapter we identify and classify soil types and then look at their global distribution. Knowledge of soil types will be important in our discussion of world vegetation in Chapter 27.

SOIL CLASSIFICATION

We human beings delight in making lists. Aware of the infinite variety and apparent chaos of nature, we try to describe it as a regular, organized system with all the appropriate labels. We have certainly tried to develop such a system for soils. For much of the past hundred years, in fact, soil scientists have been trying to construct a universally meaningful and useful soil classification. All the while, knowledge of world soils, especially in the tropical lands, has been rapidly increasing. This ever widening knowledge combined with strong feelings of nationalism has prevented the acceptance of a universal system. It is not hard, then, to understand why we have yet to agree on soil classification. With this consideration in mind, let us narrow our sights and concentrate on the American experience with soil classification.

A History of Soil Classification in the United States

Before the 1930s, soil scientists in the United States stressed the nature of parent rock in interpreting soil development. Geology was the basis for classification. This geological view rapidly waned as C. F. Marbut (then chief of the U.S. Soil Survey) translated the work of V. Dokuchaiev, a Russian soil scientist, into English in 1927.

The Russian system of classification took an evolutionary approach to soil development. It stressed the close relationships between climate, vegetation, parent material, topography, and time. Because it was then believed that climate and vegetation occurred in broad units or zones, soil was also considered to be "zonal" in distribution.

Applying a wealth of field experience, Marbut produced an American classification system based, in part, on the Russian model. He recognized 30 or so "Great Soil Groups," which he subdivided into families, series, and types. The Marbut system, with some modification, was adopted by the U.S. Soil Conservation Service in 1938 and became the government standard. It was also favored by geographers and still continues to be seen in some texts.

Soil Classification Today

By the early 1950s, it was becoming increasingly evident that Marbut's aging classification, for all its pioneering merit, could no longer accommodate the new soil findings worldwide. A more up-to-date system was clearly required.

Begun in the United States as a cooperative venture by both government and university pedologists, the effort proved to be tedious and contentious. By 1960, after it had been revised at least seven times, the new classification was born; it was christened ***The 7th Approximation.*** (Minor revisions appeared as supplements in 1964, 1967, and 1968.) The U.S. government officially adopted the system in 1965.

This new system continues to acknowledge the importance of climate and vegetation. Unlike the older system, however, it treats soils as a continuum—continuously variable from one location with one set of conditions (climate, vegetation, rock type, etc.) to another with a different set of conditions.

The Comprehensive Soil Classification System treats soils as a continuum.

Rather than using certain processes of soil formation (laterization, calcification, and podzolization) as the basis for classification, the ***Comprehensive Soil Classification System*** (as it is officially called) is based on observable soil characteristics, especially surface and subsurface ***diagnostic horizons*** (Fig. 25-1). The classification has several levels, from most general to most specific:

1. ***Orders,*** of which there are 10 (Fig. 25-2)
2. ***Suborders*** (47)
3. ***Great Groups*** (225)
4. ***Subgroups*** (970)
5. ***Families*** (4500)
6. ***Series,*** of which there are more than 13,000 in the United States alone

SOIL ORDERS AND WORLD PATTERNS

When you compare the world distribution of soil orders (Map VI) to world climates (Map I) and world vegetation

Surface horizons (epipedons)

Ochric epipedon (from Greek, *ochros,* pale). A light-colored (gray to brown) horizon, low in humus; present in alfisols, aridisols and ultisols.

Mollic epipedon (from Latin, *mollis,* soft). Dark brown to nearly black horizon, very high humus content, rich in plant nutrients. Characteristic of mollisols.

Subsurface horizons

Albic horizon (from Latin, *albus,* white). Light-colored (whitish to light gray) horizon, sandy in texture, with clay and iron oxides leached out. Underlain by a spodic horizon in spodosols.

Argillic horizon (from Latin, *argilla,* clay). Enriched with illuvial clay, usually the B horizon. Common to ultisols and many other soils.

Calcic horizon (from Latin, *calcis,* lime). Horizon with accumulation of calcium carbonate. Common in aridisols and some mollisols.

Oxic horizon (from French, *oxide,* oxide). Severely weathered horizon with iron and aluminum oxides, some clay. Characteristic of oxisols.

Spodic horizon (from Greek, *spodos,* wood ash). B horizon in spodosols, accumulation of aluminum oxide and humus, with or without iron oxide.

Fig. 25-1. ***Diagnostic horizons.*** *A selected list of surface and subsurface diagnostic horizons used in the Comprehensive Soil Classification System.*

(Map II), you see very similar patterns. In other words, many soils have "regional distribution." Yet several orders—entisols, inceptisols, and histosols—may occur in almost any climate, and alfisols are found in a variety of midlatitude climatic types. These, then, do *not* have regional distribution. Let us consider the characteristics of each of the soil orders and their worldwide distribution. For a more detailed map of soil orders in the United States, refer to Figure 25-3 and for representative soil profiles, see color insert.

Oxisols

Oxisols are characteristic of old upland surfaces in the tropics and subtropics where vegetation is broadleaf evergreen (commonly rain forest) and climatic conditions are always warm and humid (Figs. 25-4, 25-5). Rapid decomposition leaves little accumulation of litter, duff, and humus. As a result, soils in this order are primarily mineral.

Oxisols are characteristic of old upland surfaces in the tropics and subtropics.

Weathering in oxisols extends to great depths. Profiles often extend downward for 33 feet (10 m) or more, making these the deepest of world soils. The most outstanding feature of oxisols is the oxic horizon, which occurs within about 2 meters of the surface and contains mainly iron and aluminum oxides. These oxides, together with the low organic content, result in the typical red to yellow coloring. They may also indicate the presence of plinthite, which, as mentioned in Chapter 24, forms an irreversibly solid ironstone if allowed to become completely dry after the destruction of protective vegetation.

Extensive areas of oxisols occur in South America, especially in Brazil, and in nearby countries in the north and in Central Africa. Farmers in both localities have long used

Soil Order	*Source of Root Word*	*General Characteristics*
Oxisols	French *oxide,* oxide	Highly weathered, old soils of low latitudes
Ultisols	Latin *ultimus,* ultimate	Weathered soils of the middle latitudes with low nutrient content
Alfisols	*Al* for aluminum, *f* for iron	Weathered soils of the middle and low latitudes with high nutrient content
Spodosols	Greek *spodos,* wood ash	Soils of cool, humid forests with a spodic and albic horizon
Mollisols	Latin *mollis,* soft	Grassland soils of the middle latitudes with a mollic epipedon
Aridisols	Latin *aridus,* dry	Desert soils, often with a calcic horizon
Inceptisols	Latin *inceptum,* beginning	Soils just beginning to show horizons, no specific climate
Vertisols	Latin *verto,* to turn	Clay soils with surface cracks when dry, tend to "invert" with time
Entisols	*Ent* from re*cent*	Youthful soils lacking horizons, no specific climate
Histosols	Latin *histos,* tissue	Highly organic soils of bogs and swamps

Fig. 25-2. ***Soil orders.***

Fig. 25-3. Generalized map of soil orders in the United States.

them in the traditional agricultural practice of ***shifting cultivation.*** First, they partly clear and burn a small section of forest. They then plant crops at random among charred trees, allowing them to benefit from nutrients released by burning. After two to three years of harvesting and continued leaching from heavy rains, nutrient supplies are so exhausted that the farmers must find a new site with sufficiently mature forest to begin the process again.

Under modern agriculture, where farmers permanently clear large areas of forest, these soils can remain productive for tropical crops like coffee, rice, sugar cane, bananas, and pineapples but only with generous applications of nitrogen, phosphorus, and potassium. Unfortunately, the cost of fertilizer is often beyond the means of local farmers who are then faced with declining yields, erosion, and ironstone.

In the United States, oxisols are limited to the state of Hawaii where they are typical on the older islands like Kauai, but they are also seen in Puerto Rico and the Virgin Islands. Worldwide, oxisols account for about 10 percent of the area occupied by soil orders.

Ultisols

Ultisols are the soils of old land surfaces (dating back to the Pleistocene Age or earlier) and are confined primarily to the Humid Subtropics. Although they are the most weathered midlatitude soils, they are not sufficiently weathered to be oxisols. At the same time, their long history of leaching has left them too low in nutrient content to be alfisols or mollisols.

In addition to severe leaching, ultisols are characterized by clay accumulation in the B layer (***argillic horizon***) resulting from the downward movement of clay particles from the A layer as water percolates through the soil. This eluviation and illuviation of clays is simply a mechanical washing; the clays are not broken down and reformed at depth, as may be the case with oxisols.

Ultisols are the soils of old land surfaces and are found primarily in the Humid Subtropics.

Fig. 25-4. ***Tropical rain forest oxisols.*** *Oxisols develop under tropical rain forest vegetation and a warm, wet climate.*

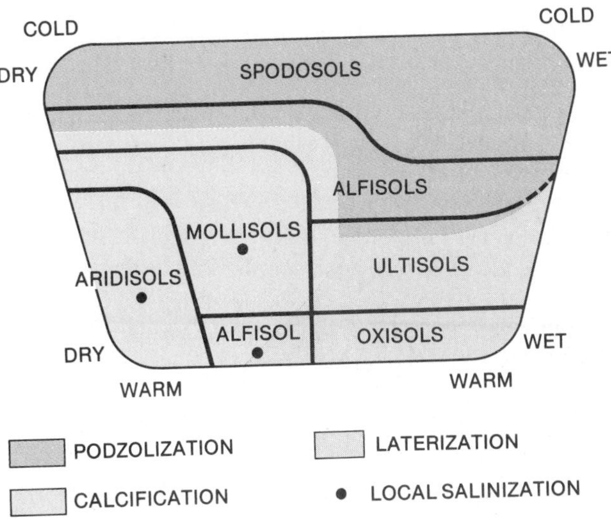

Fig. 25-5. ***Soil-forming processes and soil order distribution.*** *This diagram shows the general relationship between soil-forming processes and distribution of soil orders with a regional distribution. Soil orders lacking a regional distribution (histosols, vertisols, inceptisols, and entisols) are not included here, although they may be influenced by soil-forming processes.*

Natural vegetation on ultisols is usually a closed forest of pines and hardwoods or a more open mixture of trees and grasses (savanna). Organic matter decomposes rapidly but, aside from a dark, humus-rich A_1 layer, these soils tend to be reddish and even yellowish owing to the presence of iron and aluminum oxides.

Given their moderate acidity and advanced weathering, ultisols would not seem likely to be first-class agricultural soils. On the contrary, with proper management they can be among the world's most productive, competing favorably with alfisols and even mollisols. Having the proper climate helps, specifically a long frost-free period, adequate rainfall to maintain sufficient soil moisture, and abundant surface water for supplemental irrigation when needed. The inherently low nutrient reserve in ultisols can be overcome by a combination of liming (mixing in powdered limestone) to reduce acidity, frequent application of fertilizers, and crop rotation including the use of legumes. In addition, when a shot of insecticide is applied now and then to control the insect pests that abound in warmth and moisture, crops like cotton, soybeans, and even pines for wood and paper do very well.

In the United States, ultisols dominate the area from New Jersey to Texas, and coastal and interior locations in California, Oregon, and Washington. Ultisols occur over broad areas in South America (especially in Colombia and southern Brazil), western Africa, Southeast Asia, and northeast Australia. In total, they account for 9 percent of world soils.

Alfisols

Alfisols have the most diverse distribution of major orders. They occur in humid and subhumid locations in the midlatitudes and the tropics, as well as under tallgrass prairie, savanna, and deciduous forests. The overall theme to the pattern of alfisols is one of transition, however. In the tropics, they occupy a transitional position between desert aridisols and rain forest oxisols. In the midlatitudes, they occupy a transitional position between the semiarid mollisols and humid ultisols and spodosols.

Alfisols occur in humid and subhumid locations in the midlatitudes and tropics and occupy a transitional position.

Because of their transitional position, alfisols share certain characteristics with adjacent orders. Like the ultisols, they have illuvial clay in the B horizon, but they have only moderate leaching, with good supplies of calcium and magnesium more typical of mollisols and aridisols. Yet they do not share the mollic epipedon with mollisols. Instead, they have a gray-to-brown surface horizon with little humus, called an ***ochric epipedon.***

Although the clay layer is not favorable to plant growth, particularly if the A layer has been eroded away, their nutrient status can be maintained to make alfisols productive agricultural soils. Crop yields compare favorably with ultisols and mollisols, and they do not have to be as closely managed as ultisols.

Comprising 15 percent of world soils, alfisols are found on every continent. In the United States they occur over large areas south of the Great Lakes and in Texas, Colorado, and California. In Europe they occupy a broad band closely related to the original broadleaf deciduous forest, and in Africa, a broad band south of the Sahara.

Spodosols

Spodosols occur from the tropics to the margins of polar tundra but develop best in the cold and humid forested lands of the Northern Hemisphere (Fig. 25-6). A typical profile includes:

- Superficial accumulations of litter and duff.
- A thin, dark organic layer (A_1).
- A coarse-textured, heavily leached, whitish albic horizon (A_2).
- A B layer with accumulations of suspended humus and oxides of aluminum and iron (spodic horizon).

Spodosols occur from the tropics to the margins of polar tundra.

Spodosols typically develop under the process of podzolization, which is encouraged by the presence of coniferous trees and sandy parent material. The trees provide additional acids for leaching and breakdown of clays; the parent material allows for easy percolation and the downward movement of humus. The fine bits of humus that collect in the B layer resist further decomposition owing to the persis-

Fig. 25-6. ***Spodosols.*** *Coniferous trees strongly affect the soils they grow in by adding acids to percolating soil water. This spruce forest along the Athabasca River in Jasper National Park has helped to produce spodosols typical of such cold and humid lands.*

tently cold temperatures and the lignin, wax, and resin they contain.

Although they are not naturally fertile, spodosols can be used for cultivation. Acid-loving crops like blueberries do well, but potatoes, vegetables, grains, and pasture grasses—cold-adapted crops that are grown with success on these soils—require generous amounts of lime to reduce acidity and fertilizers to provide plant nutrients. The coarse texture of spodosols allows easy root penetration but only little water retention. As a result, these soils require irrigation in addition to other supplements.

Early settlers in the United States rarely foresaw the difficulties of using spodosols for agriculture. For example, after loggers had removed most of the valuable red and white pine in the southern edge of the boreal forest extending from Minnesota to the coasts of New England, farmers moved in to plant crops. They soon discovered that the costs of production were prohibitive, and so they had to abandon the land. Today, with synthetic fertilizers and large urban markets, these spodosols are used for high-value crops like vegetables and sugar beets, for milk production, and even for the famous potatoes of Maine. Yet worldwide, spodosols continue to support mainly forest vegetation managed for the harvest of timber for construction and pulp for paper.

In addition to Minnesota, Michigan, and the New England states, large tracts of spodosol soils occur in central and eastern Canada, the Netherlands, and a broad band extending from Sweden into the Soviet Union. In total, spodosols make up 6 percent of world soils.

Mollisols

Mollisols, the dark-colored soils of grasslands and some hardwood forests, are the best agricultural soils. Their deep, dark mollic epipedon remains soft and crumbly even when dry, and contains both abundant humus and nutrients. Leaching is at a minimum, though in sufficient quantity to remove the highly soluble salts of sodium and potassium from the profile and to move some of the less soluble calcium carbonate to lower levels where it can precipitate (the process of calcification).

Mollisols are the dark-colored soils of grasslands and some hardwood forests.

Like alfisols, mollisols occupy transitional locations: the midlatitudes between arid and humid lands where the original vegetation varied from widely spaced bunchgrass and shortgrass steppe to tallgrass prairie (Fig. 25-7). The largest body of mollisols in the world is located in a nearly continuous band from Hungary through the heartland of the Soviet Union to mainland China. In South America, they are the soils of northern Argentina (the Pampas), Uruguay, and

Fig. 25-7. Grassland mollisols. The rich mollisols of western Nebraska, as in this view of Scotts Bluff National Monument, developed under a cover of perennial grasses.

Paraguay. In North America, their largest unbroken area occurs in the Great Plains, extending north to Canada and south almost to the Mexican border. Mollisols also occupy part of the northern West (between the mountains) and the interior Pacific Northwest. Accordingly, they are the most extensive of U.S. soil orders. On a world scale, they account for 9 percent of all soils.

Although they are naturally fertile and can sustain high yields, mollisols were among the last soils used in agriculture. Part of the problem was lack of technology. Prior to the 1850s wooden plows could not penetrate the thick sod of the grassland without breaking or becoming mired down. Moreover, at least in the North and South American grasslands, the presence of hostile Indians made ground breaking hazardous. It was better to cut down forests and to plant crops in the more easily tilled ultisols and alfisols of friendly regions.

All this changed, when the all-steel moldboard plow became available and the Indians were brought under control. Thousands of settlers came to the grasslands and soon began to take advantage of the rich mollisols for the production of grains (especially wheat, but also rye, barley, and oats), corn, and sorghum. In the process, they helped establish the foundations for rapid population growth in the twentieth century. Technology continues to play an important role in the use of mollisols. Synthetic fertilizers are applied for even higher yields, large tractors and combines work and harvest fields, and center-pivot irrigation systems reduce the threat of drought.

Fig. 25-8. ***Salinization.*** *Soil in this playa basin in Utah is enriched with salt from surface evaporation. Only salt-tolerant halophytes can persist in such salty soils. Sagebrush, a plant well known for its aversion to high salinity, grows on the surrounding well-drained aridisols.*

Aridisols

Aridisols are the soils of the world's deserts. Because deserts account for nearly one-fifth of total land surface, they are the most widespread of soils with a regional distribution.

Low rainfall in deserts means that aridisols have little or no water leaching and weak horizon development. Profiles are shallow:

- The upper layer is light in color and low in organic matter (ochric epipedon).
- An illuvial B layer is usually absent.
- There is generally an accumulation of calcium carbonate (calcic horizon) in the C layer.

Low rainfall in deserts means that aridisols have little or no water leaching and weak horizon development.

In very old aridisols, the calcic horizon may be several centimeters or more thick with a dense, rocklike quality. This is called ***caliche***.

On all but the lowest positions in desert landscapes, aridisols develop by calcification. In the lowest positions—valley bottoms with poor drainage—aridisols develop by salinization. The resulting soils are easy to locate, for they will support only the most salt-tolerant desert plants called ***halophytes*** (Fig. 25-8).

Under natural conditions, aridisols are not prized for agricultural use. Grazing animals require vast tracts for support—one steer may need over 100 acres. Often, the soil surface is covered with rock and gravel—termed ***desert pavement***—left after the wind has removed fine particles (Fig. 25-9). Yet, if the soil can be irrigated, rocks removed, organic matter mixed in, and nitrogen fertilizer applied, aridisols can be among the most productive soils in the world.

Fig. 25-9. ***Desert pavement.*** *Where strong desert winds have removed fine particles from the soil surface, only rocks and coarse gravel may remain. The resulting surface, known as desert pavement, is shown here in central Australia.*

The costs are high, however. Water must come from deep wells or expensive diversion projects, fertilizer and organic matter must be brought in, and irrigation itself may lead to a buildup in soil salinity and other problems (see Chapter 24).

Aridisols occupy a large area in western North America extending from northern Mexico and including much of eastern California, Nevada, Arizona, and New Mexico. Vast areas of aridisols are found in the Sahara, central and southern Asia, southern Africa, southern South America, and Australia, bringing the world total for this soil order to 20 percent.

Inceptisols

Inceptisols are the soils that form on young geomorphic surfaces such as glacial deposits, flood-plain alluvium, and deposits of volcanic ash (Fig. 25-10). As their name implies they are just beginning. They show some horizon develop-

Fig. 25-10. ***Inceptisols on glacial deposits.*** *The soil profile is just beginning to form on this stratified drift in New England. Swallows favor exposed sand strata for nesting sites.*

ment through chemical alteration but not enough to exhibit the diagnostic layers present in mature soils.

Inceptisols do not show enough horizon development to exhibit the diagnostic layers present in mature soil.

The inceptisol order also includes:

- Soils of the tundra that lack sufficient development owing to extreme cold.
- Soils slowly forming in weathering-resistant quartz sand.
- Soils with poor drainage because of a water table that stands close to the surface at some time during the year.

All this variety means that inceptisols are not confined to any particular climate or vegetation type. They occur on all continents, mostly in humid locations, usually under trees, sometimes under grass. Inceptisols are common on volcanic ash deposits in Oregon, Washington, and Idaho; on young basalt flows in Hawaii; on the tundra of Alaska; on the flood plain of the Mississippi River; and on gentle-to-steep slopes in the Middle Atlantic states. Outside the United States, inceptisols occur in the cold northern regions of Canada and the Soviet Union and along major rivers like the Amazon and the Ganges. They make up 16 percent of world soils.

The agricultural value of inceptisols varies from excellent to poor. Flood-plain soils can be among the most productive, especially in warmer regions like the lower Mississippi where soybeans do well. The volcanic inceptisols of Hawaii produce high yields of sugar cane. Cold tundra soils, however, hold little agricultural potential.

Vertisols

Vertisols contain primarily shrinking and swelling clays. As they dry out, cracks appear that may become more than 1.5 inches (4 cm) wide and may extend to depths of 3 feet (1 m) or more. Loose material from the surface—soil particles and litter—falls into and accumulates in the cracks, leading to a sort of "inversion" of the soil. When the rains return, water running down the cracks wets the clays, which then expand causing some vertical and horizontal movement of the soil. Miniature hills and valleys then form on the surface. Most cracks close completely during the wet season.

Vertisols contain primarily shrinking and swelling clays.

The production of vertisols requires only the right kind of clay and a climate with strongly contrasting periods of wet and dry. The clay may come from weathered limestone or basalt. Suitable climates are those of deserts that receive an occasional drenching, tropical and subtropical monsoon climates, and Mediterranean climates that have a cool, moist winter and hot, dry summer (Fig. 25-11).

For obvious reasons, vertisols do not generally make good building sites, and their use in agriculture can be a little tricky. Clay soils are very difficult to break with a plow during the dry season. They are no easier in the wet season when they become very heavy and sticky, and collect on equipment such as tractor tires, which then spin without moving forward. Only during a short period of a week or so following the first rains of the wet season can the soil be

Fig. 25-11. ***Desert vertisol.*** *A mosaic of surface cracks appear as clay-rich vertisols dry. Short-lived annual plants like this member of the sunflower family may find sufficient moisture within cracks.*

worked easily. If large tractors are available, as they are in the United States, large fields can be plowed and planted in time. Otherwise, fields must be small to accommodate the human and animal effort required. Even with these limitations, vertisols are widely used for the production of sorghum, cotton, corn, and grains.

On a global scale, vertisols make up only 2 percent of all soils. Yet they are well represented in the Sudan, northwest India, and eastern Australia. In the United States, they occur primarily in coastal and central Texas, as well as in Alabama, Mississippi, and California.

Entisols

Entisols lack even the slightest hint of distinct horizons, making them the least developed of world soils. Either there has been insufficient time for horizons to form, or the physical environment is, in some way, keeping them from forming. For example, entisols are found on active flood plains where frequent alluvial deposition leaves little time between additions of sediment for soil to develop. Entisols are also associated with quartz sands of dunes and beaches; cold arctic permafrost; frequently flooded tidal marshes, deltas, and lake margins; eroding mountain slopes; and even areas of human disturbance where mature soils of other orders have had their horizons homogenized by deep plowing or excavation for urban development.

Like inceptisols, entisols are distributed according to local conditions rather than to major climate or vegetation patterns. Yet they are widespread. Entisols account for 13 percent of world soils, and they are well represented in the shifting sands of the Sahara and Namib deserts in Africa, in the thin soils of the Andes Mountains of South America, and in the uplands of central Asia from Turkey to Tibet. In the United States, entisols are widely distributed. In addition to numerous locations along major rivers and in coastal areas, they are found in the Rocky Mountains and in the sandy areas of Florida, Georgia, Alabama, and Nebraska (Sand Hills).

Like inceptisols, entisols are distributed according to local conditions.

It is impossible to generalize about the agricultural value of entisols. Here again much variety is in evidence. Some active flood plains and sandy areas are very productive with crops from corn to peanuts, whereas beaches and areas of permafrost do not fit into the current agricultural picture at all.

Histosols

Histosols are the highly organic soils of bogs and swamps (Fig. 25-12). They develop in depressions where water accumulates and where plants such as cattails, reeds, rushes, and willows thrive. Because so little oxygen is available under saturated conditions, litter, consisting of leaves and stems, together with dead roots, decomposes very slowly. This results in the gradual accumulation of thick, dark brown or black organic deposits called ***peat*** if the original plant parts can still be recognized, or ***muck*** if they cannot.

Histosols are the highly organic soils of bogs and swamps.

Fig. 25-12. ***Histosol fuel.*** *The thick layer of peat in some histosols is an important source of fuel. Here at Magillicuddy's Reeks in Ireland, peat has been cut and set out to dry before being used for home heating.*

Surprisingly, histosols are potentially of considerable agricultural value. They can be very productive, especially for vegetables, providing specialized management techniques are used. Farmers must first drain them with a series of ditches or drain tiles and plow them deeply. The farmers must then plant and smooth the seed bed with a roller to keep the light soil from blowing away.

The initial drying and planting create a new problem—fire. Farmers must take great care in the use of machinery and protect fields from brush fires. Even smoking may be hazardous to histosol health. Fires can smolder in peat and muck for months or even years. Drainage also increases the rate of decomposition, which in turn causes the surface to subside—as much as 10 feet (3 m) in 40 years has been recorded.

Histosols occur in all climatic regions from the high latitudes to the tropics, but they are most widespread in cool, humid areas where the topography inhibits drainage. Histosols are well represented in northwest Canada and south of Hudson Bay. In the United States, these organic soils occur over wide areas in Florida's Everglades, Georgia's Dismal Swamp, Louisiana's coastal wetlands, and Minnesota's Lake of the Woods. Outside North America, they are common in the British Isles and Finland. Histosols account for 1 percent of world soils. Thus, although they are found on every continent, most occurrences are too small to show on the world map.

CONCLUSION

The Comprehensive Soil Classification System described in this chapter emphasizes that soils are continuously variable from place to place and yet may still be closely related to patterns of climate, hydrology, landforms, and vegetation. The processes of soil formation discussed in the previous chapter work over time to produce soils typical of major soil orders in the classification (oxisols, ultisols, alfisols, spodosols, mollisols, aridisols, and inceptisols). Local factors including poor drainage, high clay content, and frequent deposition of new sediment override the major processes to form soils in such orders as histosols, vertisols, and entisols.

To be sure, knowledge of soil orders gives us insight into the agricultural use of soils and how they must be treated to produce crops. Most important to our study of physical geography, however, is the insight gained by classifying and relating soil types to other environmental variables. In the next chapter we see that vegetation, like soil, is closely related to other physical variables.

KEY TERMS

The 7th Approximation
Comprehensive Soil Classification System
diagnostic horizons
orders
suborders
great groups
subgroups
families
series
oxisols
shifting cultivation
ultisols
argillic horizon
alfisols
ochric epipedon
spodosols
mollisols
aridisols
caliche
halophytes
desert pavement
inceptisols
vertisols
entisols
histosols
peat
muck

REVIEW QUESTIONS

1. How does the present-day system of soil classification differ from that developed by C. F. Marbut?

2. Describe the three soil types that do not have regional distribution. How are they alike? How are they different?

3. Why do farmers in South America use "shifting cultivation" in traditional agriculture?

4. How are ultisols similar to and different from oxisols and mollisols?

5. Describe the two soil orders that occupy "transitional" positions. How are they alike and how are they different?

6. How is acid important in the development of spodosols?

7. Why were mollisols the last soil order to be used for cultivation?

8. What is the difference between aridisols developed by calcification and those by salinization?

9. What is the difference between inceptisols and entisols?

10. Why can vertisols be plowed only during a very short period in the year?

APPLICATION QUESTIONS

1. What type of soil primarily occurs in your area? How does it affect the types of farming and building that are done near your university? Does it require any special management to be productive?

2. Name two different soils of which the minerals iron and aluminum are important components. Compare and contrast them in terms of these minerals.

3. Compare and contrast the behavior of humus in three different soil orders.

4. Compare and contrast the ways in which improved technologies have made farming easier and more productive in three soil orders. Can these improved technologies *damage* the soil?

5. Using a good world atlas, look up three locations with very diverse climates. See if you can deduce the type of soil each has based on climatic conditions, natural vegetation, and crops. Does soil type influence the prosperity of these areas?

6. Find out how soil type has affected or been affected by the recent famines and droughts of Subsaharan Africa.

CHAPTER 26

UNDERSTANDING NATURAL VEGETATION

OBJECTIVES

After studying this chapter, you will understand

1. How natural vegetation is related to human habitation.
2. What the concept of holism means in the study of natural vegetation.
3. How various climatic factors affect patterns of vegetation.
4. How various topographical factors affect patterns of vegetation.
5. How biotic relationships affect patterns of vegetation.
6. How soil affects patterns of vegetation.
7. How plant succession changes patterns of vegetation over time.

Vegetation on the land is almost universal. Even in the forbidding environment of a mid-latitude desert, where aridity and bitter temperatures discourage casual propagation, numerous plant communities have a permanent home. Bailey's yucca appears healthy and confident against a background of hearty, drought-resistant shrubs.

INTRODUCTION

Virtually the entire land surface of the earth supports some sort of vegetation. Ice fields, barren rocky outcrops, sterile sand dunes, and alkaline salt flats occasionally have no growing plants. Often, however, a close examination will reveal that seemingly barren rocks support mosses and lichens, salt-tolerant plants grow in highly saline soils, and the driest desert will, after an infrequent shower or even heavy fog, display a rapid flowering of long-dormant but still viable seeds.

This natural vegetation gives a certain distinction to our landscape. We therefore think not only in terms of terrain or climatic variation from place to place but of vegetational variation as well. Grassy plains, forested hills, and impenetrable jungle are part of our standard vocabulary in describing the character of a region.

HUMANS AND NATURAL VEGETATION

The earth as a human habitat is the prime concern of geography, and natural vegetation is an important element in our physical environment. For example, when the Spanish first attempted to occupy central Chile, they found familiar Mediterranean, open grassy, and low brush country that reminded them of home. They understood a grazing economy based on this type of vegetation and discovered that the vines, grains, and olives they brought from Spain flourished here. As they pushed south into the increasingly forested southcentral Chile, however, the Spanish were repelled by the unfamiliar character of the country (as well as by the Indians occupying it).

Many years later, German immigrants, familiar with a for-

Fig. 26-1. European antecedents are revealed in the architecture of central Chile.

est environment and possessing the skills to cope with it, moved readily into the accessible portions of southern Chile and developed its latent resources. With their potato and dairy economy and their knowledge of forestry, they felt very much at home in a region that appeared extremely inhospitable to the Spanish. Today, the high-gabled frame homes in forest clearings contrast strongly with the tile-roofed stucco and adobe dwellings in the villages of central Chile (Figs. 26-1 and 26-2).

This is not to say that other factors are not as important as vegetation in controlling where and how humans live. But vegetation cannot be ignored, for it is a factor everywhere, and it is a highly visible part of the total physical environment.

In attempting to analyze and classify the natural vegetation of the world, we are forced to make a somewhat artificial distinction inasmuch as very little truly *natural* vegetation remains. Everywhere humans have lived they have changed the original vegetation. They have cut down forests, plowed prairies, drained wetlands, and introduced new or alien species. In sizable regions of China, India, and Europe dense human occupation has been of such long standing that the only way we can discern the original vegetation is from the types of pollen left in bogs and lake sediments. Even in lightly populated areas, the balance of nature has been upset by the destruction of animals that were a part of the ecosystem or by the introduction of nonnative species. (Australia's rabbits are a well-known example.) Fire, too, frequently of human origin, has caused noticeable variations.

Everywhere humans have lived they have changed the original vegetation.

Nonetheless, we will attempt to describe and classify natural vegetation because it is an important element in physical geography and because it illustrates the linkage between climate, water, landform, and soil. Indeed, focusing on natural vegetation will allow us to bring unity to the various topics we have already considered.

The Concept of Holism

In bringing together so many diverse elements, we will rely on the concept of ***holism***—the idea that nature is a whole, that everything in nature is connected to everything else. Figure 26-3 captures the myriad interactions among environmental factors and plants. Some interactions directly involve plants (for example, those between plants and solar radiation or plants and topographic position), and other interactions do not directly involve plants (for example, those between soil and atmosphere or solar radiation and temperature).

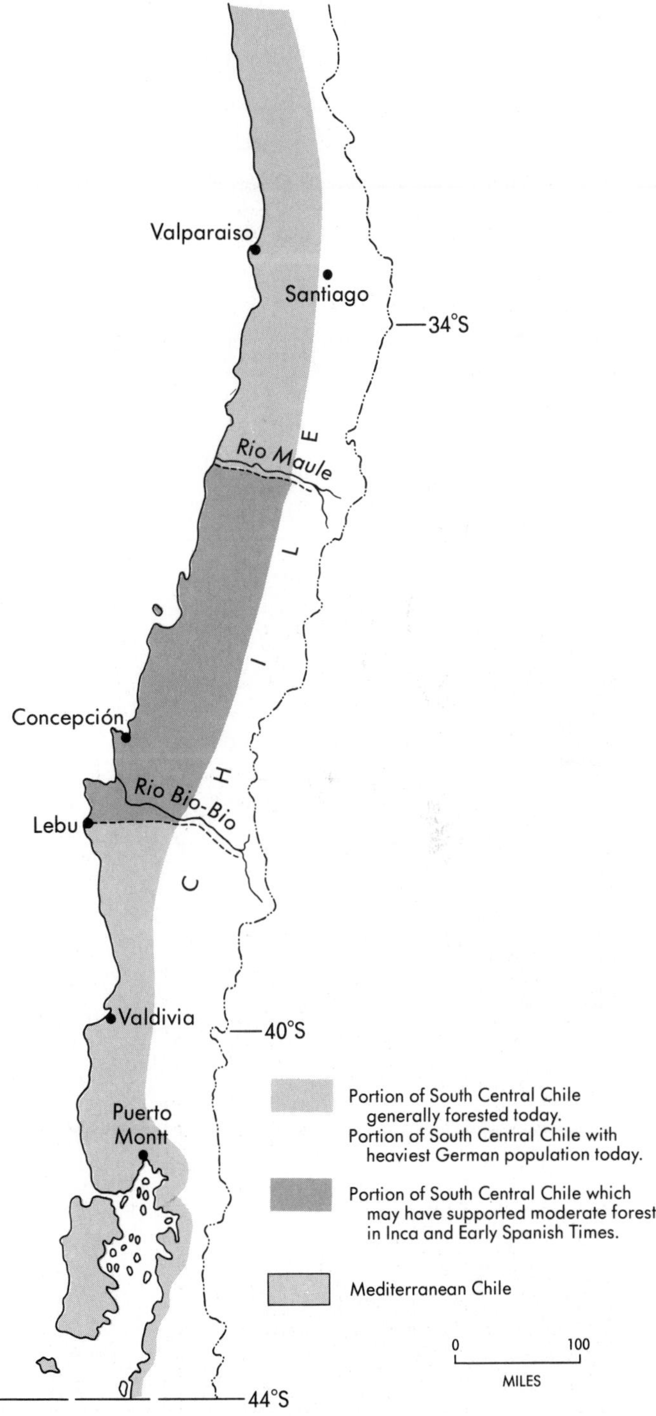

Fig. 26-2. ***Population and forest—southcentral Chile.***

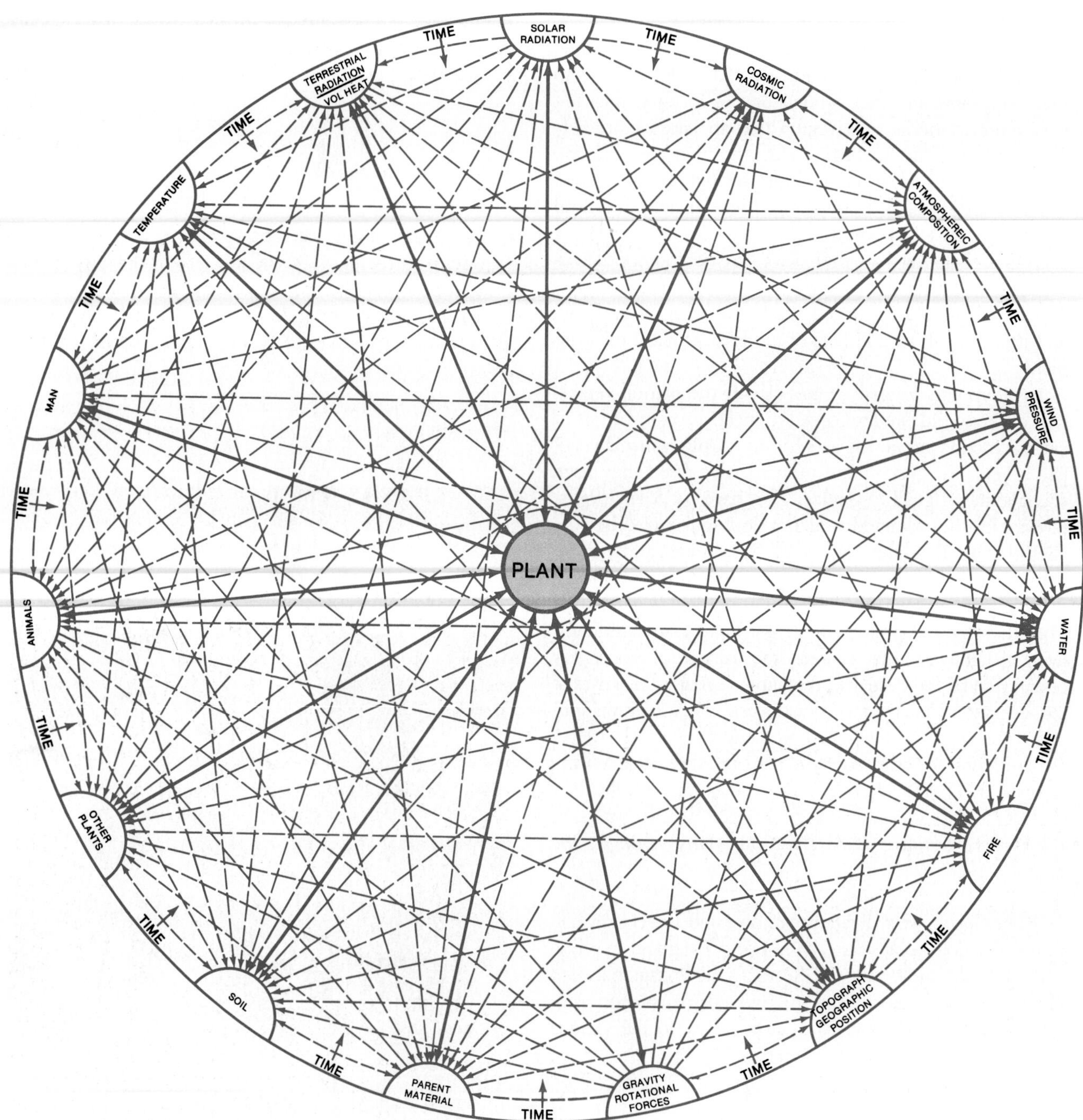

Fig. 26-3. ***A holistic view.*** ***The individual plant is strongly influenced by environmental factors, including other plants. Yet all parts of nature are interrelated in some way. This holistic view is shown in diagrammatic form. Time, not an interacting part, adds an important environmental dimension.***

Time, though not an environmental factor, is part of the concept of holism. Plant–environment relationships may change with time as ice ages come and go, as fire destroys old-growth forest allowing light to reach the soil surface, as continents collide forming new mountain ranges, and as humans alter the mixture of atmospheric gases. Let us keep this big picture in mind as we consider the relationships between vegetation and its environment.

VEGETATION AND THE ENVIRONMENT

Vegetation patterns—the distribution of plants and plant groups—are directly related to environmental factors. In a sense, vegetation reflects its environment—its particular mix of climate, topography, biological interaction, and soil. To appreciate this notion, remember that most plants are not adapted to a wide variety of environmental conditions. Each species tends to specialize for a certain restricted set of conditions. This set of conditions is the species' ***habitat*** and its place, or ***niche,*** within the overall scheme of natural vegetation. Some plants do well only in moist areas shielded from the sun, whereas others are found only on dry hillslopes receiving direct solar radiation.

The great diversity of the world's plant life reflects the great habitat diversity of the earth's surface, with its mountains, sea coasts, deserts, and wind-swept plains. Some habitats are unique, but most are repeated from place to place and from one continent to another. Moreover, similar habitats tend to be occupied by the same or similar plants. For example, broadleaf evergreen trees occupy moist, low-elevation, equatorial areas, and drought-resistant shrubs characterize the rainshadow sites of the midlatitudes.

Once we understand the relationship between environment and vegetation, therefore, we can use vegetation as an indicator of environment. Because vegetation is far more visible than the conditions under which it grows—long-term temperature patterns, soil-water regimes, seasonal grazing patterns, and the like—it is a valuable indicator indeed. Say, for example, that a certain group of plants—sedges and rushes—tend to grow only in low areas where soils are waterlogged much of the year. The presence of these plants can therefore tell us that the area has poor drainage conditions even when it appears dry. In other words, these plants indicate a set of environmental conditions. Such information can be very valuable, for instance, when we are searching for a good site to build a home.

We can use vegetation as an indicator of environment.

Let us now explore some of the vegetation–environment relationships that let us use plants as environmental indicators.

Climate, the Main Control

Among the environmental factors, climate exerts the greatest control on the distribution of plants. Some authors have gone so far as to say that "climate is the master." As we have already seen, climate is the sum of atmospheric conditions including light, temperature, moisture, and wind acting over a long period. Let us look at each of these conditions in turn.

Climate exerts greater control on the distribution of plants than any other environmental factor.

LIGHT Plant species vary widely in their tolerance to and requirements for light. To see just how widely, try putting your favorite philodendron or other house plant in full sun or growing a dandelion indoors in indirect light. In general, plants fall into one of two categories based on light requirements:

1. ***Heliophytes—helio*** for sun, *phytes* for plants—do well under conditions of full sun. Dandelions fall into this category.
2. ***Sciophytes***—*scio* for shade—prefer conditions of low light (Fig. 26-4). Philodendron falls into this category.

Fig. 26-4. ***Light and forest structure.*** *An outstanding characteristic of forests is the stratification of plant species with respect to light intensities. In this interior view of a western Oregon forest, the sciophytes oregongrape, huckleberry, and fern adorn the shady forest floor, while Douglas fir, with its massive boles and insatiable appetite for light, forms the highest strata. Mid-level species are western white pine and western hemlock.*

To complicate matters somewhat, we also recognize *obligative* heliophytes—plants that grow only in sites with full sun—and *facultative* heliophytes—as sciophytes capable of growing in the sun but preferring shade. Similarly, there are obligative and facultative sciophytes. Douglas fir, an important forest tree of the Pacific Northwest, is an obligative heliophyte. Its seedlings cannot grow in the shade of its parent tree. The facultative sciophyte western hemlock, on the other hand, does quite well in the shade of older trees as a seedling until it grows tall enough to reach full sun.

Forests, but also to some degree other vegetation types, usually have a mix of heliophytes and sciophytes with differing growth heights to match light requirements. This vertical distribution is called ***stratification***, and each individual layer is a ***stratum*** (pl. ***strata***). Equatorial rain forest may have up to five strata—three successive tree layers, a shrub layer, and an herbaceous (nonwoody) layer. Interestingly, leaf size tends to vary from upper to lower layers. Small leaves are characteristic of heliophytes extending into the forest canopy; sciophytes of the forest floor often have much larger leaves. This larger leaf size is not surprising because light intensity may be reduced over 90 percent before reaching the lowest levels, and there it takes a large leaf with plenty of chlorophyll to capture enough light for photosynthesis.

TEMPERATURE Most plants have relatively narrow temperature requirements for growth and reproduction. Little or no growth is possible below 32° F (0° C) when the movement of fluids between roots and leaves is difficult. Similarly, most plant growth is inhibited above 122° F (50° C). Certain plants may exceed these limits. Lichen, for example, is capable of growth to around 0° F (−18° C), and hot spring algae (Fig. 26-5) thrive in water at temperatures as high as 173° F (78° C).

Temperature exerts perhaps the most important control on the distribution of world vegetation patterns. As we have seen, temperatures decrease from low to high latitudes. Broad vegetation zones consisting of plants with varying temperature requirements grow along this gradient.

Temperature exerts perhaps the most important control on the distribution of world vegetation patterns.

The general pattern of vegetation from the equator to the pole, then, is one of ***latitudinal zonation*** (Map II). Using the northern half of the Western Hemisphere as an example, we see a progression of zones from the Amazon rain forest at the equator to savanna grassland, desert, middle- and high-latitude forest, and tundra in northern Canada and Alaska.

On a smaller scale, but also related to temperature, is ***altitudinal zonation***. We can easily see a series of vegetation zones when ascending mountain ranges. For example, at San Francisco Peaks, an isolated dormant volcano near Flagstaff, Arizona (Fig. 26-6), the sequence begins with desert at low elevation and then proceeds through several forest zones to alpine meadow and snow fields.

There are two exceptions to altitudinal zonation. One is

Fig. 26-5. ***Hot spring algae.*** *Algae at Emerald Spring, Yellowstone National Park, can grow in water temperatures exceeding 170° F (76° C). As a rule, lighter colored algae have the greatest heat tolerance.*

termed a ***thermal belt*** (Fig. 26-7A), a zone of mild temperatures located on valley sides above the normal winter surface inversion (see Chapter 4). Cold air settles in valleys at the foot of mountain ranges, leaving the nearby hillslopes in warmer air. These sites are often favored for orchards and vineyards because they have a lower incidence of late spring frost. The pear orchards of the Columbia Gorge in Oregon and the apple orchards in southern California are excellent examples.

The ***frost pocket*** is the second exception to the normal decrease of temperature with altitude (Fig. 26-7B). In this case interior valleys within mountain ranges receive cold air drainage from slopes, dropping temperatures below those of surrounding higher sites. The cooling is sometimes so severe that treeline species normally occurring at much higher elevations will dominate the pocket. For example, lodgepole pine is common in frost pockets at moderate elevations in the Cascade Range of western Oregon.

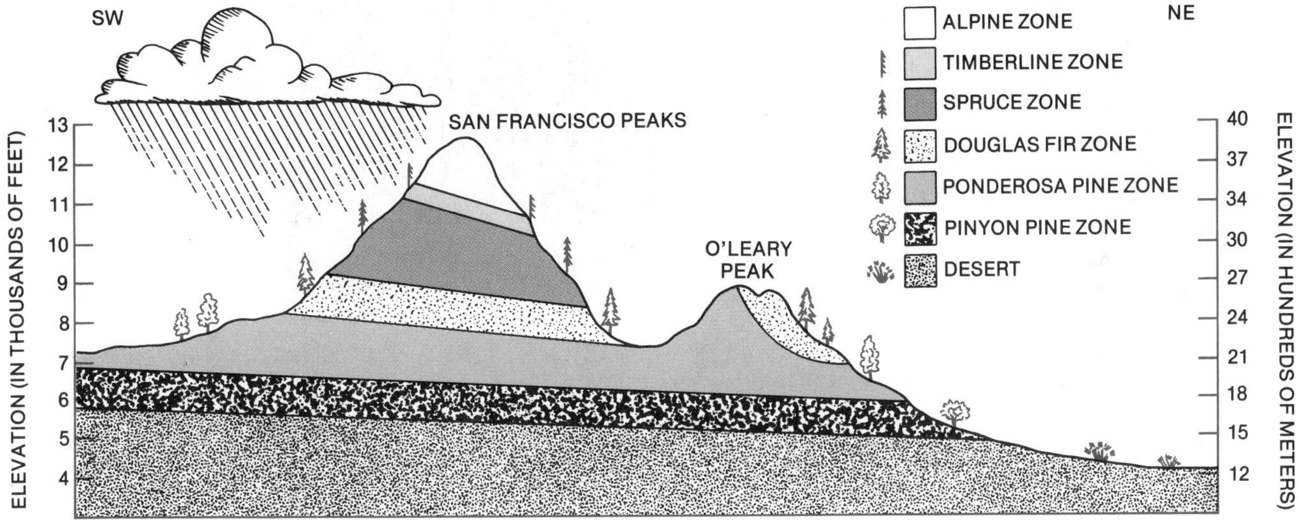

Fig. 26-6. ***Dormant volcano at San Francisco Peaks.*** *This isolated mountain of volcanic origin near Flagstaff, Arizona, rises to over 12,000 feet (3700 m) above sea level. Vegetation zones are closely related to elevation but dip to the wetter, less exposed northeast (poleward) slope.*

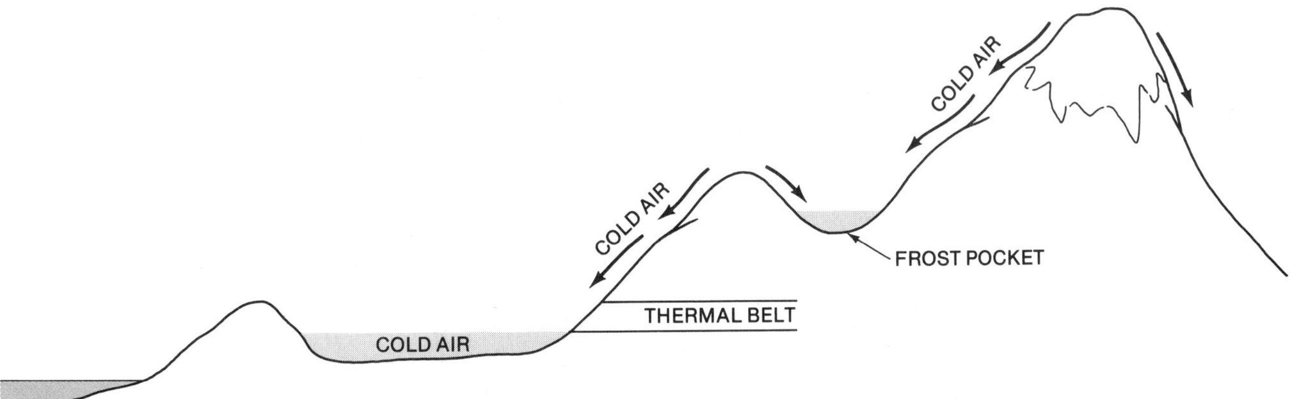

Fig. 26-7. ***Thermal belts and frost pockets.*** *During the cool months in the midlatitudes, cold air from surrounding hillsides settles into lowland valleys (A) and the depressions within mountain ranges (B). Thermal belts, located above the cold air on valley sides, are warmer than would be expected given the normal decrease of temperature with elevation. Depressions within mountain ranges, often described as frost pockets, do not follow normal temperature patterns either and often contain plants more characteristic of higher sites.*

MOISTURE Moisture is of utmost importance to plants. Water is an essential raw material to photosynthesis. Soil water contains dissolved nutrients for plant growth. Water-inflated cells give leaves rigidity, and, because water heats and cools slowly, the water content of plants, which may be greater than 70 percent, helps to protect them from rapid atmospheric temperature change.

Water is constantly moving through plants. Usually entering through the roots and bringing nutrients such as nitrate and phosphate with it, water travels up the stems to the leaves where nutrients are left behind and excess water exits by transpiration. In a sense, plants function as pumps, withdrawing moisture from the soil and speeding its return to the atmosphere. In this way, vegetation, and especially dense forests, contribute greatly to local atmospheric moisture. The removal of plants may thus have an adverse effect on rainfall. Indeed, one of the major concerns about forest removal in the Amazon Basin is the potential for reducing the amount of water vapor reaching inland agricultural areas. Without transpiration, most rainwater stays in the soil or is lost to runoff.

Plants function as pumps, speeding the return of moisture to the atmosphere.

Fig. 26-8. ***Marine hydrophyte.*** *This brown algae, dislodged from its grip to the sea floor by storm waves and stranded on a California beach, is a typical hydrophyte with large cavities for gas storage and buoyancy.*

Plants fall into three basic groups based on moisture requirements:

1. Hydrophytes
2. Xerophytes
3. Mesophytes

Hydrophytes. At one extreme are the ***hydrophytes,*** plants that have adapted to what we might consider an oversupply of water. This group includes the aquatic plants such as water hyacinth, water lily, and pond weed, as well as the wetland plants such as cattail, bulrush, and mangrove.

Plants require oxygen for metabolism, and most obtain it through root uptake. Oxygen in water bodies and saturated muds, however, is in dissolved form and often in short supply. Although most plants cannot grow under such conditions, aquatic hydrophytes can use dissolved oxygen, and many have large intercellular cavities for gas storage, which gives them a spongy character (Fig. 26-8). Some wetland hydrophytes, like mangrove, obtain oxygen through roots that extend into the air (aerial roots).

Xerophytes. Xerophytes occupy the opposite extreme. These plants are adapted to drought, that is, to the lack of available water in the normal root zone (upper 24 in.—60 cm) at some time during the growing season. In this group are most of the desert plants and those in dry sites elsewhere that are equipped to endure or escape recurrent drought. Xerophytes of the desert come in various shapes and sizes, but we can identify four general types:

1. Drought evaders
2. Evergreen shrubs
3. Phreatophytes
4. Succulents

Xerophytes are adapted to lack of available water in the normal root zone.

Drought evaders grow only when there is enough root zone water. Most of these are the desert wildflowers—termed ***ephemeral annuals***—that appear shortly after a period of heavy showers and grow rapidly, investing most of their energy in above-ground parts, especially leaves and flowers (Fig. 26-9A). By the time the soil moisture is gone, they will have produced new seeds to await the next rainy period.

Some perennial shrubs—called ***drought deciduous perennials***—are also drought evaders, but, instead of starting

Fig. 26-9. ***Desert xerophytes.*** *Xerophytes are of four general types as seen in North American deserts: (A) small, drought-evading ephemeral annuals, like these goldpoppies, (B) evergreen shrubs, like creosote bush, (C) phreatophytes, like this smoke tree growing at the edge of a dry river course, and (D) succulents, including prickly pear (foreground) and saguaro (background) cacti.*

anew from a seed, they grow new leaves when soil water is replenished and drop them when water is no longer available. Both types of drought evaders are capable of a very high rate of photosynthesis. Ephemeral annuals tend to be most common in deserts with unpredictable rainfall, whereas drought deciduous perennials are more frequent where rainfall is predictable.

It is surprising to see ***evergreen shrubs*** in the desert, but plants like creosote bush and sagebrush occupy wide areas in the drylands of North America (Fig. 26-9B). They draw water from the intermediate depths below the normal root zone, but not without a struggle. By exerting tremendous pressures (often greater than 50 atmospheres—one atmosphere is about 15 lb/sq. in.), they wrest moisture from soil particles. Abnormally thick cell walls sustain pressures of this magnitude without collapsing. Root systems are larger than the above-ground growth, extending outward and downward for 60 feet (18 m) or more, whereas leaves tend to be small and shiny to minimize overheating.

Phreatophytes have "wet toes"; their roots grow into the groundwater. This permanent source of moisture can support trees like palm, cottonwood, and salt cedar. Phreatophytes are usually limited to the desert's dry watercourses and rock outcrops where groundwater is within reach (Fig. 26-9C).

Succulents have yet another strategy to secure enough moisture for survival (Fig. 26-9D). By growing a very extensive but shallow root system, they draw in water from the rains and daily dew. Water is stored in succulent tissue and is used slowly through periods of drought. Succulents can also limit gas exchange to nighttime hours when deserts are cool, thereby greatly reducing water loss from transpiration. In all plants, however, photosynthesis takes place during daylight hours, using carbon dioxide from the atmosphere. Cacti have solved this dilemma by *storing* carbon dioxide during the night and using it during the day when they keep leaf openings tightly closed. Cacti have also reduced leaves to spines and have shifted photosynthesis to the fleshy stems. Succulents are most prevalent in warm deserts and may achieve a tree form as in the saguaro of the Sonoran Desert.

Mesophytes. Most plants are ***mesophytes*** with little or no adaptation to recurrent drought or to the waterlogged conditions enjoyed by hydrophytes. The mesophyte category is so broad that it is helpful to view it as a continuum. At one end are the ***xeric*** species with some drought tolerance. At the other are ***mesic*** species found in areas of abundant soil moisture.

Most plants are mesophytes, poorly adapted to recurrent drought or to waterlogged conditions.

WIND Wherever high-velocity wind persistently blows from one direction, it has an effect on the growth and appearance of vegetation. High winds are to be expected along the borders of the arctic tundra, at upper elevations in mountainous areas, along coasts, and in deserts. Wind tends to increase the rate of transpiration, especially on the windward side of trees, resulting in drying and uneven growth. Often branches extend only in the downwind direction, giving trees a typical flag form (Fig. 26-10). Wind at treeline on high mountains may be so drying as to keep trees permanently dwarfed, a condition referred to as ***krummholz*** (Fig. 26-11). Along coasts, onshore winds carrying salt spray prune seaside trees into distorted shapes. High winds associated with storms litter forests with downed trees—*windfalls*—leaving an enduring mark on the mosaic of species. Along coasts and in deserts where sand is available, the wind often forms and moves dunes that destroy vegetation.

Topography

In itself, topography has little direct effect on vegetation patterns, but it does strongly influence climate and soils, both of which exhibit strong control on plants. One way to consider the role of topography is to focus on five contrasts:

1. North–south slope contrast
2. Hill–valley contrast
3. Windward–leeward contrast
4. Altitudinal contrast
5. Ridge–watercourse contrast

NORTH–SOUTH SLOPE CONTRAST Because the angle of the sun's rays is always less than 90° N of the Tropic of Cancer and south of the Tropic of Capricorn, sun-facing slopes poleward from the tropics receive greater insolation than those facing away. In the Northern Hemisphere, especially in the midlatitudes, south-facing slopes tend to be warmer and drier and have an earlier onset of the growing season than the cooler and wetter north-facing slopes. In fact, north-facing slopes in steep terrain may experience direct sun during the day for only a month or two each summer.

Sun-facing slopes poleward from the tropics receive greater insolation than those facing away.

Such sharply contrasting habitats mean that vegetation on the opposing slopes is often markedly different. In the Coast Ranges of California, for example, south slopes are frequently dominated by grasses, often to the exclusion of trees, whereas forests of evergreen oak characterize the

Fig. 26-10. ***Testimony to cold winter winds is this flag-form evergreen beech at the southern tip of South America (Tierra del Fuego).***

Fig. 26-11. ***Krummholz.*** *Stunted, gnarled, and misshapen tree forms, described by the German term* krummholz, *are common at the upper treeline on high mountains.*

TIMBERLINE

Bristlecone pine at Timberline in the Colorado Rockies.

The transition from trees to low-growing shrubs and herbaceous plants on a mountain slope is one of the most striking of all vegetation changes. Timberline is the all-inclusive word for this transition, and here within a few meters the forest ends and the tundra begins. The whole character of the vegetation changes through timberline, and hiking trails that climb through it, though often strenuous, bring the reward of distant vistas unhindered by forest trees.

As a general rule, timberline is highest in the tropics and lowest in the high latitudes. It ranges from around 15,000 feet (4500 m) above sea level in the moist tropics (slightly higher in the dry subtropics) to sea level in polar regions.

Most of the timberline trees of Northern Hemisphere mountains are members of the pine family such as pine, spruce, fir, hemlock, and larch. All are needle-leaved and, except for the larch, are evergreen. In some areas, especially the Himalayas, Scandinavia, and eastern Asia, broadleaf deciduous trees like birch, alder, aspen, and beech also occur. Broadleaf evergreen species, most notably evergreen beech, dominate tropical and Southern Hemisphere mountains. Snow gum, a type of eucalyptus, dominates timberline in Australia.

The general pattern at timberline in all latitudes is for the forest trees gradually to decrease in density and height and then to occur only in small groups surrounded by low-lying vegetation. These tree islands often become established around a single tree that, as it grows, slightly modifies the environment. Its dark color causes snow to melt sooner, and its heat absorption warms nearby soil for growth of new tree seedlings. Persistent winds may so desiccate the windward side of tree islands that flag-form trees are common. At the upper limit of the timberline transition, such islands and scattered individual trees display *krummholz*—the stunted, gnarled, misshapen tree forms of high mountains and high latitudes.

Many researchers have sought to explain the cessation of tree growth on a mountainside, but the ultimate answer is still unknown. Such environmental factors as snow depth, the drying and chilling effects of wind, exposure to intense sunshine (especially ultraviolet waves), the effects of animals that eat young seedlings, and human use of fire for hunting are likely to contribute to the abrupt vegetation change. However, because of the overall similarity in the occurrence of timberlines most investigators believe that some universal ecological principle is at work.

We know that temperature decreases with altitude and latitude and that arctic and alpine tundra are very cold places. Perhaps there is a point at which the temperature is too cold for trees to survive. The warmth of the summer, however, is far more critical to tree growth than the severity of winter cold. Summer is the time of most photosynthesis, growth, and storage of energy. Observations in both arctic and alpine locations indicate that the upper limit of timberline occurs at the 50° F (10° C) isotherm for the warmest month. Apparently, lower summer temperatures do not allow trees to meet their metabolic needs and store sufficient energy for the winter. Recent research also suggests that short cool summers do not allow shoot tissues to ripen and grow.

Even though most investigators agree that temperature exerts a strong control on tree growth, a full explanation for the upper timberline remains elusive.

---

north slopes (Fig. 26-12). Even in western Oregon, with its nearly continuous forest, south slopes near hill tops are "balds" dominated by grasses.

Although the vegetation contrast is not always as sharp as that between grassland and forest, sun-facing slopes usually have more xeric plants and shady slopes more mesic plants. Not surprisingly, humans tend to favor the sunny slopes. In the European Alps, population concentrations are greater on the *adrets* (sunny slopes) than on the *ubacs* (shady slopes).

HILL–VALLEY CONTRAST The hill–valley contrast is largely one of soil and moisture differences. As seen in Chapter 24, valley soils develop on transported parent materials with

Fig. 26-12. ***North-south slope contrast.*** *The north-south slope contrast is striking in this view of Wyoming's Black Tail Butte. The north-facing slopes are covered by evergreen forest; the warmer and drier south-facing slopes support only grasses and a few shrubs.*

mixed mineral content. Valley soils, especially in more arid regions, tend to be wetter for longer periods than hillside soils. Consequently, valleys are often dominated by tall trees or even hydrophytes, if drainage is poor. In contrast, hillsides often contain more xeric species, and the size of its vegetation may be smaller (Fig. 26-13).

Valley soils tend to be wetter for longer periods than hillside soils.

WINDWARD–LEEWARD CONTRAST Where mountain ranges are located across the prevailing wind, and especially if the wind is off the sea, windward slopes tend to receive greater amounts of precipitation than the leeward slopes. Greater precipitation means a more moist environment on the windward slope, and, consequently, vegetation tends to be denser with more mesic species. The leeward slope may rapidly descend into a pronounced rainshadow, with species from the wet side poorly represented or absent. Most of the vegetation is characterized by drought-tolerant plants (Fig. 26-14).

Excellent examples of the windward–leeward vegetation contrast occur in the Andes of South America, the South Island of New Zealand, the Sierra–Cascade Ranges of western North America, and the Hawaiian Islands. Northeast slopes of the island of Maui, for instance, receive up to 400 inches (1000 cm) of precipitation and support tropical forest; southeast slopes, with 30 inches (75 cm) or less, have widely spaced dryland plants like kiawe, a close relative of mesquite.

ALTITUDINAL CONTRAST We have already discussed the decrease in temperature and the corresponding zonation of vegetation with increasing elevation. We might add, however, that the windward–leeward contrast influences this zonal pattern. Working together, the two contrasts often result in a windward dip of the vegetation zones (Fig. 26-6). That is, the greater moisture of the windward slope may allow plants of higher elevations to extend to sites lower than we would otherwise expect.

The greater moisture of the windward slope may allow plants of higher elevations to extend to lower sites.

Fig. 26-13. ***Hill-valley contrast.*** *Forested slopes are in sharp contrast to the valley meadow in this view of the Centennial Mountains, Idaho/Montana.*

Fig. 26-14. ***Windward-leeward contrast.*** *Sharp contrasts in vegetation are seen when traversing the San Bernardino Mountains.*

RIDGE–WATERCOURSE CONTRAST Differences in soil moisture between the ridges and watercourses occurring on the mountainsides may further distort the pattern of altitudinal zonation. The thin soils of ridges shed water rapidly; the watercourses tend to collect water and hold it for longer periods. The zones bend to accommodate these conditions (Fig. 26-15). In mountains with arid lands at the base, as in the southwestern United States, the tree zone reaches its lowest elevation in or next to watercourses, but trees first appear at higher elevations on ridges. In moist tropical areas, the pattern may be reversed, with lowland rain forest ascending to its highest position along watercourses rather than ridges.

A COMBINATION OF FACTORS Although topography has a strong effect on climate and thus a strong influence on vegetation patterns, what we see in the vegetated landscape is the result of many factors working together. The result is likely to be very complex. The windward–leeward contrast, for example, may be quite obvious, but on closer inspection, altitude, differences between ridges and watercourses, slope orientation, and so on, all play a part. Indeed, one of the real challenges in describing and understanding vegetation is to make sense out of what often appears to be chaos. Yet climate and topography are only part of the picture. To these factors we must add the effects of plant and animal interactions and soil parent materials.

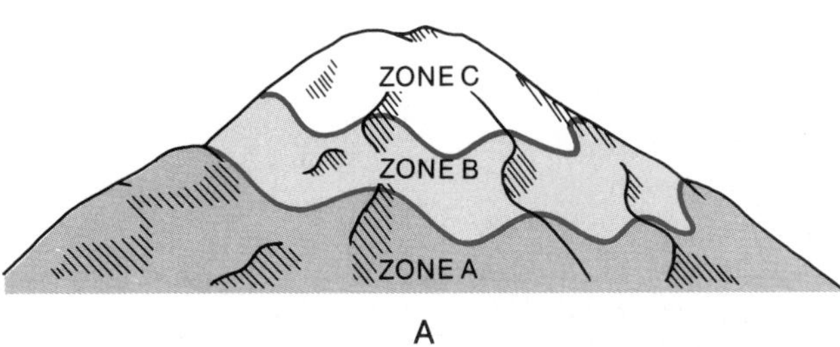

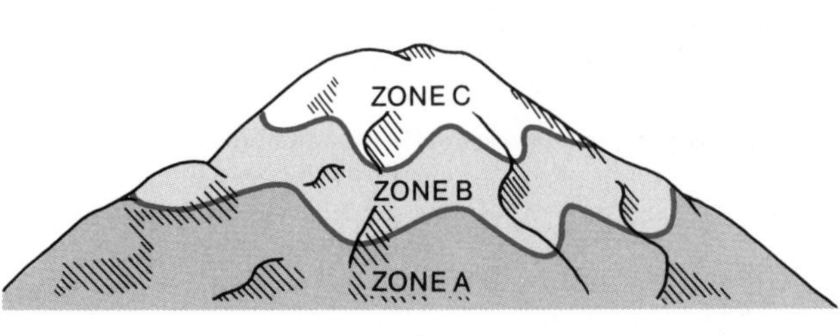

Fig. 26-15. ***Ridge–watercourse influence on vegetation zones.*** *This influence is seen in A for regions where lowlands are arid (the Rocky Mountains, for example) and in B where lowlands are wet (as in low-latitude mountains flanked by rain forest).*

Biotic Relationships

To this point we have focused on the mosaic of environments established by climate and influenced by topography. We should point out, however, that habitats provide only the site for plant and animal interactions, and these interactions ultimately result in vegetation patterns.

PLANT–PLANT INTERACTIONS Competition is central to interaction between plants. Plants have limited space in which to grow, and plants of the same species or of different species compete with one another for it. Most plant species specialize to minimize direct competition. For example, the stratification of the tropical rain forest is probably the result of plants specializing for certain light intensities. Some plants, such as creosote bush (common in the Mojave Desert) and chamise (common in the chaparral of California), take competition a bit further by using ***allelopathy,*** a sort of chemical warfare involving the liberation of toxic chemicals that block the germination of seeds around them, even their own.

Plants of all species compete with one another for the limited space available in which to grow.

Plant parasites influence the competitive struggle by affecting some plants and not others. Chestnut blight, for example, removed chestnut trees from their position of importance in the forests of the eastern United States ***Epiphytes***—nonparasitic plants like certain ferns, orchids, and bromeliads of the equatorial rain forest that grow on other plants, usually trees—have little overall effect on competition but represent a plant–plant interaction that contributes to the overall character of vegetation patterns (Fig. 26-16).

ANIMAL–PLANT INTERACTIONS Just as the composition of vegetation is influenced by the plants themselves, animals interact with plants in ways that are important to the success and distribution of plants. On the positive side, animals such as bees, ants, and butterflies aid in pollination. Birds carry seeds internally or in mud on their feet and deposit them at great distances. Animals unwittingly transport seeds in hair and fur, and, when such seeds are deposited, they are likely to do well because animals tend to visit similar habitats.

Animal–plant interactions are important to the success and distribution of plants.

Fig. 26-16. ***Epiphytes.*** *Epiphytes are not parasitic but they do use other plants for support. Here in Louisiana, the common Spanish moss hangs from live oak.*

On the negative side of animal–plant interactions, bark beetles have killed thousands of lodgepole pine in forests of the western United States, and locusts periodically threaten plant life in northern Africa. Grazing animals, if not confined, select only the most palatable forage leaving the rest uneaten (Fig. 26-17). With time the palatable species may be practically eliminated. Such a condition existed in Texas after cattle and sheep converted the grassland to shrubland. Large animals trample vegetation. Caribou on the Alaskan North Slope, for instance, regularly use the same trails between pasture and water. Even small burrowing animals, like prairie dogs and gophers, destroy much vegetation and influence the pattern that remains; beavers, by dam building, flood valleys and drown sections of forest.

The destructive action of animals pales in contrast to that of humans. Any consideration of world vegetation patterns, as we noted at the beginning of this chapter, must take into account modifications of human origin.

The Soil Influence

Major climatic patterns influence soil-forming processes that result in a typical set of horizons. Vegetation contributes organic material and influences soil pH. What about the reverse situation—do soils help to explain vegetation patterns? The answer is yes.

Fig. 26-17. ***Effects of grazing.*** *Much of eastern Oregon was originally grassland with scattered shrubs, but heavy grazing by cattle and sheep has eliminated most of the native grasses and allowed less palatable shrubs like sagebrush to spread and dominate the land. The barbed wire fence in this view separates grazed land with mostly sagebrush on the right from ungrazed highway right-of-way on the left. Note the abundant perennial grasses on the ungrazed side.*

Soils help to explain vegetation patterns.

Vegetation patterns associated with serpentine and pumice are examples. Serpentine, an ultramafic rock rich in iron and magnesium, outcrops in numerous locations in the mountains of western California and Oregon. Residual soils developed in serpentine parent material have a great deal of magnesium but only low amounts of calcium; that is, they have a low calcium/magnesium ratio. Serpentine soils may also contain large amounts of chromium and nickel. Although both calcium and magnesium are important plant nutrients, the low ratio inhibits the growth of most plants otherwise suited to the climate and topography. Vegetation growing in such soils is usually stunted and very different in composition. In California, Jeffrey pine and grasses are most prominent, but some serpentine outcrops are simply barren openings in the forest.

Pumice is fine volcanic glass that often collects in thick deposits downwind from volcanic vents. For example, eruptions of Mount Mazama (the present site of Crater Lake National Park) left broad pumice deposits on the eastern flanks of the Oregon Cascades. Pumice soils drain rapidly, have little water-holding capacity, and represent abnormally dry habitats. Instead of the ponderosa pine forests that one would expect to see east of Crater Lake, the pumice deposits are thinly populated with lodgepole pine.

Lodgepole pine and Jeffrey pine may thus indicate special environmental conditions, and vegetation patterns, in general, help us to understand the whole complex of environmental interactions. Yet we know that the earth system is in constant change, and with this change come modifications in environmental conditions, environmental interactions, and resulting vegetation patterns.

LONG-TERM CHANGE IN VEGETATION

We have seen that continents shift position, mountain ranges come and go, and ice sheets advance and retreat. It is not surprising that, over the long term in earth history, vegetation patterns have not been fixed. The fossil record contains ample evidence of shifting patterns: tropical forests in Great Britain and Alaska, northern forests in Arizona and southern California, and grasslands in the Sahara and Mojave. Pollens in lake deposits in the Great Lakes region tell us of the revegetation of lands stripped clean by glaciers and subsequent change with gradual climatic warming.

Against this backdrop of gradual vegetation change seen in the fossil and pollen record is a more rapid, short-term change associated with plant succession.

PLANT SUCCESSION

Each plant species is best suited to a certain set of environmental conditions. These conditions shift over space and time, and the plants themselves help to change the site in which they grow. Established plants may so alter the site as to create conditions best suited to other plant species. New species invade the site only to be replaced, in turn, as this process of ***plant succession*** continues.

Established plants may so alter the site as to create conditions best suited to other plant species.

Phases of Plant Succession

Plant succession is an orderly process marked by six phases (Fig. 26-18):

1. *Deposition, nudation*—New surface becomes available for plant growth. This phase may take place through, for example, the deposition of alluvium on a flood plain, lava on the flanks of a shield volcano, or

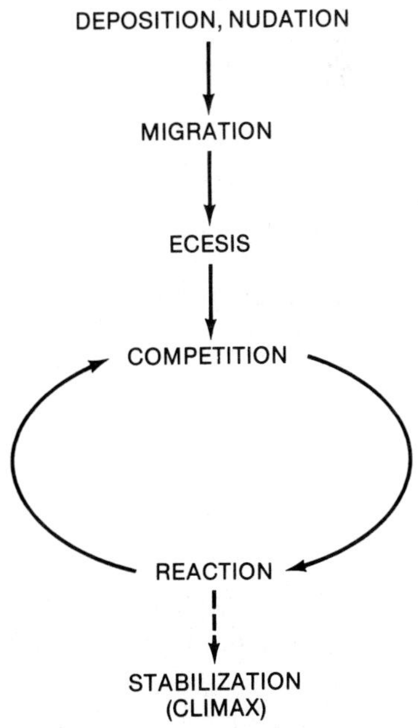

Fig. 26-18. ***Plant succession phases.***

sand blown inland from a beach; or existing vegetation may be removed by fire, logging, or storm winds.

2. *Migration*—Seeds and other plant parts capable of growth, together called *propagules,* travel to the vacant site by wind, animal, or water transport.
3. *Ecesis*—Only a few of the great number of propagules reaching the site will be of species suited to growth under the conditions there. For example, they may need to be heliophytes and be able to grow without a large nutrient supply. This phase marks the successful germination and growth of the first or ***pioneer species***—mostly weeds. Nonetheless, it is like the beginning of a symphony after the dissonance of tuning up.
4. *Competition*—Following the successful establishment of the pioneers, the site rapidly fills in as species compete for light, water, and other resources.
5. *Reaction*—Successful species change the site. (The site "reacts" to the presence of species.) As the pioneers grow, they contribute to the organic matter in the soil, retard evaporation by shading the surface, and otherwise gradually create conditions to which they are not suited or in which they are not able to compete successfully. New species now invade the site and displace those already there (returning to the competition phase). Reaction continues as the new species change the site.
6. *Stabilization*—The whole process of competition–reaction continues until the species present are able to persist without being replaced by other species. This phase, often called the ***climax,*** may take 200 years or more to achieve. It will continue until a major disturbance or a change in the climate occurs.

Dating and Tracking Plant Succession

Plant succession is a fascinating process. We have all seen weeds come into an untended garden, but none of us will live long enough to observe all the changes in plant life the garden space will go through on the road to stabilization. So how do we know about plant succession if we cannot follow it through? The secret is being able to date the surface on which the plants grow. By identifying surfaces of various ages, we can study the vegetation on each surface and piece the time puzzle together. For instance, much of what we know about succession near the Great Lakes was revealed by studying the sand dunes at the south end of Lake Michigan where the dunes increase in age from the lake shore inland. Grasses, wild cherry, and cottonwood on the younger dunes gradually give way to a climax beech–maple forest on the oldest dunes (Fig. 26-19).

Plant succession at Glacier Bay, Alaska, is also well understood because glaciers there have been retreating and exposing new surface for plant growth and soil development for more than 200 years (Fig. 26-20). Researchers have identified five successional stages or ***seres*** on sites of increasing age. Shortly after the ice retreats, exposed rock (mostly marble) and glacial outwash are colonized by moss, fireweed, dryas, and arctic willow (Fig. 26-21). All the plants in this pioneer sere are low-growing, moisture-tolerant heliophytes. They persist for several years but are gradually replaced by a shrub stage characterized by barclay willow

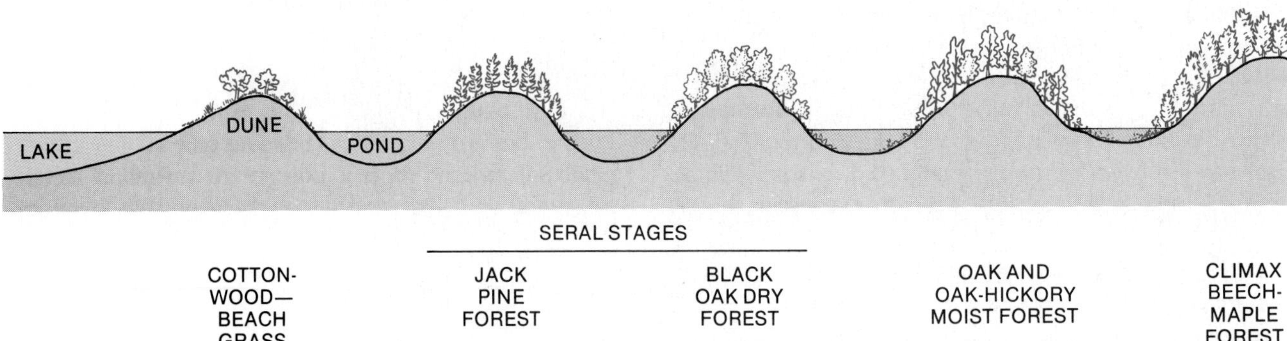

Fig. 26-19. ***Plant succession along Lake Michigan's southern shores.*** *One of the best places to view plant succession in North America is along the southern shores of Lake Michigan. Here beach sands have accumulated in a series of dunes that increase in age with distance from the lake. Cottonwood and beach grass are the first to occupy young dunes, but with time they are replaced by other seral stages. Vegetation on the older dunes, now located well away from the lake shore, has progressed through each seral stage to the climax forest of beech and maple. According to recent estimates, it may take 1000 years for succession on these dunes to reach the climax forest.* (Source: *V. E. Shelford,* Animal Communities in Temperate North America, *University of Chicago Press, Chicago, Il., 1913.)*

Fig. 26-20. **Retreating glaciers.** *At Glacier Bay on Alaska's southeast coast, ice has retreated more than 60 miles (96 km) up the fjord complex over the last 200 years. As new land is exposed, plants become established and soils begin to develop. Because the rate of ice retreat at Glacier Bay is well-documented, this area offers an excellent opportunity to chart the course of plant succession from pioneer to climax communities. Less than 50 years ago, this inlet leading to Riggs Glacier was covered by ice several thousands of feet thick.*

Fig. 26-21. **Pioneer vegetation.** *Following glacier retreat in coastal Alaska, new land is colonized by moss, fireweed, dryas, arctic willow, and other pioneer plant species. In approximately 200 years this site near Portage Glacier will support a climax hemlock forest.*

and sitka willow. The taller willows cut off the light from lower growing pioneers. The willows, in turn, are shaded out by closely spaced tree species—mostly alder but also black cottonwood (both of which are deciduous broadleaf).

This alder stage dominates sites for 10 to 80 years. Then a transition sets in: alder and cottonwood are gradually replaced by sitka spruce, a needleleaf evergreen tree. At first the sitka spruce grows in thick, pure stands, but it is eventually joined and later replaced after about 250 years by other evergreens—western hemlock and mountain hemlock. The hemlock stage is the climax on well-drained hillsides, and there it will persist indefinitely. However, on more level sites, the hemlock forest is gradually invaded by thick growing sphagnum moss. The moss retains large amounts of water—up to 20 times its weight—creating a very soggy forest floor called a ***muskeg.*** Trees are killed as their roots are unable to obtain oxygen, and sphagnum moss takes over on these sites as the climax species.

We know these successive changes in the vegetation at Glacier Bay take place, but what conditions allow the alder to become so dominant only to be replaced by sitka spruce? What signals the shift from spruce to hemlock? Fortunately, we have much of the answer. It lies in reaction—the changes in the soil generated by the plants themselves. The alder waits until the pioneer species and willows increase soil organic matter and establish a pool of soil nutrients. Once on the site, the alder, which has nitrogen-fixing bacteria living on its roots, begins to prepare the way for sitka spruce and thus its own demise. The nitrogen-fixing bacteria contribute greatly to available soil nitrate. Decaying leaves from the deciduous alder yield acids that, together with percolating rainwater and snowmelt, greatly increase soil acidity. Leaves also drastically increase the amount of organic matter on the forest floor and in the upper soil.

The shift from spruce to hemlock is signaled by reaction—the soil changes generated by the plants themselves.

After 80 years the organic, nitrate-rich, acidic soil is ideal for sitka spruce. The spruce seedlings become established in the protection of the alder and use the soil resources to grow tall enough to totally dominate the site. With the loss of alder and its bacteria, however, the high nitrate levels are not maintained. As the nitrate supply in the soil is used, spruce finds it increasingly difficult to hold its own with the hemlocks, which have lower nitrate requirements. The hemlocks are, therefore, ultimately the victors.

CONCLUSION

We know now that plants tend to specialize. Hydrophytes grow in wet places, heliophytes demand full sun, phreatophytes search out deep groundwater, and succulents persist on lands with infrequent rain and occasional dew. In many ways, the earth's richness in plant life reflects a richness in variety of habitats. Our goal in this chapter has been to understand the rudiments of the plant–habitat relationship as it contributes to the holism of physical geography.

We have seen that climate, topography, and soil come together in various combinations. In any particular place, however, plant success or failure is at least partly the result of competition with other plants and of relationships with animals including human beings.

One of the ironies in success, at least in successional stages before stabilization or climax, is that established plants tend to modify their own habitat. In time they lose out to others better suited to the modifications. Plants therefore have *a place in space as well as a place in time.* As habitats change so do the types of plants that occupy them.

Plants have much to tell us about the conditions under which they live, about their habitat and the current stage in plant succession. In the next, final chapter, we expand our treatment of plant–habitat relationships to a general discussion of world vegetation patterns and how they integrate all aspects of physical geography.

KEY TERMS

holism
habitat
niche
heliophytes
sciophytes
stratification
stratum (a)
latitudinal zonation
altitudinal zonation
thermal belt
frost pocket
hydrophytes
xerophytes
drought evaders
ephemeral annuals
drought deciduous perennials
evergreen shrubs
phreatophytes
succulents
mesophytes
xeric
mesic
krummholz
allelopathy
epiphytes
plant succession
deposition
nudation
migration
ecesis
pioneer species
competition
reaction
stabilization
climax
seres
muskeg

REVIEW QUESTIONS

1. Describe three different patterns of vegetation. What do they indicate about the conditions in which they grow?

2. How does the amount of available light affect the pattern of vegetation in forests?

3. How does windward–leeward contrast affect altitudinal zonation? Can you think of one other climatic, biotic, or topographic factor that influences another? What is the effect of such a relationship on natural vegetation?

4. Describe how three different kinds of plants adapt to too much or too little moisture.

5. Why do the north slopes of the California Coast Ranges have forests of evergreen oak whereas the south slopes have mainly grasses? Where might you expect to find the *opposite* pattern?

6. Describe two ways in which plants that occupy the same habitat minimize direct competition with any other plant by specializing.

7. Describe two ways in which animals can help their environment and two ways in which they can hurt it.

8. Describe two ways in which soil types help to explain patterns of vegetation. Can you think of one other possibility?

9. Describe the six phases of plant succession.

10. Describe two ways in which soil scientists can find out about the plant succession in a particular area.

APPLICATION QUESTIONS

1. Who were the original European settlers of your area (either your university area or your hometown)? What kind of natural vegetation did they find when they first arrived? How did they change the pattern of vegetation? Did their experiences in the "Old Country" help to determine their approach to the "new land"?

2. When people come to a new place, they often bring new plants and animals with them. Sometimes these nonnative species have a damaging effect on the natural vegetation. Research and describe one instance in which the introduction of a new ***animal*** species proved disastrous for a local environment. Do the same with one ***plant*** species. Then find out how your state protects its natural environment against such introductions.

3. Choose a plant that occurs naturally in your area. (Be careful—many plants that *look* "natural" were really planted by local landscapers!) Describe the conditions that allow this plant to grow in your area. Then describe three different areas where it could *not survive*.

4. Find three pictures of distinctive natural vegetation patterns. See if you can deduce the environmental conditions from the patterns you see. For each one, explain why you would or would not want to (a) build a house there, (b) go backpacking there, and (c) plant tomatoes there.

5. Choose one form of *non*natural plant species that people plant in your area—such as roses, kiwi trees, dahlias, soybeans, or Kentucky bluegrass. Find out what kind of climatic, topographic, and biotic factors affect the decision to plant this species and the kind of maintenance required. Would such a plant be able to survive *naturally?* Why or why not?

6. Choose one plant that occurs naturally in your area and create a holistic diagram (like the one in Figure 26-3) to show its interrelationships with various environmental factors. Try to include at least *five* factors. Explain your reasoning.

C H A P T E R 27

WORLD PATTERNS: VEGETATION AND PHYSICAL GEOGRAPHY

O B J E C T I V E S

After studying this chapter, you will understand

1. The levels of classification used to describe vegetation patterns.
2. The distribution, characteristics, and global importance of
   a. Tropical rain forest
   b. Tropical deciduous forest and scrub
   c. Savanna
   d. Shrub desert
   e. Mediterranean scrub forest
   f. Subtropical coniferous forest
   g. Middle-latitude coniferous forest
   h. Middle-latitude broadleaf and mixed forest
   i. Middle-latitude prairie and steppe
   j. Taiga
   k. Tundra

This is the fate of most of the world's natural vegetation. As our numbers swell, we cut the ancient forests to feed an insatiable appetite for wood, fuel, and building space. Everywhere we push aside natural plant life and replace it with sprawling cities, roads, and planted fields, or drown it beneath hundreds of hydroelectric reservoirs. Even though geographers persist in describing natural *vegetation, to see it in an unspoiled state is becoming increasingly difficult.*

INTRODUCTION

Physical geography, as we know, embraces five components that together comprise the physical environment: weather and climate, water, landforms, soils, and vegetation. Each is a study in itself; yet the overall theme in physical geography is that all components are tightly interwoven into the grand global fabric. In this final chapter, we conclude our discussion of vegetation with the theme of ***integration*** in mind, for world vegetation patterns do not stand alone. In many ways they reflect the total physical environment and its complex interrelationships.

World vegetation patterns reflect the total physical environment.

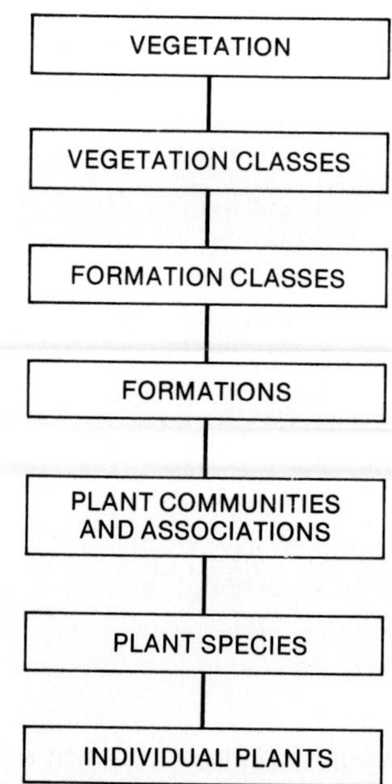

Fig. 27-1. ***Vegetation classification.*** *Each level contains those below; thus many individual plants together constitute a particular species, many different species make up plant communities and associations, and these in turn form particular formations, and so on. Higher levels—formations and above—are based on forms of dominant plants, whereas individuals, plant species, and communities and associations are in reference to plant names. Formations and formation classes are especially useful in describing world vegetation patterns.*

CLASSIFICATION

Before we can describe the great variety in world vegetation and relate it to other physical patterns, we must develop a classification scheme and choose the appropriate level of detail to serve our needs. We start with the term ***vegetation,*** which means plant life in general. This term is too general, however. It does not tell us if we are dealing with trees, shrubs, loblolly pines, Kentucky bluegrass, or some combination thereof. We need more specific terminology to break the term down into more useful components (Fig. 27-1).

Classes of Vegetation

At the first level of classification are four ***vegetation classes:*** (1) ***forest,*** (2) ***grassland,*** (3) ***desert,*** and (4) ***tundra.*** Each class is based on a dominant *form* of plant life. Forests are noted for their woody plants with easy-to-recognize trunks, branches, and leaves. In other words, trees are the dominant form of plant life in forests. Some forests, with a mixture of stunted trees and tall shrubs, are called *scrub forests.* Grasslands have mainly nonwoody plants with grasslike appearance. Deserts have a variety of plant forms, but the form that characterizes deserts above all others is the shrub. Shrubs are woody plants, but unlike trees they have multiple stems instead of a single trunk, and their branches and leaves often extend nearly to the ground. Tundra is similar to grassland in appearance, but it also contains lichen and moss forms and even shrubby trees. The overall emphasis on form here is important. At this level we are not at all concerned with the actual names of the dominant plants or of the other constituent plant species.

Forests are noted for their woody plants.

Formation Classes

The second level of classification is the ***formation class.*** Here we subdivide each general vegetation class, giving more attention to basic forms. The forest vegetation class contains five main formation classes:

1. ***Broadleaf evergreen forest***—nonseasonal forest; trees have broad, flat leaves that are retained year round. Because it occurs only within the tropical zone with rain all year, this formation class is often called *tropical rain forest.*
2. ***Broadleaf deciduous forest***—seasonal forest; trees

have broad leaves but do not retain them through the cold or dry season. The two varieties of this formation class, using the names most commonly applied, are *tropical deciduous forest and scrub* and *middle-latitude broadleaf forest.*

3. ***Needleleaf evergreen forest***—seasonal forest; trees have long, slender, rounded leaves that remain on the branches all year. Varieties are *subtropical coniferous forest, middle-latitude coniferous forest,* and *taiga.*
4. ***Mixed forest***—seasonal forest; broadleaf deciduous trees and needleleaf evergreen trees. This formation class is called *middle-latitude mixed forest.*
5. ***Mediterranean forest***—seasonal forest; trees are broadleaf evergreen with leathery, drought-resistant leaves. Because trees often share dominance with tall shrubs, this formation class is sometimes called *mediterranean scrub forest.*

The grassland vegetation class has three main formation classes:

1. ***Savanna***
2. ***Prairie***
3. ***Steppe***

All three classes have predominant grass forms, but *savanna* sticks to the low latitudes in areas of extended seasonal drought and often contains scattered trees. Prairie and steppe occur in the middle latitudes (hence *middle-latitude prairie* and *middle-latitude steppe*). Prairie is the moister and taller growing of the two; trees, when present in prairie or steppe formation classes, are usually confined to watercourses.

Desert has two formation classes:

1. ***Shrub steppe*** occupies semiarid lands on the transition between steppe and true desert. It is a combination of shrubs and grasses.
2. ***Shrub desert*** is found in true desert locations and has predominantly shrub forms and few grasses.

Tundra also has two formation classes:

1. ***Arctic tundra*** occurs beyond the northern limit of tree growth.
2. ***Alpine tundra*** occurs above the altitudinal limit of tree growth.

Formations

The third classification level contains ***formations.*** These are subdivisions of formation classes and are actual vegetation units that occur in a specific, though often extensive, area of the globe. For example, the *South American Tropical Rain Forest Formation* is one of the many formations in the tropical rain forest formation class, and the *Asian Steppe Formation* is one of many formations in the middle-latitude steppe formation class. You can see that, even though there are only a few formation classes, the number of formations is much larger. Larger still by many magnitudes is the number of plant communities or associations that make up the formations.

Formations are subdivisions of formation classes and occur in a specific area of the globe.

Communities, Species, and Individuals

Communities or ***associations*** within a single formation may number in the tens or even hundreds. These are synonymous terms referring to a recognizable and recurrent group of plant species that tend to occur together in a particular habitat. Thus each formation, with its hills and valleys, north- and south-facing slopes, wet and dry sites, and areas of varying animal influence, supports a diversity of habitats and, in response, a diversity of recurrent plant communities.

Each formation supports a diversity of habitats and, in response, a diversity of recurrent plant communities.

Consider the Douglas fir/Oceanspray community, for instance. Common to coastal Washington and Oregon and typical of the driest forested sites, this community of the tree Douglas fir, the shrub oceanspray, and several other plant species is one of many that make up the North American Coniferous Forest Formation.

Although there are thousands of plant communities the world over, each one contains numerous ***plant species,*** and each species in turn is represented by countless ***individual plants.***

Obviously, the further we descend in the vegetation hierarchy, the more we need the tools of the ecologist, botanist, and plant taxonomist. That is, we need to know the plants by name and something about their individual habitat preferences. Since we are focusing on the global aspects of physical geography, however, we need not concern ourselves with such matters here. Indeed, geographers have long specialized in the broad view of world vegetation patterns from the perspective of formation class and formation. In an introductory text, this level of integration will serve us quite well.

As we proceed, you will find it helpful to refer frequently to Maps I–VI.

Geographers have long viewed world vegetation patterns from the perspective of formation class and formation.

WORLD PATTERNS

Our description of world vegetation patterns begins with the formation classes and formations of the low latitudes and continues poleward. This sequence is similar to that followed in the description of world climates in Chapters 7–12.

Tropical Rain Forest (Selva)

Tropical rain forest or ***selva*** is located in virtually the identical region as the Tropical Wet climate (compare Maps I and II). The forest straddles the equator wherever land is available at low altitude. Three main formations are easily recognized: (1) The South American Formation centers on the vast Amazon Basin; (2) the African on the Congo (Zaire) basin; and (3) the Indo-Malayan on a series of large and small islands including Sri Lanka, Malaya, the East Indies, and Melanesia.

Together these three major Tropical Rain Forest Formations constitute one of the largest zones of climax vegetation anywhere in the world and certainly the greatest area of dense forest vegetation. In addition, the east coasts of continents and islands in the tradewind zone and minor, smaller regions are also clothed with this formation class. (In these minor regions, the typical Tropical Wet climate is not always absolutely necessary for the growth of a true tropical rain forest. The galeria forest is of this type and is discussed a little later in this chapter.) Parts of the Guinea coast of west Africa, southeast coastal India, and the west coast of Burma and adjacent Bangladesh also contain tropical rain forest. Although these regions have a dry season and are classified as Tropical Wet and Dry, their rainy season is so very wet (200 or more in. or 508 cm of rain each year) that the soil never dries out. This condition allows the growth of selva despite a definite rainless period.

The three major tropical rain forest formations constitute one of the world's largest zones of climax vegetation.

THE FOREST STRUCTURE Stimulated by constantly high temperatures and heavy precipitation, the vegetative growth of the tropical rain forest is unceasing, and the resulting forest is of a particular type duplicated nowhere else on earth. Alexander von Humboldt, the noted nineteenth-century geographer, described it as a "forest above a forest." Indeed, several strata are present (Fig. 27-2). Tall, straight

Fig. 27-2. ***Vertical structure of the tropical rain forest.*** *The highest trees form the emergent layer of noncontinuous, individual trees that are fully exposed to the sun. The canopy layer consists of medium-height and shorter trees. Their joining crowns absorb much of the sunlight, leaving only scant amounts for herbaceous plants and new growth at the forest floor.*

trees with their crowns fully exposed to sunlight extend to 150 feet (46 m) or more, creating a low-density *emergent layer.* Below the emergents, somewhat shorter trees grow closer together, and, reaching for the light, their foliage nearly joins to form the *canopy layer.* Still lower and completing the canopy is a dense stratum of small trees. One of these trees is the common cacao from whose beans chocolate is made. When cacao is planted in commercial plantations, great care must be exercised to ensure continual shade by retaining many of the taller trees. Finally, in the shadowy half-light below the canopy is an undergrowth of treelets, saplings, and herbaceous plants.

Vegetative growth of the tropical rain forest is unceasing, and the resulting forest is unique.

Branches and foliage of most trees persist only at levels with enough light. Thus trunks are featureless shafts. Trunk bases, however, are often buttressed or fluted, presumably to support the tall growth. Lacking climatic seasons of any kind and supplied abundant moisture, these trees are broadleaf evergreens that continually regenerate their foliage and drop dead leaves throughout the year. The forest floor is remarkably clean because the hot, moist climate fosters microbiotic decomposition at a rapid rate. Even fallen timbers are disposed of with dispatch by decay and a large array of termites.

Such decay results in a wealth of minerals that, when carried to plant roots by infiltration, feed the unending cycle of plant growth. But percolating waters, as we know from the discussion of tropical oxisol soils (Chapter 25), leach away valuable minerals, making the presence of the forest mandatory for the continuation of natural soil fertility.

Decay results in a wealth of minerals that feed the unending cycle of plant growth.

At the ground, where the shade is deepest, few plants can survive. There is no real undergrowth as we know it in the middle-latitude forests, and there are considerable expanses of bare soil. Rattan, a woody vine that runs along the ground, and a few low, fernlike plants are typical in the sparse vegetation of the forest floor.

Lianas, or climbing vines, can cling to trees to make their way upward to the light of the upper canopy. One feature of the interior is the many hanging lianas (and pythons) draped like great ropes among the trunks of the tall trees. Lianas are long lived and grow to considerable length; several have been reported in excess of 700 feet (213 m) and some over 1000 feet (300 m).

Epiphytes, too, are a common plant form in the tropical rain forest as they drape many of the canopy branches in great variety. Some are ferns, others delicate orchids, and still others massive shrubby bromeliads. All depend on rainfall and dew for needed water. Some, the so-called ***tank epiphytes,*** hold several liters in a basin of leaves that serves as water storage and as habitat for protozoans, insects, frogs, worms, and even small hydrophytes. Ready access to water and decay products, it would seem, is the reward for maintaining this small ecosystem in the branches.

Notwithstanding this overall frenzy of continued vegetational growth, the interior of the tropical rain forest is relatively open country and a distinctive world unto itself, with perpetual gloom, a dank and moldy odor, and still, moist air. The active life zone is remote in the high canopy where tree-dwelling animals and birds spend their entire life cycles and the trees flower and fruit. At the ground, where deep shade inhibits growth, there is no great difficulty in moving about from place to place, and the great myth of thousands of square miles of impenetrable jungle is just that, a myth.

The interior of the tropical rain forest is a distinctive world unto itself.

JUNGLE A jungle is a real entity, and it is highly impenetrable. It is merely a phase of rain forest, however, limited to certain favored localities and by no means dominant. ***Jungle*** comes about when light is allowed to reach the forest floor. If a storm or fire removes some trees and causes a break in the canopy or, as is more frequent, if humans open a clearing to plant a garden, the light brings countless dormant seeds to life. Grasses, saplings, bamboo, and low brush take over very rapidly, and within just a few years the region is a dense tangle of interlocking vegetation. Elements of the original planted garden, abandoned by migratory agriculturalists as the secondary growth becomes too aggressive to cope with, often compete with the other species for survival.

Jungle develops when light is allowed to reach the forest floor.

Left to itself, such a jungle will eventually revert to the original selva. However, if fire is repeatedly used to clear a given plot, it will usually develop into a semipermanent, rank, grass savanna. Many parts of Southeast Asia and Africa are characterized by many of these grassy openings in the forest. The dense grass strongly suppresses the establishment of tree seedlings, but presumably even this growth will finally revert to the climax vegetation as the rain forest slowly encroaches from all sides.

This jungle is only temporary, but, given a situation such as a fairly steep slope where the trees at various levels do not have interlocking crowns, more light than normal will find its way to the forest floor and a permanent jungle will result. More common than the hillside jungles are riverside jungles that develop along the banks of streams wide enough to cause a rift in the overall canopy. Such dense vegetation, merely fringing the stream course for many miles, has given rise in the past to reports of dense jungle over extensive areas. Early explorers, who used the waterways as the easiest means of penetrating the interior, observed nothing but the densest jungle from their craft and quite naturally assumed that the back country was of the same character.

Even greater misconceptions came about when these travelers, still clinging to the streams, journeyed well beyond the margins of the rain forest proper. Here a dry season each year results in a much lighter, more open vegetation. Fringing the river, where water is available to tree roots even during the annual rainless period, is a continuation of the riverine jungle called a ***galeria forest.*** A true picture of the situation can be seen from the air during the dry season. The brilliant green of the galeria, only a narrow band faithfully following each turn of the river, pushes long fingers into the more open, brown-colored vegetation of the surrounding countryside. Voyagers on the river have no means of knowing that they have left the tropical rain forest far behind.

The brilliant green of the galeria forest pushes into the brown-colored vegetation of the surrounding countryside.

There are also permanent coastal jungles where a dense wall of trees and underbrush faces out to sea at the inland margin of the beach sands. Many of the usual forest species are represented, supplemented by vegetational varieties whose seeds can remain viable for long periods of time when immersed in seawater as they are transported longshore or whose root systems can cope with some mild soil salinity, such as a number of palms and casuarina.

A variant on the coastal jungle is the frequently encountered, amphibious ***mangrove.*** This hydrophyte actually requires brackish water, preferably a few inches to 3 feet (1 m) deep. Thus it flourishes offshore, making it difficult to determine the exact point where land meets water (Fig. 27-3).

A variant on the coastal jungle is the frequently encountered amphibious mangrove.

SPECIES DIVERSITY Aside from its amphibious habits, the mangrove community is atypical of the tropical rain forest in another respect—it is made up of a single plant species. This characteristic may not seem particularly odd to those of us familiar with middle-latitude forests where a single type of tree is often strongly dominant. In the selva, however, the number of plant species in a community is likely to be in the thousands.

Probably no other widespread plant association has such species diversity as the tropical rain forest. One of the many difficulties in attempting to exploit the great economic potential of the selva — be it rubber, Brazil nuts, or mahogany — has been the fact that none of these valuable trees grows in even small groves, much less forest. Indeed, it is not unusual to find more than 30 different tree species growing in a single acre of forest.

Probably no other widespread plant association has such species diversity as the tropical rain forest.

THREATENED FOREST The greatest area of untouched virgin tropical rain forest is found in the still lightly populated and inaccessible upper Amazon Basin. Elsewhere in the world, its occurrence is somewhat more spotty. Southeast Asia and much of Africa display the effects of human hands to an increasing extent. Generally, where population pressures are heavy or along the great river valleys and accessible coastlines, secondary growth or permanent plantation development has changed the original vegetation. Even in the backcountry of interior New Guinea, upper Amazonia, or Borneo, however, the virgin rain forest is threatened. Recent estimates place the global destruction of tropical rain forest by humans at the incredible rate of 75,000 acres (30,000 ha) per day.

The global destruction of tropical rain forest by humans is estimated at 75,000 acres per day.

Tropical Deciduous Forest and Scrub

The climatic, and therefore geographic, limit of the tropical rain forest nearly always begins where a dry season intrudes along the poleward margins of the wet tropics. This dry season occurs during the respective hemisphere's winter

A

B

C

Fig. 27-3. ***Aboveground roots in the tropical rain forest.*** *(A) A Virgin Island kapok tree has plank buttress roots common to trees growing on well-drained sites. (B) Pandanus grow along tropical beaches and use stilt roots to withstand frequent flooding by brackish water. (C) And if a land plant intends to march out to sea, as does the mangrove, its stilt roots must be equipped with "knees" for breathing.*

months, since the rains of the Equatorial Low follow the sun across the equator into the opposing summer hemisphere. When it becomes dry enough to interrupt plant growth for even a short period each year, the character of the natural vegetation changes. Widespread tropical grasslands are the usual result of an extended annual dry period or erratic and unreliable rainfall from year to year. Where the moisture receipt is too great for continuous grass but still insufficient to support a true selva, a variety of intermediate vegetation types develop as a transitional phase, grouped rather loosely here into ***tropical deciduous forest and scrub.***

The outstanding features of this broad formation class are

1. The continuing dominance of woody trees.
2. A tendency toward more open country than in the tropical rain forest.
3. A loss of foliage during the dry season, indicating dormancy.

FOREST REGIONS In large parts of Southeast Asia and to a lesser extent in South America, Africa, and Australia, where rainfall is heavy and reliable, considerable forests of large trees occur. They are more widely spaced than in the tropical rain forest, and since their crowns do not join into a continuous canopy, there is a fair development of bushy undergrowth.

Brown and barren during the dry season and admitting light to the forest floor, these deciduous woodlands appear altogether different from the nearby selva. In addition, even though it has some of the same species as are found in the selva, the tropical deciduous forest tends to display large stands of a single species. An outstanding example is teak in Burma, Thailand, and adjacent regions. Because teak is a valuable cabinet and construction timber, widespread forestry has caused some alteration of the climax plant community.

The tropical deciduous forest tends to display large stands of a single species.

Because decomposition of fallen leaves and the like is retarded during the dry season, dry debris tends to accumulate. Fire is thus an annual hazard. Many believe that frequent burning is an important factor explaining the occurrence of this type of forest cover and of the formation class in general.

SCRUB REGIONS In regions of lower precipitation or an extended dry period, true forest gives way to greater variety of scrub associations. To dignify such vegetation by the term *forest,* despite the dominance of trees is to evoke an image of something much more lordly than actually exists, for the trees are stunted and seldom reach heights of more than 25 feet (8 m). Most species display the deciduous character in response to the dryness of their environment, but others have developed some xerophytic characteristics, such as thick bark, small pulpy leaves, and protective thorns, and maintain their leaves the year round as evergreens.

In drier regions or in dry periods, true forest gives way to greater variety of scrub associations.

Some trees, such as the flat-topped acacia of east Africa, spread out rather thinly with grass between and no bushy undergrowth (Fig. 27-4). Some group themselves in groves with sizable grassy openings separating one grove from another. Others, though fairly widespread as individuals, develop a dense thicket of thorny undergrowth. In northeastern Brazil, such thorny vegetation, called ***caatinga,*** extends for hundreds of miles and is virtually impenetrable. The remarkable evergreen eucalyptus family of Australia (of which there are several hundred varieties adapted to nearly every climate on the continent) is found, along with acacia, in a virgin scrubby woodland that encompasses the whole of the north coastal region.

A TRANSITIONAL FORMATION CLASS Altogether, the entire category of tropical deciduous forest and scrub represents a gradual change from the dense selva of the Tropical Wet climate region to the savanna grasslands that occupy the heart of the Tropical Wet and Dry climate region. Tree size and density of undergrowth decrease as the length of dry season increases. It would therefore appear that various types of deciduous tree and scrub vegetation would be arranged in bands parallel to the outer margin of the selva. This pattern holds true to a certain extent but is less evident than might be expected. Variables that affect this pattern include:

1. Total amount of annual precipitation (not merely length of the dry season).
2. Reliability of rainfall from year to year and season to season.
3. Terrain.
4. Soils.
5. Human activities, particularly with respect to fire and conversion of lands to agriculture.

The tropical deciduous forest and scrub category represents a gradual change to the savanna grasslands.

What remains of tropical deciduous forest and scrub is now threatened by rapid population growth. This empha-

Fig. 27-4. **Baobab.** *Dominating the ground-hugging thorny scrub is a baobab. These distinctive trees are slow growers but drought-resistant. They usually appear as individuals spotted among the deciduous thickets.*

sizes the difficulties involved in assessing what the climax vegetation of large areas might actually be once humans have occupied them. Southeast Asia, India, West Africa, and South America all have populations expanding at rates well above the world average. Strong dependence on agriculture and forest products to support increasing numbers means continued alteration and destruction of natural vegetation.

Savanna

In the savanna formation class are tropical grasslands representing the drier phase of the Tropical Wet and Dry climate. Where there is not enough moisture to support woody trees and bushes as the dominant plants the grasses take over. Only in a few places, however, are trees entirely absent; the savanna normally features scattered acacia, palm, brush, or even giant cactuslike plants along the desert margin (Fig. 27-5).

Without enough moisture to support woody trees and bushes, the grasses take over.

In this respect, and several others, the tropical grasslands are somewhat different from the more familiar middle-latitude prairie and steppe. The grass is taller and coarser, sometimes reaching heights of 8 to 10 feet (2 to 3 m), and during the long dry season, these sharp-edged blades become parched and harsh to the touch. Nor does the savanna form a turf; it is more typically arranged in tufts or bunches with patches of soil visible in between. The sharply contrasting wet and dry seasons control a vegetative rhythm that alternates between the tall, rank, brownish mature grass and rapidly growing, new green shoots with the onset of summer rains. If adequate surface drainage is lacking, this same rhythm is also reflected in an annual change from a dry, drab landscape to one of flood and seasonal marsh.

Toward the end of the dry period comes the fire season. Graziers usually deliberately set the fires to encourage a full and early sprouting of the new grasses. The common use of fire has made it difficult to determine precisely the boundary between the savannas and the forest scrub country—no two distribution maps are exactly alike. Authorities are in rather general agreement, however, that the grasslands are expanding at the expense of the trees.

Fig. 27-5. ***African savanna.*** *The savanna grassland in Africa is seldom an endless sea of grass—spindly flat-topped acacias seem always to be about either singly or in open groves. Acacias are less affected by fire than other woody plants.*

The greatest extent of unbroken savanna occurs in a wide latitudinal band in northern Africa between the Zaire basin (where in places it meets the selva) and the southern margin of the Sahara where short bunchgrass begins to grade into desert. Comparable zones, though somewhat smaller, are found in much of the veldt country of southern Africa, the campos of interior Brazil, the llanos of central Venezuela and adjacent Colombia, and a broad belt just south of the north coast of Australia. Lesser, widely scattered pockets of tropical grasses, including even the Florida Everglades, are difficult to show at the scale of most maps.

The greatest extent of unbroken savanna occurs between the Zaire basin and the southern margin of the Sahara.

ANIMALS OF THE SAVANNA Savanna grassland supports the greatest number and variety of hoofed (ungulate) grazing animals of any formation class. These include more than 40 species and such common animals as antelope, wildebeest, zebra, giraffe, and hippopotamus. Birds are common in the wet season, and reptiles in the dry season. The animals do not consume all the annual growth of grass; most of the grassy bulk is decomposed by countless termites who often live in elevated earthen mounds.

DESERTIFICATION Though highly variable, savanna soils are generally nutrient-poor alfisols that do not take kindly to plowing. Mechanized agriculture often leads to increased soil erosion and rapid nutrient loss. In Africa and South America some farmers still practice traditional shifting cultivation or "bush-fallowing" to avoid these problems. By far the greatest human use of savanna is for livestock grazing. Because many indigenous cultures measure family wealth in herd size, there is a tendency toward overgrazing. In some places, the Sudan–Sahel of North Africa for example, overgrazing of grasses coupled with prolonged drought is so severe that grass cover has been destroyed and the land resembles a desert. This process is called ***desertification,*** and once set into motion it is difficult to correct.

The greatest human use of savanna is for livestock grazing.

TRANSITIONAL CHARACTER In comparing the map of world vegetation distribution with the map of world climate regions (Maps I and II), note that, although only one type of vegetation (selva) coincides with the Tropical Wet climate, two distinctive vegetation complexes (tropical deciduous forest and scrub, and savanna) occupy the Tropical Wet and Dry climate. This illustrates the transitional character of this region between the very wet and the very dry within the tropics.

Shrub Desert

All desert plants must be equipped to withstand a harsh and difficult environment. As a result, they do not grow in great numbers; individuals are widely scattered, nowhere forming a continuous cover. It is amazing that vegetation cover and plant variety exist at all throughout the world's desert re-

Fig. 27-6. ***Desert halophytes.*** *What could possibly grow in the poisoned ground of Utah's Great Salt Desert? Obviously not much, but these halophytic shrubs are alive and well.*

gions. As we have seen in Chapter 26, nature has provided these plants with a number of special mechanisms to survive in an almost waterless habitat, and only very limited desert locations fail to support some kind of vegetative cover (Fig. 27-6).

All desert plants must be equipped to withstand a harsh environment.

Shrub desert (true desert) begins where grass ceases to dominate. In the slightly better watered regions and in regions where precipitation is reliable, bunchgrass persists, intermingled with shrubby vegetation to form shrub steppe.

The largest number of desert plants are xerophytic (ephemeral annuals, drought deciduous perennials, evergreen shrubs, phreatophytes, and succulents). They have special adaptations to conserve moisture or to grow only when moisture is plentiful. The natural vegetation of middle-latitude deserts must also be able to survive severe winter frost.

Most of the desert supports some kind of plant cover, often in much richer variety than is generally appreciated. It ranges from the spectacular saguaro cactus and Joshua trees that may grow up to 50 feet (15 m) tall and appear from a distance as forests, to the simple lichens giving color to rock faces (Fig. 27-7).

Fig. 27-7. ***Joshua Tree.*** *Among the tallest of desert plants, the Joshua Tree in California's Mojave Desert is now becoming scarce as subdivisions take over its habitat.*

Most of the desert supports some kind of plant cover.

DESERT SOIL Lack of dense vegetative cover and general dryness mean that little organic matter from plant decay finds its way into desert aridisols. Thus humus must be added if desert soils are to produce agricultural crops.

DESERT ANIMALS Desert animals, like desert plants, are adapted to a lack of water. Some can even exist without ever

COPING IN DEATH VALLEY

In Death Valley, plants must adapt to the most radical of environmental extremes. Since weather records have been kept at Furnace Creek, there has been one year with no rainfall at all, one year when the shade temperature reached 134° F (57° C), one year when there were 134 successive days over 100° F (38° C), and one year when the minimum temperature reached 15° F (−9° C). The difficulties of survival in such adverse conditions are immense for any vegetative life.

A salt marsh now covers most of the valley, which is below sea level, but there is evidence that a 116-mile long (187 km) and a 600-foot deep (183 m) lake filled the entire depression as recently as 20,000 years ago. A cooler, moister period fostered a lush and varied vegetation, but this picture would change as current desert conditions began to assert themselves. Some plants simply died and were replaced by invading Mexican arid-land species, some adapted in a variety of ways to cope, and others found sanctuary in shady canyons with dripping springs. Today we find little enclaves of temperate vegetation, responding to the cooling effects of altitude, stranded atop mountainous islands in a desert sea, for example, the bristlecone pines on Telescope Peak. Within Death Valley proper are 21 endemic plants, most of them endangered and all of them rare. Included among them are

Death Valley Sandpaper Plant
Death Valley Monkey Flower
Panamint Locoweed
Napkin Ring Buckwheat
Hollyleaf Four Pod Spurge
Rockalady
Rattleweed

A plant can survive primarily by developing a xerophytic character, that is, seeking out the moisture that does exist and/or conserving what it finds. Tiny leathery or hairy leaves and thick bark inhibit transpiration; deep roots tap groundwater; a mass of widespread surface roots utilize runoff. The *succulents,* such as cactus, can store up to a year's water supply in their tissue.

Halophytes show yet another type of xerophytic adjustment—the ability to use water with a high salt content. Often surrounding a Valley salt pan will be a concentric vegetative pattern with several plant communities arranged according to salinity tolerance. The succulent pickleweed will concentrate nearest the center, for it can withstand up to 6 percent salt; then comes a zone of arrowweed where the soilwater is only about 1 percent salt; and, finally, mesquite or saltbrush grow, whose salt tolerance is still less.

Annuals may have a lifespan of only a few weeks or months as they sprout, flower, and broadcast their seed during the cooler late winter and early spring. The new seed is varnished with a growth-inhibiting substance that washes off in the first rain of the new year, and a temperature-activated growth enzyme ensures that frost will not damage the delicate new seedlings.

---

consuming water directly. Desert beetles and certain rodents like kangaroo rats and jerboas, for instance, can obtain water from dry seeds through metabolic breakdown of carbohydrates—so called metabolic water. Woodrats chew on cactus for its stored water. Snakes, on the other hand, obtain all the water they need by eating an assortment of these juicy rodents.

REGIONS OF DESERT The Sahara and adjoining deserts extend through Arabia and the Middle East into Central Asia to form the largest dry zone in the world supporting shrub desert vegetation. Other, but lesser, regions include all continental west coasts in the tradewind belt, Patagonia (southern Argentina), and the Great Basin of North America.

The Sahara and adjoining deserts form the largest dry zone in the world supporting shrub desert vegetation.

Mediterranean Scrub Forest

The rainfall of the Mediterranean climatic regime, heavily concentrated during the winter season when evaporation is low, can support fairly dense vegetation, even sizable trees.

Fig. 27-8. Cork oak. The thick bark of this tree common to Mediterranean shores is an adjustment to resist seasonal drought. Bark is periodically harvested for use in beverage bottles.

In contrast, the summer dry season is long and often quite warm so that plants and trees taking advantage of the winter moisture must adapt themselves to extended drought.

DROUGHT RESISTANCE The deciduous habit and complete dormancy are means of combatting seasonal aridity in the tropical scrub forest, but the *mediterranean scrub forest,* which is broadleafed and evergreen, cultivates a variety of its own adjustments to resist drought. The cork oak with its thick bark (Fig. 27-8), the eucalyptus with its deep tap root, and the California oak with its tiny leathery leaves are examples.

Mediterranean scrub forest cultivates its own adjustments to resist drought.

Most trees and shrubs have drought-resistant leathery leaves, thereby giving this formation class the alternate name of *sclerophyll forest.* The entire plant assemblage has a gray and dusty appearance during much of the year as it continues to grow, very slowly, during the long summer. With the coming of the winter rains, new shoots appear and the landscape takes on a somewhat greener aspect as the rate of plant growth accelerates. Winter frosts are not unknown, but they are usually mild and short and seldom damage the tender new buds.

TREES Trees, though common in the mediterranean scrub forest, are not dominant. They are usually widely spaced, or, in certain specially favored areas such as draws or seasonal watercourses where deep roots can reach a higher water table, they occur in limited groves. Moderate height and gnarled trunk and branches are typical. Oaks of many kinds are native to both North America and Europe and have been introduced successfully into the Southern Hemisphere. Conversely, the Chilean pepper tree with its grotesque trunk and weeping character and several species of the Australian eucalyptus have been very popular in California and many parts of the Mediterranean Basin. These trees grow taller, straighter, and more rapidly than most Mediterranean trees and are an important source of timber in western Australia.

A few of the mediterranean scrub forest trees, notably the olive, fig, and chestnut, have been removed from their native context and cultivated as important sources of food. Still another representative of the trees, found in small numbers in this formation class, is the needleleaf evergreen. These trees, such as the digger pine and Lebanon and Aleppo cedars, usually grow along the slightly wetter foothill margins. Others, like Monterey pine and cypress and the

Fig. 27-9. ***Maquis (matorral in spanish) along the south coast of Spain.*** *Although a widespread element in the mediterranean scrub forest, maquis is scrub rather than forest. Trees were cut long ago for ships, charcoal, and home construction.*

coast redwood, require the low temperature and high humidity of foggy coasts to survive in the Mediterranean climate.

CHAPARRAL, MAQUIS, MALLEE Despite the wide occurrence of trees in the mediterranean scrub forest, the dominant plant assemblage is a woody brush, generally called ***chaparral*** (United States) or ***maquis*** (Europe and South Africa) (Fig. 27-9). The many shrub varieties that make up this complex are beautifully adapted to their environment. Their extensive root systems take advantage of water at various levels in the soil, and thick-walled cells exert the high osmotic pressures required to remove water tightly held by soil particles. They also spread themselves through sucker propagation from the roots and thus expand their domain in all directions.

The dominant plant assemblage of the mediterranean scrub forest is chaparral.

In the ***mallee*** (scrub eucalyptus) country of southern Australia, some shrub species have proven almost impossible to eradicate short of the monumental task of digging out every involved root system. Scrub vegetation, often a tangled intertwined mass, is very susceptible to summer fires. Dead leaves and other litter build up over the years and burn vigorously when dry. However, shrubs are remarkably fire tolerant, with new growth appearing rapidly from undamaged roots. Moreover, recent research suggests that shrubs can persist only with recurrent fire. Otherwise shrub vegetation is gradually replaced by trees.

Sun-facing slopes, too dry to support shrubs or trees, are

often covered with grass. Large areas within the California chaparral, for example, are in grass—brown in summer and green in winter with the colorful California poppy and blue lupine intermixed.

ANIMALS Mediterranean animals, like the vegetation, tend to be dull colored. Birds are especially common, and ground squirrels, woodrats, mule deer, rabbits, and coyotes are supported by abundant food and good cover.

SOILS Soils, highly variable, are frequently assigned to the alfisol category.

DISTRIBUTION AND HUMAN IMPACT Because the Mediterranean climate is limited to narrow coastal regions of the Mediterranean Basin and tiny west coast exposures poleward of the tropical deserts, the mediterranean scrub forest is not a widespread formation class. Humans have lived for so long in the Mediterranean climate regions that a typical scrub forest in its virgin state is difficult to reconstruct or encounter today.

The mediterranean scrub forest is not a widespread formation class.

Fig. 27-10. ***Loblolly pine in coastal North Carolina.*** *Despite their age (about 75 years), these are not large trees and the forest is open. The chief virtue of the southern pine as a commercial timber tree is its ability to occupy land that is unproductive for agriculture and to reproduce itself rapidly.*

Subtropical Coniferous Forest

In only one place in the world are coniferous trees the dominant species within the subtropics—in the southeastern coastal United States. There an extensive region is forested almost exclusively by several varieties of pine.

Stretching from Chesapeake Bay in the north, along the coast in an ever-widening band into eastern Texas, the ***subtropical coniferous forest*** appears to coincide with the sandy soils of the low-lying coastal region (Fig. 27-10). The deep alluvia of the Mississippi bottoms and the heavier ultisols of the interior, on the other hand, support a mixed or broadleaf forest that develop under a climate identical to that of the coniferous forest (Humid Subtropic). The texture of the soils is almost certainly the reason for this difference. In other words, the vegetation is controlled by the soil rather than by the climate. Although the Humid Subtropic is found in every continent, only in a couple of these areas is there any hint of a true coniferous forest. Southern Brazil contains some limited stands of araucaria pine that may have been more extensive at one time. Today they are found only at slight elevations that partially moderate the subtropical climate. South of this region, the grasslands of Uruguay and much of the Argentine Pampas are sufficiently wet to support trees, although they have not supported them within historic time. Very probably, they did before repeated fires wiped them out, but it is not known whether these imagined forests were coniferous.

Within the pine forest area of the southeastern United States, loblolly, yellow, and shortleaf pine are all represented, often mixed with one another. They grow fairly widely spaced, and they mature into medium-height trees, less than 100 feet (30 m) tall. Grass and occasional evergreen, shade-loving broadleaf shrubs such as rhododendron occupy a remarkably clean forest floor where heat and moisture promote the rapid decay of fallen needles and other litter.

One important feature of the subtropical coniferous forest is its reproductive ability after the trees have been removed. Southern pine forest is able to regrow mature timber trees within 30 years after cutting (less for pulpwood), a quality that has given the southern lumbering operation a real advantage over that of competitors.

Southern pine forest can regrow mature timber trees within 30 years after cutting.

Much of this area has been in crops at one time, but improved soils and increasing demand for timber have led to extensive plantings of farm woodlots and reforestation of eroded slopes. Today, more of this region is probably in timber than at any time in the past 100 years.

Middle-Latitude Coniferous Forest

Like the subtropical coniferous forest, the ***middle-latitude coniferous forest*** occurs only in North America. It appears to be a product of the Marine West Coast climate and is limited to a narrow coastal strip from northern California to southern Alaska. Although there are other forested areas in every continent except Africa that inhabit a Marine West Coast climate, conifers nowhere else constitute the dominant species. Probably the short summer dry season with its accompanying forest fires, which shows up only in North America, is responsible.

Fig. 27-11. ***Ancient redwoods.*** *Many of these ancient redwoods in northern California are over 200 feet (61 m) high. One holds the world record at 368 feet (112 m). But these are* Sequoia sempervirons. *The fabled "big" trees of the western Sierra slopes are* Sequoiadendron giganteum *and, although not quite as tall, are much longer lived and of huge girth.*

Whatever the cause, the Pacific Northwest and adjacent Canada and Alaska support some of the world's heaviest forests. Not only are the trees huge—coast redwood and Douglas fir reach over 200 feet (61 m) in height and 30 feet (9 m) in girth—but they grow very close together (Fig. 27-11). On the west side of the Olympic Peninsula and Vancouver Island, what might well be described as a middle-latitude rain forest exists, with as heavy a vegetation growth as any tropical rain forest. With 150 to 200 inches (381 to 508 cm) of rainfall and mild year-round temperatures, the giant Douglas fir grows mightily, often in stands with trees all about the same age.

The Pacific Northwest and adjacent Canada and Alaska support some of the world's heaviest forests.

As mentioned in Chapter 26, Douglas fir is a successional species that comes in after natural fires and persists until replaced by a climax western hemlock forest. This is not a rapid process, however, for magnificent stands of old-growth Douglas fir may be 600 or more years old. Yet even in these ancient forests western hemlock makes up a significant part of the understory. In some places the groundcover of middle-latitude coniferous forest is bracken fern that occasionally reaches 6 to 8 feet (2 m) in height. Or scattered fern may be interspersed with huckleberry, rhododendron, and other low herbaceous growth. In the less well-drained areas, western red cedar or red alder supplants the Douglas fir; beneath cedar/hemlock communities, devil's club and skunk cabbage are common. This tangle of growing vegetation is made even more impenetrable by the accumulation of moss-covered downed trees that fail to decompose as rapidly as in the tropics.

MIDDLE-LATITUDE CONIFEROUS FOREST SOIL Alfisols and ultisols are the main soil orders, with some spodisols occurring locally.

VARIATIONS ON THE TYPICAL PATTERN By no means does all the North American northwest coast support this heavy growth or even all these specific plants and trees, but only minor variations are the rule. For instance, northern California features the redwood, whereas the Douglas fir fades out north of Oregon and Washington and spruce and hemlock take over. In addition, since the mountains very closely ap-

Fig. 27-12. ***Forest cutting.*** *A snow-capped monarch looms over an extensive stand of old-growth Douglas fir. Logging in the foreground will pay some salaries and local taxes, but it certainly detracts from the pristine character of a classic natural landscape.*

proach the coast throughout most of the region, the forest changes its character in a regular sequence with gain of altitude. Remember, however, that we are not classifying mountain vegetation and must draw an upper limit to the typical coastal forest, probably well below 1000 feet (305 m).

LOGGING Certain portions of the middle-latitude coniferous forest have been removed permanently as in the Willamette Valley and Puget Sound country. In some less accessible regions and on some steeper hillsides, however, millions of board feet of timber remain despite commercial cutting and frequent summer fires (Fig. 27-12).

Some experts are increasingly concerned that the current high rate of logging will leave little of the old-growth Douglas fir forest intact by the turn of the century. As much of the ancient forest is converted to tree farms, animal species that are dependent on old growth (examples are the northern spotted owl and flying squirrel) face an increasing threat of extinction.

The current high rate of logging may leave little of the old-growth Douglas fir forest intact by the year 2000.

Middle-Latitude Broadleaf and Mixed Forest

Here we combine two formation classes—***middle-latitude broadleaf forest*** and ***middle-latitude mixed forest***—because they have similar structure and geographic location. This combined forest formation class, displaying many variations, occupies three different climatic regions—Humid Subtropic, Humid Continental, and Marine West Coast. All these regions have relatively heavy precipitation that allows trees to become the dominant vegetation.

BROADLEAF DECIDUOUS REGIONS OF THE UNITED STATES In the United States, much of the South, inland from the sandy coastal plains, supports a deciduous broadleaf forest dominated by oak and chestnut. In the Middle Atlantic states and the Middle West this forest merges into a quite similar oak/hickory or walnut/poplar forest. These trees are of considerable diameter but only moderate

Fig. 27-13. ***Broadleaf deciduous forest.*** *Deciduous trees allow a good deal of light to reach the forest floor although generally at the wrong season to encourage heavy growth. But in the spring life is everywhere as the delicate new leaves impart a limpid quality to the gently filtered sunshine.*

height, and although they grow quite close together, they do not form a canopy as in the broadleaf selva.

A fair amount of light can reach the forest floor, especially since this is a deciduous forest, and the growth of saplings and young trees is thus encouraged (Fig. 27-13). Bushy undergrowth is not dense, but windfalls and the accumulation of several inches of litter and duff over alfisols and ultisols are typical.

In poorly drained areas, swamp communities such as the cypress/red gum of the southern Mississippi River flood plain replace dryland forest and may cover many square miles with their junglelike tangle of supporting brush and vines.

MIXED REGIONS OF THE UNITED STATES North of this broadleaf deciduous forest, in the region of the Great Lakes and New England, roughly coinciding with the Humid Continental–Short Summer climate, is a gradual change to a mixed forest. Conifers begin to appear as individuals or in sizable groves or clumps among the broadleaves. Birch, beech, and maple take over from oak and hickory, and various pines, spruce, and fir show darkly among them. Sandy or rocky soils are often responsible for extensive stands of conifers, as in northern Michigan, New Jersey, and Maine. Elsewhere they are much more generally intermingled.

EUROPE As in North America, European climatic regions tend to correspond to vegetative type. However, in Europe the Marine West Coast climate supports broadleaf deciduous forest, and the Humid Subtropic climate does not exist.

Unlike the Marine West Coast climate of North America where coniferous forests predominate, western coastal Europe from northern Spain to southern Norway has an original deciduous forest similar to that of the eastern United States. Conifers are often seen today at some elevation, on rocky or sandy soil, or in a planting intended to replace the slower growing hardwoods.

Western coastal Europe has an original deciduous forest similar to that of the eastern United States.

The Humid Continental–Long Summer climate of northern Italy and much of the Danube basin produces a broadleaf deciduous forest, and the Humid Continental–Short Summer regions of Northern Europe are mantled by a mixed forest.

ASIA China, Korea, and Japan demonstrate quite clearly the general progression from broadleaf deciduous and semideciduous in the south to mixed forests in the north, with increasingly severe winters and a shorter growing season.

SOUTHERN HEMISPHERE Broadleaf and mixed forests are represented in the Southern Hemisphere to a more limited degree. The narrow coastal strip of southern Chile, with its Marine West Coast climate, displays a dense broadleaf deciduous forest where beech is the dominant species. Along the Humid Subtropic coast of southeastern Australia, the ever present eucalyptus—broadleaf but evergreen—is the most common tree.

HUMAN IMPACT Without a doubt, humans have destroyed the middle-latitude broadleaf and mixed forest more thoroughly than any other vegetative type in the world. In places, the character of the original climax vegetation is mere conjecture. In northern China, where for centuries the forests have been cut again and again and the roots systematically grubbed out for fuel, not a vestige of the original forest remains.

Humans have destroyed the midlatitude broadleaf and mixed forest more than any other vegetative type in the world.

Soil erosion is, of course, inevitable unless some sort of crop or substitute vegetation is introduced to intercept the surface runoff. Reforestation is being pushed with considerable success in many parts of the world, not only to minimize soil erosion, but also to make steep slopes and rocky or sandy areas productive.

Northeastern United States, Europe, and China are heavily populated and timber-short and have long had to import wood products from other parts of the world or go without. Since the original forest was slow growing, it has been replaced with more rapidly maturing conifers.

Fig. 27-14. Steppe grassland in Wyoming.

Middle-Latitude Prairie and Steppe

In the middle latitudes as in the tropics, grasslands show up as transitional vegetation between climates that are moist enough to support vigorous tree growth and moisture-deficient deserts. The ***middle-latitude prairie*** with its undulating expanse of tall grass is next to the forest. It often merges with fingers of tree-lined streams pushing well out into the sea of grass, and grassy openings penetrate the forest margin. The prairie features perennial grass vegetation averaging 2 feet (0.6 m) high, although in places it reaches well over that, with the root systems merging into solid turf or ***sod***. Many other herbaceous plants are also represented in the prairie—some of them annuals with showy flowers—but woody growth is distinctly lacking. A prairie landscape is therefore one of endless vistas of waving grass.

As they near the margin of the desert, the grasses decrease in height to 6 or 8 inches (15 to 20 cm) and the species change, but the sod continues. This short-grass formation class is commonly called ***middle-latitude steppe***[1] (Fig. 27-14).

In the middle latitudes, grasslands show up as transitional vegetation.

Generally, when bunchgrass and woody scrub with bare soil between make their appearance and the continuous sod is no longer in evidence, the dry margin of the steppe has been reached. Such grasses as the distinctive American buffalo grass are typical of the steppe. If overgrazing is allowed, they are often gradually replaced by a less nutritious secondary growth that does not change the appearance of the grassland but is no longer the climax vegetation.

The dry margin of the steppe has been reached when continuous sod is no longer in evidence.

REGIONS OF PRAIRIE AND STEPPE In North America the prairie coincides with the drier western part of the Humid Continental climate region, whereas the adjoining steppe is found in the moister eastern fringe (chiefly east of the Rockies) of the Middle-Latitude Dry climate. This combined area represents the world's largest continuous middle-latitude grassland from Alberta and Saskatchewan south almost to the Gulf. Minor representatives occur in the Washington Palouse country and in eastern Texas.

Almost as large a region stretches from Hungary through the Ukraine and vicinity and north of the Caspian Sea in a

[1] In the tropics the height of the grass is also lower as precipitation lessens and a kind of steppe develops. Since the term *steppe* was first applied in Russia to describe a middle-latitude short-grass grassland, its use has been confined to that particular type of vegetation.

long narrow band far into central Siberia. Sizable outliers occur in Asia Minor, Iberia, and Manchuria.

The situation in Argentina, Uruguay, and southern Brazil has already been discussed, but this has been prairie and steppe as long as western people have known it and must be so classified. Australia exhibits a large region of middle-latitude intermixed scrub and grassland in the interior southeast (chiefly New South Wales and Queensland). Small areas of grassland occur in the Orange Free State of South Africa and in New Zealand.

THE INFLUENCE OF AGRICULTURE The prairie with its rich mollisols has proved to be highly productive for agriculture since farmers have acquired the implements to break the heavy turf. As might be expected, the food grasses, chiefly corn and wheat, predominate. The steppe, on the other hand, with its lower total moisture and cyclic precipitation pattern has traditionally been a grazing area, although with improved techniques including advanced irrigation systems, small-grain agriculture is at least partially successful.

The prairie with its rich mollisols has proved to be highly productive for agriculture.

This means that there are few areas of steppe or prairie in the world where either the plow or overgrazing has not wrought considerable change. Long gone are most of the original grasses, and gone too are the vast herds of grazing animals like bison and antelope that depended on them.

Taiga

The *taiga*, also called the great north woods and boreal forest, is the world's greatest forest, at least in terms of area. It stretches in a wide band across the entire breadth of North America from Alaska to Labrador and across the much greater breadth of Eurasia from Scandinavia to Kamchatka. It is virtually a virgin forest, partially because of its remoteness from major population centers and partially because of the inferior quality of the trees for saw-timber. Only along its periphery have commercial lumbering (and Christmas tree plantations) modified the climax forest to any significant degree.

The taiga is virtually a virgin forest.

Home to wolves, bears, moose, and migrating caribou, the taiga is basically a coniferous evergreen forest made up of a relatively small number of tree species. Extensive stands of spruce, fir, and pine are common in North America, whereas pine dominates the Eurasian forest except for a large area in northeastern Siberia where larch, a rare deciduous conifer, prevails.

The extremely long, cold winters and limited growing season are not an ideal habitat for trees, and those few species that do exist are simply the ones that have demonstrated their ability to survive. As a result the trees are small, even stunted, seldom reaching over 50 feet (15 m) in height and 8 to 10 inches (20 to 25 cm) in diameter.

Widely spaced, the pointed conifers do not form a high canopy, but, because they receive light at lower levels, they tend to develop branches and foliage well down the bole. These skirts tend to join and form what amounts to a low canopy, effectively cutting out light at the forest floor. Limited by lack of light and cold temperatures, undergrowth is very sparse. A thick carpet of needles and many dead branches and downed trees collect above the spodosol soils and demonstrate the very slow rate of decay of organic material in the absence of high temperatures.

FIRE AND SECONDARY GROWTH This surface accumulation as well as the resinous character of the trees makes the taiga highly susceptible to fire, and each summer forest fires burn over extensive areas. Secondary growth of aspen, alder, and birch appears out of place among the dark conifers and, if left undisturbed, will eventually be succeeded by them. Growth is slow, however, relegated to just a few summer months each year, and the rate of replacement in the high-latitude forests contrasts strongly with that in the tropics.

The taiga is highly susceptible to fire, and each summer forest fires burn over extensive areas.

ICE AND BEDROCK In some places, notably the Ob River valley of western Siberia, northward-flowing streams that are frozen during the winter maintain the ice at their mouths long after the upper courses have thawed. The blocked mouth causes the river to overflow each spring and seasonal swamps to form over large areas. This condition prevents the establishment of normal forest. As a result, such areas are characterized by coarse grasses and spotty low brush.

In regions that have suffered severe glaciation, as in much of Quebec, bedrock has been exposed at the surface. Although the trees display remarkable tenacity in establishing their roots in cracks and crannies where even a little soil may be available, true forest occurs only in scattered pockets

Fig. 27-15. ***Taiga in eastern Canada.*** *Much of Labrador has been heavily glaciated, the exposed rocky surfaces covered here with lichen. Thirty-foot (9 m) spruces strive mightily to form even a patchy forest.*

between the expanses of barren rock (Fig. 27-15). This is a patchy forest, if a forest at all, and locally it breaks up the general taiga landscape of endless trees extending beyond the horizon.

NORTH VERSUS SOUTH Within the taiga, the largest trees and the better developed forest grow along the southern fringe, gradually deteriorating until the far north trees fade out or persist only in gnarled flag and krummholzlike forms. This is not because the winters are colder, for the trees are dormant in any case. Whether the temperature is −30° F (−35° C) or −50° F (−46° C) makes little difference. Rather, the controlling factors are length of growing season and summer heat. Both of these factors decrease with distance north, and as roots are forced to compete with permafrost and leaves and branches with drying arctic winds, the increasingly harsh conditions are reflected in the size of the vegetation.

Within the taiga, the better developed forest grows along the southern fringe.

Tundra

Now we have gone beyond the treeline, and here, between the northern limit of forest and the permanent snow and ice of the highest latitudes, occurs the arctic tundra.[2]

The two-month summer season with its low temperatures and occasional frost is simply inadequate for trees. Botanists claim that there are trees in the arctic tundra, and as evidence they produce examples of alder or juniper—about 6 inches (15 cm) tall with a tendency to lie flat on the ground. To most of us these are not trees.

Woody plants of all kinds are rare. Tundra vegetation (Fig. 27-16) is a highly specialized mix of species that can:

1. Withstand (or produce seeds that can withstand) long bitter winters.
2. Tolerate high-velocity dry winds.
3. Survive in cold, moist, and low-nitrogen soils (mainly inceptisols, entisols and histosols).

[2] If mountain ranges at any latitude are sufficiently high, alpine tundra will be found between treeline and unvegetated barren surface or permanent ice and snow. Both formation classes are similar in many physical respects.

Fig. 27-16. ***Summer in the arctic tundra.*** *Cotton grass prepares to shed its seed before the snow and icy wind return. Permafrost is not far below the hummocky surface—vegetation size and variety are strictly limited by cold, damp soils and restricted root space.*

4. Grow in shallow soils over permafrost.
5. Mature rapidly and reproduce during a short, cool summer.

Such conditions rule out most plants. The ever present grasses are represented, as are sedges. Mosses and the simple primitive lichen are everywhere. These, with various other low herbaceous plants—some flowering briefly, and many reproducing through shoots from their perennial root systems—make up the vegetation.

Tundra is a low, ground-hugging vegetation on a flat to gently rolling terrain. Surface drainage is inadequate at best, and vertical soil drainage is entirely absent as a result of permafrost, which is often less than a foot under the surface in midsummer. Thus, despite an average of only 5 to 10 inches (13 to 25 cm) of precipitation per year, the surface is constantly wet and boggy during the growing season. Corridors of low trees may push out from the taiga along stream courses in response to better drainage and lower permafrost levels, for, paradoxically, flowing streams mean drier surface conditions.

TUNDRA WILDLIFE During the summer, seemingly from nowhere, come great clouds of gnats, biting flies, and mosquitoes that have somehow survived the killing winter cold and can literally eat alive anyone who attempts to inhabit. Even animals are not immune to harassment; insects keep the vast herds of caribou (North America) and reindeer (Eurasia) in constant movement. These large grazing animals will nearly immerse themselves in shallow lakes to avoid, even temporarily, the insect menace.

Great clouds of gnats, biting flies, and mosquitoes come during the summer.

Summer is also a time of foraging for the lemming population and of the predatory activities of jaegers and snowy owls who feed on them. Migratory waterfowl nest in vast numbers but leave with their offspring as the fall signals an end to the growing season. Winter brings bitter cold but eases human transportation, the insects are gone, and the surface has frozen solid and is no longer one great wetland.

TUNDRA REGIONS The arctic tundra is a north coastal vegetation and is found in virtually unbroken series along the north coast of North America and Eurasia. It is nonexistent in the Southern Hemisphere except in a few tiny high-latitude islands. The Falkland Islands off the coast of Argentina and the Aleutians in the North Pacific are generally included in the tundra grouping despite their relatively low latitudes, for they do not support trees, although there is more grass than in true arctic tundra. High winds and cool

The arctic tundra is found in virtually unbroken series along the north coast of North America and Eurasia.

summers with occasional frost probably keep trees from growing.

HUMAN INFLUENCE The traditional home of Eskimos and certain other Native Americans, and the seasonal habitation of Lapps herding reindeer, arctic tundra has been until recently little despoiled by the human presence. This situation is changing with the quest for mineral wealth. The Alaska North Slope, for instance, is now a vast oil field with roads, derricks, modular housing, refuse dumps, and pipes on the surface to carry crude petroleum to waiting tankers. Vegetation and wildlife have been pushed aside, although migrating caribou still come each year to cross the roads and pipeline to feed on remaining lichen and grass.

If our thirst for petroleum continues unchecked and the will of the oil industry prevails, other tundra areas like the pristine Arctic National Wildlife Refuge east of the North Slope may suffer a similar fate.

CONCLUSION

In this chapter our view of vegetation extends beyond the general relationships and dynamics considered in Chapter 26 to world patterns and the integration of physical geography seen in the distribution of plants. The vegetation classification contains decreasing levels of integration from vegetation in general through vegetation classes, formation classes, formations, associations, and communities to individual plant species. Our focus is on world patterns at the formation class/formation level. Although rather broad, this level helps us understand the complexity of world vegetation patterns and how they relate to climate, soil, geographic position, human population, and human alteration. From tropical rain forest to the arctic tundra each formation class is noted for a particular array of geographic characteristics and life processes, but all have been changed by the human presence.

Physical geography is the basis for understanding the human home.

Indeed, as we complete this final chapter we see more than ever that it is not enough to study only the forms and processes shaping the physical environment. Humans affect change. We cut the tropical rain forest and replace it with grassland. We pave vast areas as cities expand. We pollute the air and change the mix of atmospheric gases, causing a shift in global climate. We till the soil, changing its structure and allowing salts to accumulate in some places and leaching to proceed rapidly in others.

So physical geography, the study of weather and climate, water, landforms, soils, and vegetation—the topics that have occupied us in this text—is more than the study of nature, of rocks, lakes, and sky. It is the basis for understanding the human home, its fragility, its interlocking systems. Through such knowledge we see our own context within nature and appreciate the need for responsible human action, which, as we rapidly approach the end of the twentieth century, is of critical importance.

KEY TERMS

integration
vegetation
vegetation classes
forest
grassland
desert
tundra
formation class
broadleaf evergreen forest
broadleaf deciduous forest
needleleaf evergreen forest
mixed forest
mediterranean forest
savanna
prairie
steppe
shrub steppe
shrub desert
arctic tundra
alpine tundra
formations
communities
associations
plant species
individual plants
tropical rain forest (selva)
emergent layer
canopy layer
liana
tank epiphyte
jungle
galeria forest
mangrove
tropical deciduous forest and scrub
caatinga
desertification
mediterranean scrub forest
sclerophyll forest
chaparral
maquis
mallee
subtropical coniferous forest
middle-latitude coniferous forest
middle-latitude broadleaf forest
middle-latitude mixed forest
middle-latitude prairie
sod
middle-latitude steppe
taiga

REVIEW QUESTIONS

1. Identify the various levels of vegetation classification. Why do we proceed from general to specific? Why do geographers concentrate on formation classes and formations?

2. Describe the five forest types and explain how they differ.

3. Why did Alexander von Humboldt call the tropical rain forest "a forest above a forest"?

4. Explain how—and why—tropical rain forest and jungle are not the same. Why did the early explorers believe that jungle was more extensive than it really is?

5. Within the tropics is a transition from the very wet to the very dry. How does this transition affect major vegetation patterns?

6. Why is desertification such a problem in grasslands? What sets this process in motion and how is it related to cultural traditions?

7. Why is the Mediterranean climate a unique challenge to plant life, and how have plants met this challenge?

8. What is the difference between prairie and steppe? Tundra and taiga? How are they similar?

9. How have humans altered each of the major formations? Why have some been more adversely affected than others?

10. Why is the study of physical geography more than simply a consideration of weather and climate, water, landforms, soils, and vegetation?

APPLICATION QUESTIONS

1. Study the world vegetation patterns shown in Map II. For each unit, list dominant climate(s), soils(s), landforms, and water features. Choose one unit and describe it from an "integrative" point of view. (How do all elements work together?)

2. What vegetative formation class characterizes the area around your college or university? How have human beings altered its original appearance? Have any human activities been harmful? How can further harm be avoided?

3. Find a mail order nursery catalogue and either tear out or photocopy the map of the United States, divided into zones. (It *should* be there!) Then draw a map of vegetative formation classes *over* the zone map. How are the divisions similar and how are they different? Could the map of vegetative formation classes tell gardeners anything that the zone map could not? What and why?

4. Choose a tree or other plant on your university grounds and find out what it is (say, a ginkgo tree or a rhododendron). Look the plant up in the encyclopedia to find out its original habitat. In what vegetative formation does it grow? Is it different from the university grounds? How? What special treatment, if any, does it require to *simulate* its natural growing conditions?

APPENDIX A

SELECTED METRIC EQUIVALENTS

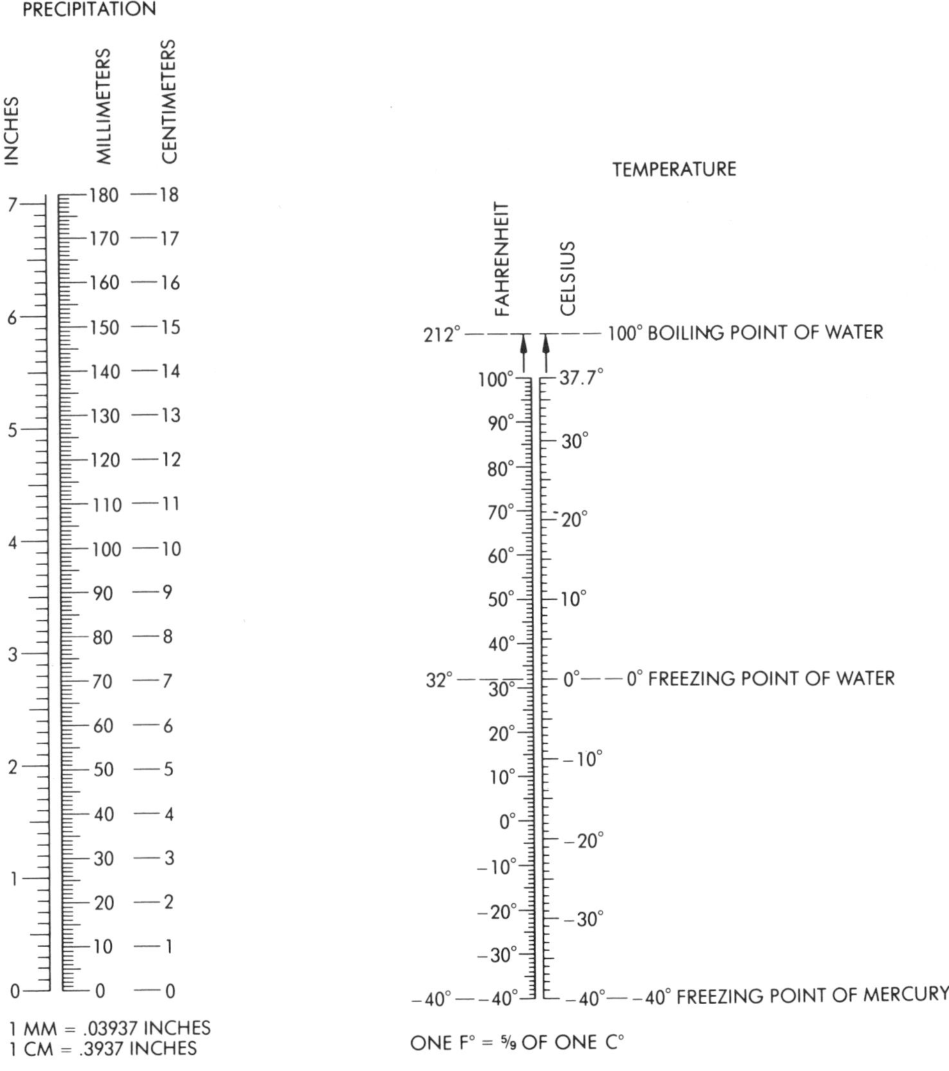

LINEAR DISTANCE

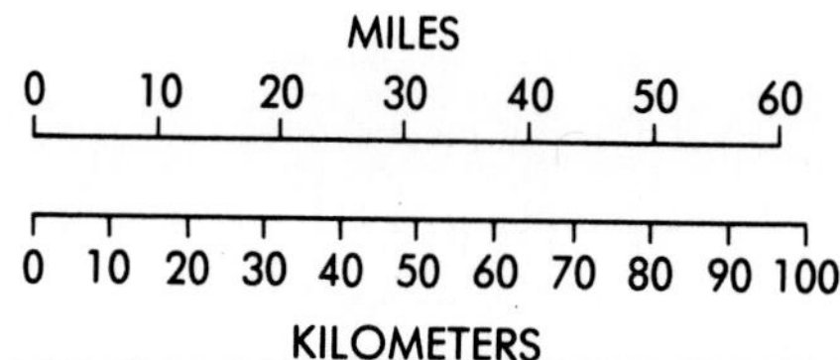

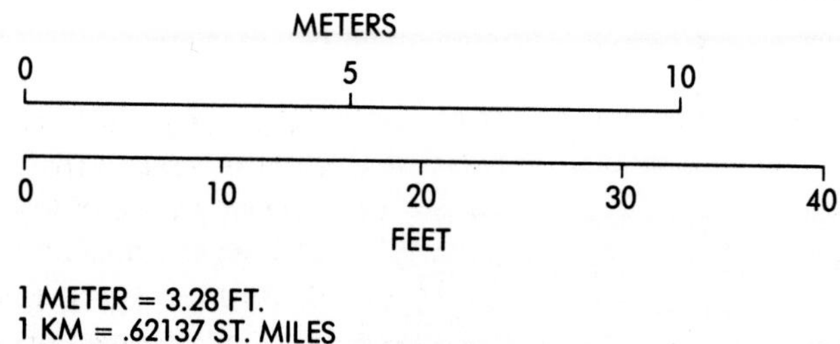

1 METER = 3.28 FT.
1 KM = .62137 ST. MILES

AREA

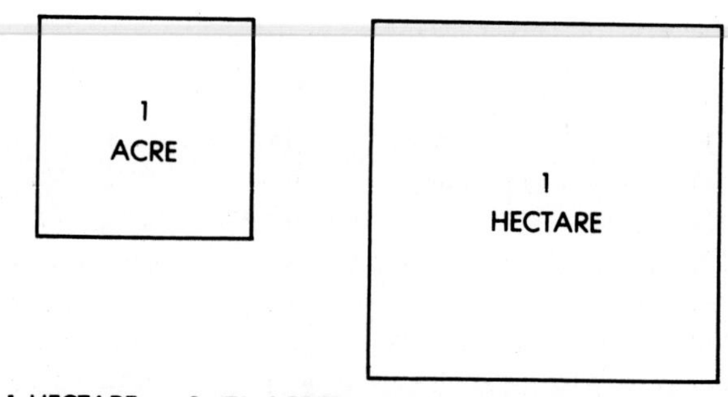

1 HECTARE = 2.471 ACRES

VOLUME

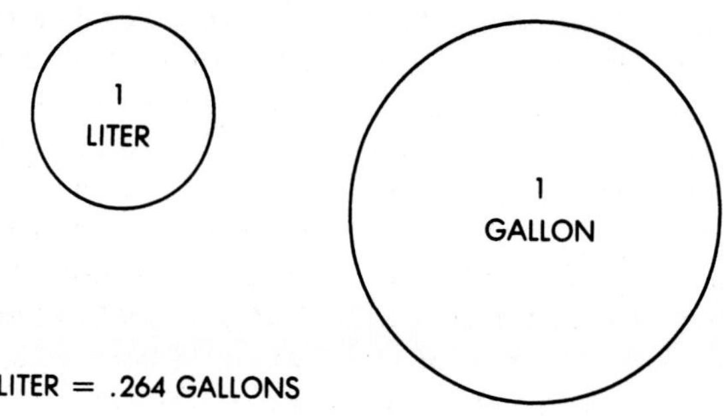

1 LITER = .264 GALLONS

APPENDIX B

MAPS

INTRODUCTION

A *map is merely a symbolized representation of all or part of the earth's surface on a flat piece of paper.* As such, it is a highly useful geographical tool, for in attempting any description or analysis of the earth's surface or of the human activity on it, relative location is essential. Maps help to solve the problem of the individual on the ground who is unable to observe more than a very small part of the earth at any one time. With a map it is possible to visualize large segments or even the whole earth and relationships that are not immediately evident from the restricted local view. And maps are capable of much more than mere location. With the proper selection of data and symbolism, a great variety of concepts may be rendered in visual form.

The urge to map is almost intuitive in all of us. How often do we grab a piece of paper to draw a crude map of road directions? How often the cave dweller must have scratched the same sort of information with a stick in the dirt, long before any means of preserving artwork for posterity was conceived. The oldest known map is a fragment of clay tablet from Mesopotamia dating back to 2500 B.C., but nobody suggests that this was the first map ever made. Maps are such obviously useful tools that they have surely been around continuously from the earliest time. We have, of course, improved on early mapping practices. The Greeks recognized the sphericity of the earth and introduced a system of coordinates for measuring. Later, with the great voyages of exploration, the distribution of the continents became known. Now ***cartography,*** the science of mapping, has become of necessity a world-encompassing affair and is no longer restricted to the immediate local neighborhood. Today, with satellite imagery, computer-assisted cartography, and highly sophisticated techniques of drawing and reproduction, maps are more widely available than at any time in history—cheaper, better, and in greater variety (Figs. B-1 and B-2). But the average person often does not know how to use them except in a perfunctory way. Knowledge of a number of basic practices is necessary to achieve the maximum understanding of this important geographic tool. In this appendix we focus on the various projections that cartographers use to show the earth's surface on a map, and we describe such important map characteristics as scale, direction, and symbolism that must be understood for effective map interpretation.

PROJECTIONS

If we assume the earth to be a perfect sphere, its only true representative is a globe, not a map, for *it is simply a physical impossibility to transfer a curved surface onto a flat one without distortion and error (Fig. B-3).* In other words, *there is no such thing as a perfect map.* Maps of very small areas, where the curvature of the earth is minimal, come close to perfection, but the larger the region the map is attempting to show, the greater the error involved, until it reaches its maximum in world maps. So why not forget maps and use only the globe? Cost is one factor. A good globe is many times more expensive than a map. However, of more importance is the fact that the globe just cannot do the job that a map can in certain instances. Imagine the size that a globe would have to be in order to show real detail of the country of Belgium. A map could handle this easily, but the globe would scarcely fit in the room. And a world map is a necessity in many situations, but only half the globe is visible at any one time. Add to this the ease of carrying a map around in your hip pocket, the simple printing of multiple copies from an original drawing, and the ease of storing in a flat case, and it becomes apparent that, despite their distortions of the areas they purport to show, maps have many advantages over the globe. This does not mean that you should throw away your globe, for it too is useful in other ways. Only on the globe can great circles be properly understood,

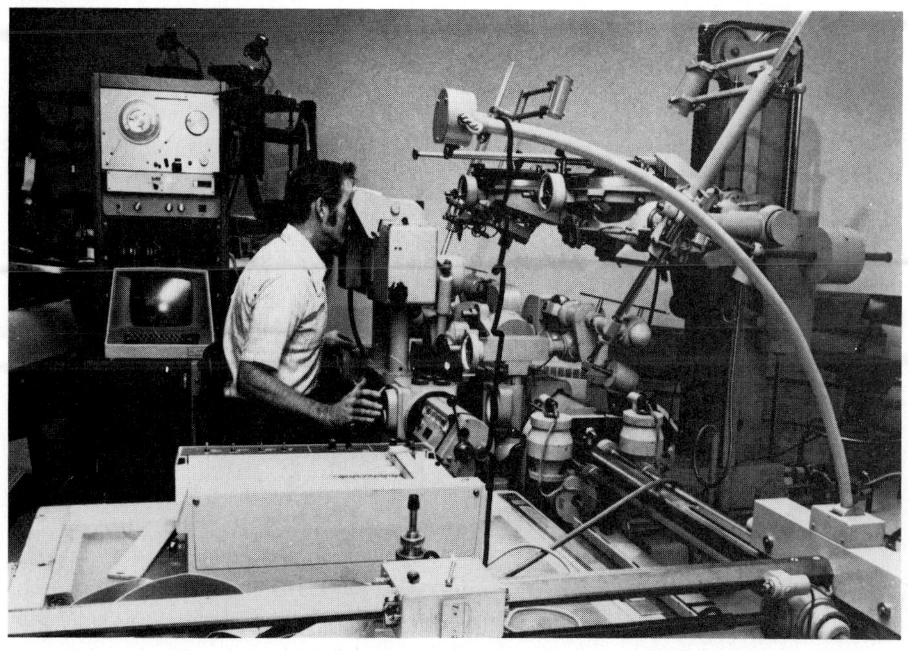

Fig. B-1. ***Modern cartography.*** *A cartographer with pen in hand has virtually gone the way of the maker of buggy whips, shoe buttons, or basketball laces. The machine and computer are the wave of the future. Here is a DPROS (Digital Profile Recording and Output System) at the U.S. Geological Survey's mapping center. The instrument digitizes and records terrain profiles from information spewed out of a diverse array of stereoplotters. Digital profile data then are stored on magnetic tape to be processed into terrain models by a computer.*

as well as true direction or earth/sun relationships. Maps and globes are complementary, not competitive, although they both attempt to represent the earth at a reduced scale.

If maps are necessary and we must accept their inherent defects, the problem of control arises. How can we keep general distortion to a minimum and limit specific error to that part of the map where it will do the least harm? This is the function of ***projection,*** the process of transferring the latitude/longitude grid from the earth's or globe's surface to a flat piece of paper.

The earth and globe have a number of important properties that a cartographer would hope to retain on the map. The most vital of these are:

1. True direction.
2. True distance (scale).
3. True shape (conformality).
4. Equivalence of area.

Unhappily, no map can successfully emulate the globe and have all these properties, but depending on the method of projection, certain critical ones can be retained at the ex-

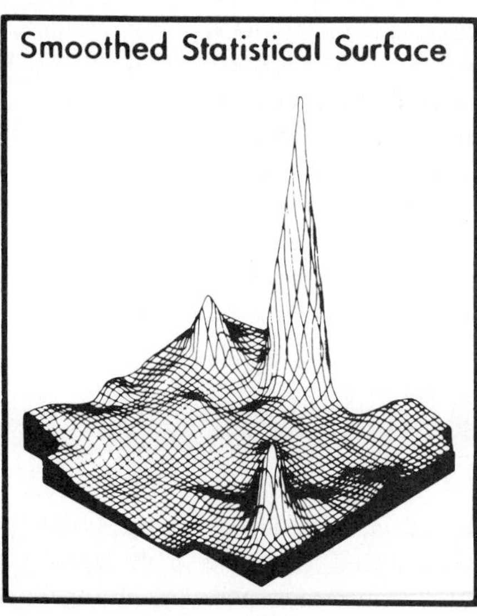

Fig. B-2. ***Cartographs.*** *Strictly speaking, these strange looking cartographs are not maps at all, but rather statistical materials presented in quasi-map form.*

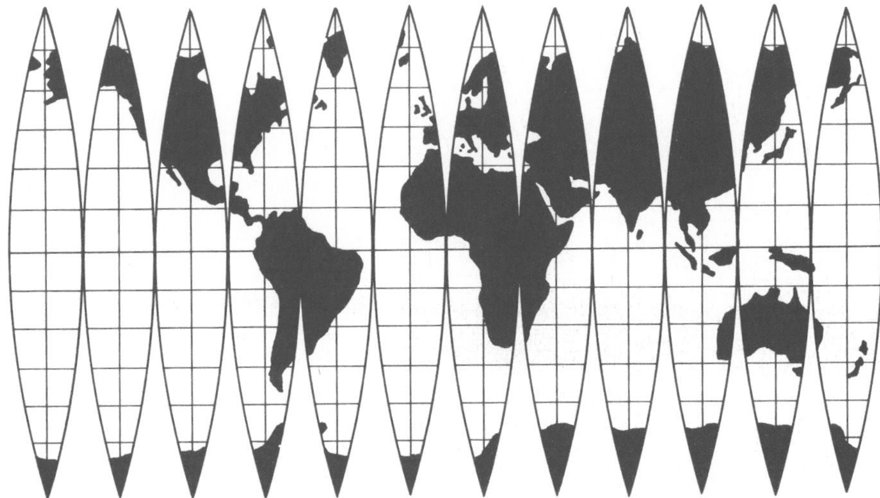

Fig. B-3. **Globe gores.** *Stripped off the globe, where they have formed a perfect picture of the earth, globe gores leave a great deal to be desired as a flat map.*

pense of others. So this matter of which projection to select relates very closely to the purpose of the map. A navigational chart must be conformal above all else, but this means that it cannot be equal area too, for no map can be both (Fig. B-4). If true direction and distance can be maintained—and they frequently can be, at least along certain lines—so much the better, for these are useful in navigation. On the other hand, if the map is to be for classroom use to show distribution of terrain or population and the like, equivalence of area is more important than true shape and again, one must be sacrificed to gain the other. Some maps are compromises, achieving none of these qualities but coming close enough in all so as not to destroy an effective visual image. But, in every case, the map user must select the projection exhibiting the properties that are needed. Military maps, for instance, often distributed free by the government, will not always work well in the classroom any more than atlas maps can be used for navigation.

The graphic and mathematical methods by which the earthly grid may be transferred onto a plane surface are endless and limited only by the ingenuity of the mapper and the useful properties of the end-product. For example, there is a group of related projections that might be termed "true projections." Although they can be arrived at mathematically or graphically, we can also place surfaces (the plane, cone, and cylinder) tangent to a model globe and actually project its grid pattern.

Azimuthal Projection: Plane

Imagine a globe made up entirely of selected latitude and longitude lines represented by wires—an open hollow wire cage. In the center of this we will put a light. Now place a flat piece of paper tangent anywhere on the globe and turn on the light. The shadow pattern of the wire grid will be *projected* onto the paper and can be traced, giving us a latitudinal/longitudinal base on which to draw in the details of continental outline. Here we have a map on our paper actually projected from the globe. Since this was done onto a plane, it is an ***azimuthal*** projection; the name is derived from the fact that all projections constructed in this manner

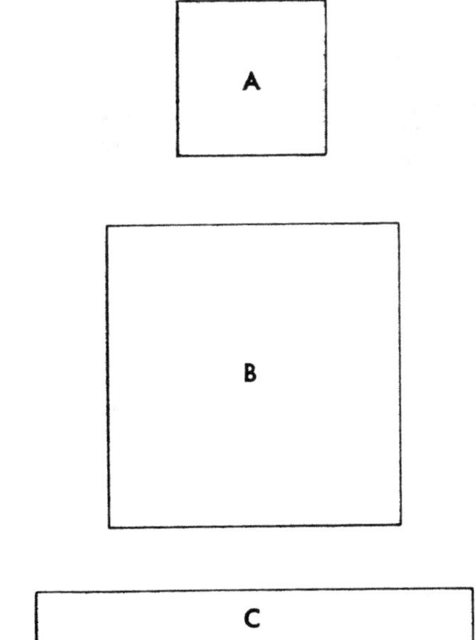

Fig. B-4. **Conformality and equivalence of area.** *A and B are the same shape (conformal) but obviously different in area. A and C are equal in area but far from conformal. These are the choices that a mapmaker must face, for no map can be* both *conformal and equal area.*

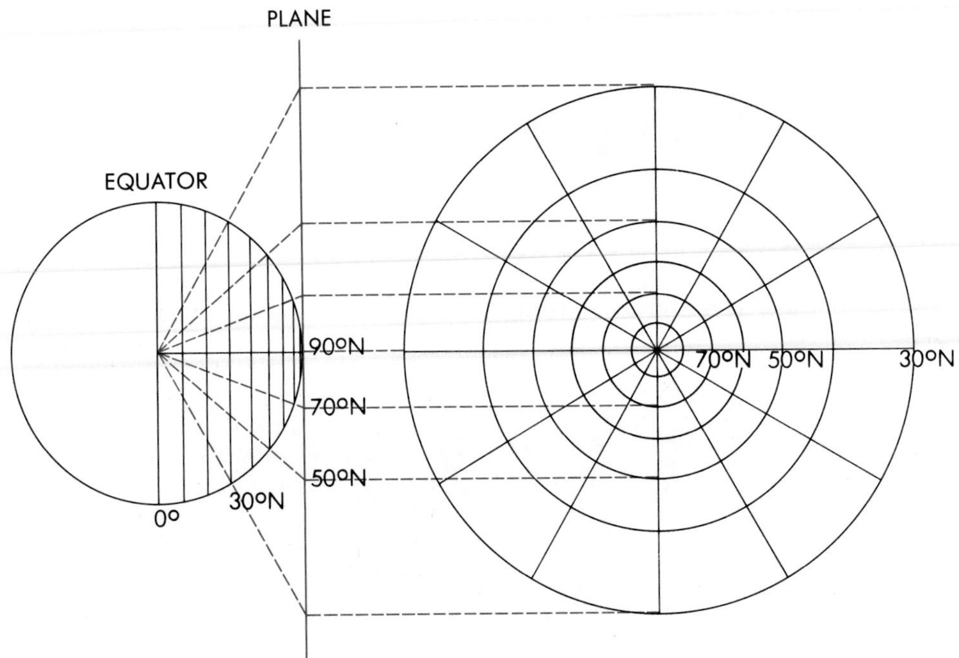

Fig. B-5. ***Azimuthal projection.*** *An azimuthal projection is derived from a tangent plane.*

have the property of true direction (azimuth) from their center or point of tangency.

If we turn the globe so that the point of tangency is the North Pole, this true direction becomes apparent (Fig. B-5). The grid pattern now appears as a series of concentric circles (parallels) about the pole at their center and radiating straight lines (meridians) from the pole. These meridians are azimuths or true compass headings, and if we follow any one of them out from the center we are going south. But notice the spacing of the concentric parallels. Although they were evenly spaced on the globe, on the projection they are increasingly farther apart with distance away from the pole—evidence that the projection is not a true picture of the globe. Land areas drawn on this projection will exhibit increasing distortion in size and shape toward the margin at the same rate.

Conic Projections: Cone

Now take our wire cage representing the earth and place a cone over the top of it with its peak above the North Pole. The cone will be tangent at a particular line of latitude, latitude that depends on the angle of the cone. If we construct a very low-angle cone, a coolie hat, it will be tangent at a latitude not far from the pole, whereas a high-angle cone, a dunce cap, will be tangent near the equator. Obviously, no one could be so flattened as to be tangent at the pole or it would have become a plane, and in order for a cone to be tangent at the equator, it would have to be transformed into a cylinder. Therefore *conic* projections are especially suited to regions in the middle latitudes between these two extremes.

Let us construct a conic projection on which to map the United States. We would select a cone that would be tangent at a latitude running through the middle of the country, say 40° N. Turn on the light at the center and trace the shadow outline on the cone. Slit the cone up the back and lay it flat and we have a base on which to draw our map (Fig. B-6). The lines of latitude will be curved but unequally spaced parallel lines crossed at right angles by equally spaced straight-line meridians converging toward the tip of the original cone. At the critical central 40° parallel, scale, shape, and area are all true. But since the other parallels, though representing equally spaced lines of latitude on the globe, become increasingly farther apart with distance from the 40° parallel, those portions of the map will exhibit greater and greater distortion.

Cylindrical Projections: Cylinder

Here we fit a cylinder over a globe. If the globe is upright with the North Pole at the top, the cylinder will be tangent at the equator. Then having turned on the light at the center and traced the grid, we cut the cylinder down one side and

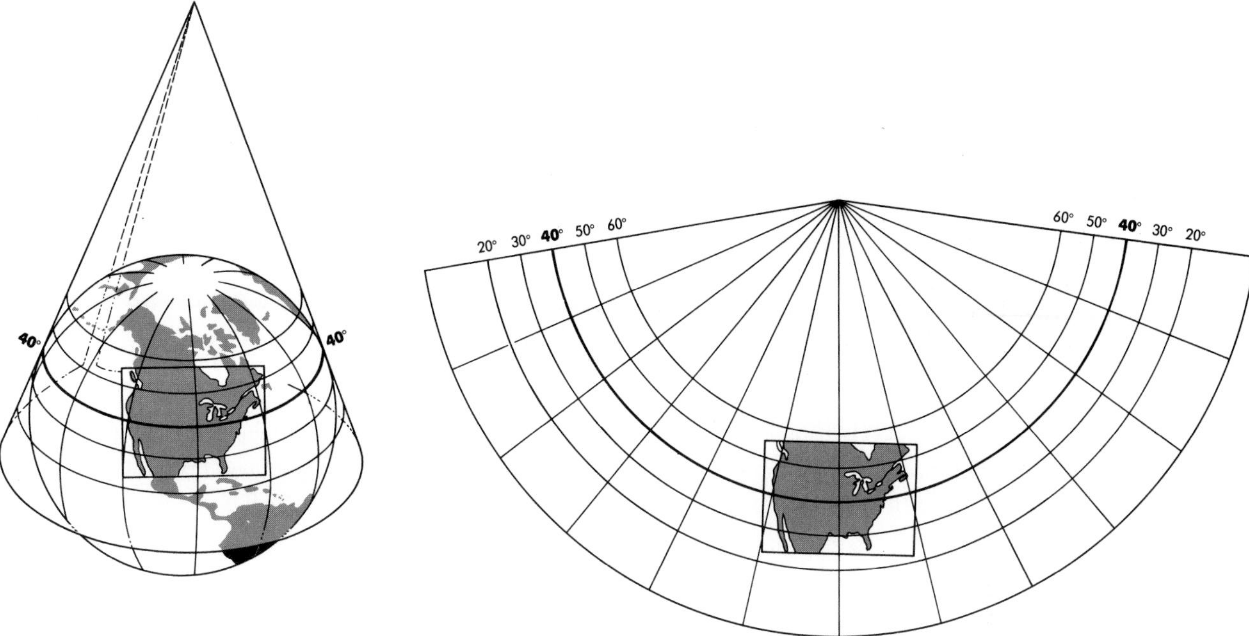

Fig. B-6. Conic projection.

lay it out flat (Fig. B-7). Once again, we have constructed a projection, this time of virtually the whole world. One weakness of both the azimuthal and the conic projections was their inability to reproduce a great deal more than a hemisphere, but now we have a grid that, except for the polar regions, represents the entire earth.

Offhand it looks pretty good. The parallels are nice straight lines (though spaced unequally again) instead of those disconcerting curves, and the meridians are also straight and at right angles to the parallels.

But once we draw the continental outlines onto the grid, it becomes manifest that there are inherent problems. Only at the equator do things look right. Elsewhere shapes and sizes get stretched in all directions. The meridians are

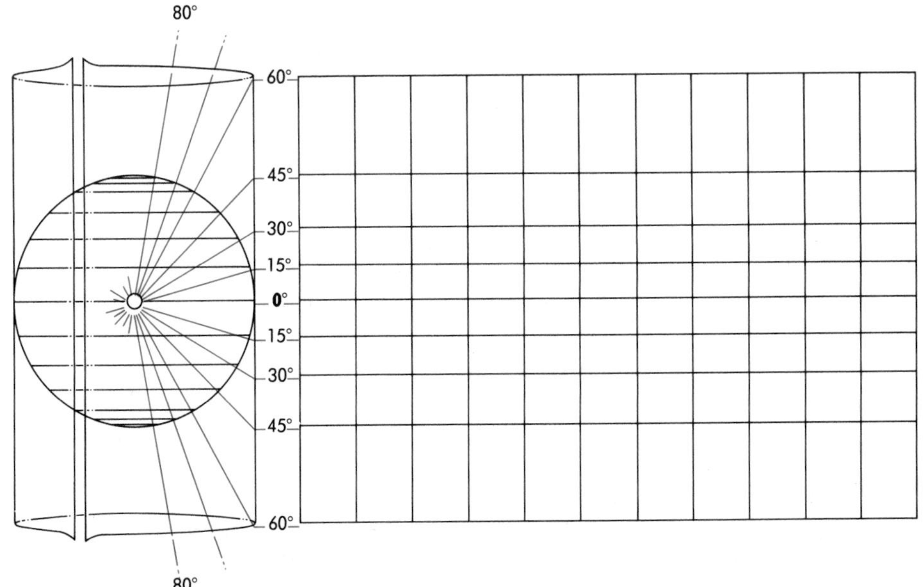

Fig. B-7. Cylindrical projection.

Fig. B-8. ***Mercator projection.*** *This famous conformal projection, derived from a cylindrical base, is still a standard navigational chart. But it contains the obvious flaw of gross size distortion in the high latitudes, which limits its use for other purposes.*

primarily at fault; they are parallel to each other and do not converge at all. This means that at the higher latitudes where the meridians should be closer together, they are as far apart as at the equator and tremendous east/west distortion results.

Because the parallels are spaced increasingly farther apart, some of this meridional stretching is compensated for, but not enough. If only the parallel spacing exactly equaled the error in meridian spacing, we would have a conformal projection. This is exactly what an early mapmaker by the name of Mercator did; he adjusted one to compensate precisely for the other. So the well-known ***Mercator*** map is a world map on a ***cylindrical*** projection, slightly doctored to achieve conformality (Fig. B-8). As it happens, any straight line on a Mercator projection is a true compass heading, and this quality plus conformality has made it one of the most widely used of all navigational charts.

Somehow the Mercator map found its way into the classroom and the atlas and was out of its element there. Proportions were all wrong. Greenland near the pole was true shape but was so badly distorted in size that it appeared to be larger than South America, which, being astride the equator, came much closer to its true size. Obviously, a map explicitly constructed for navigational use was not of universal utility. And this brings us back to the original proposition regarding projections—their purpose is to emphasize those qualities required for a particular use and to relegate the errors and distortions to the regions where they will do the least harm. The map is too useful an instrument to abandon simply because it cannot be perfect, so we accept it, warts and all. The choice of projection is our control mechanism.

SCALE

All maps are reductions of the region they are depicting; there are no enlargements. This means that once a portion of the earth's surface has been rendered onto a flat projection, there must be some indication of how much reduction has occurred in the process so that the map reader can measure distance and properly interpret relative size. *Scale, then, is simply the index of reduction* and is written at the bottom of each map in one of three ways:

1. *Graphic scale.*
2. *Stated scale.*
3. *Representative fraction.*

The simplest and easiest to use, especially for the general map reader, is the graphic scale. It is the common bar scale marked off in tens or hundreds of miles so that we can measure simply with a straightedge the distance between any two places.

Graphic Scale

Scale of Miles

It may also be in kilometers or feet—whatever is most convenient. Subdivisions for more finite measurements are frequently introduced at the left end of the bar. But this scale, like all others, cannot be applied with impunity all over a map, for no map is perfect, especially if it is of a large area. The general rule is: *Trust the scale only at the center of the map and regard it as a mere generalization near the margins.* The graphic scale has one advantage over all others: if the map is to be enlarged or reduced, its scale changes, but the length of the bar changes at the same rate and so remains true. Any other scale must be reworked to conform to the varied size of the map.

Stated Scale

The stated scale is a prose sentence that merely states the number of miles (or kilometers) that is the equivalent of 1 inch (or 1 centimeter), so that again, with a ruler, distances can be readily determined.

Stated Scale: 1 inch equals 8 miles

Representative Fraction

The representative fraction is in many ways the most versatile of the lot and is in very common use, but since it is merely a number, it does not appear very useful to the uninitiated. It is shown as a fraction, or more frequently a ratio, and is applicable in many situations where other scales are not by carefully avoiding any reference to inches, miles, or feet.

Representative Fraction (R.F.): 1:316,800

The ratio, alone, simply states that *1 unit on the map is equal to 316,800 of the same units on the earth.* This means that such a scale is just as applicable in a country using the metric system, or any other, as it is where we use inches and miles, and tells us how much reduction has occurred. The large number, 316,800, is the number of times the map has been reduced. Obviously, if a map of the world were reduced sufficiently to fit on a piece of paper of manageable size, the number would be even larger, on the order of 1:10,000,000 (Fig. B-9).

It is all well and good to know that a map has undergone reduction several million times, but our minds simply do not grasp exactly what this means. A snapshot of a person is a considerable reduction from the original too, but we have seen that person and know how big he or she is, and our minds take care of the reduction problem automatically. But we have never seen the United States at one glance or probably even our county, so a simple notation of how many times it has been reduced does not help much. Somehow we must make better sense out of that number. Let us come back to our original scale of 1:316,800 and use inches as our unit: one inch on the map equals 316,800 inches on the earth. Still not much help; 316,800 inches is hard to visualize. Let us change the inches to miles, and then we will have something useful. If one inch on the map is equal to so many miles, measurement has suddenly become easy. The critical key to all this is 63,360, *the number of inches in a mile.* This number will make sense out of any R.F. scale when it is applied to the larger number. In this case, 316,800 ÷ 63,360 = 5, or 1 inch on the map is equal to 5 miles.

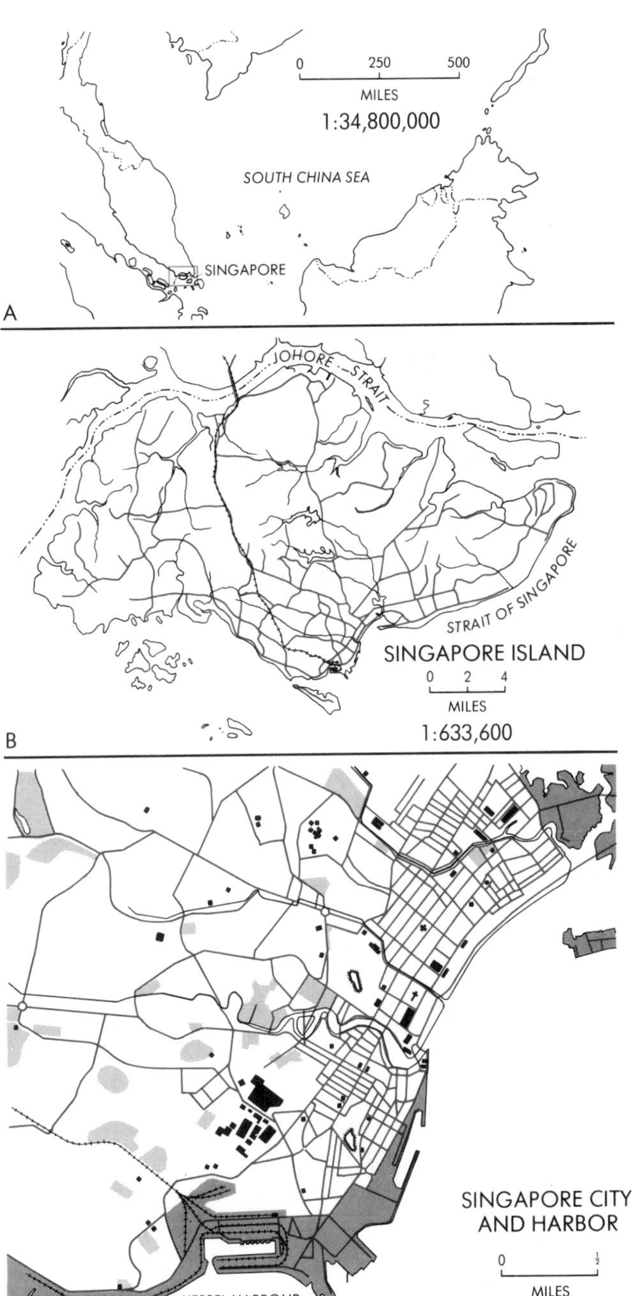

Fig. B-9. ***Scale and detail.*** *The map of Southeast Asia (A), has been reduced many more times than that of Singapore harbor (C), and thus shows much less detail.*

Without a scale, maps are not much more than pictures, and no intelligent interpretation can be made. Therefore the first point of business on attempting to read any map is to check its scale.

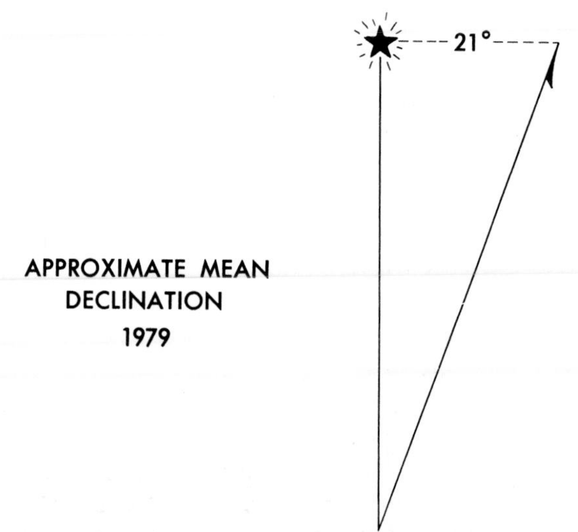

Fig. B-10. **Compass variation.** *The star represents Polaris or true north; the arrow, the direction of the magnetic compass needle. A symbol such as this frequently appears at the margin of a map to indicate the proper correction to be made when the map is used with a compass in the field.*

DIRECTION

On a map with latitude and longitude lines, determining direction is a simple matter, for they are by definition east/west and north/south lines. But on the occasional large-scale map without these lines and on any map that might be taken out into the field, there will appear an arrow pointing to true north and often a second arrow pointing to magnetic north (Fig. B-10). Their direction may differ by a significant amount (termed "declination" or "variation"), for the North Magnetic Pole is over a thousand miles from the North Pole at Prince of Wales Island, Canada.

If one were to stand near the southern tip of Lake Michigan and observe a magnetic compass, it would point to true north because this particular point is on a great circle passing through both the true and magnetic poles. This is called an ***agonic line.*** Anywhere on the agonic line the two poles are in line, and a magnetic compass will point true north; anywhere else there will be variation of some degree. Oddly, the earth's lines of magnetic force change from time to time in an unpredictable manner, so that the agonic line wanders a bit and at any given moment will display waves and therefore is never quite a true great circle. Consequently, the map arrows showing the amount of compass variation are usually dated to call attention to the fact that they may be in slight error if the date is not recent.

SYMBOLISM

A map in its entirety is merely an assemblage of symbols. Even a line representing a coastline is a symbol, for it attempts to demark mean sea level, a line that does not exist in nature. Or if it shows the broken Norwegian coast on a small-scale map, only the major fiords are at all accurate, and the rest is merely generalized to indicate a ragged character. Roads, if they are to show clearly, may be actually represented as 5 miles (8 km) wide, and railroads display a single rail 3 miles (5 km) wide with 10-mile-long (16-km-long) ties 50 miles (80 km) apart. These are all deliberate exaggerations to call attention to features of importance; other major features that are not deemed useful in what that particular map is attempting to show are left out. A map can show political boundaries, latitude and longitude, etc., all of which are symbols representing features invisible in nature.

It has been said that an aerial photo is essentially a map. It is not. For one thing, it shows too much. Forest foliage may hide a road, and the tremendous amount of minutiae detract from the important things. A map is selective; a map is intelligently generalized; a map is often exaggerated to illustrate a particular synthesis—and it does this through the use of selected symbols.

Through long use, most basic symbols have become standardized so that even foreign maps can be read rather easily, although the printed words may be unintelligible. To a considerable degree, symbols are pictorial, especially on ***large-scale*** maps, that is, those depicting a small area in great detail. Schools, churches, mines, and the like are denoted by symbols that even the novice can recognize immediately (Fig. B-11). With the increasing use of color on maps, symbolism becomes even simpler, for here too international standardization of basic colors aids in recognition. Blue, of course, is always used for hydrographic symbols, rivers, lakes, springs, and swamps; black and red are used for cultural features such as roads, buildings, and political borders. Any symbol illustrating terrain or topography is shown in brown, while green refers to vegetation, both natural and cultivated.

But symbolism must be adjusted to conform to scale. A perfectly legitimate symbol on a world map simply will not do on one of a county. Cities, as an example, are merely a dot on a ***small-scale*** map, and even that dot may have to be exaggerated in size in order to show up readily. But on a large-scale map, the entire street pattern of the city may be called for with major buildings, parks, and airports indicated in their exact location. The important requirement for map symbols is that within the limitations of scale, they should be readily recognizable if at all possible. Obviously, the occa-

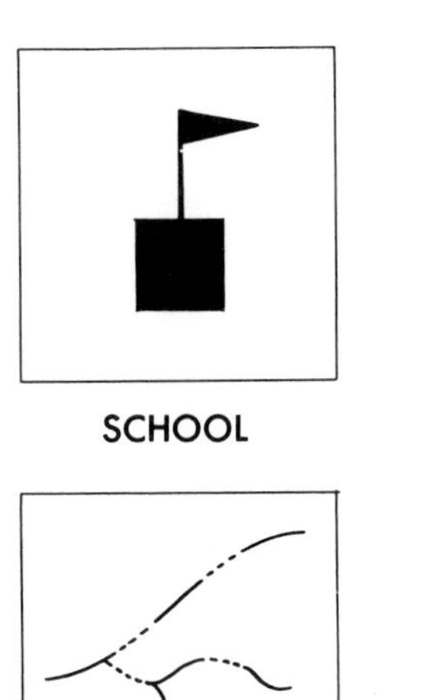

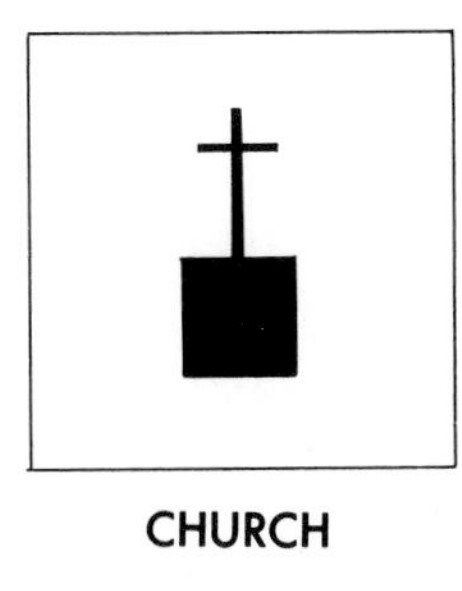

MARSH

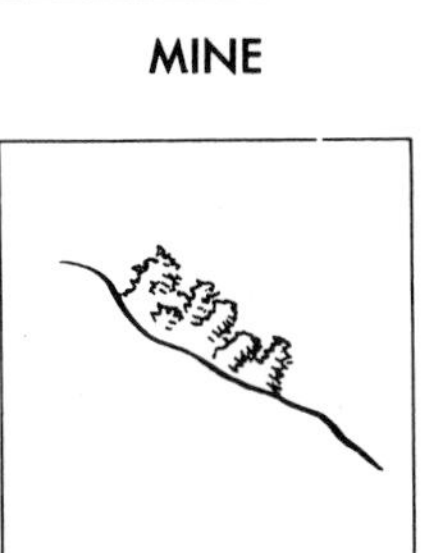

Fig. B-11. ***Standardized symbols.*** *Examples of internationally standardized symbols for use on medium to large-scale maps.*

sion arises now and again for the use of unusual symbols, and a legend explaining their meaning is appended to the map. This is perfectly all right, and most people have no difficulty in understanding legends if their notations are clear and succinct. But the less explanation required, the better the map.

Thus far, we have been dealing with symbols that are indicative of location alone, and there is no question that this is the prime function of map symbolism. But if a symbol can do double duty and show something else in addition to location, so much the better. We have such symbols, and they are highly useful; as a general term, let us call them ***value symbols.***

Value Symbols

The common dot map is an example of a value symbol. The location of the dots immediately tells us *where*, for instance, tomatoes are grown in California; but also, if we consult the legend, we find that each dot is given a value of so many acres, and such a map then tells us, essentially, *how many* tomatoes are grown in a certain locale (Fig. B-12). Or take, for example, a map locating cities with circles for those with over 1 million population, and triangles, squares, or smaller circles that indicate a population of 100,000 to 500,000 and so on. These are value symbols. So are bales of cotton, stacks of dollars, or rows of little men—the variations are endless, but all show location plus.

One particular kind of value symbol that is applicable to many situations is the ***isopleth*** or ***isarithm, a line connecting all points of equal value.*** Within this group are many applications, some of which are undoubtedly already familiar to even the most casual map reader. Examples include:

Isotherm—a line connecting all points of equal temperature.

Isobar—a line connecting all points of equal barometric pressure.

Isohyet—a line connecting all points of equal precipitation.

Isagon—a line connecting all points of equal compass variation.

Contour—a line connecting all points of equal elevation.

There are many more, often without specific names. Corn production in a given county may be plotted by the use of these lines. All areas with over 80 percent of each farm planted to corn will be enclosed by one line, over 50 percent by another, and so forth. The use of color or shading, with the darkest shade representing the heaviest production, will make the picture very effective. Whether it is political affiliation, soup consumption, or the percentage of men wearing derby hats that is to be mapped, the isopleth can do the job (Fig. B-13).

Contours and the Third Dimension

One of the most difficult problems faced by mapmakers is how to show nature's third dimension on a flat map. All sorts of methods have been tried, from crude pen strokes to

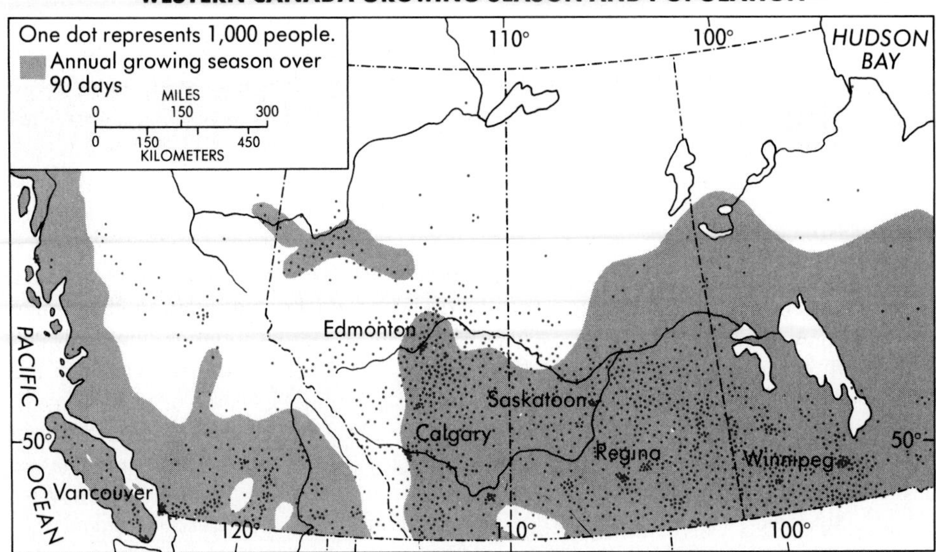

Fig. B-12. ***Value symbols.*** *Dot map with isopleths.*

indicate a mountain range to highly artistic sketches of mountains and valleys. Recently, a process called ***plastic shading*** has become widespread, wherein—assuming a low sun—usually in the northwest, the artist shades deeply those sides of the ridges that would be in deep shadow and highlights the sunny side. Actual construction of a plastic model to be photographed can give much the same effect. This is very striking, and the third dimension is strongly suggested. However, the weakness in most of these methods, not including exorbitant cost and a high degree of artistic ability required on the part of the cartographer, is that the symbolism merely represents terrain character with-

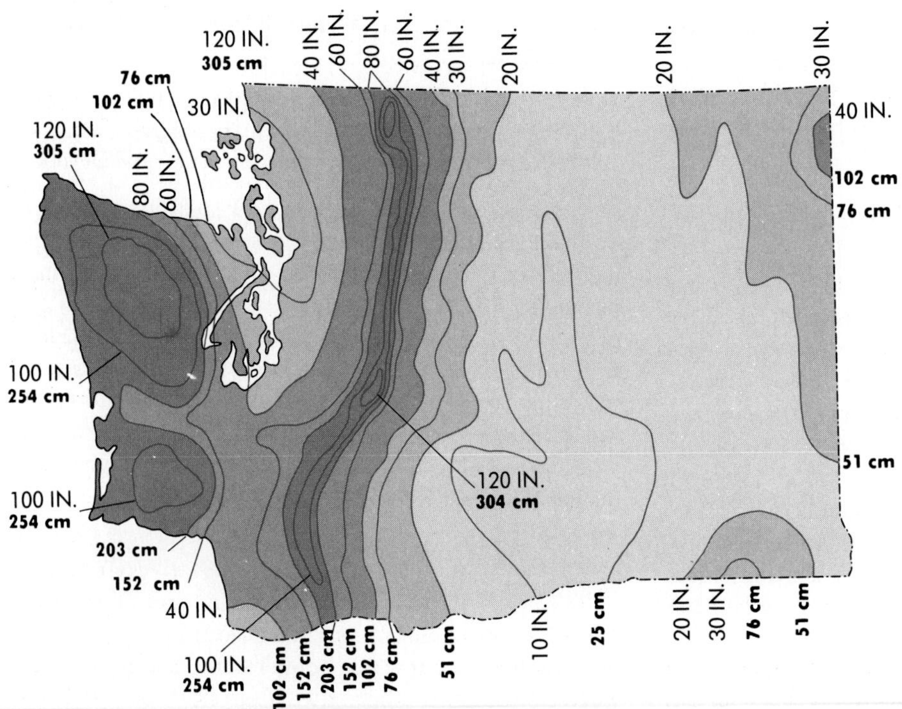

Fig. B-13. ***Value symbols.*** *Isohyet map of Washington State.*

out allowing the map reader to pick off specific elevations anywhere on the map except perhaps certain prominent peaks that may be labeled (Fig. B-14).

All things considered, the map using ***contour lines*** is the most accurate from the point of view of depicting actual elevations throughout. These maps have their drawbacks too in that they are costly to construct, are most applicable only as relatively large-scale maps, and are not easy to read without some practice and training. But the general map reader is likely to come across them, especially via the widely distributed U.S. Geologic Survey sheets that cover a large part of the United States and should know a few basic rules so that such maps can be meaningful.

Atlases and wall maps of continents very frequently represent topographic variation by the use of graduated colors—greens grading through yellows and browns indicating lowland versus highland. These are contour maps, albeit highly generalized. The legend states that everything colored green is under 500 feet (152 m). The margin of green then is a contour 500 feet (152 m) above sea level. [The green color does not mean flat, merely under 500 feet (152 m)]. Similarly, the limit of each of the other colors is a contour line. But the usual contour map is of a much smaller area than a continent, and much more detail is shown.

The secret to reading contours (usually drawn in brown) is to recognize that each line represents a certain vertical distance; therefore the closer they are together, the steeper the slope. Always at the bottom of the map a ***contour interval*** will be indicated in feet, the elevation differential between any two contour lines. As an illustration, let us examine a round but asymmetrical hill rising from sea level with a contour interval of 100 feet (31 m) (Fig. B-15). Since elevations are virtually always reckoned from mean sea level, we measure a series of points all the way around the hill 100 feet (31 m) above the sea. When these are connected by a line, we will have drawn a contour line completely encircling the hill 100 feet (31 m) above sea level. Now since the contour interval specifies that contour lines must represent 100-foot (31-m) differences of elevation, our second line will be constructed above the first in the same manner at 200 feet (60 m) above sea level—a smaller circle since the hill tapers to a peak, but parallel to the sea and the contour line below it. Similarly, the 300-foot (90-m) and 400-foot (122-m) contours are drawn on the hill. Since the top of the hill is above 400 feet (122 m) but under 500 feet (152 m), its exact elevation is often written in at, say, 450 feet (137 m).

If we were to fly over this hill and look down on it from the top, we would see these contour lines as a series of concentric circles, but they would not be evenly spaced. On the steep side of the hill the foreshortening of the overhead view would make them appear close together, while the gentle slope would be represented by wider spacing. We know that all these contours have been accurately measured and are at the proper distance above sea level, yet the plane or map view from above gives us a direct relationship between slope and line spacing (Figs. B-15 and B-16).

Let us assume that a road on a map is 1 inch long, that 1 inch equals 1 mile, and that during the course of that mile, the road crosses five 100-foot (31-m) contours. This means a change of elevation of 500 feet (152 m) in 1 mile—quite a steep slope. But elsewhere on the same map, another 1-mile road might cross only two contours and thus climb only 200 feet (60 m)—a much gentler slope. A quick glance at the average contour map can immediately tell us the essence of the terrain. In the areas that appear brown because of great crowding of many contour lines close together, the country must be very rough (or steep), while in the white regions where only a few contours show up, the topography is subdued. Just how rugged the mountains are and what the exact elevations might be will, of course, require somewhat closer scrutiny.

The contours themselves are labeled—often only every fifth line, which is drawn slightly darker—and any place falling directly on a line can be assigned an accurate elevation. Also, prominent peaks, road junctions, lakes, and the like, will usually be marked with their exact height. Locations between contour lines must have their elevations interpolated within the limits of the contour interval. The selection of the contour interval is determined by the scale of the map and particularly by the character of the terrain. If a section of the Rocky Mountains is to be mapped, a large contour interval, 100 feet (31 m) more, must be used, since there are huge changes of elevation within very short distances. If in such country a 10-foot (3-m) contour interval were used, there would be so many brown lines on the map attempting to show every 10-foot (3-m) change of altitude that they would merge and be indecipherable. On the other hand, a map of almost flat terrain must utilize a very small contour interval in order to bring out the minor undulations. A 100-foot (31-m) interval on such a map would result in no contour lines at all if there were no 100-foot (31-m) change in elevation on the plain.

Theoretically, every contour line must close, that is, eventually come together to form a circle, for all land masses in the world are islands, and since we are measuring from sea level, closure becomes mandatory. In actual practice, however, only limited areas are represented on a single map, and a single closed line or a series of concentric circles means a hill. This is a good general rule to keep in mind when we attempt to interpret contour maps, but there is one

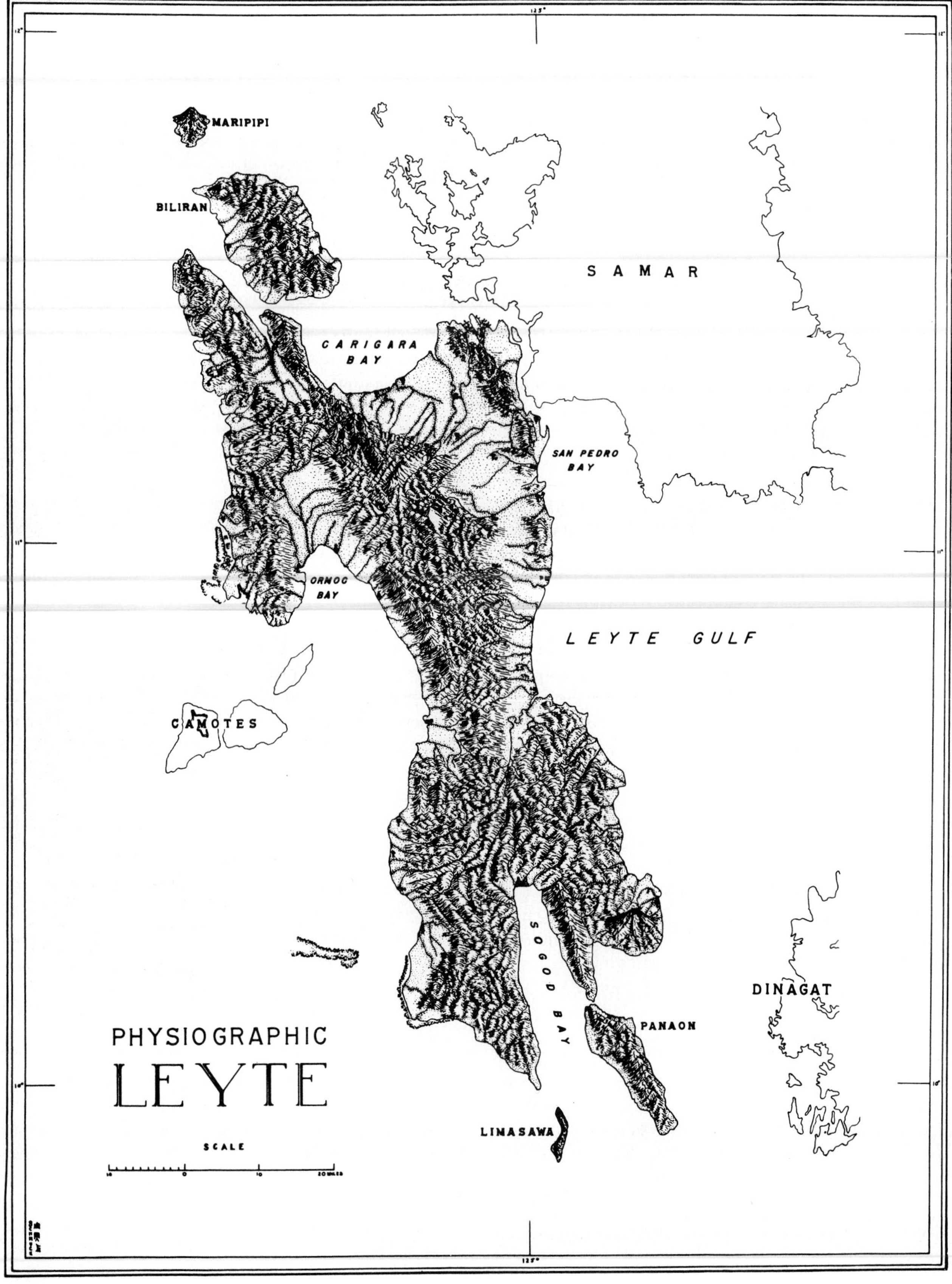
125°
12°
MARIPIPI
BILIRAN
SAMAR
CARIGARA BAY
SAN PEDRO BAY
11°
ORMOG BAY
LEYTE GULF
CAMOTES
SOGOD BAY
DINAGAT
PANAON
PHYSIOGRAPHIC
LEYTE
10°
LIMASAWA
SCALE
10
0
10
20 MILES

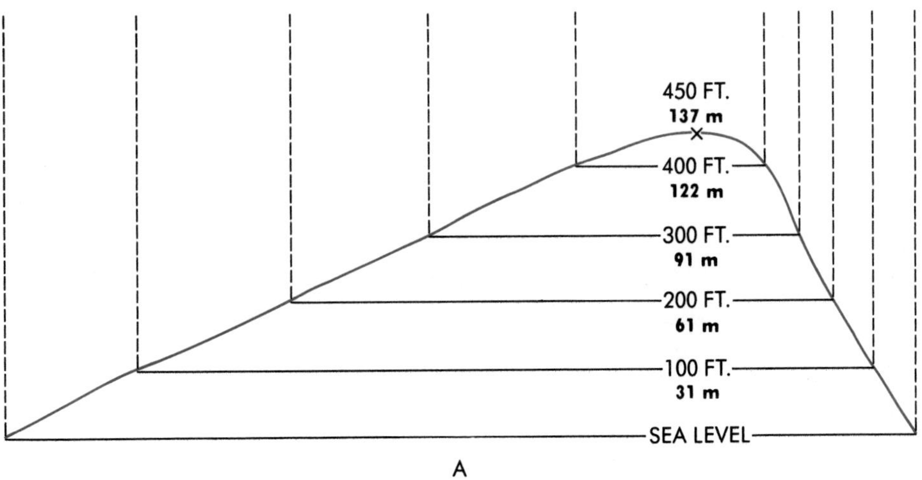

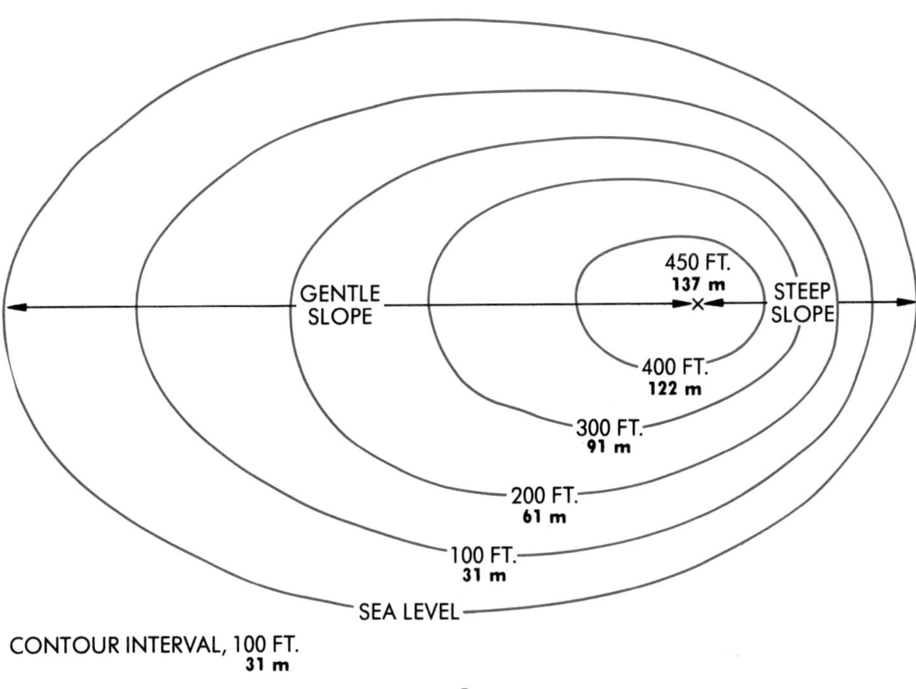

*Fig. B-15. **Contours.** (A) All contour lines represent a 100-foot (30.5-meter) elevation differential as is obvious from side view. (B) They are closer together on the steep slope than on the more gentle one.*

---

*Fig. B-14. **Plastic shading.** This realistic pen and ink rendering is not only time consuming and costly to construct, but is not really a map at all. A map, by definition, is a vertical or plan view, whereas this pictorial resembles an oblique aerial photo.*

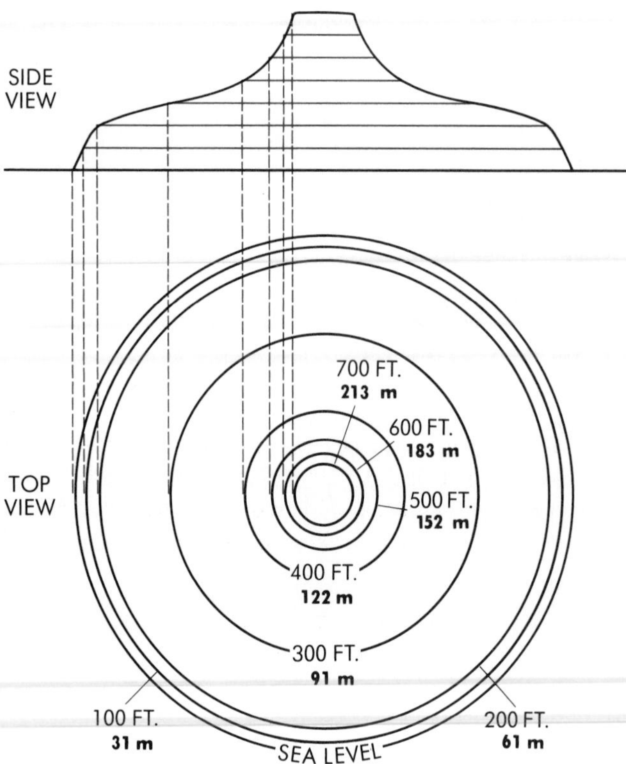

Fig. B-16. ***Contour spacing.*** *The map reader should be able to visualize the actual profile of the island from simple contour spacing.*

exception. A hole in the ground must also be represented by closed contours. In order to distinguish one from the other so that the map can be read intelligently, the depression contours have little ***hachures*** on the downslope side. For instance, we might encounter a volcano with a substantial crater at the top. The mountain would be represented by a series of closed contours, but the crater, if its depth were more than the contour interval, would also be closed, and the only way we could tell that it was a depression rather than a simple extension of the cone would be by the hachures (Fig. B-17).

One of the best guides to interpreting terrain slope is an assessment of the stream pattern, for streams flow in valleys and therefore slope will inevitably be up as we move away from the stream. But streams must flow downhill and therefore must cross contours that represent slope. It may appear that contour lines parallel stream valleys, indicating the steepness of the valley sides, but if the map reader carefully follows a given line, it will be discovered that it eventually crosses the stream. Normally, however, each contour line will run far upstream before it crosses, for the elevation of the stream bottom some distance upstream is equal to the height of the bank below.

Remember, then, *contours must cross streams but they will point upstream.* This concept is useful in determining the direction of stream flow on the frequent maps where a stream enters on one side and flows off the other with no other indication of which way it is going.

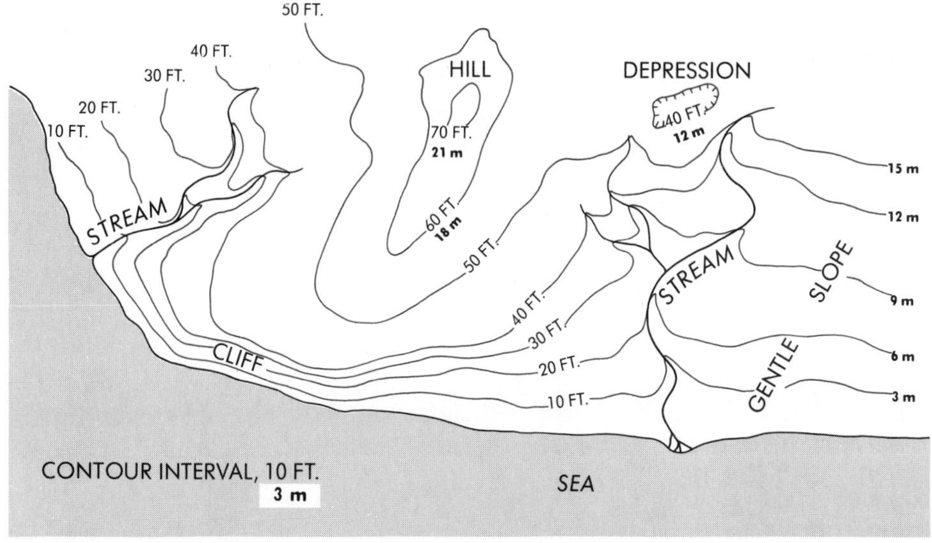

Fig. B-17. ***Simplified contour map.***

APPENDIX C

THE KÖPPEN CLIMATIC CLASSIFICATION

There are five basic climatic categories in the Köppen system, symbolized as A, B, C, D, and E. These are further subdivided by adding lower-case letters to indicate lesser variations of temperature and moisture within the major groupings.

A. TROPICAL HUMID CLIMATES

Coolest month must be above 18° C (64.4° F).

Af—Tropical Rain Forest (f—feucht or moist) No dry season. Driest month must attain at least 6 cm (2.4 in.) of rainfall.

Aw—Tropical Savanna (w—winter) Winter dry season. At least one month must attain less than 6 cm (2.4 in.) of rainfall.

The following lower-case letters may be added for clarification in special situations:

m (monsoon)—despite a dry season, total rainfall is so heavy that rain forest vegetation is not impeded.

w′—autumn rainfall maximum.

w″—two dry seasons during a single year.

s (summer)—summer dry season.

i—annual temperature range must be less than 5° C (9° F).

g (Ganges)—hottest month occurs prior to summer solstice.

B. DRY CLIMATES

No specific amount of moisture makes a climate dry. Rather, the rate of evaporation (determined by temperature) relative to the amount of precipitation dictates how dry a climate is in terms of its ability to support plant growth. This is reckoned through the use of formulas that are not included here. See Selected Bibliography for Section 3, Köppen and Geiger.

BW (W—wuste or wasteland)—desert.

BS (S—steppe)—semiarid.

The following lower-case letters may be added for clarification in special situations:

h (heiss or hot)—average annual temperature must be above 18° C (64.4° F).

k (kalt or cold)—average annual temperature must be under 18° C (64.4° F).

k′—temperature of warmest month must be under 18° C (64.4° F).

s—summer dry season. At least three times as much precipitation in the wettest month as in the driest.

w—winter dry season. At least ten times as much precipitation in the wettest month as in the driest.

n (nebel or fog)—frequent fog.

C. TEMPERATE HUMID CLIMATES

Coldest month average must be below 18° C (64.4° F), but above −3° C (26.6° F). Warmest month average must be above 10° C (50° F).

Cf—no dry season. Driest month must attain at least 3 cm (1.2 in.) of precipitation.

CW—winter dry season. At least ten times as much rain in the wettest month as in the driest.

Cs—summer dry season. At least three times as much

rain in the wettest month as in the driest. Driest month must receive less than 3 cm (1.2 in.) of rainfall.

The following lower-case letters may be added for clarification in special situations:

a—hot summer. Warmest month must average above 22°C (71.6° F).

b—cool summer. Warmest month must average below 22° C (71.6° F). At least 4 months above 10° C (50° F).

c—short cool summer. Less than four months over 10° C (50° F).

i—see A climate.

g—see A climate.

n—see B climate.

x—maximum precipitation in late spring or early summer.

D. COLD HUMID CLIMATES

Coldest month average must be below −3° C (26.6° F). Warmest month average must be above 10° C (50° F).

Df—no dry season.

Dw—winter dry season.

The following lower-case letters may be added for clarification in special situations:

a—see C climate.

b—see C climate.

c—see C climate.

d—coldest month average must be below −38° C (−36.4° F).

f—see A climate.

s—see A climate.

w—see A climate.

E. POLAR CLIMATES

Warmest month average must be below 10°C (50° F).

ET (T—tundra)—Warmest month average must be above 0° C (32° F).

EF—All months must average below 0° C (32° F).

APPENDIX D

THE GEOLOGIC TIME SCALE

Era	*Period*	*Epoch*	*Beginning of Each Period*
Cenozoic	Quaternary	Recent (Holcene)	10,000
		Pleistocene (ice age)	2,000,000
	Tertiary	Pliocene	5,000,000
		Miocene	24,000,000
		Oligocene	37,000,000
		Eocene	58,000,000
		Paleocene	66,000,000
Mesozoic	Cretaceous		144,000,000
	Jurassic		208,000,000
	Triassic		245,000,000
Paleozoic	Permian		286,000,000
	Carboniferous	Pennsylvanian	320,000,000
		Mississippian	360,000,000
	Devonian		408,000,000
	Silurian		438,000,000
	Ordovician		505,000,000
	Cambrian		570,000,000
Precambrian	Late Precambrian		1,000,000,000
	Middle Precambrian		2,500,000,000
	Early Precambrian		3,800,000,000
Formation of the Earth			4,600,000,000
Age of the Universe			18,000,000,000

GLOSSARY

absolute age Age given in units of time, usually years.

absorption The change of solar radiation to heat energy as it interacts with the earth's atmosphere and surface.

abyssal plain A deep, sea-bottom plain.

accordance All ridge tops at the same elevation; an indicator of the original surface before erosion.

adiabatic lapse rate *See* lapse rate (adiabatic-dry); lapse rate (adiabatic-wet).

adiabatic (process) Change of sensible temperature of a gas resulting from compression or expansion without any heating or cooling from outside.

adits Horizontal mine shafts.

advection Heating or cooling resulting from horizontal air movement.

advection fog *See* fog (advection).

aeolian Refers to landforms resulting from wind erosion and deposition.

aggradation The deposition of eroded material; part of the gradation process.

air drainage Dense, cold surface air flowing downslope in response to gravity.

air mass A great body of air that displays a singular homogeneity of both temperature and moisture.

albedo The reflectivity of light waves from an unpolished surface.

albic horizon Light-colored horizon, sandy in texture, with clay and iron oxides leached out. Found in the profile of spodosols.

Aleutian Low The small but intense (especially in winter) center of low atmospheric pressure in the North Pacific. It is a generator of middle-latitude storms.

alfisols Highly weathered, old soils of humid and subhumid climates. They have an ochric (pale) epipedon and an argillic (clay) horizon.

allelopathy The liberation of toxic chemicals from certain plants that blocks germination of seeds around them.

alluvial fan Depositional feature at the mouth of a mountain canyon, usually in arid or semiarid regions. As stream flow loses its velocity exiting the canyon, stream load deposits in a gently sloping, fan-shaped accumulation of alluvium.

alluvial piedmont Large, coalescing fan deposits of alluvium at the base of mountains.

alluvium Stream-transported and deposited material, usually unconsolidated deposits of gravel, sand, silt, and clay.

alpine tundra Formation class of low-growing plants found at high mountain elevations above treeline.

altitudinal zonation The arrangement of plant communities or formations with increasing altitude on mountainsides. Closely related to decreasing temperature.

Altocumulus Middle-level cloud with many small cumulus masses together in a regular pattern. Indicator of fair weather.

Altostratus A smooth, grayish, blanket-like cloud that indicates the approach of a storm and through which the sun shows as a bright spot.

andesite Grayish, moderately silica-rich lava and fine-grained rocks.

anemometer Device with rotating cups used to measure wind speed.

aneroid barometer An instrument for measuring air pressure. A metal diaphragm enclosing a partial vacuum compresses or expands in response to differing pressures.

angle of incidence The angle at which the sun's rays strike the earth's surface.

Antarctic Circle Latitude 66 1/2° S. The northernmost latitude in the Southern Hemisphere that experiences at least one 24-hour period of light and one 24-hour period of dark each year.

ante meridiem (A.M.) The period before the sun has transited a given meridian, hence morning.

anticyclone Center of high atmospheric pressure with outward flowing air.

Antitrades Poleward flowing upper air wind currents directly above the Trades. *See also* Trades.

aphelion The greatest distance of the earth to the sun each year, on or about July 3.

aquifer A permeable and porous rock mass or layer from which flowing water may be obtained.

Arabian (desert) Area of Tropical Dry climate between the Red Sea and the Persian Gulf. Part of the dry climatic belt extending from northern Africa to western India.

arctic, A (air mass) Extremely cold and extremely dry, reflecting its source region, Antarctica or the frozen north polar seas.

Arctic Circle Latitude 66 1/2° N. The southernmost latitude in the Northern Hemisphere that experiences at least one 24-hour period of light and one 24-hour period of dark each year.

arctic front The somewhat theoretical line of contact, in the high latitudes, between arctic and polar air.

arctic tundra Formation class of low-growing plants occurring poleward of tree growth and mainly on the shores of the Arctic Sea. Associated with the Tundra climate.

argillic horizon Soil horizon (usually the B horizon) enriched with illuvial clay. Common to ultisols and many other soils.

aridisols Desert soils, often with a calcic horizon. These soils are too dry for mesophytic plants.

arroyos Desert water courses, usually dry, may carry large water volumes during heavy rainfall. Also called wadis.

artesian (well) A flowing well. One that taps an aquifer whose water entry point is higher than the well head.

association (plant) Group of plants that occur together in a certain habitat within a specific formation. Characterized by one or more dominant species. Also known as a plant community.

asthenosphere Soft, plastic zone within the upper mantle of the earth, below the lithosphere.

Atacama Region of Tropical Dry climate and desert vegetation in northern Chile and southern Peru.

Atlantic currents System of ocean currents in the North and South Atlantic. Important to transport of equatorial heat poleward.

aureole Baked zone where the combination of heat and pressure, magmatic fluids, and gases have formed a concentration of minerals.

Australian desert Most of interior Australia with Tropical Dry climate and desert vegetation.

autumnal equinox First day of the autumn (fall) season; in the Northern Hemisphere about September 21. The noon sun is directly overhead at the equator.

Azores High The North Atlantic Horse Latitude cell of high atmospheric pressure featuring generally dry sunny weather and sinking air. Also called the Bermuda High.

back slope The gentle slope on the windward side of a sand dune.

baguio Local name for a hurricane in the Philippines. *See also* hurricane.

banks Shallow plateaus atop the continental shelves. Sunlight and plankton supply make them excellent fisheries.

bannwalds Carefully protected forest plots on slopes above Swiss villages. Maintained to impede or block snow avalanches in winter.

barchan dunes *See* dunes, barchan.

barrier islands Long, narrow offshore islands paralleling the coast. Formed from sand built up on broad coastal plain and elongated by longshore drift.

basalt A dark, silica-poor lava and fine-grained rock.

baselevel The level of the ocean surface extended through the continent. Theoretically, the lower limit to which a stream can erode.

batholith A large intrusive plutonic mass, affecting a surface area of at least 40 square miles (104 km^2).

baymouth bar Sand spit that has closed off the entrance to a bay when tide action and/or stream flow is not strong enough to maintain an opening to the sea.

bed load Sediment (usually gravel, pebbles, boulders, etc.) carried along a stream bottom by rolling, sliding, and skipping. The heaviest particles a stream can transport.

Bora A cold, high-velocity gravity wind descending the Julian Alps at the head of the Adriatic.

braided stream A stream that has a broad course and multiple, intertwining channels. During low flow the stream is choked with bar deposits; subsequent high water flow breaks up into many channels diverted around deposits.

broadleaf deciduous forest One of the forest formation classes. Seasonal forest of the middle latitudes with broadleaf trees like oak and elm that lose all leaves in the fall and grow new leaves in the spring.

broadleaf evergreen forest One of the forest formation classes. Nonseasonal forest of the low latitudes. Broadleaf trees retain leaves year round. Also called tropical rain forest.

caatinga A dense thorn thicket occurring in semiarid northeastern Brazil. Part of the tropical scrub formation class.

calcic horizon Soil horizon with accumulation of calcium carbonate. Common in aridisols and some mollisols.

calcification The process of soil formation in arid and semiarid locations. Minimal leaching results in deposition of calcium carbonate (calcic horizon) in the B or C horizon.

caldera A large craterlike depression in a volcanic mountain formed by collapse as the magma chamber below the mountain is suddenly emptied.

caliche Name given to a thick, rocklike calcic horizon in aridisol soils of the southwestern United States.

Campos Region of southern Brazil with Tropical Wet and Dry climate and savanna vegetation.

canopy layer Nearly continuous layer (or stratum) of tree leaves and branches in a forest. In the tropical rain forest, the canopy layer is below the emergent layer and above the undergrowth of treelets, saplings, and herbaceous plants.

carbon dioxide Atmospheric gas with molecules consisting of one atom of carbon and two atoms of oxygen. One of two basic raw materials for photosynthesis. (The other is water.) Given off during respiration. Readily absorbs long-wave terrestrial radiation. Increases in carbon dioxide in the atmosphere may lead to world climate warming.

chaparral Dense vegetation of tall drought-resistant shrubs and low trees that dominates regions of Mediterranean climate. Known also as maquis (Europe) and mallee (Australia).

chemical weathering Chemical processes that decompose surface rock exposed to water and the atmosphere. Breaks down minerals through oxidation, hydration, and direct solution.

chernozem A black, high-humus soil developed in the middle latitudes under 15 to 20 inches (38 to 51 cm) of annual precipitation and short grass cover. First identified and named in the Ukraine. Now included within the mollisol soil order.

chinook Warm, dry wind descending lee mountain slopes.

cinder cone A relatively small and symmetrical steep-sided volcanic mountain of unconsolidated ejecta.

circle of illumination The boundary between the sunlit half and the dark half of the earth. It always divides the earth into a solar illuminated hemisphere and a hemisphere that is not illuminated.

cirque A semicircular, rocky indentation in high mountains at the head of a glaciated valley. Glacial plucking and quarrying enlarge the bowllike structure.

cirriform One of the three major cloud forms. High-level clouds thought to be made of ice crystals may contain supercooled liquid water droplets. Appear fragile and feathery. Often described as a high-level haze.

Cirrocumulus Cirriform clouds. High-level layer of delicate cloud masses in a geometric pattern. Called "mackerel sky."

Cirrostratus Cirriform clouds. High-level, light clouds making an even layer that produces a halo around the sun or moon.

Cirrus Highest of the cirriform clouds. Forms wispy trails that may show the position of the jet stream. Known as "mare's tails."

clay-humus complex Minute, negatively charged particles of clay and humus in soil that attract, hold, and exchange cations from soil water.

climate Generally, a longtime, overall aggregate of weather observations for an area.

climax The end or final stage in plant succession marked by equilibrium with environment and long-term persistence of the same community of plant species.

climograph A graph on which can be plotted the average of longtime monthly temperatures and precipitation records.

clouds of vertical development Clouds with little horizontal but much vertical spread. *See* Cumulus and Cumulonimbus.

cold front A segment of the Polar Front, on the westerly side of middle-latitude cyclones, where local circulation causes cold air to advance against warm.

community (plant) *See* association (plant).

competition Interaction between two or more plants, and between two or more animals, attempting to use the same scarce resources. Important to plant succession and vegetation patterns.

composite cone A steep-sided volcanic peak constructed of alternating pyroclastic eruptions and lava flows.

Comprehensive Soil Classification System Soil classification system officially adopted by the U.S. government in 1965. Known also as the 7th Approximation.

compression When a parcel of atmosphere descends in altitude and is forced to occupy a smaller volume. Results in heating, lower humidity, and cloud dissipation.

condensation The change in form of water vapor to liquid water, usually as a result of lowered temperature.

condensation nuclei Minute dust particles that serve as the initial site for condensation of water vapor and formation of clouds.

conduction The transfer of energy through contact.

confined aquifer Aquifer with impervious rock layers above and below it. *See also* aquifer.

conglomerate (rock) A sedimentary rock. Pebble size or larger particles cemented together by any of a variety of materials.

continental glaciers Massive sheets of ice overlying whole or parts of continents and flowing out in all directions from their centers, or region of origin.

continental polar, cP (air mass) Cold and dry reflecting its source region, the interior of high-latitude continents.

continental shelf Relatively shallow, submerged margins of the continent extending into the ocean, usually for several miles before reaching the continental slope.

continental slope Steep slope that extends from the continental shelf into the deeper ocean.

continental tropical, cT (air mass) Hot and dry, reflective of its source region, the tropical deserts.

controls of weather and climate Those factors that work together to influence weather and climate patterns such as latitude, nearness to large water bodies, ocean currents, and storm tracts.

convection A rising, warm central column of matter and sinking, colder side columns triggered by heat at the base.

convection in the mantle Motion of mantle rock; heating expands and causes flow toward the lithosphere. Thought to be the force behind the movement of lithospheric plates.

convectional precipitation Violently rising moist air, triggered by heat, expands and cools to a sub-dew point temperature. *See also* convection.

converging boundaries Lines along which lithospheric plates moving toward one another meet.

core (earth's) The extremely dense, superheated center of the earth; thought to have a mostly solid inner core and a liquid outer core.

Coriolis force The earth's rotation causes an apparent lateral drift of horizontally moving objects. The force increases with latitude from nil at the equator to maximum at the poles. *See also* Ferrel's Law.

crevasses Deep stress fractures in flowing valley glaciers. Movement of glacier over resistant rock and more rapid flow at the top and center cause ice to open in crescent-shaped crevasses with their concavity facing upslope.

crop line Line related to latitude and elevation beyond which a particular crop plant cannot be grown economically. In North America, the crop line for corn divides the region of Humid Continental climate into Long-Summer and Short-Summer types.

crustal warping Tectonic deformation by compressional forces that results in broad areas of upwarped and downwarped crust.

cumuliform General category for relatively deep and fleecy, fluffy, or cottony clouds. Cumuliform clouds indicate turbulent air movement but never cover the entire sky.

Cumulonimbus Tall, billowing cloud of vertical development that brings thundershowers and may extend to the top of the troposphere.

Cumulus Fleecy, fluffy, cottony clouds that often indicate fair weather but may develop into towering thunderclouds. *See also* Cumulonimbus.

current meter Instrument with rotating cups for measuring stream water velocity.

cutbank Steep erosional streambank on the outside of a river bend or meander curve.

cycle of erosion (geomorphic cycle) Progressive erosion of a landscape through youthful, mature, and old-age stages. A conceptual model for landscape evolution.

cyclone Generally, any low-pressure center. Specifically, in India, Australia, Pakistan, a violent hurricane/typhoon type storm. *See also* middle-latitude cyclone.

cyclonic (frontal) precipitation Light moist air ascending over a denser air mass, expands and cools to a sub-dew point temperature.

deflation The ability of strong prevailing winds to remove loose material, such as sand, and form a surface depression.

degradation General lowering of land surfaces by weathering and erosion processes.

delta Depositional feature built at the mouth of rivers emptying into oceans or lakes. Sediment deposits are sorted according to size as the stream loses its velocity.

demersal Refers to deep sea bottoms.

deposition (geomorphology) The accumulation of rock debris, sediment, organic material. Usually the laying down of eroded rock material by agents such as wind, streams, or glacial ice.

deposition (plant succession) The first phase in plant succession in which newly deposited surface becomes available for plant growth. *See also* nudation.

depression Low atmospheric pressure usually related to a storm system.

desert Vegetation class characterized by thinly dispersed, drought-resistant shrubs, grasses, and herbaceous plants, but very few trees. *See also* true desert.

desert pavement Rocky surface from which finer particles have been removed by wind.

desertification Expansion of desertlike conditions onto grasslands and other regions by overgrazing and land clearing especially during periods of drought.

dew Condensation on a cold surface above 32° F (0° C), a result of air cooling through contact.

dew point temperature (dew point) Temperature at which cooled air becomes saturated with water vapor. Air cooled below this temperature will produce condensation.

diagnostic horizons Certain surface and subsurface horizons that are the basis for categorization of soils in the Comprehensive Soil Classification System.

diastrophism Buckling, warping, folding, faulting of the earth's crust by tectonic forces.

diffuse daylight Radiated short-wave solar energy arrives at the earth after scattering by atmospheric dust or clouds.

dike A relatively small intrusive volcanic mass that forms a thin vertical sheet cutting across horizontal strata.

dilatancy The opening of small cracks along a fault plane under extreme pressure. A possible indicator of an impending earthquake.

discharge Volume of water moving through a cross-section of stream channel in a given time period. Usually expressed as cubic feet per second.

distributaries Branching channels flowing through a delta.

diurnal Daily occurrence. The earth's rotation causes a daily progression from night to day and day to night.

diverging boundaries Line along which two lithospheric plates are moving apart.

Doldrums An equatorial belt of high temperature, low pressure, heavy rainfall, and generally fickle breezes or calm.

drainage basin Total land area drained by a major river and its tributaries.

drought deciduous perennials Certain plants, mostly shrubs, growing in deserts and other recurrently dry habitats that grow rapidly when shallow soil water is plentiful, then drop all leaves and become dormant during periods of drought.

drought evaders Certain desert plants that avoid drought by growing only when shallow soil water is plentiful. *See also* drought deciduous perennials; ephemeral annuals.

dry adiabatic lapse rate *See* lapse rate (adiabatic-dry).

duff Partially decayed organic matter on the soil surface from previous growing seasons. Grades from distinguishable plant and animal parts to shapeless particles.

dunes, barchan Crescentic sand dunes with gentle slope upwind on the convex side and steep slip face on the leeward, concave side. Horns of crescent point downwind.

dunes, longitudinal Ridgelike elongated sand dunes formed parallel to the direction of the prevailing wind.

dunes, transverse Ridgelike dunes formed at right angles to prevailing winds.

dust (atmospheric) Solid particles in atmospheric suspension (excluding ice). *See also* condensation nuclei.

dynamic equilibrium The concept in geomorphology that tectonic and gradational processes approach a balance in shaping landforms so that a change in the intensity of a process (e.g., erosion, tectonic uplift) will result in an alteration in the characteristics of the landform (e.g., slope). In time, the landform will adjust to the changed geomorphological conditions, and a balance will be achieved. Since changes in process intensity continue to occur, a balance, or true equilibrium, is rarely achieved.

Easterlies *See* Polar Easterlies.

easterly wave Intertropical storm that occasionally occurs in the streamlines of the Trades. Some become hurricanes.

ecesis Phase in plant succession marked by the successful germination and growth of the first or pioneer plants.

ejecta Rock fragments and other material thrown out of an erupting volcano.

elastic rebound Along an active fault rocks deform as they sustain increasing pressure and then snap back as sudden movement occurs (earthquake).

El Niño Episodic warming of the usually cold water off the coast of Peru. Related to changes in atmospheric pressure and wind patterns, and associated with worldwide weather anomalies.

elements of weather and climate Conditions that describe the instantaneous or long term state of the atmosphere including temperature, pressure, wind, and moisture.

emerged shorelines Occur along coasts where recent relative vertical upward movement of land has exposed the sea floor. Characterized by a simple, smooth shoreline with few embayments or headlands.

emergent layer Uppermost stratum of trees in the tropical rain forest.

ENSO Acronym for El Niño and the Southern Oscillation. The usually cold waters off the coast of Peru are episodically warmed by a change or oscillation in which the normal low-and high-pressure patterns switch places across the western and southeastern Pacific. *See also* El Niño.

entisols Order in the Comprehensive Soil Classification System with youthful soils lacking horizons. No specific climate.

ephemeral annuals Short-lived plants (wildflowers) of deserts and other dry habitats. They grow rapidly, flower, and produce seed with available soil moisture, and then persist as seed during prolonged drought. *See also* drought evaders.

epicenter The point on the earth's surface directly above the focus of an earthquake.

epiphytes Nonparasitic plants that grow on other plants, usually trees. Especially common in tropical rain forest.

equator A great circle whose plane is at a right angle to the earth's axis of rotation, therefore bisecting the earth into a Northern Hemisphere and a Southern Hemisphere.

equatorial, E (air mass) Located in equatorial regions, it is a hot and very moist air mass.

Equatorial Low A belt of low atmospheric pressure astride the equator and extending roughly 10 degrees of latitude on either side. Caused by surface heating from high-angle solar rays.

equinox When the noon sun is directly overhead at the equator. Occurs twice a year on or about March 21 and September 21. Day and night are of equal length at all latitudes.

erg "Sand sea." Large desert region of deep, shifting sand with many dune forms.

erosion A general term for the wearing away and transport of soil and surface rock.

erratics Large rocks transported by glacial ice into areas where the rock does not match the local geology.

evaporation The change in form of liquid water to vapor, usually as a result of heat.

evergreen shrubs In general, shrubs that retain leaves year round. In deserts, a type of xerophyte with a very extensive, deep root system and thick cellular structure that allow it to obtain water from intermediate depths below the normal root zone for most plants.

exfoliation Peeling away of surface rock layers as the result of unloading and pressure release.

exotic streams Streams that flow in, and often through, an arid or semiarid region that have their main source of water in a humid region. Example: the Colorado River in the Sonora Desert of Arizona and California, which brings water from rain and snowmelt in the humid Rocky Mountains through the desert to the Gulf of Mexico.

expansion (rising air) As a parcel of air rises over a topographic barrier, front, or from surface heating, its barometric pressure decreases, its volume increases (the parcel expands), and its temperature decreases. *See also* adiabatic (process).

extratropical storms Storms that occur primarily in the middle latitudes (middle-latitude cyclones and tornadoes).

extrusive vulcanism *See* vulcanism, extrusive.

eye (hurricane) The warm, windless, cloudless center of a hurricane.

families (soil) Fifth level in the Comprehensive Soil Classification System.

fault A deep-seated crustal fracture along which there is movement of blocks of crust.

fault, normal An inclined crustal fracture with the upper block moving down and the lower block moving up.

fault, overthrust A very low-angle crustal fracture along which the upper block is thrust over the lower.

fault plane Surface along which blocks of crust slip during fault movement.

fault, reverse An inclined crustal fracture with the upper block thrust up and the lower block down.

fault, strike-slip A crustal fracture along which there has been horizontal movement parallel to the direction of the fault.

fault, transform Where two crustal blocks move horizontally past one another between two offset segments of an ocean ridge (rise).

felsenmeer Extensive ground cover of sharp-edged, angular rock fragments caused by cold temperatures and ice action at high altitudes and high latitudes.

felsic General term for rock of the earth's continental crust. Felsic-type rocks have high silica content (silicates), and relatively low density, and are light in color.

Ferrel's Law Any horizontally moving object in the Northern Hemisphere will exhibit an apparent right-handed deflection. Left-handed in the Southern Hemisphere. *See also* Coriolis force.

fissure flows Highly liquid flows of lava emanating from cracks or fissures in the earth's crust. The igneous rock formed from these flows arranges itself in extensive horizontal layers.

fjords "U"-shaped, ice-deepened valleys intruded by the sea as glacial ice retreated at the end of the Pleistocene era.

flood plain The flat, frequently flooded area adjacent to a stream channel.

focus (earthquake) Point of maximum energy release in an earthquake.

foehn *See* chinook.

fog A cloud at ground level.

fog (advection) A transported radiation fog, usually the result of a gentle breeze.

fog (radiation) Tiny water particles in suspension, usually near the ground, a result of air cooling below the dew point via a temperature inversion.

folding Tectonic deformation of the crust by compressional forces that produces folded structures in a formerly level surface.

forest Vegetation class marked by the dominance of trees that grow close together with branches and leaves forming a more or less continuous canopy.

formation Fourth level in the classification of vegetation, a subdivision of formation class. An actual vegetation unit based on dominant plants occurring in a specific, though often extensive, area of the globe.

formation class Third level in the classification of vegetation, a subdivision of vegetation class giving more attention to basic forms such as size, shape, structure, and seasonality of the plants that dominate the vegetation.

fossil water Groundwater that has accumulated in an aquifer over a long period of time—thousands, perhaps tens of thousands of years; may take just as long to be naturally replenished after withdrawn for use.

friable A term pertaining to the ease of crumbling of soils.

Frigid Zone In the ancient Greek classification of climate, the zone of cold winds and snowy forests north of the Mediterranean basin.

front The line of contact between two differing air masses.

frost Ice crystals forming on cold surfaces as moist air is cooled below the dew point by contact.

frost heave The lifting of soil and rock particles by water freezing below the surface.

frost pocket A deviation in the normal decrease of temperature with altitude. Interior valleys within mountain ranges that receive cold air drainage from slopes dropping temperatures below surrounding higher sites.

fumarole A natural steam jet contaminated with magmatic gases and emitting an odor.

galeria forest A narrow riverine forest almost completely dependent for its moisture on the watercourse.

geoid Earthlike. Refers to the earth's theoretical shape based on gravitational measurements.

geomorphology The study of landforms and the processes that shape them.

geyser A spasmodic erupter of hot water and steam.

global energy budget The balance of heating and cooling of the earth system.

graben Crustal block between two high-angle faults that has slipped downward forming a steepsided structural valley.

gradational processes Leveling of land surfaces by weathering, mass wasting, and erosion.

graded stream A theoretical condition in a stream where the load supplied from the drainage area is equal to the capacity of the stream to transport it.

Grand Stairway A series of south-facing cliffs that run north from the Grand Canyon exposing the rock strata of the Colorado Plateau.

granite Coarse-grained, igneous rock formed through slow cooling of magma within the crust.

grassland A vegetation class of mainly nonwoody plants with a grasslike appearance. *See also* prairie; savanna; and steppe.

great circle The largest circle that can be drawn on a globe; a circle whose plane bisects a globe. The line described on the surface of a globe by a plane passed through its center.

great groups Third level in the Comprehensive Soil Classification System.

greenhouse effect Short-wave solar energy, allowed easy ingress through the glass walls, is transformed into long heat waves by absorption. The glass is not transparent to the heat waves, which are trapped within. Applied to the global atmosphere, some gases like water vapor, carbon dioxide, and methane are transparent to short-wave but absorb long-wave radiation. Increases in these gases may contribute to climate warming.

Greenwich (England) The prime meridian (0° longitude) passes through Greenwich. Also known as the Greenwich meridian, this meridian is one-half of a great circle running from the North Pole to the South Pole and passing through Greenwich.

Greenwich mean time (GMT) Local time at Greenwich, England. Time used to determine longitudal position, also known as universal time (UT).

groins A short, low jetty constructed perpendicular to the shore, across the beach into the surf zone. Intended to check longshore drift and to build up beach sand deposits behind groin. *See also* jetty.

groundwater Subterranean water in saturated soil and rock, subject to movement by gravity.

groundwater table (water table) The upper limit of groundwater. *See also* groundwater.

groundwater zone Subterranean zone saturated with water.

Gulf air, gT (air mass) Warm and moist air mass over the Gulf of Mexico. Frequently invades the United States east of the Rocky Mountains and brings moisture for heavy rainfall and high humidity.

habitat Home of a plant species, includes a particular combination of environmental factors such as light, temperature, slope orientation, and soil type.

halophytes Plants with the capacity to utilize soil water with a high content of salt.

hamada wind-swept upland desert of exposed and scoured bedrock, sometimes thinly covered with coarse material.

hanging valleys Ice-truncated tributary valleys that once intersected a master stream that was subsequently widened and deepened by a valley glacier. Since glaciation, some tributary valleys now display waterfalls dropping into the glacial valley.

Hawaiian High The eastern end of the North Pacific Horse Latitude belt. Typically, dry sunny weather, sinking air, and high pressure.

heat gradient The increase in heat with increase in depth below the earth's surface.

heliophytes Plants that grow best in full sun.

high clouds Clouds that occur in a layer

above 20,000 feet (6000 m) altitude. Include Cirrus, Cirrostratus, and Cirrocumulus.

high-latitude circulation cell Theoretical atmospheric circulation poleward of the Polar Front in both hemispheres with easterlies toward the front at ground level and upper tropospheric flow from the front to the pole.

histosols Highly organic soils of bogs and swamps.

holism Idea that nature is a whole, that everything in nature is connected to everything else.

horizons (soil) Succeeding parallel layers in soil with increasing depth, each distinctive from the other by color, texture, structure, or chemical characteristics.

Horse Latitudes A subtropical region of high temperatures, high pressure, minimal rainfall, and generally calm conditions. Encountered especially at sea in both hemispheres at approximately 30° latitude.

horst Crustal block that has been upthrust between two high-angle faults.

hot, dry rock system (geothermal energy) A method to produce steam for power generation by injecting water into geothermally hot rocks and recovering the steam to drive generators.

Humid Continental–Long Summer Climate of continental interior and east coasts of Northern Hemisphere middle latitudes with long, hot, and moist summers and cold winters with snow. Rain occurs all year with 20 to 50 inch annual precipitation. Traditionally, that part of the Humid Continental climate region south of the crop line that will support the commercial growth of corn and winter wheat.

Humid Continental–Short Summer Similar to Humid Continental–Long Summer but with longer and colder winters, more northern location (north of the crop line), and spring wheat rather than winter wheat. Corn does not do well.

Humid Subtropic Climate of subtropical east coasts with warm, humid summers, mild winters with cold spells, and moist all year with rainfall annual average 40 inches.

humidity A measure of water vapor in the atmosphere.

humus Partially decomposed organic matter in soil, imparting a dark color as carbon is released. Significant to soil friability and retention of moisture and nutrients.

hurricane An intense summer/fall storm of the tropical seas—originating as an easterly wave in the Trades. Normally encountered in the southwest corners (Northern Hemisphere) and northeast corners (Southern Hemisphere) of major ocean basins. The South Atlantic is the exception.

hydrograph A graph showing stream discharge over time at a particular stream location, usually where water velocity data are recorded.

hydrologic cycle The loss of oceanic water via evaporation; the return of that moisture to the earth's surface via precipitation; and finally, the restoration of surface water to the sea via rivers, springs, and icebergs.

hydrophytes Plants adapted to growth in water or saturated soil. Includes aquatic and marsh/swamp plants.

hygrograph (hygrothermograph) A graph of relative humidity over time. A hygrothermograph shows relative humidity and temperature over time. Also instrument(s) that measure and record such data.

Icelandic Low The small but intense North Atlantic Subpolar Low cell. A major generator of middle-latitude storms.

igneous rock Rock that has cooled and solidified directly from the molten state.

inceptisols Soils just beginning to show horizons.

individual plants Lowest category in the classification of vegetation.

infiltration The absorption and movement of water from rainfall and snowmelt into the soil.

insolation The total solar energy received at any given point on the earth's surface.

integration The idea that all parts of physical geography—weather and climate, landforms, water, soils, and vegetation—are inseparable and that patterns in each are influenced by patterns in the others.

interfluves The dividing high ground or ridges between streams.

International Date Line (IDL) Roughly the 180° meridian (the line accommodates political boundaries). A day is lost crossing from east to west, and gained crossing from west to east.

Intertropical Convergence Zone (ITC) Zone of convergence of air masses carried by the trade winds, coincident with the Doldrums.

intertropical storms Storms of the low latitudes, including weak equatorial lows, easterly waves, and hurricanes/typhoons.

intrusive vulcanism *See* vulcanism, intrusive.

isobar A line on a map connecting all points of equal barometric pressure.

isostasy The theory of buoyant equilibrium that suggests that an uplift of crust in one region must be compensated by a depression in another.

isotherm A line on a map connecting all points of equal temperature.

jet stream High-speed air flow in the upper troposphere, usually westerly but in some cases easterly.

jetty Long, narrow wall constructed at entrances to bays to protect them from excessive sand deposition. *See also* groins.

jungle Dense tropical forest of herbs, shrubs, and trees. Common along river courses where light can penetrate below rain forest canopy and where rain forest has been disturbed or removed.

Kalahari Tropical Dry climate region of southern Africa (mostly western Botswana).

karst Landscape that results from the solution of limestone. Characterized by sinkholes, solution valleys, and karst towers.

katabatic (wind) Cold air flowing downhill at high velocity responding to gravity. An extreme form of simple air drainage.

Köppen system (climate) A quantitative system of climate classification. First introduced in 1918, it has been modified many times.

krummholz The stunted, gnarled, misshapen tree forms of high mountains and high latitudes.

laccolith A relatively small intrusive igneous mass fed from below by a magma conduit cutting across strata to form lense-shaped masses. Surface is uplifted in a small dome.

lagoons Brackish water bodies between mainland and barrier islands on wide coastal plains.

lake Large water body surrounded by land, may have fresh or salt water.

Lake Agassiz Ancient glacial meltwater lake covering much of what is now Manitoba, Minnesota, and North Dakota.

Lake Bonneville Large ancient lake that formed in the basins of northwestern Utah as glacial ice melted at the end of the Pleistocene era.

land breeze Light wind blowing from land to sea, usually at night and early morning when the sea is warmer than the land.

land subsidence Sinking ground surface resulting from removal of groundwater and subsequent compaction of the aquifer. Any sinking or lowering of a level surface resulting from geomorphological processes.

La Niña Abnormally cold water and intense upwelling off the coast of Peru. Alternates with El Niño.

Lapps People of northern Norway, Sweden, Finland, and northwestern Soviet Union, who keep large reindeer herds. Formerly migratory with herds; now mostly settled in towns.

lapse rate The change in air temperature upward through the troposphere, usually a loss of about 3.6° F (2° C) per 1000 feet (305 m). Also called normal lapse rate.

lapse rate (adiabatic-dry) A mass of air impelled aloft will experience a loss of 5.5° F per 1000 feet (3° C per 305 m) as a result of expansion.

lapse rate (adiabatic-wet) If condensation is occurring within it, a mass of air impelled aloft will experience a loss of less than 5.5° F (3° C) per 1000 feet (305 m). 5.5° F (3° C) will be lost due to expansion, but a modicum of heat will be returned through condensation.

latent heat of condensation Evaporation requires heat, which means that each molecule of water vapor in the atmosphere has an increment of latent heat. Condensation, the reverse of evaporation, releases this heat.

lateral moraines *See* moraines, lateral.

laterization The soil-forming process of warm, humid regions. Mildly acidic rainwater percolates through soils, removing silicon from the A layer and leaving it in the B layer where clays may form. Nutrients are rapidly removed by leaching.

latitude Distance north and south of the equator measured in degrees of surface arc from 0° (equator) to 90° north and south (the geographic poles).

latitudinal zonation The arrangement of formations (major vegetation units) in a sequence from the equator to the poles in relation to the decreasing temperature.

leaching The process by which nutrient ions and other soluble minerals are washed downward in the soil by percolating water.

liana General term for vines that grow from the forest floor into the tree canopy. Common in tropical rain forests.

limestone A sedimentary rock typically formed by the long-term compression of lime-rich sediments accumulated on the sea floor. Includes coral and sea shells cemented into coarse rock.

lithosphere The rigid outer shell of the earth. Includes continental and oceanic crust as well as a portion of the upper mantle above the asthenosphere. Also a term referring generally to the solid earth as one of the "spheres"—atmosphere, hydrosphere, biosphere, and lithosphere.

litter The uppermost organic layer above the soil, composed of recently deposited leaves, branches, and dead animals. Some decomposition, but parts are still recognizable.

Llanos Region of Tropical Wet and Dry climate in South America (Colombia)

loess Wind-borne and deposited silt. The foundation of deep, fertile soils in many parts of the world.

longitude Distance east and west of the prime meridian (Greenwich) measured in degrees of surface arc from 0° (prime meridian) to 180° (its opposing limb).

longitudinal dunes *See* dunes, longitudinal.

longshore drift Transport of sand and other sediments parallel to shore.

low clouds Cloud family with clouds that usually occur below 6500 feet (2000 m) altitude. Included are Stratus, Nimbostratus, and Stratocumulus.

low (pressure) An area of relatively low atmospheric pressure, characterized by converging winds, ascending air, and precipitation.

mafic Rock that has high proportion of iron and magnesium and is relatively low in silica. Rock of the oceanic crust and below the continents; darker and denser than most continental rock.

magnetosphere The influence of the earth's magnetic field far beyond the outer margin of the atmosphere. *See* Van Allen radiation belts.

mallee Scrub eucalyptus thickets in Australia. Coincident with Mediterranean climate.

mangrove A low tropical maritime tree whose habitat is shallow seawater necessitating "knees" on its roots for aeration. It grows aerial roots for aeration and develops extensive thickets.

mantle The bulk of the earth's mass; located between the core and the crust.

maquis *See* chaparral.

marble The metamorphic rock formed from limestone.

Marine West Coast Climate of middle-latitude west coasts with mild winters, cool summers, cloudiness, frequent light rain and drizzle, and precipitation of 20 to 60 inches (50 to 150 cm) per year.

maritime polar, mP (air mass) Cold and moist reflecting its source region, the high-latitude oceans.

maritime tropical, mT (air mass) Hot

and moist, reflective of its source region, the tropical seas.

marshes Wetlands with nonwoody plants (grasses, sedges, and rushes). Soils are saturated, and water may or may not be present on the soil surface.

mass wasting Downslope movement of rock debris on slopes, banks, and cliffs as a result of gravity; involves no flowing agent of transport such as water, wind, or ice.

mechanical weathering Physical break-up of rock into smaller particles by such processes as ice action, root growth, exfoliation, or rapid temperature change.

mediterranean forest *See* mediterranean scrub forest.

mediterranean scrub forest Vegetation formation class, a seasonal forest with trees and tall shrubs, both of which are broadleaf evergreen with leathery, drought-resistant leaves. Coincident with Mediterranean climate. Also called sclerophyll forest after the leathery leaves.

mercury barometer An instrument measuring air pressure. A column of mercury in a glass tube responds to differing pressures.

meridian One half, or a limb, of a great circle passing through the poles. A line of longitude. Also defined simply as a great circle passing through the poles.

mesic Moist. Plants classed as mesophytes that grow in moist but not saturated habitats.

mesopause The narrow zone in the atmosphere separating the mesosphere from the thermosphere (ionosphere)—60 miles (97 km) altitude.

mesophytes Plants with little or no adaptation to recurrent drought or to waterlogged conditions. Most plants fit this category. *See also* mesic; xeric.

mesosphere That atmospheric layer extending outward from the stratosphere to about 60 miles (97 km).

metamorphic rocks Rocks transformed to a much different type of rock by the long-term application of pressure and heat—normally denser, harder, and displaying different molecular structure.

middle clouds Family of clouds found between 6500 and 20,000 feet (2000 and 6000 m) altitude. Includes Altostratus and Altocumulus clouds.

middle-latitude broadleaf forest Forest formation class of the middle latitudes dominated by broadleaf deciduous trees.

middle-latitude circulation cell Atmospheric circulation cell with part of the sinking air in the subtropical highs flowing poleward to the Polar Front, rising air at the front, and flow aloft back to the subtropical highs.

middle-latitude coniferous forest Forest formation class of the middle latitudes dominated by needleleaf evergreen trees.

middle-latitude cyclone A large, slow-moving, frequent and generally mild storm along the Polar Front, usually with well-defined warm front, cold front, and low-pressure center.

Middle-Latitude Dry Climate of leeward (eastern) side of mountain chains in the midlatitudes. Similar to Tropical Dry with hot summers but with cooler winters and snow. Precipitation 10 inches (25 cm) or less (true desert), or 10 to 20 inches (25 to 50 cm) (semiarid).

middle-latitude mixed forest Forest formation class of the middle latitudes with broadleaf deciduous and needleleaf evergreen trees.

middle-latitude prairie Grassland formation class found in the middle latitudes in areas of 20 to 40 inches (50 to 100 cm) precipitation. Grasses are tall, averaging 24 inches (60 cm) or more.

middle-latitude steppe Grassland formation class found in the middle latitudes in areas with 10 to 20 inches (25 to 50 cm) precipitation. Grasses are short, usually less than 12 inches (30 cm).

migration (plant succession) Second stage in plant succession marked by movement of seeds and other plant reproductive parts to a new open site.

millibar A force of 1000 dynes per square centimeter. Unit for measuring atmospheric pressure.

mineral soils Soils with little or no organic matter.

mixed forest Seasonal forest of the middle latitudes with a mixture of broadleaf deciduous and needleleaf evergreen trees.

moder soils Soils with moderate amounts of litter, duff, and humus.

Mobo discontinuity The boundary, based on composition and density, between the rock of earth's crust and the rock of the mantle.

mollic epipedon Soil surface horizon, dark brown to nearly black, very high in humus, rich in plant nutrients. Characteristic of mollisols.

mollisols Grassland soils of the middle latitudes with a mollic epipedon.

monadnock A resistant, erosional remnant hill or mountain remaining in an old age landscape.

monsoon The alternating summer onshore winds (rainy season) and winter offshore winds (dry season), basically a result of differential seasonal heating of land versus water.

moraine lake Small lake dammed by a terminal moraine.

moraine, recessional Ridges of till left behind as a glacial ice margin pauses during its retreat.

moraine, terminal A ridge of till deposited at the terminus of advancing glacial ice. Marks the farthest advance of the glacier.

moraines, lateral Ridges of glacial till deposited on either side of a retreating valley glacier. Accumulation of rock material from erosion and mass wasting of valley sides.

mor soils Soils with large amounts of litter and duff on the surface, but little humus in the A horizon.

mound springs Artesian springs of mineralized water in arid regions that build up mineral deposits in mounds.

muck Poorly drained, highly organic soil of marshes and swamps. with nonfibrous, shapeless organic matter.

mull soils Soils with small amounts of litter and duff on the surface, but abundant humus in the A horizon.

muskeg Thick water-saturated accumulations of sphagnum moss, climax of plant succession in some northern coniferous forests on level terrain.

natural levees Higher ground (stream banks) on either side of an old-age stream built up by deposition of sediments from frequent flooding.

neap tides Tides with low tidal range occurring at two-week intervals when the moon is in its first-or third-quarter phases.

needleleaf evergreen forest Forest formation class. Trees have long, slender, rounded leaves remaining on branches all year. Varieties of this formation class are subtropical coniferous forest, middle-latitude coniferous forest, and taiga.

niche The place or role taken by a species in the overall scheme of nature. Sometimes referred to as the species' "occupation" in an ecosystem.

Nimbostratus Type of low cloud, appears as a featureless layer from which gentle precipitation falls. Rain cloud of warm fronts.

normal fault *See* fault, normal.

normal lapse rate *See* lapse rate.

Northeast Trades Persistent winds of the Northern Hemisphere blowing from the subtropical high-pressure cells in a northeasterly direction to the Intertropical Convergence Zone.

nudation First phase in plant succession. Destruction of former vegetation by such agencies as fire, logging, or overgrazing provides a site for new plants to grow.

nuée ardente A massive superheated cloud of volcanic ash and gases that flows rapidly downslope out of an erupting volcano. Literally "glowing cloud."

nutrient ions Food substances in the soil water taken in by plants. These are positively charged molecules like potassium, phosphate, and nitrate that are attracted to minute clay and humus particles and removed from the soil by leaching.

occluded front The front formed when a rapidly advancing cold front in a middle-latitude cyclone overtakes the warm front, leaving only cold air at the ground.

occlusion When a rapidly advancing cold front in a middle-latitude cyclone overtakes the warm front.

ochric epipedon Soil surface layer, light colored, low in humus. Present in alfisols, aridisols, and ultisols.

one atmosphere The average pressure on the earth's surface exerted by a column of atmosphere, about 15 pounds per square inch.

orders The first level in the Comprehensive Soil Classification System, 10 in all.

orographic precipitation Wind forcing a moist air mass up a mountain slope, thereby causing sub-dew point cooling via expansion and condensation.

outwash plain An alluvial apron of stream-worked glacial debris laid out in front of a receding continental ice sheet.

overland flow A form of surface runoff over sloping ground, not confined to a channel. Occurs when precipitation exceeds the rate of infiltration.

overthrust fault *See* fault, overthrust.

oxbow lake Water body formed when a meander of an old-age stream is cut off; the abandoned channel becomes a lake.

oxic horizon Subsurface soil horizon, severely weathered with iron and aluminum oxides, some clay. Characteristic of oxisols.

oxisols Highly weathered, old soils of the low latitudes.

Pacific currents System of ocean currents in the North and South Pacific Ocean. Important to the movement of tropical heat poleward.

paleomagnetism Remanent magnetism in rock. When molten rock cools and solidifies, iron minerals are aligned with earth's magnetic field. This record in the rock is used to determine the polarity and intensity of the earth's magnetic field in the geologic past and to help determine the relative age of rock.

parallelism The inclination of the earth's axis always in the same direction. The axis of rotation is always parallel to itself throughout the earth's orbit around the sun.

parallels Lines of latitude. Circles of latitude whose planes are at right angles to the axis of rotation, hence parallel to the equator.

parent material Mineral matter from the disintegration of rock through weathering that is basic material in which soil develops.

peat Partially decomposed plant material that accumulates in the soils of a saturated environment like a marsh or bog. A soil in the histosol order.

pedology The study of soils.

peds Natural aggregates of soil made up of many particles held together by mucus from soil organisms, iron oxides, carbonates, or clays. Basis for soil structure classification.

peneplain Erosional plain in the old-age stage of the cycle of erosion. Gently rolling surface near baselevel.

percolation The downward movement of water through soil and rock material in response to gravity.

percolation pond A shallow reservoir that confines water so that it may percolate into the surface to replenish groundwater.

perihelion The nearest approach of the earth to the sun each year—on or about January 3.

permafrost Permanently frozen ground below the surface.

permeability The free movement of a liquid through porous rocks and soil.

pH scale A range of values from 1 to 14 that expresses the acidity or alkalinity of a solution based on the weight of hydrogen ions per liter.

phreatophytes Plants with deep roots that obtain moisture directly from groundwater. Type of xerophyte.

phytoplankton Microscopic plant life in the sea.

pioneer species In plant succession, the first plants to occupy a new site.

plane of the ecliptic The plane whose edge is described by the earth's orbit around the sun.

plant species The sixth level in the vegetation classification, type of plant made up of many individual plants.

plant succession The orderly sequence of plant communities that evolves over time beginning with the growth of pioneer species on a bare or disturbed surface and proceeding toward stabilization (climax).

plate tectonics The dynamic model of

the lithosphere. Lithospheric plates are set in motion by forces within the mantle, and they interact along their boundaries to produce major global landforms.

playa Dry lake bed in the interior-drained basins of arid regions. Occasional stream runoff forms shallow lakes in basin bottom, evaporates, and leaves behind light-colored alkaline lake flat, or playa.

Pleistocene The epoch in geologic time from about 2 million years ago to about 10,000 years ago. A time of global climate cooling and the advance and recession of continental scale ice sheets in the Northern Hemisphere. The Ice Age.

plinthite An oxic horizon in some oxisol soils that is very high in iron and aluminum oxides and that, with repeated wetting and drying, may become nearly irreversibly hardened.

plucking The process by which moving glacial ice in contact with rock surfaces pulls away and transports rock fragments by freezing them into the ice.

plug volcano An active, often craterless, volcanic peak that features thick pasty lava pushing up a dome or spire. Trapped gases make this type of volcano subject to violent explosions.

pluton An intruded volcanic rock mass in the crust such as a batholith or laccolith.

plutonic Refers to coarse-grained igneous rock mass that has cooled slowly below the surface.

podzol Russian word for ash-colored soil. *See also* spodosols.

podzolization A soil-forming process resulting from strongly acid percolation. Found mainly in regions of cool, humid climate and coniferous forest vegetation.

Polar Easterlies Surface winds of the high latitude circulation cells bringing cold polar air to the Subpolar Low.

Polar Front The line of contact, in the middle latitudes, between polar and tropical air masses.

Polar Ice Cap High-latitude climate with severely cold winters, cold summers, and precipitation of 10 inches (25 cm) or less. Found in Antarctica and Greenland.

Polaris The North Star. The star visible almost directly overhead at the North Pole. From the earth it appears virtually stationary as the earth rotates, unlike any other heavenly body.

polar front jet stream High-velocity, narrow, westerly currents near the top of the troposphere in both hemispheres. They are roughly above the Polar Front and appear to act as middle-latitude storm tracts.

polar night jet stream High-latitude, high velocity air currents active only during the six months of polar darkness.

pond Body of water, smaller than a lake.

pores Spaces between soil peds; many contain soil air or water.

porosity The degree of empty space in soils, rocks, or sediment. May be related to water-holding if open spaces are interconnected.

post meridiem (P.M.) After the sun has transited a meridian, hence afternoon.

prairie *See* middle-latitude prairie.

precipitation Liquid water droplets or ice falling out of the atmosphere.

pressure (atmospheric) The weight of the total atmosphere at a given location on the earth's surface.

pressure gradient The pressure differential expressed by the spacing of isobars.

prime meridian *See* Greenwich (England).

pyroclastic Meaning "broken by fire." A violent volcanic eruption of broken solid particles, from dust to jagged boulders.

quartzite The metamorphic rock formed from sandstone.

radiation The transfer of energy through space.

radiation fog Tiny water particles in suspension, usually near the ground, a result of air cooling below the dew point via a temperature inversion.

radiogenic heat Heat energy from decay of radioactive elements.

radiometric dating Determination of the absolute age of matter by the rate of decay of certain natural radioactive elements.

rainshadow The dry lee of a mountain where air descends and warms adiabatically.

reaction Repetitive phase of plant succession in which plants, by growing on a site, change their environment and in the process make it more suitable to other species. Literally, the plant's environment "reacts" to its presence.

recessional moraine *See* moraine, recessional.

recycling The recovery of materials and resources for reuse by humans.

reflection (solar energy) A mirrorlike reversal of selected wave lengths by atmospheric dust, water, gas, or the earth's surface into outer space.

reg A stony desert plain swept clean of fine particles by wind erosion.

regolith Loose inorganic material above bedrock; parent material for soil development.

relative age The age of a geographic feature or event in relationship to other features or events rather than its age in units of time such as years.

relative humidity The amount of water vapor in the air relative to that which it is capable of holding at a given temperature.

residual (soil parent material) Soil parent material that comes from the disintegration of rock in place. It accumulates gradually and overlies decomposing bedrock.

reverse fault *See* fault, reverse.

reverse osmosis The method by which fresh water flows from a plastic, semipermeable membrane as pressure is applied to seawater on the other side of the membrane.

revolution (earth's) The movement of the earth around the sun once each year.

rhyolite A fine-grained igneous rock of the same chemical composition as granite but differing in the size of its crystals because of rapid cooling.

rift valley A down-faulted valley resulting from the tensional forces along diverging plate boundaries.

rip currents Strong, narrow currents that flow seaward from a beach after water from successive waves has built up on the shore.

rotation (earth's) The motion of the earth on its axis once each day.

runoff Water flow along the surface.

Sahara Region of Tropical Dry climate in northern Africa.

salinization Process of soil formation that occurs locally in arid and semiarid regions resulting in accumulation of salts in the upper soil and soil surface.

salted out Describes a fresh water well near a salt water body that has withdrawn so much groundwater that subsurface salt water intrudes into the aquifer and the well.

sand spit A low sandy bar that extends from a headland across an embayment. Spits are the result of longshore drift.

sandstone A sedimentary rock consisting of loose sand particles cemented together by any variety of materials. Normally high in silica and permeable.

Santa Ana (wind) A downslope, hot, dry wind blowing occasionally from the Great Basin to southern California.

saturation (humidity) The condition that exists when the air is holding all the water vapor it is capable of holding at a given temperature. Also called dew point.

savanna Grassland formation class of the low latitudes in areas of seasonal drought. Grasses are the predominant plant form but scattered trees may also be present.

sciophytes Plants that grow best in shaded locations.

sclerophyll forest *See* mediterranean scrub forest.

sea arch An arch formed by wave action eroding through either side of a headland.

sea breeze Light wind blowing from the sea to the land, usually during the day when the sea surface is cooler than the land.

sea cliff Vertical slope facing the sea undercut and eroded by wave action.

sea floor spreading The movement of ocean crust away from a spreading axis usually found at midocean. Thought to be caused by convection in the mantle that produces a midocean ridge where new crust is being formed.

sedimentary rock Rock formed from the recombination of the accumulated unconsolidated deposits through pressure and cementation. Includes rocks of organic origin such as coral or coal as well as evaporative salt deposits.

semiarid Refers to a region with 10 to 20 inches (25 to 50 cm) annual precipitation, a subdivision of Middle-Latitude Dry climate.

sensible temperature The temperature that the human body senses. A result of both actual temperature and humidity.

seres Stages in plant succession.

series The basic unit of soil classification consisting of soils that are essentially alike except for minor profile differences.

The 7th Approximation *See* Comprehensive Soil Classification System.

shale A sedimentary rock of accumulated fine muds compressed into a solid. Normally displays foliation at right angles to exerted pressure.

shield volcano A volcano with gentle slopes and shield-like profile constructed by relatively quiet flows of highly liquid lava.

shifting cultivation Traditional agricultural practice in tropical rain forests. A small plot is cleared leaving tall trees, debris is burned to unlock nutrients, and crops are planted for one to three years until nutrient supply is exhausted and weeds take over. A new site is chosen and the process repeated. Old plots are not used again for 20 or more years.

shrub desert Desert formation class found in true desert locations; has predominantly shrub forms and few grasses.

shrub steppe Desert formation class; occupies semiarid lands on the transition between steppe and true desert. It is a combination of shrubs and grasses.

sill A relatively small intrusive volcanic mass. It is fed from below by a vertical conduit cutting across strata. Magma then intrudes between horizontal strata in a thin sheet to form a sill.

sinks Deep surface depressions caused by solution and collapse of rock.

Sirocco A strong southerly wind bringing hot, dry continental tropical air from the Sahara Desert to Mediterranean Europe.

slate The metamorphic rock formed from shale. Typically displays more distinctive foliation than the parent shale.

sling psychrometer An instrument for measuring relative humidity utilizing two thermometers, one with a piece of wet cloth wrapped about the bulb. The different temperature readings reflect the evaporative capacity of the air.

slip face Steep leeward side of a sand dune. Sand blowing over the crest falls into the lee of the dune and slips down a steeply formed slope.

slip-off slope Gentle depositional slope on the inside of a river bend or meander curve.

sod Intertwined roots and lateral stems of closely spaced grasses in a grassland, also called turf.

soil erosion The physical removal of soil by wind and runoff, often initiated by human action.

soil profile A cross-section of soil to a depth of several feet, normally revealing distinctive horizons.

soil structure The arrangement and grouping of soil particles, described on the basis of shape, size, and distinctiveness of peds.

soil texture A soil characteristic based on the percentage composition of sand, silt, and clay.

soil-water zone Shallow zone of developed soil in which water is held as a film around particles (peds) and in small passageways between them (pores).

solar constant The energy received at a plane placed at right angles to the solar radiation stream at the outer edge of the atmosphere; equal to about 2 calories per square centimeter per minute.

solfatara *See* fumarole.

solution load Dissolved components in a stream derived from chemical solution of minerals in rock.

Sonoran (Desert) Region of Tropical Dry climate in northern Africa.

Southeast Trades Persistent winds of the Southern Hemisphere blowing from the subtropical high-pressure cells in a southeasterly direction to the Intertropical Convergence Zone.

Southern Oscillation *See* ENSO.

specific heat The number of calories needed to raise one cubic centimeter of a substance one degree Celsius. Almost five times as much energy (calories) is required to heat water as to heat an equal amount of dry land.

spectrum of electromagnetic radiation The range of wavelengths forms very short gamma rays to very long radio waves.

spodic horizon B horizon in spodosols, accumulation of aluminum oxide and humus, with or without iron oxide.

spodosols Soils of cool, humid forests with spodic and albic horizons.

spring tides Tides with high tidal range occurring at two-week intervals when the moon is in new or full phases.

stabilization (plant succession) *See* climax.

stack (sea) An erosional remnant of a headland that forms a small, high island offshore.

steppe *See* middle-latitude steppe.

storm Generally, any low pressure center. A cyclone.

storm surge Excessively high water along the coast resulting from the passage of a hurricane.

stratification (vegetation) The arrangement of plants in a plant community into a series of layers or strata based on light requirements.

stratification of the atmosphere The arrangement of various components of the atmosphere into a series of concentric layers about the earth. The layers have characteristic temperature lapse rates and they are, from the bottom up: troposphere, stratosphere, mesosphere, and thermosphere.

stratiform One of the three basic cloud forms (the others are cirriform and cumuliform). Clouds of this type cover the entire sky, are usually gray (because they cut out much of the light), and show little vertical development. They indicate minimal turbulence.

Stratocumulus Low, fluffy, grayish clouds in the low cloud family with spaces or "sun breaks" between them.

stratopause The narrow zone of atmosphere separating the stratosphere from the mesosphere—about 35 miles (56 km) above the earth.

stratosphere That atmospheric layer extending outward from the tropopause to about 35 miles (56 km).

stratum, strata (vegetation) A layer of plants of a certain average height within the vertical structure of a plant community. The tropical rain forest, for example, may have three tree strata, a shrub stratum, and an herbaceous stratum.

stratus A low cloud that covers the entire sky. One of the members of the low cloud family.

stream flow Surface water flow confined to a channel, as opposed to unconfined overland flow.

striations Parallel scratches and grooves on bedrock gouged by angular rocks frozen in a passing glacier.

strike slip fault *See* fault, strike-slip.

subduction Process in which one lithospheric plate angles below another, usually forming a deep ocean trench.

subgroups Fourth level in the Comprehensive Soil Classification System.

sublimation The direct change from water vapor (gas) to ice crystals (solid) and vice versa.

submarine canyons Deep narrow canyons cut into the continetnal shelf.

submerged shorelines Occur along coasts where recent relative vertical movement of land is down; results in an irregular and indented shoreline.

suborders Third level in the Comprehensive Soil Classification System.

Subpolar Low Low pressure zone in both hemispheres at 60 degrees latitude. Also called the Polar Front.

subsidence inversion A sinking air mass, heated via compression, results in a warm layer aloft. An increased temperature with altitude gain results. Also called an upper air inversion.

subsolar point (SSP) Point on the earth's surface where the sun's rays are directly overhead, or perpendicular to the surface.

subtropical climates Transitional climates located between the tropical and middle-latitude climates. They are Mediterranean and Humid Subtropic.

subtropical coniferous forest Variety of the needleleaf evergreen forest formation class found only in the southeastern United States on sandy soils. Trees are mostly pines with an understory of broadleaf shrubs and an herbaceous layer of grasses.

Subtropical Highs Centers of air subsidence, clear sky, and high pressure located at approximately 30° latitude in the Northern and Southern Hemispheres (Horse Latitudes). *See also* Azores High; Hawaiian High.

succulents Xerophytic plants like cacti that survive drought by using water stored in soft tissue and by limiting atmospheric gas exchange to cool nighttime hours.

Sudan-Sahel Region of Tropical Wet and Dry climate in north Africa, south of the Sahara. Much of its grass cover has been destroyed by overgrazing and drought.

summer solstice Northern Hemisphere—the day when the sun at noon is directly overhead at latitude 23 1/2° N (late June); for the midlatitudes, the longest day when the noon sun appears highest in the sky and it is the longest day of the year. Southern Hemisphere—the day when the sun at noon is directly overhead at 23 1/2° S (late December); for the midlatitudes, the day when the noon sun appears highest in the sky, and it is the longest day of the year.

supercooled The condition in which cloud droplets are cooled below the freezing point but remain unfrozen. Such droplets tend to freeze rapidly and precipitate if dry ice or silver iodide crystals are applied.

surface inversion A shallow increase in temperature with altitude from the ground up caused by radiational cooling of the earth's surface at night.

suspension load Fine material carried by a stream, small and light enough to remain suspended.

swamps Wetlands with woody plants (trees or shrubs) growing in water-saturated soils, with or without standing water on the surface.

swells Regular, long wavelength, ocean waves generated by storms at sea. They travel great distances over the sea.

Taiga (climate) Climate of high-latitude, continental interiors of the Northern Hemisphere. Winters are long and cold with snow; summers are short and cool. Precipitation averages 5 to 15 inches (13 to 38 cm).

taiga (vegetation) Variety of needleleaf evergreen forest formation class found in a wide band across the entire breadth of North America from Alaska to Labrador and across the much greater breadth of Eurasia from Scandinavia to Kamchatka (regions of Taiga climate). Trees are closely spaced with few understory plants and accumulations of organic matter (litter and duff) cover the forest floor.

talus The accumulation of rock debris at the base of a weathered cliff or slope. Delivered by gravity, talus is arranged in an apron according to size.

tank epiphyte Type of epiphyte found in the tropical rain forest. Its basin of leaves holds several litres of rainwater that provides the plant with water but serves as habitat for protozoans, insects, frogs, worms, and even small hydrophytes.

tarn Small lake in an ice-abandoned cirque at the head of a glaciated mountain valley.

tectonic processes Folding, faulting, crustal warping, and vulcanism—the processes that build major landforms.

Temperate Zone In the ancient Greek classification of climate, the zone of moderate temperatures centered on the Mediterranean basin.

temperature inversion An increase in air temperature with altitude. An anomaly. The opposite of normal lapse rate.

terminal moraine *See* moraine, terminal.

Thar Region of Tropical Dry climate in northwest India and Pakistan.

thermal belt A zone of mild temperatures located on valley sides above the normal winter surface inversion that results in an exception to the normal altitudinal zonation of plant communities. Favored for orchard crops.

thermosphere (ionosphere) The atmospheric layer that extends outward from the mesopause to the far periphery of the atmosphere—perhaps 6000 miles (9656 km).

tidal bores Channeled incoming tidal flow rushing up the lower course of a river as a wall of water a foot or more in height.

tidal inlet An opening through a barrier island maintained by tidal action.

till Unsorted rock debris deposited directly by moving ice.

tombolo A sand spit connecting a small offshore island to the mainland. The sand is deposited by converging waves between the island and the shore.

tomography The technology that uses computers to generate three-dimensional interior views from multiple images. By using seismic data, this technology can be applied to the earth's interior.

tornado The most violent storm in nature. It is small, swiftly moving, lethal, and almost exclusively American—normally a product of the Polar Front.

Torrid Zone In the ancient Greek classification of climate, the hot, dry zone including the Sahara Desert south of the Mediterranean basin.

Trades (winds) Constant brisk breezes, encountered especially at sea in both hemispheres at 10° to 30° latitude. *See also* Northeast Trades; Southeast Trades.

transform boundaries The lines along which two lithospheric plates slip horizontally by one another rather than collide (convergent boundaries) or pull apart (divergent boundaries).

transform fault *See* fault, transform.

transmission (of energy) The process by which radiation passes through atmospheric gases without being absorbed.

transpiration The evaporation via leaf surface of the cellular moisture in plants.

transported (soil parent material) Soil parent material that has been eroded and transported from another site by gravity, moving water, wind, or ice. Thus, unlike residual parent material, it does not come from the disintegration of underlying bedrock.

transverse dunes *See* dunes, transverse.

triple junction Where three lithospheric plate boundaries meet.

Tropic of Cancer Latitude 23 1/2° N. The northernmost latitude that experiences the noon sun directly overhead. (This occurs on or about June 21.)

Tropic of Capricorn Latitude 23 1/2° S. The southernmost latitude that experiences the noon sun directly overhead. (This occurs on or about December 21.)

tropical circulation cell (Hadley cell) Circulation cells in the troposphere of both hemispheres with air rising near the equator in the Doldrums, flowing north and south as the antitrades, descending in the Horse Latitudes into the subtropical high-pressure cells, and returning equatorward as the Trades.

tropical deciduous forest and scrub Variety of the deciduous forest formation class characterized by trees, some of which are stunted (scrub) in appearance, that lose their leaves during the dry season. Such forest and scrub lands are found on the transition between the tropical rain forest and savanna grasslands.

Tropical Dry Climatic type located astride the Tropics of Cancer and Capricorn on west coasts and continental interiors. Hot and dry climate dominated by air descending and warming in subtropical high-pressure cells. Contains the world's driest deserts.

tropical easterly jet stream High-velocity air current in the upper troposphere flowing easterly over northern Africa, India, and Southeast Asia during the summer (high-sun) season. It is considered to play a major role in the summer monsoon.

tropical rain forest (selva) Forest formation class that straddles the equator and occurs on east coasts of continents and islands in the tradewind zone. Dense, multilayered forest of evergreen broadleaf trees and shrubs, including numerous epiphytes and lianas.

Tropical Wet Climatic type of equatorial regions (Doldrums) and tropical east coasts (onshore trade winds). Warm temperatures, heavy and frequent precipitation, and rain forest vegetation are characteristic.

Tropical Wet and Dry Transitional climatic type of the low latitudes located between regions of Tropical Wet and Tropical Dry. Climate is warm year round. Monsoon circulation and the annual shift in solar heating cause distinct wet and dry seasons.

tropopause The narrow atmospheric zone separating the troposphere from the stratosphere, 8 to 9 miles (13 to 15 km) above the earth.

troposphere The atmospheric layer that is closest to the earth extending outward about 8 to 9 miles (13 to 15 km).

true desert Division of the Middle-Latitude Dry climate with precipitation 10 inches (25 cm) or less.

tsunami Shock waves generated by earthquakes, volcanic eruptions, or undersea slides transmitted through the ocean as sea waves. Seismic sea waves (*tsunami*) from earthquakes travel rapidly across oceans to expend their energy on coasts thousands of miles away.

Tundra (climate) Climate of high-latitude coastal regions primarily in the Northern Hemisphere with cold winters (but warmer than taiga winters) and short, cool summers. Annual precipitation is 5 to 15 inches (13 to 38 cm).

tundra (vegetation) Formation class of low-growing plants occurring poleward of tree growth and mainly on the shores of the Arctic Sea (arctic tundra) or above timberline on high mountains (alpine tundra). Associated with the Tundra climate.

turbidity currents Dense, muddy flows of seawater across the ocean bottom, usually in submarine canyons or deep-ocean trenches.

typhoon *See* hurricane.

ultisols Weathered soils of the middle latitudes with low nutrient content.

ultramafic Dark, dense rock with very high proportion of iron and magnesium minerals. Describes rock of the earth's mantle.

undertow Water retreating down a steep beach face and into the surf zone.

unloading The erosion and removal of overlying soil and rock material or the retreat of glacial ice that releases pressure on the exposed surface.

upper air inversion *See* subsidence inversion.

upwelling Cold water from ocean depths ascends as the surface water is moved aside by winds and currents.

valley glaciers (alpine glaciers) Tongues of ice flowing in mountain valleys.

valley train Alluvium from glacial meltwater streams that is sorted and deposited along the valley stretching beyond the terminal moraine.

Van Allen radiation belts Concentration of highly charged particles from the sun trapped by the earth's magnetic field. One is about 10,000 miles (16,000 km) above the earth, and a second at about 23,000 miles (37,000 km).

varve Seasonal stratified clay/silt deposit on the bottom of a glacial meltwater lake. Useful as a dating tool.

vegetation Plant life in general. First level of the vegetation classification.

vegetation classes Second level in the vegetation classification and based on major plant forms. The four vegetation classes are forest, grassland, desert, and tundra.

Veldt Region of Tropical Wet and Dry climate in southern Africa between the Congo basin and the Kalahari.

vernal equinox First day of the spring season. In the Northern Hemisphere, around March 21. The noon sun is directly overhead at the equator; equal day and night for all latitudes.

vernal ponds Small water bodies occupying shallow depressions and containing water only for a few months in the spring.

vertisols Clay soils with surface cracks when dry. Surface particles fall into and collect in cracks causing these soils to "invert" with time.

volcanic neck Eroded, remnant vent or neck of an ancient volcano. The solidified magma in the neck is more resistant to erosion than the extruded materials composing the rest of the volcano.

vulcanism, extrusive Lava welling out onto the surface of the earth by way of volcano or fissure.

vulcanism, intrusive An invasion of the earth's crust by magma from below that fails to reach the surface.

wadis *See* arroyos.

warm front A segment of the Polar Front usually on the eastern side of middle-latitude cyclones, where warm air advances against cold.

water control The survey, harnessing, and distribution of water for irrigation and other uses, especially important in regions of seasonal drought like that of Mediterranean climates.

waterlogging Saturation of the root zone of poorly drained agricultural soils.

water vapor The gaseous form of water.

wave-built terrace Extension of the wave-cut bench by deposition of fine sediments carried offshore.

wave-cut bench Shallow rock platform eroded by wave action and extending seaward from a wave-cut cliff.

wave refraction Bending of ocean wave fronts as they approach shallower zones offshore; directs wave energy at headlands.

weak equatorial low Intertropical storm centered on the Intertropical Convergence Zone that once developed tends to move westward. Rainfall occurs from numerous individual convective cells.

weather The day-to-day or even hour-to-hour condition of the atmospheric elements (temperature, pressure, wind, and moisture).

Westerlies Winds of the middle latitudes (35° to 60°) that blow from the west. Also a zone covering the same latitudes.

wet adiabatic lapse rate *See* lapse rate (adiabatic-wet).

willy willy Local name for a hurricane/typhoon in Australia.

wind Air movement that is horizontal or nearly horizontal (not ascending or descending air).

winter solstice Northern Hemisphere—the day when the sun at noon is directly overhead at latitude 23 1/2° S (late December); the noon sun is at its lowest point in the Northern Hemisphere sky and it is the shortest day of the year. Southern Hemisphere—the day when the sun at noon is directly overhead at 23 1/2° N (late June); the noon sun is at its lowest point in the Southern Hemisphere sky, and it is the shortest day of the year.

xeric Dry. Plants classed as mesophytes that grow in relatively dry habitats but not the excessively dry habitats of xerophytes.

xerophytes Plants adapted to growth in soils that lack available water sometime during the growing season. This includes most desert plants and those in dry sites elsewhere that are equipped to endure or escape recurrent drought.

zone of active deposition The area of deposition of glacial till near the outer margins of continental ice sheets.

zone of active erosion The area of scoured and eroded bedrock within a few hundred miles of the center of origin of continental ice sheets.

zone of eluviation Soil layer, usually the A horizon, from which minerals are removed by leaching.

zone of illuviation Soil layer, usually the B horizon, that is the site of accumulation of minerals leached from the A horizon.

zone time Time throughout a zone that is defined generally by 15 degrees of longitude; it is the time of the central meridian for the zone. Each zone is one hour different from its neighbors.

zooplankton Microscopic animal life in the sea.

BIBLIOGRAPHY

GENERAL REFERENCES

Carter, D. B., T. H. Schmudde, and D. M. Sharpe. *The Interface as a Working Environment: A Purpose for Physical Geography.* Commission on College Geography Technical Paper No. 7, Association of American Geographers, Washington, D.C., 1972.

Dassman, R. F. *Environmental Conservation.* 5th ed. John Wiley & Sons, Inc. New York, 1984.

Davidson, D. A. *Science for Physical Geographers.* John Wiley & Sons, Inc. New York, 1978.

Doerr, A. H. *Fundamentals of Physical Geography.* Wm. C. Brown Publishers, Dubuque, Iowa, 1990.

Gabler, R. E., R. J. Sager, S. M. Brazier, and D. L. Wise. *Essentials of Physical Geography.* 3rd ed. Saunders College Publishing, Philadelphia, 1987.

Haines-Young, R. H. *Physical Geography: Its Nature and Methods.* Harper Collins, New York, 1986.

Huber, T. P., R. P. Larkin, and G. L. Peters. *Dictionary of Concepts in Physical Geography.* Greenwood Press, Westport, Conn., 1988.

McKnight, T. L. *Physical Geography: A Landscape Appreciation.* 3rd ed. Prentice Hall, Englewood Cliffs, N.J., 1990.

Oberlander, T. M., and R. A. Muller. *Essentials of Physical Geography Today.* 2nd ed. Random House, New York, 1987.

Strahler, A. N., and A. H. Strahler. *Elements of Physical Geography.* 4th ed. John Wiley & Sons, Inc. New York, 1989.

Whittow, J. B. *The Penguin Dictionary of Physical Geography.* Penguin, Harmondsworth, England, 1984.

World Resources Institute. *World Resources 1990–91: A Guide to the Global Environment.* WRI Publications, Washington, D.C., 1990.

PART I/THE PLANET EARTH

CHAPTER 1/EARTH AS A SPHERE

Bowditch, N. *New American Practical Navigator.* Reprint Services Corp., Irvine, Calif., 1989.

Espenshade, E. B., Jr. and J. L. Morrison, eds. *Goode's World Atlas.* 18th ed. Rand McNally & Co., Chicago, 1990.

Howse, D. *Greenwich Time and the Discovery of Longitude.* Oxford University Press, 1983.

McCarthy, D. D., and J. D. Pilkington, eds. *Time and the Earth's Rotation.* Kluwer Academic Publishers, Norwell, Mass., 1979.

Smith, J. R. *Basic Geodesy: An Introduction to the History and Concepts of Modern Geodesy Without Mathematics.* Landmark Enterprises, Rancho Cordova, Calif., 1988.

CHAPTER 2/EARTH AND SUN RELATIONSHIPS

Bowditch, N. *New American Practical Navigator.* Reprint Services Corp., Irvine, Calif., 1989.

Friedman, H. *Sun and Earth: A Scientific American Book.* W. H. Freeman & Co., San Francisco, 1987.

Scientific American Editors. *The Solar System: A Scientific American Book.* W. H. Freeman & Co., San Francisco, 1975.

PART II/EARTH'S ATMOSPHERE

Boucher, L. *Global Climate.* John Wiley & Sons, Inc., New York, 1975.

Breuer, G. *Weather Modification: Prospects and Problems.* Cambridge University Press, London, 1979.

Budyko, M. I. *The Earth's Climate.* Academic Press, New York, 1982.

Clayton, H. H. *World Weather Records.* Smithsonian Miscellaneous Collection, Vol. 79, Smithsonian Institution, Washington, D.C., 1944.

Critchfield, H. J. *General Climatology.* 4th ed. Prentice Hall, Englewood Cliffs, N.J., 1983.

Hoskins, B., and R. Pearce, eds. *Large-Scale Dynamical Processes in the Atmosphere.* Academic Press, New York, 1983.

Lee, A. *Weather Wisdom: Facts and Folklore of Weather Forecasting.* Congdon & Weed, New York, 1990.

Lutgens, F. K., and E. J. Tarbuck. *The Atmosphere: An Introduction to Meteorology,* 3rd ed. Prentice Hall, Englewood Cliffs, N.J., 1986.

Schaefer, V. J., and J. A. Day. *A Field Guide to the Atmosphere.* Houghton Mifflin, Boston, 1981.

Skinner, B. J., ed. *Climates Past and Present,* William Kaufmann, Los Altos California, 1981.

Trewartha, G. T. *The Earth's Problem Climates,* University of Wisconsin Press, Madison, Wis., 1981.

Trewartha, G. T., and L. H. Horn. *An Introduction to Climate.* McGraw–Hill, New York, 1980.

U.S. Oceanic and Atmospheric Administration. *Climates of the States.* 2 vols. U.S. Government Printing Office, Washington, D.C., 1974.

Walls, J. *Desertification.* Pergamon, New York, 1980.

Whittow, J. *Disasters: The Anatomy of Environmental Hazards.* University of Georgia Press, Athens, 1980.

CHAPTER 3/THE ATMOSPHERE'S COMPOSITION AND STRUCTURE

Allen, O. E., and the editors of Time-Life Books. *Atmosphere.* Time-Life Books, Alexandria, Va., 1983.

Anthes, R. A., et al., *The Atmosphere.* 2nd ed. Merrill, Columbus, Ohio, 1981.

Hanwell, J. D. *Atmospheric Processes.* Allen & Unwin, Winchester, Mass., 1980.

Idso, S. B. *Carbon Dioxide: Friend or Foe: An Inquiry into the Climatic and Agricultural Consequences of the Rapidly Rising CO_2 Content of the Earth's Atmosphere.* IBR Press, Tempe, Ariz., 1982.

Meetham, A. R., et al. *Atmospheric Pollution.* 4th ed. Pergamon, Oxford, England, 1981.

National Research Council. *The Upper Atmosphere and Magnetosphere.* National Academy of Sciences, U.S. Government Printing Office, Washington, D.C., 1977.

Pasquill, F. *Atmospheric Diffusion: The Dispersal of Windborne Material from Industrial and Other Sources.* 3rd ed. John Wiley & Sons, Inc. New York, 1983.

Riehl, H. *Introduction to the Atmosphere.* McGraw–Hill, New York, 1978.

CHAPTER 4/HEATING AND COOLING THE ATMOSPHERE: TEMPERATURE

Behrman, D. *Solar Energy: The Awaking Science.* Little, Brown, Boston, 1976.

Gribbin, J. *Future Weather and the Greenhouse Effect.* Delacorte Press, New York, 1982.

Jones, P. D., and T. M. Wigley. "Global Warming Trends." *Scientific American,* vol. 263, no. 2, pages 84–91, 1990.

Miller, D. H. *Energy at the Surface of the Earth.* Academic Press, New York, 1981.

Robinson, N. *Solar Radiation.* Elsevier, New York, 1966.

Weinberg, C. J., and R. H. Williams. "Energy from the Sun." *Scientific American,* vol. 263, no. 3, pages 146–155, 1990.

CHAPTER 5/PRESSURE AND WINDS

Corby, G. A., ed. *The Global Circulation of the Atmosphere.* Royal Meteorological Society, London, 1970.

Eldridge, F. S. *Wind Machines.* Van Nostrand Reinhold, New York, 1980.

Franklin Institute. *Wind Energy.* Franklin Institute Press, Philadelphia, 1978.

Palmen, E., and C. W. Newton. *Atmospheric Circulation Systems: Their Structure and Physical Interpretation.* Academic Press, New York, 1969.

Panofsky, H. A., and J. Dutton. *Atmospheric Turbulence.* John Wiley & Sons, Inc. New York, 1983.

Ramage, C. S. *Monsoon Meteorology.* Academic Press, New York, 1971.

CHAPTER 6/ATMOSPHERIC MOISTURE

Battan, L. J. *Cloud Physics and Cloud Seeding.* Doubleday, New York, 1962.

Gokhale, N. R. *Hailstorms and Hailstone Growth.* State University of New York Press, Albany, 1975.

Kirk, R. *Snow.* Morrow, New York, 1978.

Ludlam, F. H. *Clouds and Storms: The Behavior and Effect of Water in the Atmosphere.* Pennsylvania State University Press, University Park, 1980.

Mason, B. J. *Clouds, Rain, and Rainmaking.* 2nd ed. Cambridge University Press, Cambridge, England, 1975.

Perry, A. H., and J. M. Walker. *The Ocean-Atmosphere System.* Longmans, London, 1977.

Thornthwaite, C. W., and J. R. Mather. *The Moisture Balance.* Princeton University Press, Princeton, N.J., 1949.

CHAPTER 7/AIR MASSES, FRONTS, AND STORMS

Battan, L. J. *The Thunderstorm.* Signet Science Library, New York, 1964.

Eagleman, J. R. *Severe and Unusual Weather.* Van Nostrand Reinhold, New York, 1982.

Flora, S. C. *Tornadoes of the United States.* Revised ed. University of Oklahoma, Norman, 1973.

Harman, J. R. *Tropospheric Waves, Jet Streams, and United States Weather Patterns.* Commission on College Geography Resource Paper No. 11, Association of American Geographers, Washington, D.C., 1971.

Jennings, G. *The Killer Storms, Hurricanes, Typhoons, and Tornadoes.* Lippincott, New York, 1970.

Reiter, E. H. *Jet Streams.* Doubleday, Garden City, N.Y., 1967.

Simpson, R. H., and H. Riehl. *The Hurricane and Its Impact.* Louisiana State University Press, Baton Rouge, 1981.

Whipple, A.B.C., and the editors of Time-Life Books. *Storm.* Time-Life Books, Alexandria, Va., 1982.

CHAPTER 8/CLIMATE CLASSIFICATION

Köppen, W., and R. Geiger. *Hanbuch der Klimatologie.* 5 vols. Verlagsbuchhandlung Gebruder Borntaeger, Berlin, 1930.

CHAPTER 9/THE TROPICAL CLIMATES

Ayoade, J. O. *Introduction to Climatology for the Tropics.* John Wiley & Sons, Inc. New York, 1983.

Niewolt, S. *Tropical Climatology: An Introduction to the Climates of the Low Latitudes.* John Wiley, Chichester, England, 1982.

Riehl, H. *Tropical Meteorology.* McGraw–Hill, New York, 1954.

CHAPTER 11/MIDLATITUDE CLIMATES

Golany, G. S. *Design for Arid Regions.* Van Nostrand Reinhold, New York, 1983.

Landsberg, H. E. *The Urban Climate.* Academic Press, New York, 1981.

Lowry, W. P. *Weather and Life: An Introduction to Biometeorology.* Academic Press, New York, 1967.

CHAPTER 12/THE HIGH-LATITUDE CLIMATES

Orvig, S., ed. *Climates of the Polar Regions.* American Elsevier, New York, 1970.

Williams, P. J. *Pipelines and Permafrost.* Longmans, London, 1980.

PART III/EARTH WATER

CHAPTER 13/GLOBAL WATER DISTRIBUTION AND CIRCULATION

Bowen, R. *Groundwater.* Halsted Press, New York, 1980.

Charlies, R. H. *Tidal Energy.* Van Nostrand Reinhold, New York, 1982.

Gaskell, T. F. *The Gulf Stream.* John Day, New York, 1973.

Gross, M. G. *Oceanography: A View of the Earth.* Prentice Hall, Englewood Cliffs, N.J., 1982.

Ingmanson, D. E., and W. J. Wallace. *Oceanology: An Introduction.* Wadsworth, Belmont, Calif., 1973.

Leopold, L. B., K. S. Davis, and the editors of Time-Life Books. *Water.* Time-Life Books, New York, 1969.

Melchioir, P. J. *The Tides of the Planet Earth.* Pergamon, N.Y., 1978.

Miller, D. H. *Water at the Surface of the Earth: An Introduction to Ecosystem Hydrodynamics.* Academic Press, New York, 1982.

Neumann, G. *Ocean Currents.* Elsevier, New York, 1968.

Pedlosky, J. "The Dynamics of the Oceanic Subtropical Gyres." *Science,* vol. 248, pages 316–322, 1980.

Raghunath, H. M. *Groundwater Hydrology.* John Wiley & Sons, Inc. New York, 1982.

Rand McNally Atlas of the Oceans. Rand McNally Chicago, 1977.

Stowe, K. *Essentials of Ocean Science.* John Wiley & Sons, Inc. New York, 1987.

Ward, R. C. *Principles of Hydrology.* 2nd ed. McGraw–Hill, New York, 1975.

Whipple, A.B.C., and the editors of Time-Life Books. *The Restless Oceans.* Time-Life Books, Alexandria, Va., 1984.

CHAPTER 14/THE EARTH'S FRESH WATER

Baldwin, H. L., and C. I. McGuinness. *A Primer on Groundwater.* U.S. Geological Survey, Washington, D.C., 1963.

Bitton, G., and C. P. Gerbs. *Groundwater Pollution Microbiology.* John Wiley & Sons, Inc. New York, 1984.

Blair, P. B., T.A.V. Cassel, and R. H. Edelstein. *Geothermal Energy.* John Wiley & Sons, Inc. New York, 1982.

Bowen, R. *Geothermal Resources.* 2nd ed. Elsevier Science Publishing Co., New York, 1989.

Bowen, R. *Groundwater.* 2nd ed. Elsevier Science Publishing Co., New York, 1986.

Driscoll, F. G. *Groundwater and Wells.* Signal Environmental Systems, Johnson Division, New Brighton, Minn., 1986.

Dunne, T., and L. B. Leopold. *Water in Environmental Planning.* W. H. Freeman & Co., San Francisco, 1978.

Geothermal Resources Council. *The Future of Geothermal Energy.* Geothermal Resources Council, Davis, Calif., 1987.

Graf, W. L. *The Colorado River: Instability and Basin Management.* Resource Publications in Geography, Association of American Geographers, Washington, D.C., 1985.

Howe, C. W., and K. W. Easter. *Interbasin Transfers of Water: Economic Issues and Impacts.* Johns Hopkins University Press, Baltimore, Md., 1971.

Leopold, L. B. *Water: A Primer.* W. H. Freeman & Co. San Francisco, 1974.

Mather, J. R. *Water Resources: Distribution, Use, and Management.* John Wiley & Sons, Inc. New York, 1984.

National Science Teachers. *Earth: The Water Planet.* National Science Teachers Association, Washington, D.C., 1989.

O'Mara, G. T. ed. *Efficiency in Irrigation: The Conjunctive Use of Surface and Groundwater Resources.* The World Bank Publications, Washington, D.C., 1988.

Parker, R. H. *Hot Dry Rock Geothermal Energy: Phase 2B Final Report of the Camborne School of Mines.* 2 vols. Pergamon Press, Elmsford, N.Y., 1989.

Price, M. *Introducing Groundwater.* George Allen & Unwin, Boston, 1985.

Reisner, M. P. *The Cadillac Desert: The American West and Its Disappearing Water.* Viking Penguin, New York, 1986.

Shainberg, I., and J. Shalhevet, eds. *Soil Salinity Under Irrigation.* Springer–Verlag, New York, 1985.

Smith, Z. A. *Groundwater in the West.* Academic Press, San Diego, Calif., 1988.

Water Pollution Control Federation Staff. *Water Reuse.* Water Pollution Control Federation, Alexandria, Va., 1983.

Willis, R., and W. Yeh. *Groundwater Systems Planning and Management.* Prentice Hall, Englewood Cliffs, N.J., 1987.

CHAPTER 15/OCEANIC RESOURCES

Borgese, E. M., and Norton Ginsburg, eds. *Ocean Yearbook No. 6.* University of Chicago Press, Chicago, 1987.

Carson, R. *Sea Around Us: Special Edition.* (Introduction by A. H. Zwinger). Oxford University Press, New York, 1989.

Charlies, R. H. *Tidal Energy.* Van Nostrand Reinhold, New York, 1982.

Clark, R. B. *Marine Pollution.* 2nd ed. Oxford University Press, New York, 1989.

Couper, A., ed. *The Times Atlas and Encyclopedia of the Sea.* Harper Collins, New York, 1989.

Cronin, D. S. *Underwater Minerals.* Academic Press, New York, 1980.

Davis, R. A., Jr. *Oceanography: An Introduction to the Marine Environment.* Wm. C. Brown Publishers, Dubuque, Iowa, 1987.

Gross, M. G. *Oceanography: A View of the Earth.* Prentice Hall, Englewood Cliffs, N.J., 1982.

Heezen, B. C., M. Tharp, and M. Ewing. *The Floors of the Oceans.* Geological Society of America, Special Paper No. 65, New York, 1959.

Johnson, D. *The Origin of Submarine Canyons.* Columbia University Press, New York, 1939.

King, C.A.M. *Oceanography for Geographers.* Arnold, London, 1975.

Knox, M., and M. Rodriguez. *Steinbeck's Street, Cannery Row.* Presidio Press, San Rafael, Calif., 1980.

Scott, J., ed. *Desalination of Seawater by Reverse Osmosis.* Noyes Data Corp., Noyes Publications, Noyes Press, Park Ridge, N.J., 1981.

Shepard, F. P. *Geological Oceanography: Evolution of Coasts, Continental Margins and Sea Floors.* Crane, Russak, New York, 1977.

Shepard, F. P., and F. Dill. *Submarine Canyons and Other Sea Valleys.* Rand McNally & Co., Chicago, 1966.

Shepard, F. P., and K. O. Emery. *Submarine Topography Off the California Coast: Canyons and Tectonic Interpretation.* Geological Society of America Special Paper No. 31, New York, 1941.

Whittaker, J. H., ed. *Submarine Canyons and Deep-Sea Fans.* Academic Press, San Diego, Calif., 1976.

PART IV/EARTH'S LITHOSPHERE AND LANDFORMS

CHAPTER 16/INTRODUCTION TO LANDFORMS

Chorley, R. S., S. A. Schumm, and D. E. Sugden. *Geomorphology.* Methuen, New York, 1984.

Cloud, P. *Oasis in Space: Earth History from the Beginning.* W. W. Norton, New York, 1988.

Garver, J. B., H. A. Curran, P. S. Justus, and D. M. Young. *Atlas of Landforms.* 3rd ed. John Wiley & Sons, Inc. New York, 1984.

Gerrard, J. *Rocks and Landforms.* Unwin Hyman, Winchester, Mass., 1987.

Davis, W. M. *Geographical Essays.* Ginn, Boston, 1909 (reprinted Dover, New York, 1954).

Hamilton, E. I., and Farquhar, R. M., eds. *Radiometric Dating for Geologists.* Books on Demand, Ann Arbor, Mich.

Hsu, K. J. *Mountain Building Processes.* Academic Press, New York, 1983.

Hunt, C. B. *Natural Regions of the United States and Canada.* W. H. Freeman & Co., San Francisco, 1974.

Piper, J. D. *Paleomagnetism and the Continental Crust.* Halsted Press, New York, 1987.

Mabbutt, J. A. *Desert Landforms.* M.I.T. Press, Cambridge, Mass., 1977.

Melhorn, W. N., and R. C. Flemal. *Theories of Landform Development.* Allen & Unwin, Winchester, Mass., 1981.

Ritter, D. F. *Process Geomorphology.* Wm. C. Brown, Dubuque, Iowa, 1978.

Shelton, J. S. *Geology Illustrated.* W. H. Freeman & Co., San Francisco, 1966.

Smith, D. G., Editor-in-Chief. *The Cambridge Encyclopedia of Earth Sciences.* Cambridge University Press, New York, 1981.

Tarbuck, E. J., and Lutgens, F. K. *Earth Science.* 5th ed. Merrill Publishing Co., Columbus, Ohio, 1987.

Tarling, D. T., ed. *Paleomagnetism: Principles and Applications in Geology.* Geophysics and Archaeology, Methuen, New York, 1983.

CHAPTER 17/EARTH'S STRUCTURE AND PLATE TECTONICS

American Geological Institute. *Dictionary of Geological Terms,* 3rd ed. Doubleday Publishing Co., Garden City, N.Y., 1984.

Brown, G. C., and A. E. Musset. *The Inaccessible Earth.* Allen & Unwin, Winchester, Mass., 1981.

Condie, K. C. *Plate Tectonics and Crustal Evolution,* 3rd ed. Pergamon, New York, 1989.

Cox, A., and R. B. Hart. *Plate Tectonics: How It Works.* Blackwell Scientific Publications, Palo Alto, Calif., 1986.

Cox, K. G., J. D. Bell, and R. J. Pankhurst. *The Interpretation of Igneous Rocks.* Allen & Unwin, Winchester, Mass., 1979.

Fry, N. *The Field Description of Metamorphic Rocks.* John Wiley & Sons, Inc. New York, 1983.

Hamblin, W. K. *The Earth's Dynamic Systems.* 4th ed. Burgess Publishing, Minneapolis, Minn., 1985.

Le Grand, H. E. *Drifting Continents and Shifting Theories: The Modern Revolution in Geology and Scientific Change.* Cambridge University Press, New York, 1989.

Miller, R., and the editors of Time-Life Books. *Continents in Collision.* Time-Life Books, Alexandria, Va., 1983.

Nolet, G., ed. *Seismic Tomography: With Applications in Global Seismology and Exploration Geophysics.* Kluwer Academic Press, Norwell, Mass., 1987.

Piper, J. D. *Paleomagnetism and the Continental Crust.* Halsted Press, New York, 1987.

Raymo, C. *The Crust of Our Earth: An Armchair Traveler's Guide in the New Geology.* Prentice Hall, Englewood Cliffs, N.J., 1984.

Shea, J. H., ed. *Plate Tectonics.* Van Nostrand Reinhold, New York, 1985.

Smith, D. G., editor-in-chief. *The Cambridge Encyclopedia of Earth Sciences.* Cambridge University Press, Cambridge, 1981.

Tucker, M. E. *The Field Description of Sedimentary Rocks.* John Wiley & Sons, Inc. Somerset, N.J., 1982.

Van der Voo, R., C. R. Scotese, and N. Bonhommet, eds. *Plate Reconstruction from Paleozoic Paleomagnetism.* American Geophysical Union, Washington, D.C., 1984.

Wegener, A. *The Origin of the Continents and Oceans* (translated by J. Byram). Dover, New York, 1966.

Wilson, J. T., ed. *Continents Adrift and Continents Aground.* W. H. Freeman & Co., San Francisco, 1976.

CHAPTER 18/TECTONIC PROCESSES

Bertero, V., ed. *Reducing Earthquake Hazards: Lessons Learned from the Mexico Earthquake.* Earthquake Engineering Research Institute, El Cerrito, Calif., 1989.

Burton, I. R., R. W. Kates, and G. White. *The Environment as Hazard.* Oxford University Press, New York, 1978.

Coffman, J. L., C. A. von Hake, and C. W. Hover, *Earthquake History of the United States.* U.S. Department of Commerce, Boulder, Colo., 1982.

Collett, L. W. *The Structure of the Alps.* R. E. Krieger, Huntington, N.Y., 1974.

Decker, R., and B. Decker. *Volcanoes.* 2nd ed. W. H. Freeman & Co., New York, 1989.

Dudley, W. C., and M. Lee. *Tsunami.* University of Hawaii Press, Honolulu, 1988.

Eiby, G. A. *Earthquakes.* Van Nostrand, New York, 1980.

Erickson, J. *Volcanoes and Earthquakes.* TAB Books, Blue Ridge Summit, Pa., 1988.

Harris, S. L. *Fire and Ice: The Cascade Volcanoes.* 2nd ed. Mountaineering and Pacific Search Press, Seattle, 1980.

Heppenheimer, T. A. *The Coming Quake: Science and Trembling on the California Earthquake Frontier.* Random House, New York, 1988.

Howell, D. G. *Tectonics of Suspect Terranes: Mountain Building and Continental Growth.* Routledge, Chapman & Hall, New York, 1989.

Hsu, K. J. *Mountain Building Processes.* Academic Press, New York, 1983.

Jaroszewski, W. *Fault and Fold Tectonics.* Halsted Press, New York, 1985.

Keller, S. A., ed. *Mount St. Helens: Five Years Later.* Eastern Washington University Press, Cheney, Wash., 1987.

Novosti Press Agency Staff. *Armenian Earthquake Disaster.* Sphinx Press, Madison, Conn., 1989.

Ollier, C. *Volcanoes.* Blackwell, Basil, Cambridge, Mass., 1988.

Ramsay, J. G. *Folding and Fracturing of Rocks.* McGraw–Hill, New York, 1967.

Richter, C. E. *Elementary Seismology.* W. H. Freeman, San Francisco, 1958.

Rosenfeld, C., and R. Cooke. *Earthfire: The Eruption of Mount St. Helens.* M.I.T. Press, Cambridge, Mass., 1982.

Sheets, P. D., and D. K. Grayson. *Volcanic Activity and Human Ecology.* Academic Press, New York, 1979.

Simkin, T., R. Fiske, and S. Melcher, eds. *Krakatau 1883: The Volcanic Eruption and Its Effects.* Smithsonian Institution Press, Washington, D.C., 1983.

Stern, R. M. *Tsunami.* W. W. Norton, New York, 1988.

Unesco. *Tsunami Research Symposium, 1974: Proceedings of an International Symposium of Tsunami Research.* UNIPUB, Lanham, Md., 1974.

U.S. Congress. *Loma Prieta Earthquake: Lessons to Be Learned.* House of Representatives Hearings, U.S. Government Printing Office, Washington, D.C., 1990.

U.S. National Academy of Sciences. *Panel on Earthquake Prediction, A Scientific and Technical Evaluation with Implications for Society.* N.A.S., Washington, D.C., 1976.

Wallace, R., and the editors of Time-Life Books. *The Grand Canyon.* Time-Life Books, Alexandria, Va., 1973.

Weyman, D. *Tectonic Processes.* Allen & Unwin, Winchester, Mass., 1981.

CHAPTER 19/GRADATIONAL PROCESSES

deBoodt, M., and D. Gabriels, eds. *Assessment of Erosion.* John Wiley & Sons, Inc. New York, 1980.

Ford, D. C., and P. W. Williams. *Karst Geomorphology and Hydrology.* Unwin Hyman, Winchester, Mass., 1989.

Holy, M. *Erosion and Environment.* Pergamon, New York, 1980.

Jennins, J. N. *Karst Geomorphology.* Blackwell, Basil, Cambridge, Mass., 1985.

Ollier, C. *Weathering.* 2nd ed. John Wiley & Sons, Inc. New York, 1986.

Rayner, J. N. *Conservation, Equilibrium and Feedback Applied to Atmospheric and Fluvial Processes.* Association of American Geographers, Washington, D.C., 1972.

Selby, M. J. *Hillslope Materials and Processes.* John Wiley & Sons, Inc. New York, 1983.

Small, R. J., and M. J. Clark. *Slopes and Weathering.* Cambridge University Press, New York, 1982.

Sweeting, M. *Karst Landforms.* Columbia University Press, New York, 1973.

Trudgill, S. T. *Weathering and Erosion.* Butterworth Publishers, Stoneham, Mass., 1982.

Zarubia, Q., and V. Mendel. *Landslides and Their Control.* 2nd rev. ed. Elsevier Science Publishing, New York, 1982.

CHAPTER 20/STREAMS AND LANDFORMS

Bremer, H., ed. *Fluvial Geomorphology.* Coronet Books, Philadelphia, Pa., 1985.

Chorley, R. J., Jr., ed. *Water, Earth, and Man.* Methuen & Co., London, 1969.

Clark, C., and the editors of Time-Life Books. *Flood.* Time-Life Books, Alexandria, Va., 1982.

Fraser, G. S., and L. J. Suttner. *Alluvial Fans and Fan Deltas: An Exploration Guide.* Prentice Hall, Englewood Cliffs, N.J., 1988.

Hey, R. D., J. C. Bathurst, and C. R. Thorne. *Gravel Bed Rivers: Fluvial Processes, Engineering and Management.* John Wiley & Sons, Inc. Somerset, N.J., 1982.

Leopold, L. B., M. G. Wolman, and J. P. Miller. *Fluvial Processes in Geomorphology.* W. H. Freeman & Co., San Francisco, 1964.

Morisawa, M. *Fluvial Geomorphology.* Allen & Unwin, Winchester, Mass., 1981.

Morisawa, M. *Streams, Their Dynamics and Morphology.* McGraw–Hill Book Co., Planetary Science Series, New York, 1968.

Petts, G. E. *Rivers.* Butterworth, Woburn, Mass., 1983.

Pitty, A. F., ed. *Geographical Approaches to Fluvial Processes.* University of East Anglia, Norwich, 1979.

Rachocki, A. H. *Alluvial Fans: An Attempt at an Empirical Approach.* John Wiley & Sons, Inc. New York, 1981.

CHAPTER 21/GLACIATION

Agassiz, L. *Geological Sketches.* Reprint Services, Irvine, Calif., 1985.

Bahn, P. G. *Images of the Ice Age.* Facts on File, New York, 1989.

Bailey, R. H., and the editors of Time-Life Books. *Glacier,* Time-Life Books, Alexandria, Va., 1982.

Chorlton, W., and the editors of Time-Life Books. *Ice Ages.* Time-Life Books, Alexandria, Va., 1983.

Coats, D. R. *Glacial Geomorphology.* Allen & Unwin, Winchester, Mass., 1981.

Colbeck, S. C., ed. *Dynamics of Snow and Ice Masses.* Academic Press, New York, 1980.

Denton, G. H., and T. Hughes, eds. *The Last Great Ice Sheets.* John Wiley & Sons, Inc., New York, 1981.

Embleton, C., and C.A.M. King. *Glacial and Periglacial Geomorphology.* 2nd ed. John Wiley & Sons, Inc. New York, 1975.

Fitzgerald, D. M., and P. S. Rosen. *Glaciated Coasts.* Academic Press, San Diego, Calif., 1987.

Flint, R. F. *Glacial and Quaternary Geology.* John Wiley & Sons, Inc. New York, 1966.

Gilbert, G. K. *Glaciers and Glaciation.* Vol. 3. Kraus Reprint and Periodicals, Millwood, N.Y., 1910.

Goldthwaite, R. P., ed. *Glacial Deposits.* Dowden, Hutchinson & Ross, Stroudsburg, Pa., 1975.

Gray, M. *The Quaternary Ice Age.* Cambridge University Press, New York, 1985.

Mathes, F. E., and F. Fryxell, eds. *The Incomparable Valley: A Geological Interpretation of Yosemite.* University of California Press, Berkeley, 1950.

Muir, J. A., and W. E. Colby, eds. *Studies in the Sierra.* Rev. ed. Sierra Club, San Francisco, 1960.

Paterson, W.S.B., *The Physics of Glaciers.* 2nd ed. Pergamon Press, Elmsford, N.Y., 1981.

Plummer, C. C., and D. McGeary. *Physical Geology.* 3rd ed. Wm. C. Brown Publishers, Dubuque, Iowa, 1985.

Price, R. J. *Glacial and Fluvioglacial Landforms.* Books on Demand, Ann Arbor, Mich.

Sharp, R. P. *Living Ice: Understanding Glaciers and Glaciation.* Cambridge University Press, New York, 1988.

CHAPTER 22/WIND AND AEOLIAN LANDFORMS

Bannan, J. G. *Sand Dunes.* Carolrhoda Books. Minneapolis, Minn., 1989.

Brookfield, M. E., and T. A. Ahlbrandt, eds. *Eolian Sediments and Processes.* Elsevier Science Publishing, New York, 1983.

Eden, D. N., and R. J. Furkert, eds. *Loess: Its Distribution, Geology and Soils: Proceedings of an International Symposium.* New Zealand, 1987. Gower Publishing Co., Brookfield, Vt., 1987.

Greeley, R., and J. D. Iversen. *Wind as a Geologic Process.* Cambridge University Press, Cambridge, 1985.

Jennings, J. N., and H. Hagedorn, eds. *Dunes: Continental and Coastal.* Coronet Books, Philadelphia, Pa., 1983.

Johnson, V. *Heaven's Tableland: The Dust Bowl Story.* DeCapo Press, New York, 1974.

Lai, R. J. *Wind Erosion and Deposition Along a Coastal Sand Dune.* University of Delaware Press, Newark, Del., 1978.

Leopold, A. S., and the editors of Time-Life Books. *The Desert.* Time-Life Books, New York, 1967.

Liu, T. *Loess in China.* Springer–Verlag, New York, 1988.

Mutel, C. F. *Fragile Giants: A Natural History of the Loess Hills.* University of Iowa Press, Iowa City, Iowa, 1989.

Nickling, W. G., ed. *Aeolian Geomorphology.* Unwin Hyman, Winchester, Mass., 1986.

Peure, T. L. *Desert Dust: Origin, Characteristics, and Effect on Man.* Geological Society of America, Boulder, Colo., 1981.

Sears, P. B. *Deserts on the March.* 3rd ed. University of Oklahoma Press, Norman, Okla., 1959.

Smalley, I. J. *Loess: Lithology and Genesis.* Halsted Press, New York, 1975.

CHAPTER 23/WAVES AND COASTAL LANDFORMS

Bascom, W. *Waves and Beaches.* Rev. ed. Doubleday Anchor Books, New York, 1980.

Coates, D. R. *Coastal Geomorphology.* Allen & Unwin, Winchester, Mass., 1981.

Fox, W. T. *At the Sea's Edge: An Introduction to Coastal Ocean-*

ography for the Amateur Naturalist. Prentice Hall, Englewood Cliffs, N.J., 1983.

Kauffman, W., and O. Pilkey. *The Beaches Are Moving: The Drowning of America's Shoreline.* Doubleday, New York, 1979.

King, C.A.M. *Beaches and Coasts.* 2nd ed. St. Martin's Press, New York, 1972.

Komar, P. D. *Beach Processes and Sedimentation.* Prentice Hall, Englewood Cliffs, N.J., 1976.

Komar, P. D., ed. *Handbook of Coastal Processes and Erosion.* CRC Press, Boca Raton, Fla., 1983.

Leatherman, S., ed. *Barrier Islands: From the Gulf of Saint Lawrence to the Gulf of Mexico.* Academic Press, San Diego, Calif., 1979.

Magoon, O. T., ed. *Barrier Islands: Process and Management.* American Society of Civil Engineers, New York, 1989.

Mei, C. C. *The Applied Dynamics of Ocean Surface Waves.* John Wiley & Sons, Inc. New York, 1982.

Schwartz, M. L. *Spits and Bars.* Dowden, Hutchinson & Ross, Stroudsburg, Pa., 1972.

Smith, D. E., and A. G. Dawson. *Shorelines and Isostasy.* Academic Press, New York, 1984.

Snead, R. E. *Coastal Landforms and Surface Features: A Photographic Atlas and Glossary.* Van Nostrand Reinhold, New York, 1982.

Webber, B., and Margie Webber. *Bayocean: The Oregon Town That Fell into the Sea.* Webb Research Group, Medford, Oreg., 1989.

PART V/SOILS AND VEGETATION

CHAPTER 24/UNDERSTANDING SOILS

Bridges, E. M. *World Soils.* Cambridge University Press, London, 1970.

Buckman, H. O., and N. G. Brady. *The Nature and Properties of Soils.* 7th ed. Macmillan, New York, 1969.

Donahue, R. L., R. W. Miller, and J. C. Shickluna. *Soils: An Introduction to Soils and Plant Growth.* Prentice Hall, Englewood Cliffs, N.J., 1977.

Eyre, S. R. *Vegetation and Soils: A World Picture.* Aldine, Chicago, 1968.

Gerrard, J. *Soils and Landforms.* Allen & Unwin, Winchester, Mass., 1981.

Kirby, M. J. *Soil Erosion.* Halsted Press, New York, 1981.

Young, A. *Tropical Soils and Soil Survey.* Cambridge University Press, Cambridge, England, 1976.

CHAPTER 25/CLASSIFICATION AND SOIL REGIONS

Basile, R. M. *A Geography of Soils.* Wm. C. Brown, Dubuque, Iowa, 1971.

Bridges, E. M., and D. A. Davidson, eds. *Principles and Applications of Soil Geography.* Longmans, London, 1982.

Finkle, C. W., ed. *Soil Classification.* Hutchinson Ross, New York, 1982.

FitzPatrick, E. A. *Soils, Their Formation, Classification, and Distribution.* Longmans, London, 1980.

Foth, H. D. *Soil Geography and Land Use.* John Wiley & Sons, Inc. New York, 1980.

Soil Conservation Service, Soil Survey Staff. *Soil Taxonomy.* U.S. Department of Agriculture, Agriculture Handbook No 436. U.S. Government Printing Office, Washington, D.C,. 1975.

Steila, D. *The Geography of Soils.* Prentice Hall, Englewood Cliffs, N.J., 1976.

CHAPTER 26/UNDERSTANDING NATURAL VEGETATION

Billings, W. D. *Plants, Man, and the Ecosystem.* Wadsworth, Belmont, Calif., 1970.

Cox, C. B., and P. D. Moore. *Biogeography, An Ecological and Evolutionary Approach.* 3rd ed. Blackwell, Oxford, England, 1980.

Grime, J. P. *Plant Strategies and Vegetation Processes.* John Wiley, Chichester, England, 1979.

Kershaw, D. A. *Quantitative and Dynamic Ecology.* American Elsevier, New York, 1964.

Mitsch, W. J., and J. G. Gosselink. *Wetlands.* Van Nostrand Reinhold, New York, 1986.

Odum, E. P. *Fundamentals of Ecology.* Saunders, Philadelphia, 1973.

Perl, P., and the editors of Time-Life Books. *Cacti and Succulents.* Time-Life Books, New York, 1978.

Price, L. W. *Mountains & Man: A Study of Process and Environment.* University of California Press, Berkeley, 1981.

Went, F. W., and the editors of Life. *The Plants.* Time, New York, 1963.

CHAPTER 27/WORLD PATTERNS: VEGETATION AND PHYSICAL GEOGRAPHY

Collinson, A. S. *Introduction to World Vegetation.* Allen & Unwin, London, 1977.

Daubenmire, R. *Plant Geography with Special Reference to North America.* Academic Press, New York, 1978.

de Laubenfels, D. J. *Mapping the World's Vegetation: Regionalization of Formations and Flora.* Syracuse University Press, Syracuse, N.Y., 1975.

Eyre, S. R. *Vegetation and Soils: A World Picture.* Aldine, Chicago, 1968.

Humphrey, R. R. *The Desert Grassland.* University of Arizona Press, Tucson, 1968.

Larsen, J. A. *The Boreal Ecosystem.* Academic Press, New York, 1980.

Larsen, J. A. *Ecology of the Northern Lowland Bogs and Conifer Forests.* Academic Press, New York, 1982.

Lovelock, J. E. *Gaia: A New Look at Life on Earth.* Oxford University Press, Oxford, 1979.

Myers, N. *The Primary Source: Tropical Forests and Our Future.* W. W. Norton, New York, 1985.

Page, J., and the editors of Time-Life Books. *Forest.* Time-Life Books, Alexandria, Va., 1983.

Pruitt, W. O. *Boreal Ecology.* Institute of Biological Studies in Biology, No. 91. Edward Arnold, London, 1978.

Richards, P. W. *The Life of the Jungle.* McGraw–Hill, New York, 1970.

Shimwell, D. W. *The Description and Classification of Vegetation.* University of Washington Press, Seattle, 1971.

Thirgood, J. V. *Man and the Mediterranean Forest.* Academic Press, New York, 1981.

Tivy, J. *Biogeography, A Study of Plants in the Ecosphere.* 2nd ed. Longmans, London, 1982.

Vankat, J. L. *The Natural Vegetation of North America: An Introduction.* John Wiley & Sons, Inc. New York, 1979.

Walter, H. *Vegetation of the Earth.* Springer–Verlag, New York, 1973.

APPENDIX B/MAPS

Bertin, J. *Semiology of Graphics: Diagrams, Networks, Maps* University of Wisconsin Press, Madison, 1984.

Bies, J. D., and R. A. Long. *Mapping and Topographic Drafting.* South-Western, Cincinnati, Ohio, 1983.

Brown, L. A. *The Story of Maps.* Little, Brown, Boston, 1950.

Campbell, J. *Introductory Cartography.* Prentice Hall, Englewood Cliffs, N.J., 1984.

Greenhood, D. *Mapping.* 3rd ed. University of Chicago Press, Chicago, 1964.

Keates, J. S. *Understanding Maps.* John Wiley & Sons, Inc. New York, 1982.

Robinson, A. H., et al. *Elements of Cartography.* 5th ed. John Wiley & Sons, Inc. New York, 1984.

U.S. Department of the Army. *Map Reading.* Technical Manual FM 21–26. U.S. Government Printing Office, Washington, D.C., 1969.

Wilford, J. N. *The Mapmakers.* Vintage, New York, 1982.

APPENDIX C/THE KÖPPEN CLIMATE CLASSIFICATION

Köppen, W., and R. Geiger. *Hanbuch der Klimatologie,* 5 vols. Verlagsbuchhandlung Gebruder Borntaeger, Berlin, 1930.

APPENDIX D/THE GEOLOGIC TIME SCALE

Eicher, D. L. *Geologic Time.* 2nd ed. Prentice Hall, Englewood Cliffs, N.J., 1976.

PHOTO CREDITS

PART OPENERS

Part 1: NASA.
Part 2: C. F. Luhmann/Naval Photographic Center, Washington, DC.
Part 3: Owen Franken.
Part 4: French Government Tourist Office.
Part 5: U.S. Forest Service.

CHAPTER 1

Opener: Murray Alcosser/The Image Bank.
Fig. 1–1: The Bettmann Archive.
p.5 Culver Pictures.
Fig. 1–5: U.S. Navy Photo.
Fig. 1–9: Paolo Koch/Photo Researchers.

CHAPTER 2

Opener: NASA.

CHAPTER 3

Opener: Grant Heilman.
Fig. 3–2: Grant Heilman.
Fig. 3–3: Joint Task Force One.
p.37 R. Taylor/SYGMA.

CHAPTER 4

Opener: © 1989 Gordon E. Smith.
Fig. 4–3: Grant Heilman.
Fig. 4–8: NOAA.
p.56 Plasma Physics Laboratory, Princeton University.
Fig. 4–12: Peter Menzel/Stock, Boston.
Fig. 4–13: New Zealand National Publicity Photo.
Fig. 4–16: Dr. John S. Shelton.
Fig. 4–18: Florida Department of Citrus.

CHAPTER 5

Opener: Stan Goldblatt/Photo Researchers.
Fig. 5–1: NOAA.
Fig. 5–4: NOAA.
p.73 Culver Pictures.
Fig. 5–6: Peter Menzel/Stock, Boston.
Fig. 5–9: Jacques Jangoux/Peter Arnold.
Fig. 5–13: Cary Wolinsky/Stock, Boston.

CHAPTER 6

Opener: Grant Heilman.
Fig. 6–4: Robert W. Wallace/U.S. Geological Survey.
Fig. 6–6: Grant Heilman.
Fig. 6–7: J. R. Hollard/Stock, Boston.
p.96 John Asztalos/National Center for Atmospheric Research/National Science Foundation.
Fig. 6–10: Grant Heilman.

CHAPTER 7

Opener: NOAA.
Fig. 7–4: Charlie Ott/Photo Researchers.
Fig. 7–5: Georg Gerster/Comstock.
Fig. 7–8: NOAA.
Fig. 7–10: Sherwood Idso.
p.116(T) Harmit Singh/Photo Researchers.
Fig. 7–11: Gerald E. Gscheidle/Peter Arnold, Inc.
Fig. 7–13: NOAA.
Fig. 7–15: Paolo Koch/Photo Researchers.

CHAPTER 8

Opener: Christopher Brown/Stock, Boston.
Fig. 8–1: Grant Heilman.
p.130 Colorado State University.

CHAPTER 9

Opener: Fritz Henle/Photo Researchers.
Fig. 9–3: Jacques Jangoux/Peter Arnold.
p.141 Greg Smith/Gamma Liaison.
Fig. 9–5: Courtesy Arabian Oil Co.
p.147 Roy Attaway/Photo Researchers.
Fig. 9–7A: Department of Travel Tourist Industry—Philippines.
Fig. 9–7B: Barbara Alper/Stock, Boston.

CHAPTER 10

Opener: Helena Kolda/Photo Researchers.
Fig. 10–1: French Government Tourist Office.
Fig. 10–4: Georges Viollon/Photo Researchers.
p.159 Dr. John S. Shelton.
Fig. 10–6: Larry LeFever/Grant Heilman.

CHAPTER 11

Opener: Tim McCabe/USDA—Soil Conservation Service.
Fig. 11–2: Grant Heilman.
Fig. 11–4: AP/Wide World Photos.
Fig. 11–6: Ralph H. Felker/USDA—Soil Conservation Service.

Fig. 11–7: Donald Dietz/Stock, Boston.
p.176 Denver Metro Convention and Visitors Bureau.
Fig. 11–10: Canadian Government Travel Bureau.

CHAPTER 12

Opener: Steve McCutcheon/Alaska Pictorial Service.
Fig. 12–3: Swedish Information Service.
Fig. 12–5: Alaska Pipeline Service Company.
p.190 Riwkin/Photo Researchers.
Fig. 12–7: J. J. Carnevale/Naval Photographic Center.

CHAPTER 13

Opener: NASA Photo/Grant Heilman.
Fig. 13–3: Paolo Koch/Photo Researchers.
Fig. 13–7: New Zealand National Publicity Studios.
Fig. 13–10: Teledyne Gurley.
Fig. 13–12: Michael P. McIntyre.
Fig. 13–13: Jean-Claude Lejeune/Stock, Boston.
p.211 Woods Hole Oceanographic Institute.
Fig. 13–15: George Bellrose/Stock, Boston.
Fig. 13–17A: Fredrick D. Bodin/Stock, Boston.
Fig. 13–17B: Fredrick D. Bodin/Stock, Boston.
Fig. 13–18: Michel Brigaud/Phototeque EDF. French Embassy—Press and Information Division.
Fig. 13–19: Canadian Hydrographic Service.

CHAPTER 14

Opener: Peter Menzel/Stock, Boston.
Fig. 14–2: U.S. Army Corps of Engineers.
Fig. 14–3: Santa Clara County Water Conservation District.
Fig. 14–6: Gary Z. Zahm/USFWS.
p.226 E. Kotlyakov/Sovfoto.
Fig. 14–7: State of California Department of Water Resources.
Fig. 14–8: S. Jorge/Image Bank.
Fig. 14–9: New Zealand Consulate General.

CHAPTER 15

Opener: The Sardine Factory.
Fig. 15–3: H. Bristol Jr./Photo Researchers.
Fig. 15–5: International Salt Co.
Fig. 15–6: B. J. Nixon/Tenneco, Inc.
Fig. 15–8: Klaus D. Francke/Peter Arnold.
Fig. 15–9: Nova Scotia Power Corp.
Fig. 15–10: AP/Wide World Photos.

CHAPTER 16

Opener: Roy Bishop/Stock, Boston.
Fig. 16–1A: Japan National Tourist Organization.
Fig. 16–1B: Spanish Ministry of Information and Tourism.
Fig. 16–3: U.S. Geological Survey/W. J. Lee.
Fig. 16–4: C. A. Reeds/American Museum of Natural History.
p.256 Harvard University Library.
Fig. 16–7: Dr. John S. Shelton.

CHAPTER 17

Opener: John Mairs.
Fig. 17–3: American Museum of Natural History.
Fig. 17–4: Grant Heilman.
Fig. 17–5: Harry Wilks/Stock, Boston.
Fig. 17–6: National Park Service.
Fig. 17–8: Jerome Wyckoff.
Fig. 17–10: Jerome Wyckoff.

CHAPTER 18

Opener: John S. Shelton.
Fig. 18–3: National Park Service.
Fig. 18–4: NASA.
Fig. 18–5: Grant Heilman.
Fig. 18–8: Swiss National Tourist Board.
Fig. 18–10: Mary Hill/CA Division of Mines and Geology.
Fig. 18–11: Georg Gerster/Comstock.
p.293(T) Sovfoto.
Fig. 18–12: NASA.
Fig. 18–13: Dr. John S. Shelton.
Fig. 18–14: Dr. John S. Shelton.
Fig. 18–17: Grant Heilman.
Fig. 18–19: Tom Duncan, Oakland Tribune/Sygma.
Fig. 18–20: Sygma.
Fig. 18–23: French Government Tourist Office.
Fig. 18–24: Dr. John S. Shelton.
Fig. 18–26: USGS.
p.305 Ira Kirschenbaum/Stock, Boston.
Fig. 18–27: UPI/Bettmann Newsphotos.
Fig. 18–28: Grant Heilman.
Fig. 18–29: Dr. John S. Shelton.
Fig. 18–30: Georg Gerster/Comstock.
Fig. 18–31A: Dr. John S. Shelton.
Fig. 18–31B: Alain Pitcairn/Grant Heilman.

CHAPTER 19

Opener: Jerome Wyckoff.
Fig. 19–1: Tom Hollyman/Photo Researchers.
Fig. 19–2: G. K. Gilbert/National Park Service.
Fig. 19–3: Dr. John S. Shelton.
Fig. 19–4: Stock, Boston.
Fig. 19–6: G. R. Roberts.
Fig. 19–7: H. E. Stork/National Park Service.
Fig. 19–8: R. D. Brugger/Picture Cube.
Fig. 19–10: UPI.
Fig. 19–12: L. A. County Fire Dept.
p.325 Lily Solmssen/Photo Researchers.
Fig. 19–13: W. B. Finch/Stock, Boston.

CHAPTER 20

Opener: Coast and Geodetic Survey.
Fig. 20–1: Jen and Des Bartlett/Photo Researchers.
Fig. 20–4: Oregon State Highway Travel Division.
Fig. 20–5: Dr. John S. Shelton.
Fig. 20–7: Grant Heilman.
Fig. 20–10: Dr. John S. Shelton.
p.338 Ann Karen/Eastfoto.
Fig. 20–12: Grant Heilman.
Fig. 20–15: Grant Heilman.

CHAPTER 21

Opener: Steve McCutcheon/Alaska Pictorial Service.
Fig. 21–1: Swissair.
Fig. 21–2: Danish Information Office, NY.
Fig. 21–3: Grant Heilman.
Fig. 21–5: French Government Tourist Office.
Fig. 21–6: National Park Service.
Fig. 21–7: G. K. Gilbert/U.S. Geological Survey.
Fig. 21–8: Charlie Ott/Photo Researchers.
Fig. 21–9: New Zealand National Publicity Studio.
Fig. 21–11: Ira Kirschenbaum/Stock, Boston.
p.354 Culver Pictures.
Fig. 21–14: Danish Information Office, NY.
Fig. 21–15: Canadian National Railways.
Fig. 21–17: E. E. Hertzog/Bureau of Reclamation.

CHAPTER 22

Opener: Union Pacific Railroad Photo.
Fig. 22–1: Hubertus Kanns/Rapho-Photo Researchers.
Fig. 22–2: National Park Service.
Fig. 22–3: National Park Service.
p.369 New Zealand Forest Service Library.
Fig. 22–5: Georg Gerster/Comstock.
Fig. 22–6: M. Woodbridge Williams/National Park Service.
Fig. 22–7: New Zealand Forest Service Library.

CHAPTER 23

Opener: Larry Ulrich Photography.
Fig. 23–1: German Information Center.
Fig. 23–3: Verna R. Johnson/NAS—Photo Researchers.
Fig. 23–6: Washington State Department of Commerce and Economic Development.
Fig. 23–9: Bill Bachman/Photo Researchers.
Fig. 23–10A: U.S. Army Engineer District, San Francisco.
Fig. 23–10B: Port Angeles Chamber of Commerce.
Fig. 23–11: New Zealand National Publicity Studios.
Fig. 23–12: NASA Photo Research by Grant Heilman.
Fig. 23–14: New Zealand National Publicity Studios.
Fig. 23–15: G. R. Roberts.

CHAPTER 24

Opener: Almasy.
Fig. 24–2: Keith Gunnar/Photo Researchers.
Fig. 24–3: Grant Heilman.
Fig. 24–6: Grant Heilman.
Fig. 24–9: Bob & Ira Spring.
Fig. 24–10: Grant Heilman.
Fig. 24–15: Peter Menzel/Stock, Boston.
Fig. 24–16: D. Brody/Stock, Boston.
Fig. 24–19: New Zealand National Publicity Studios.
Fig. 24–20: Grant Heilman.
Fig. 24–21: USDA.

CHAPTER 25

Opener: Owen Franken/Stock, Boston.
Fig. 25–4: Owen Franken/Stock, Boston.
Fig. 25–6: Hubertus Kanus/Photo Researchers.
Fig. 25–7: Jeff Gnass.
Fig. 25–8: Grant Heilman.
Fig. 25–9: G. R. Roberts.
Fig. 25–10: Janet K. Jahoda/Photo Researchers.
Fig. 25–11: Peter Menzel/Stock, Boston.
Fig. 25–12: Ellis Herwig/Stock, Boston.

CHAPTER 26

Opener: Richard Weymouth Brooks.
Fig. 26–1A: C. Frank/Photo Researchers.
Fig. 26–1B: Owen Franken/Stock, Boston.
Fig. 26–4: Weyerhaeuser Co.
Fig. 26–5: J. P. Iddings/U.S. Geological Survey.
Fig. 26–8: Grant Heilman.
Fig. 26–9A: Grant Heilman.
Fig. 26–9B: Grant Heilman.
Fig. 26–9C: Richard Weymouth Brooks/Photo Researchers.
Fig. 26–9D: Alan Pitcairn/Grant Heilman.
Fig. 26–10: E. Aubert de la Rüe/Photo Researchers.
Fig. 26–11: U.S. Geological Survey.
p.438 Grant Heilman.
Fig. 26–12: Grant Heilman.
Fig. 26–13: Jeff Gnass.
Fig. 26–14: Peter Eilers.
Fig. 26–16: Alan Pitcairn/Grant Heilman.
Fig. 26–17: Peter Eilers.
Fig. 26–20: William E. Ferguson.
Fig. 26–21: Larry Ulrich.

CHAPTER 27

Opener: Grant Heilman.
Fig. 27–3A: J. E. Boucher/National Park Service.
Fig. 27–3B: Australian News and Information Bureau.
Fig. 27–3C: National Park Service.
Fig. 27–4: South African Information Service.

Fig. 27–5: Ira Kirschenbaum/Stock, Boston.
Fig. 27–6: Daniel S. Brody/Stock, Boston.
Fig. 27–7: Verna R. Johnston/Photo Researchers.
p.460 Grant Heilman.
Fig. 27–8: Maurice E. Landre/Photo Researchers.
Fig. 27–9: Spanish Ministry of Information and Tourism.
Fig. 27–10: U.S. Forest Service.
Fig. 27–11: Redwood Empire Association.
Fig. 27–12: Weyerhaeuser Co.
Fig. 27–13: P. Berger/NAS, Photo Researchers.
Fig. 27–14: Grant Heilman.
Fig. 27–15: Courtesy Environment Canada.
Fig. 27–16: Peter Southwick/Stock, Boston.

APPENDIX B

U.S. Geological Survey.

INDEX